GOSHEN COLLEGE
CULTURE
FOR SERVICE

Good Library

This book is placed
in the library
in memory of

Arthur Troyer '62

GOSHEN COLLEGE
CULTURE
FOR SERVICE

Good Library

This book is placed
in the library
in memory of

Arthur Troyer '62

PLATE 1   PLANET EARTH AND HER ECOSYSTEMS
(A) Desert, (B) Glacier, (C) Swamp, (D) Temperate rainforest,
(E) Coral reef, and (F) Salt marsh.

PLATE 2   CHARACTERS IN THE DRAMA
(A) Water lily, (B) Yellow-bellied marmot, (C) Chum salmon, (D) Vampire bat, (E) Climbing ferns, (F) Least tern, (G) *Neurospora crassa* (false-color SEM), (H) Green iguana, (I) Scorpion, and (J) Tree frog.

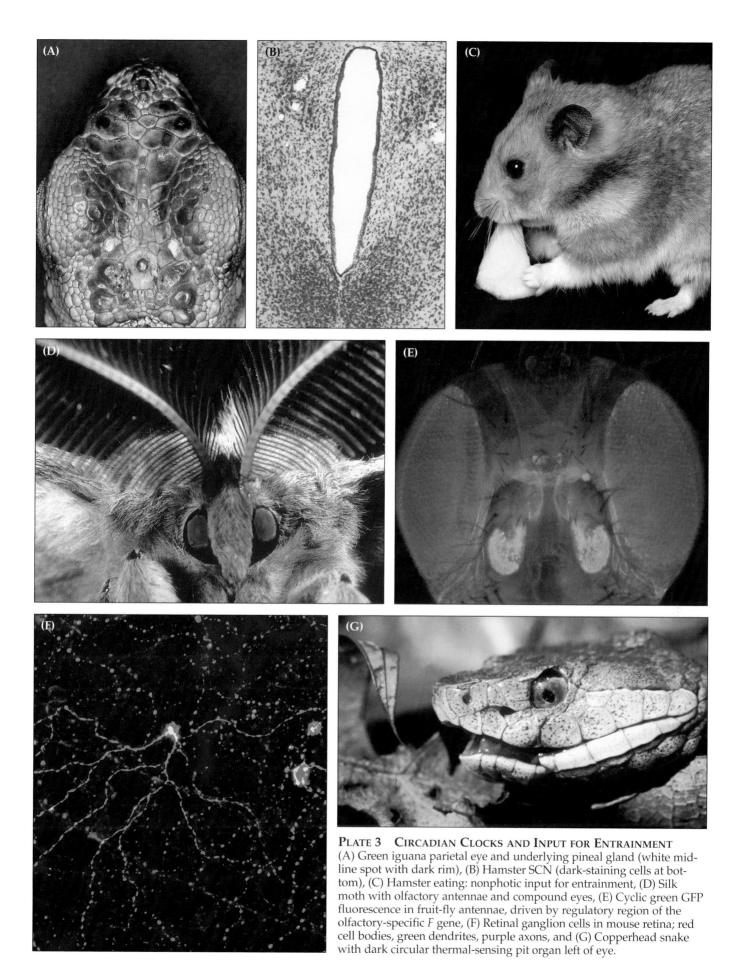

PLATE 3   CIRCADIAN CLOCKS AND INPUT FOR ENTRAINMENT
(A) Green iguana parietal eye and underlying pineal gland (white midline spot with dark rim), (B) Hamster SCN (dark-staining cells at bottom), (C) Hamster eating: nonphotic input for entrainment, (D) Silk moth with olfactory antennae and compound eyes, (E) Cyclic green GFP fluorescence in fruit-fly antennae, driven by regulatory region of the olfactory-specific *F* gene, (F) Retinal ganglion cells in mouse retina; red cell bodies, green dendrites, purple axons, and (G) Copperhead snake with dark circular thermal-sensing pit organ left of eye.

PLATE 4 CIRCADIAN OUTPUT: ACTION
(A) Hawksbill turtle, (B) Horseshoe crabs, (C) Humpback whale,
(D) Honeybee, (E) Great blue heron, (F) Emperor penguins,
(G) Human kayakers, (H) Sandhill cranes, and (I) Gray wolf.

Chronobiology

Chronobiology
Biological Timekeeping

Edited by

Jay C. Dunlap
Dartmouth Medical School

Jennifer J. Loros
Dartmouth Medical School

Patricia J. DeCoursey
University of South Carolina

Sinauer Associates, Inc. Publishers
Sunderland, Massachusetts, U.S.A.

The Cover
Light streaming through redwood trees in Jedediah Smith Redwoods State Park, Crescent City, California, © CORBIS.

Sinauer Associates
23 Plumtree Road
Sunderland, MA 01375 U.S.A.
FAX: 413-549-1118
Email: publish@sinauer.com
Internet: www.sinauer.com

Library of Congress Cataloging-in-Publication Data
Chronobiology: biological timekeeping/edited by Jay C. Dunlap, Jennifer J. Loros, Patricia J. DeCoursey.
     p. cm.
Includes index.
ISBN 0-87893-149-X
1. Chronobiology. I. Dunlap, Jay C. II. Loros, Jennifer J. III. DeCoursey, Patricia J. QP84.6.C453 2004 571.7'7—dc21
                      2003004552

Dedication

In appreciation, and in memory of
Colin S. Pittendrigh and Jürgen Aschoff

Founders of Chronobiology

For those who knew and worked with them,
they were a constant source of inspiration.

Brief Contents

Contents

3   Fundamental Properties of Circadian Rhythms 67

4   Circannual Rhythms and Photoperiodism 107

5 Functional Organization of Circadian Systems in Multicellular Animals 145

6 Cell Physiology of Circadian Pacemaker Systems in Metazoan Animals 181

7 Molecular Biology of Circadian Pacemaker Systems 213

8 Adapting to Life on a Rotating World at the Gene Expression Level 255

9 Human Circadian Organization 291

10 The Relevance of Circadian Rhythms for Human Welfare 325

11 Looking Forward 359

Preface

Time is the wisest of counselors.
—Plutarch

The purpose of *Chronobiology: Biological Timekeeping* is to provide a comprehensive text-book for use in biological rhythms courses at both the advanced undergraduate and graduate levels and to provide a nontechnical source book for scientists in other fields. Chronobiology is the study, at all levels of organization, of adaptations evolved by living organisms to cope with regular geophysical cycles in their environment. Strong intuitions of internal timing as the basis of rhythms in organisms had existed for many years in the work of several outstanding physiologists of the nineteenth century. Twentieth-century pioneers, including Erwin Bünning, Curt Richter, and others, had been fascinated with daily rhythmic phenomena in a wide variety of organisms. However, the use of the words "living clock" by the gifted German scientist Gustav Kramer in a 1952 publication, provided a unifying principal. As a result, numerous attempts have since been made to identify and measure the properties of living pacemakers, and the field of rhythms research has continued to attract adherents and attention.

Chronobiology research has arisen independently in many fields and remains one of the most interdisciplinary fields in biology. Its impact is evident at the molecular, cellular, organismal, behavioral, and societal levels. Circadian systems have become a favored neurobiology model because of the cellular simplicity of the suprachiasmatic nucleus (SCN), the daily pacemaker of mammals, in combination with high precision behavioral marker outputs. The early identification of circadian rhythms in biochemical and behavioral parameters of several microbial systems provided excellent models for research on cellular regulation. Geneticists and, later on, molecular biologists turned to biological timekeeping as one of the great unsolved unifying principals of biological systems regulation. Eventually this research led to the identification and cloning of clock genes in several species, including humans. As a result of these discoveries and applications, chronobiology is an immensely diverse and dynamic field today, relevant to all of us.

Early stages of the field were documented in a series of pivotal *Symposium Proceedings* in 1960, 1965, 1971, 1976, and 1982, and by the *Handbook of Behavioral Neurobiology: Volume IV: Biological Rhythms* in 1981. Interdisciplinary growth of chronobiology was reflected in the founding of the *Journal of Biological Rhythms* in 1986, and the establishment of The Society for Research on Biological Rhythms in 1988. Several other societies for chronobiology have been established worldwide. In November 2001, the 14 chief international societies formed a consortium called The World Federation of Societies for Chronobiology.

As the field of chronobiology expanded exponentially, the need became ever more pressing to make the information available to a broad readership of students and scientists in other fields. Although many chronobiologists teach courses on biological rhythms, the last university-level textbook on the subject was published in 1982, at the very beginning of the molecular era of rhythms research. Contributing to the lack of an updated textbook was the realization that the field was too broad for any single scholar to write such a book. A conference sponsored by the American Physiological Society entitled "Chronobiology: Understanding the Biological Clock from Genetics to Physiology," was convened at Dartmouth Medical School in 1995 and featured sessions dedicated to the development of textbook materials. Although no text emerged from that effort, the seeds for this book were sown there, and the growing need for a textbook kept the concept alive. In late 2001, the three Editors joined forces to marshal a textbook to completion.

The book's contents reflect one of the most profound generalizations about biological rhythms. All clocks must measure time accurately, then communicate the timing commands to effector systems of an organism. This dictum applies to all organisms, from primitive cyanobacteria to humans, and as a result, clocks appear highly conserved. Evolution does not seem to have produced hundreds of different clock types. What is learned about the clockworks of an insect or fungus may relate directly to the clockworks of vertebrates. Thus, the plan for the book is to survey clocks primarily by level of operation rather than by taxonomic order.

As introduction, pacemaker systems are first presented from a historical perspective and include coverage of terminology and basic foundations (Chapter 1). A part of the phenomenology of biological rhythms is next presented as rhythmic behavior of organisms (Chapter 2), followed by in-depth coverage of the practice and theory that underlie analysis of the fundamental formal properties of biological rhythms as self-sustained oscillators (Chapter 3). The conclusion of behavioral phenomenology covers circannual rhythms and photoperiodism (Chapter 4). The text then follows a progression from the organization of the circadian system in multicellular organisms at the level of organs and tissues (Chapter 5), on to the cellular level (Chapter 6), to the genetic and molecular levels (Chapter 7), and to the output of regulatory commands through gene expression (Chapter 8). Finally, coverage of the basic properties of the human circadian system has been separated out from that of other vertebrates in Chapter 9, followed by an examination of the relevance of circadian physiology to human society (Chapter 10). Chapter 11 then incorporates selected concepts from the earlier chapters into a glimpse of possible future directions for the field.

As a result of the broad scope of *Chronobiology*, it may be used in diverse programs, including neuroscience, biology, plant physiology, psychology, ecology, or medical physiology. In trying to meet the needs of students and instructors with different backgrounds, several strategies have been employed. Each chapter begins with an opening spread that guides the reader by bridging consecutive chapters and providing an overview of the chapter at hand. Many of the chapters start with a brief historical overview to help acquaint readers with the field. At the end of Chapters 1 through 10 are sections entitled Study Questions and Exercises, which have been designed to give insights, provide practical experience, and suggest ideas for additional library- or laboratory-based research into chronobiology. Many illustrations are included in the text, based on the old adage that a picture is worth a thousand words. In the cellular and molecular chapters, new methods are presented in the text and in illustrations, and are then directly related to breakthrough discoveries. The overarching themes of the book are summarized in four Plates that contain collages of visually striking, full-color photographs.

Another means of increasing the appeal of the material is the minimal use of technical terms and jargon. Readers familiar with the circadian literature may be surprised that material can be discussed in depth with limited reference to PTMs, PRCs, PTCs, FRRs, PPRCs, ZTs, GnRHs, or to terms such as *tau*, delta phi, and *zeitgeber*. Some readers, particularly those living outside North America, will need to know the scientific names of plants and animals in order to identify them. Rather than fill the pages of the book with Latin names, the book provides in an Appendix a list of organisms by both Latin and common English names. Finally, a Glossary provides working definitions for selected technical terms.

Acknowledgments

Appreciation is expressed to the many people who helped complete *Chronobiology: Biological Timekeeping*. A most gratifying aspect of the preparation of this textbook has been the collaboration of over 50 expert chronobiologists who contributed to the writing and reviewing of the component pieces. Authors and contributors are acknowledged at the end of each chapter in alphabetical order or in the order requested by the authors. We greatly appreciate the efforts of our chronobiology colleagues who kindly evaluated outlines, read early versions of the book, and provided invaluable suggestions. Especially helpful were Jo Arendt, Gerta Fleissner, Günther Fleissner, Al Lewy, Michael Menaker, Ralph Mistlberger, and David Weaver. Bill Schwartz found time in his busy life for consultation on many aspects of the book. We also thank Hildur Colot, Allan Froehlich, Kwangwon Lee, Han Cho, and Deanna Denault at Dartmouth, who carefully reviewed later versions and made suggestions for both factual and stylistic improvements. We also gratefully acknowledge Teresa Eastman for administrative support. Colleagues who were especially helpful in efforts to acquire some of the photographs include Gerta Fleissner, Günther Fleissner, Michael Grace, Michael Menaker, Ignacio Provencio, Phil Rizik, Richard Vogt, and Clint Cook. Jennifer Keller translated many of the chapter-opener ideas into visual images for the book. Donna Paquin, Sierra Jones, and Alicia Mims helped carry out the meticulous effort of obtaining figure permissions for seven of the chapters.

We are deeply grateful for the generous help of these many individuals but we realize that undetected errors may remain, and we take full responsibility for them. We are aware, also, that the field of chronobiology continues to grow daily and that we had to arbitrarily pick a point to stop additions and improvements to the text. We encourage all our readers to give us feedback about areas that can be improved.

Funding for the Dartmouth Conference came to J.C.D and J.J.L. in part from the American Physiological Society and the U.S. Air Force Office of Scientific Research, and for completion of parts of the textbook to P. DeC. from the National Science Foundation.

No single agent can add more to the production of an informative and exciting book than its publisher. We feel honored to have worked with Sinauer Associates, Inc., and express our deepest appreciation for their incomparable cooperation in production of the book. Andy Sinauer organized the team that shepherded our book through to completion; he has generously encouraged the use of extensive photographs, color plates, and the memorable color cover photo. Our Production Editor at Sinauer Associates, Kerry Falvey, has been a tower of patience, fortitude, and skill, who engineered every turn and twist with grace and charm. Our Copy Editor, the untiring Stephanie Hiebert, polished our prose to enhance its precision and appeal. Chris Small coordinated production details and Joan Gemme designed the book and created effective and attractive page spreads. David McIntyre's knowledge and skill helped make possible many of the photographic features of the book. Imagineering Media Services, Inc., redrew every illustration to insure uniform format and style. Marie Scavotto produced our announcement brochure and handled the task of organizing the book's promotion. To all of these friends at Sinauer, we extend our most cordial thanks.

Jay C. Dunlap	Jennifer J. Loros	Patricia J. DeCoursey
Hanover, NH	Hanover, NH	Columbia, SC

The Editors

Jay C. Dunlap is Chairman of the Department of Genetics at Dartmouth Medical School. An early interest in the physiology of marine organisms drew him to graduate work with J. W. Hastings at Harvard, where he first became fascinated with biological clocks and with the genetic dissection of complex biological phenomena. Dr. Dunlap has co-edited 16 books in genetics and published over 100 articles on the genetics and the molecular biology of circadian systems. A recipient of the Honma International Prize for Circadian Rhythms Research, he has served on the editorial board of the *Journal of Biological Rhythms*, and as president of the Society for Research on Biological Rhythms.

Jennifer J. Loros is Professor of Biochemistry and of Genetics at Dartmouth Medical School. A lifelong interest in horticulture led to graduate work on circadian biological clocks at the University of California, Santa Cruz and postdoctoral research at Dartmouth. She serves as Associate Editor for the journal *Genetics* and on the Advisory Board for the *Journal of Biological Rhythms*, and received the Aschoff's Rule Award for her contributions to understanding circadianly regulated gene expression. Dr. Loros's lab explores the genetic and molecular underpinning of circadian timing systems, the means through which fungal and mammalian clocks control gene expression and organismal behavior, and molecular mechanisms of the clock's response to environmental stimuli.

Patricia J. DeCoursey is Distinguished Professor of Biology in the Department of Biological Sciences at the University of South Carolina. Her early interests in ornithology led her to Cornell University for undergraduate studies in zoology. She completed a Ph.D. in zoology and biochemistry at the University of Wisconsin, Madison, and then carried out postdoctoral research with Jürgen Aschoff at the Max-Planck Institute for Behavioral Physiology in Erling-Andech, Germany. Dr. DeCoursey was a member in 1985 of the original Organizing Committees for both the *Journal of Biological Rhythms* and the Society for Research on Biological Rhythms. She served two terms as SRBR's Secretary, and has been active on the Advisory Boards of both groups. Her career-long interests in chronobiology have centered on behavioral, physiological, and ecological aspects of circadian rhythms, primarily in mammals.

Chronobiology

Before the earliest linguistic record of human awareness and interest in the passage of time, primordial architects and engineers created monumental megaliths of stone, such as Stonehenge, which is aligned to detect the solstices. Other peoples built giant labyrinths of stone to measure lunar months for use in calculating religious festivals. Sophistication in time-measurement devices increased through the pendulum clocks of the Middle Ages on to present-day precise atomic clocks. Yet humans, in spite of great fascination in determining time of day and year, had only the barest glimmering of biological timing and its significance for life on planet Earth until a half century ago.

Chapter 1 begins with an overview of human ingenuity over the ages in reading time from instruments. It then turns to basic terminology and some initial principles of biological timekeeping relative to environmental rhythms. In portraying concepts and methods, this chapter starts to unravel the mysteries and wonders of biological timing by means of internal living pacemakers.

Overview of Biological Timing from Unicells to Humans

Time is the coin of your life . . . and only you can determine how it will be spent.

—Carl Sandburg

Introduction: Structural Organization and Time Organization Are Two Pivotal Parallel Features of Living Organisms[4,5,16,31,45,62]

The rotating features of Earth have influenced the evolution of rhythmicity in all living organisms[7,23,32,37,46,51,52]

OVERVIEW. The world of living organisms has evolved around two sets of contingencies. The spatial configurations of each organism's own physical structure and environment is a familiar concept to all. Far more subtle and less frequently noticed by the untrained eye is the temporal, or time, structure of each organism in response to rhythmic elements of its environment. Only now, after several thousand years of human interest in rhythmicity in organisms, has science begun to unravel the complex but compellingly interesting biological basis and ecological usefulness of timing processes within organisms. Whereas structure in terms of cells, tissues, organisms, and organ systems gives form to living matter, appropriate timing gives relevance to behavior by relating the functions of organisms to rhythmic aspects of the physical and biotic environment.

CONCEPTS OF TIME. Jesperson and Fitz-Randolph aptly summarize time as a paradoxical riddle. They note that time is present everywhere, but occupies no space. It can be measured but not seen or touched or weighed. Although time is an everyday phenomenon, no one can define it completely. It can be spent, saved, wasted, or killed, but not destroyed or even changed.[37]

Isaac Newton and Albert Einstein, two of the greatest theoretical physicists of all time, spent much of their careers trying to define time but never were completely satisfied with their own answers. Stephen Hawking, the current guru of cosmological time, speaks of curved unified spacetime and reveals how truly complicated concepts of time can be. He attempts to simplify the enormously complex concepts of 11-dimensional super-gravity, black holes, superstrings, and p-branes with superb drawings in his new book, *The Universe in a Nutshell*.

Webster's Dictionary defines **time** in three different ways: (1) as the measured or measurable period during which an action, process, or condition exists or continues; (2) as the grouping of the beats of music; or (3) as a moment, hour, day, or year as indicated by a clock or calendar. Turn a page further to the word **timing**, and *Webster's* enlightens us with the following definition: selection for maximum effect of the precise moment for beginning or doing something. These concepts of time, timing, clock, and rhythm are key characters in *Chronobiology: Biological Timekeeping*.

Chronobiology is a word derived from three Greek stems: **chronos** for time, **bios** for life, and **logos** for study. By union of the concepts, chronobiology comprises the systematic scientific study of living timing processes in plants and animals. The field deals with living internal clocks that track and help anticipate important environmental rhythmic events, as well as read the passage of time of day. Biological clocks are thus elegant proactive strategies that avoid the problems of mere passive responses of organisms to changes in the environment.

ADAPTIVENESS OF BIOLOGICAL CLOCKS. The field of chronobiology is dynamic and fascinating. Its subjects span the biotic menagerie from the most primitive single-celled cyanobacteria to the human species (see Plate 2). Internal biological clocks play a role as temporal regulatory pacemakers in practically every function of every living species. The story exemplifies the evolutionary history of every species and the driving need to develop adaptiveness of performance. Adaptiveness is manifested in organisms as genetic fitness, the ability to place viable offspring in future generations. Success in temporal terms involves an organism doing the right thing at the right place at the right time (see Plate 4).

An explanation about organisms solving the "**right time**" challenge traverses eons of time and almost unimaginable space. The answers take the reader back through the past several million years of human conscious thought and self-awareness, back through innate, instinctive neural timing of daily activities as a response to rhythmic environments in multicellular organisms, and even further back to the ancient cyanobacteria of over 3 billion years ago. Possibly these primordial unicellular forms of life contained genetic timing messages in each single cell.

Perhaps internal timing goes back to the very origin of cellular life on our planet. In order to survive, a primordial organism must have evolved the requisite capacities for reproduction, for differential passage of materials in and out of the cell, for detecting and responding through irritability to its environment, and for pacing out time.

CHRONOBIOLOGY AS AN EXCITING ADVENTURE. Part of the magical appeal of the chronobiology saga is an element of mystery and surprise. Humans were clearly aware of seasonal and daily activities of many species millennia ago. The interest is witnessed in the awe-inspiring constellation of gigantic rocks arranged in an intriguing pattern at Stonehenge, which is pictured at the beginning of this chapter. With few exceptions, however, humans had minimal clues until 50 years ago about the marvels of living clocks in organisms. Such rhythms were viewed instead merely as direct responses of organisms to changes in their environment. Then in 1952, the simple observation and deduction of a scientist watching migrating birds, which is described later, changed the picture dramatically. A new window on animal timing was pushed open.

Chapter 1 turns first to human interest in the passage of time and the development of devices and instruments for measuring time ever more accurately. The narrative then continues in a historical sweep to the recognition of the rhythmic passage of time in plants and animals as an internal biological process allowing internal estimation of time in the external world. The second part of the chapter pivots abruptly from historical interest to the special terminology and beginning concepts of the science of biological timekeeping. In conclusion, the rationale for the progression of the remaining chapters of the book is explained.

Clocks and Calendars Have Been Important to Humans from Earliest Recorded History[9,10,27,28,33,40,60,61]

Time measurement was crucial in early cultures[9,13,15,52]

OVERVIEW. A thread of unprecedented invention runs through the history of early attempts by humans to measure and predict the time of day and the season. Few fields of scientific study have origins as colorful and compelling in their interest as chronobiology. F. Crane writes the following in his forward to Brearley's *Time Telling through the Ages*:

> The Thousand and One Tales of the Arabian Nights are not more interesting than this account of how the stars and the passing hours were through long ages of experiment finally confined into a tiny silver casket (the watch) and given, not to some prince for a fabulous sum, but to Everyman and for a dollar.[9]

Details would fill volumes, but a quick sprint through the highlights will emphasize the importance of time in human history.

Early humans used timing devices for two main purposes: telling the time of day and marking the passage of the annual cycle. Some inventions incorporated elements of both daily and seasonal timing. The earliest instance of awareness of time by humans will never be known, but the importance of the mysterious forces of sun, moon, planet, and stars in the everyday and religious life of primitive peoples even today suggests that awareness of daily, lunar, and seasonal time in organizing their activities is as old as the human race itself.

CELESTIAL BODIES FOR TIME-TELLING. No known instruments for measuring time were produced by Paleolithic humans, but the sun, moon, and stars must have been very effective for estimating daily and seasonal time. The sun was a powerful presence in the life of early humans. It dictated life by its rising and setting. The solar passage provided favorable daytime and dangerous night options for humans. Its daytime path told viewers the time of day, allowing anticipation for efficiently partitioning the hours of the day.

Paleolithic humans almost certainly read solar shadows to estimate time of day. Some tribes may even have erected slender rocks or posts in open areas specifically to cast solar shadows. At noon the shadow was always shortest. Furthermore, the seasonal transit of the noon point marked on a north–south axis gave seasonal information for planning events in the life of a tribe. The northernmost point of the arc in the Northern Hemisphere occurs on the summer solstice on June 21, when daylight is longest, and the southernmost point on December 21 is the winter solstice, or shortest day of the year. Halfway between are the spring or vernal equinox and the comparable autumn equinox, when day and night are of equal length. Similarly, the moon was also used in determining seasonal time.

CELTIC ADVANCES. At Stonehenge on the Salisbury Plain of England lies an ancient monument of megalithic stones, so huge and so precisely aligned that it defies human imagination to figure out how these stones were moved many kilometers and hoisted into place by mere mortals (Figure 1.1). The record of Stonehenge starts over 8 millennia ago when Stone Age hunters followed the retreating glaciers northward. Radiocarbon dating substantiates that three huge wooden posts were erected, perhaps as carved totems, but perhaps to cast shad-

FIGURE 1.1 A view at Stonehenge. The rising sun shines through one of the arches at the summer solstice.

ows. After about 3000 B.C., a henge, or circle, of wooden posts was built surrounded by a mound of dirt and an outer trench, with the river Avon flowing through one corner. The sacred ground may have served as a microcosm of these people's world.

Within a century a much more elaborate structure of roofed posts arose. Finally in about 2000 B.C. a final stone phase was initiated. An outer ring consisted of towering rectangular stones linked above by enormous stone lintels. Inside was a ring of much smaller bluestones. Most massive and impressive of all were the five trilathons, or arches, arranged in a horseshoe shape. The tallest was the central arch, with only a narrow slit between the verticals. Directly in front of it, in line with the rays of the setting sun at the winter solstice was the so-called sacred heel stone.

The awe-inspiring rings of giant stones were most likely used for astronomic observations and religious ceremonies. At sunset on the winter solstice, the sun shines through the slit and touches the heelstone with its blood-red rays. Possibly the unforgettable spectacle played a role in religious ceremonies, assuring the people that winter's bitter nadir had been reached and that spring's restorative rebirth was about to begin again. Other positions in the stone circle mark the rising of the sun at the summer solstice or related moon positions.

Another late Stone Age burial site in Ireland, called Newgrange, dates back to 3100 B.C. Beneath a large marker stone, a 25 m tunnel leads to a burial chamber, oriented such that a beam of light from the sun at sunrise on the winter solstice shines along the length of the tunnel. Today, some 5,100 years later, this primitive calendar still works.

The rise of science in the Western world and in China promoted astronomy and math[40,42,44,50,58]

MESOPOTAMIAN AND EGYPTIAN ASTRONOMERS AND ASTROLOGISTS. Meanwhile, humans in Asia Minor learned to work soft metals into useful objects, and Neolithic cultures were slowly transformed into a Bronze Age culture about 4,000 years ago. Agriculture and herding dictated a settled life where cities grew. On the fertile floodplain between the Tigris and Euphrates rivers of ancient Mesopotamia, the first written language, cuneiform, evolved in Sumeria with characters inscribed on mud envelopes and tablets. With a permanent record now possible, the sciences flourished. The science of astronomy was very practical for understanding the heavens and for predicting appropriate times and seasons for activities of such great and fabled cities as Babylon, Ur, and Nineveh. The heavenly bodies were revered as religious totems, and priests spent much of their time observing planetary objects and their movements.

These early astronomers documented the annual cycle using the key 12-star constellation figures as a zodiac to give a 12-part division of the year into 30-day lunar months (Figure 1.2). The zodiac functioned chiefly in astrology, to relate the supposed influence of celestial bodies on human affairs and terrestrial events by their positions and appearances. However, the early Sumerian observers also made remarkable scientific discoveries that affect time measurement even today. They recognized precisely the units of time and built accurate calendars. The early Babylonian astronomers recognized the 24 h day with hours of 60 minutes and minutes consisting of 60 seconds. The initial 360-day Babylonian calendar did not accurately mark the true 365.25 days of a year, but astronomers soon added a corrective thirteenth month every 6 years to account for the extra 30 days that had accumulated. So precise was their calendar that it lasted until the Roman emperor Julius Caesar added corrective days to some months and a leap year day. Only one further correction, made by Pope Gregory XII in 1582, was necessary to create the annual calendar in use today.

Nearby in Egypt, another great civilization was rising on the banks of the Nile River, paralleling in many ways the culture of Mesopotamia on the Tigris–Euphrates plain. As a result of native genius and cultural exchanges with Sumeria, the Egyptians made similar intellectual progress: a written language of symbols called hieroglyphics, great irrigation projects, and the flourishing of the applied sciences. In particular, the sciences of mathematics, engineering, architecture, and astronomy were cultivated. As early as 2100 B.C. the Egyptians designed star clocks to tell time at night by constructing tables for the times of rising and setting of certain stars in the sky.

FIGURE 1.2 Zodiac calendar. An 1886 painting by Alphonse Mucha shows the 12-part division of the zodiac.

One of the most actively pursued astronomic interests involved daily timing by flat sundials or vertical obelisks (Figure 1.3). The great obelisk called Cleopatra's Needle was built in 1500 B.C. at the temple of Heliopolis; today it still marks off the hours in London, where it was carried in the mid-1800s. A big improvement on the relatively inaccurate sundial was the hemicycle. This instrument was made of a shallow bowl with a horizontal pointer projecting out over the bowl. A more advanced form of hemicycle consisted of a circular courtyard with a small hole that formed a pointer of sunlight. The hemicycle circumvented some of the latitudinal and seasonal problems of the simple sundial.

Another invention was the hourglass or sand clock. This simple device used the force of gravity to pull fine sand through a tiny constriction in a vessel to mark the passage of time. The devices were first used in the 3rd century B.C. in Alexandria. The great advantage was simplicity, cheapness, and portability. Practicality was not one of its virtues. Although a 3-minute hourglass is still occasionally used to time the cooking of a breakfast

FIGURE 1.3 Obelisk with dial for telling time of day and making calendrical measurements.

est labyrinth designs are very ancient. Patterns on a clay tablet from Knossos dating back to about 1400 B.C. suggest the famous labyrinth of the Minotaur. Supposedly this legendary creature was imprisoned within its meandering walls until killed by brave Theseus.

The best-known labyrinth today lies in the floor of the great cathedral at Chartres, France. It dates back to about A.D. 1260. Most labyrinths were intended for walking in a symbolic or religious fashion. The most intriguing ones, however, for a discussion of human timing instruments had a feature called lunations. At the 11-course Chartres Cathedral labyrinth, the shallow lunation scallops around the outside circumference number 28 for each quadrant except at the entrance, where 1.5 lunations are lacking (Figure 1.5). The lunations may have served as a counting device useful in calculating the dates of festivals such as Easter.

Post-Renaissance scientific advances resulted in radical changes in clock measurement devices[1,2,10,29,37,42,43]

EUROPEAN MECHANICAL CLOCKS. A revolution in timekeeping came about with the invention of European mechanical clockworks in the 1200s. A variety of designs based on several different principles were used. The

egg, the appeal is quaint charm, not versatility. Sandglasses need to be turned over to start counting anew, for they have no element of self-sustainability.

WATER THIEF CLOCKS. The next major advance in daily time measurement was the clepsydra, or water thief, developed in Egypt. Originally it used a simple water-containing vessel with a pinhole that allowed water to drip out slowly and regularly. Level marks on the bowl indicated the passage of time. No machinery was needed, and the dripping water was totally independent of sunlight, which didn't work for sundials on cloudy days or at night. Water clocks quickly became more and more elaborate. To extend the usefulness of a water clock, a float or toothed ratchet could be used to transmit the force of the rising water to a gear train. Often water clocks were literally water towers that used the force of gravity to turn one or more wheels at a stable rate, or to announce the time of day with dials or ringing bells. A remarkable water tower, built under the direction of Su Sung in about the year A.D. 1000, was considered one of the wonders of China until it was carried off by Tartar invaders in A.D. 1126 (Figure 1.4).

LABYRINTHS. An interesting variant of the seasonal timer is the labyrinth, consisting of a spiral path without crossovers or blind endings. The path leads by continual pendular change of direction to a central endpoint and returns by the same route. Thus the route is unicursal, with various possible numbers of switchbacks or courses. Some small ones are only 3-course, but far more common are 7-course or 11-course labyrinths. The earli-

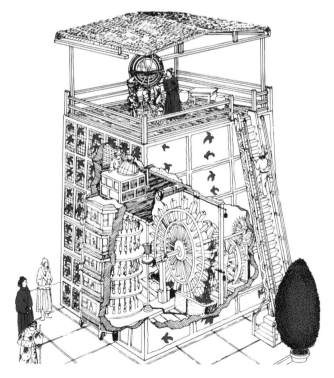

FIGURE 1.4 Su Sung's water tower clock. This clock was built circa A.D. 1000. (Illustration by John Christiansen; from Needham et al., 1986.)

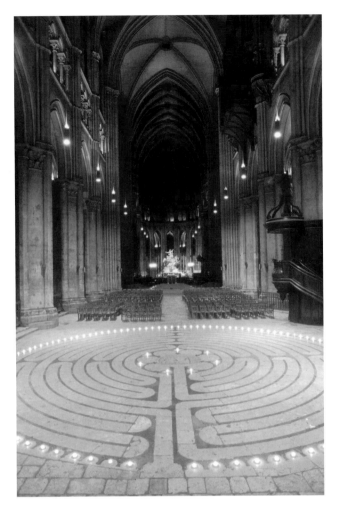

FIGURE 1.5 The labyrinth in Chartres Cathedral. This 11-course labyrinth shows paths and lunations.

interest was still primarily astrological, but the clocks also served to call the religious to worship with impressive clock faces or ringing bells. Some clocks used a drum of mercury falling regularly from one chamber to another to turn dials and hands. Others used a cable wound around large drums, with an escapement to regulate the rate of turning. Some used a falling weight and escapement connected with bells to chime the hours. The best-known clock in the world is Big Ben in London (Figure 1.6).

A major improvement in the earliest clockworks was based on a chance observation by the boy genius Galileo. In 1581, at the age of 17, he watched the swinging of an overhead lamp in the cathedral at Pisa, timed it with his own pulse rate, and formulated the principle of isochronism of the swinging of a pendulum: equal arcs in equal time. Regardless of the length of the arc, the time of each swing was the same. Pendulum clocks were soon built, and some called grandfather clocks are

still in use today. Pendulum clocks, however, are large and bulky.

Until the mid-1600s, clockworks were simply too big and too expensive for any purpose except public institutions such as cathedrals or administrative palaces. After the awkward pendulum of most clocks was replaced by a coiled spring, miniaturized domestic clocks became feasible, but even these were primarily amusements for the very wealthy (Figure 1.7).

TEMPERATURE-COMPENSATED CLOCKS TO SOLVE THE LONGITUDE PROBLEM. In spite of the improvements brought about by spring coils, clocks still had problems. They were unreliable particularly because of temperature sensitivity. A pressing need existed for a highly accurate, portable, temperature-compensated timepiece. It should be stable enough on long ocean voyages to determine longitude at sea and thus allow safe travel to a destination. In the guesswork navigational method of the day in 1707, for example, four British warships returning to harbor ran unexpectedly aground on the English coast only a few miles from port, and in that night alone 2,000 sailors drowned.

As a direct result of the continuing dire loss of ships at sea, the British crown offered a reward of 20,000 pounds sterling in 1714 for the invention of a suitable ship's chronometer. The sum was truly wealth equal to

FIGURE 1.6 Big Ben clock, London.

FIGURE 1.7 An American domestic clock from the nineteenth century. (Courtesy of Marc Desrosiers.)

FIGURE 1.8 An atomic clock. NIST-F1, a cesium atomic clock located at the National Institute of Standards and Technology in Boulder, CO. (Copyright Geoffrey Wheeler.)

a king's ransom. Thereafter, the solving of the longitude problem became a major issue in all European astronomic observatories. Space does not permit the telling of the story of John Harrison, the humble carpenter who struggled 40 years to build his three exquisite clocks (H1, H2, H3) and finally a crowning victory, his H4 watch, which is illustrated at the beginning of the chapter. He had solved at last the longitude problem.

MODERN ELECTRIC AND ATOMIC CLOCKS. In our modern-day world we rely for accurate timing on oscillations far different from mechanical pendulum swings. Electric clocks use the reliable 60 swings per second of 60-cycle AC electric current to move the clock hands and tell the time. For the even more precise timing needed by power companies, broadcasters, and telephone companies, the vibrations of quartz crystal oscillators activated by electric current are employed.

The ultimate is reached in the atomic clocks that set world timing standards (Figure 1.8). The resonance rate of the hydrogen atom, or of rubidium or cesium, is used as the most accurate timing yardstick. The examples discussed up to this point suggest the fascination of humans in all ages with the concept of time, but almost exclusively from a very practical and applied point of view. The concern had been how to measure and use information about time, not to understand the basis and function of biological timing within an organism.

Awareness of Internal Biological Clocks Began Relatively Recently[11,26]

Early knowledge of physiology was limited

The intrinsic wonders of time awareness and time measurement through the ages justify inclusion in this chapter. However, the information serves an even more important function here by providing a stunning counterpoint for the chapters to follow. Humankind was almost totally ignorant until a century ago about the existence of internal timekeeping systems in plants and animals.

Until the Renaissance, knowledge of the structure and function of living organisms, even humans, was rudimentary. Not only ignorance of facts, but also lack of any interest in inquiring into the possibility of internal biological timing, was rampant. Even though the early Greek natural historians had noted and commented on the rhythms of birdsong or leaf movements, none had considered that the response to environmental cycles was more than a passive exogenous reaction to external stimuli. In the 4th century B.C., the scribes of Alexander the Great noted daily leaf and petal movements, and another Greek naturalist, Androsthenes, reported the daily rhythms of leaves opening and closing but assumed they were a direct response to environmental stimuli.

Early circadian pioneers had some first inklings of endogenous rhythms[26]

The first ray of intellectual light comes from an enterprising French astronomer, Jean Jacques d'Ortous deMairan. He developed a side interest in the leaf movements of light-sensitive plants, particularly a species of *Mimosa* (see Figure 8.1). DeMairan went far beyond merely noticing the movements. He carried out a few experiments concerning causality. When placed in darkness without cyclic environmental information, the leaf rhythm continued for many days. DeMairan suggested other experiments for which he had no time. These included testing other plant species, or reversing the light–dark (LD) cycle. His work was published as a short communication addressed to the Royal Academy of Science in Paris. The date was 1729.

Little new information was added for another century. Then Swiss botanist Alphonse de Candolle demonstrated that the *Mimosa* daily leaf movements persisted in darkness. Moreover, in constant darkness the rhythm advanced to an earlier start each day. Here were convincing data for a free-running rhythm. Even Darwin got into the preliminary act briefly by noting that daily leaf movement in plants was an innate property.

Although several later observers noted behavioral patterns attributable to internal timing, a major obstacle to further progress was the wrong choice of words. The internal clock of bees allows their return to a feeding table at a fixed time of day, regardless of weather or even the presence of food (Figure 1.9). The process was named **Zeitgedächtnis** in a publication in 1929. The German word, which translates literally as "time memory," only perpetuated the idea of a learning response instead of innate timing.

Other scientists questioned whether persistent daily rhythms under so-called constant conditions merely reflected the failure to exclude all possible latent clues from the research facility. Because most laboratories were unable to eliminate barometric pressure cycles, minute daily temperature changes, or cosmic ray cycles from their laboratories, a definitive defense of the internal nature of rhythmicity was not possible.

Later pioneers had good insights into circadian rhythmicity[11,38,39,41]

Even the unequivocal published demonstrations of accurate free-running activity rhythms of white-footed mice in wheel cages by Maynard S. Johnson in 1926 and 1939 made no impact on the prevailing mentality of

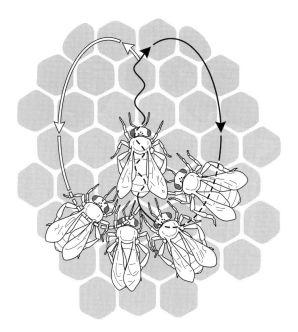

FIGURE 1.9 A forager bee dancing in the comb of a beehive. Bees communicate distance and direction of food with a waggle dance. (From Bradbury and Vehrencamp, 1998.)

exogenous rhythmicity. In a powerful statement about causation, he wrote,

> The white-footed mouse has, normally, a well-marked daily periodicity of spontaneous activity, and this periodicity persists for months when the mouse is kept in continuous darkness. The active period normally occurs at night, but in continuous darkness can be made to come at any time during solar day or night, and so can be shown not to be dependent on any unrecognized or uncontrolled daily variable in the environment.[39]

Johnson even went on to show that the rate was dependent on the intensity of the continuous light regime and increased to a longer and longer period with logarithmic increase of the continuous light intensity (Figure 1.10). He continues:

> The duration here reported for *Peromyscus*, of a well-defined activity rhythm after 18 months in absence of daily change in light or dark, seems to be a new endurance record for a diurnal rhythm. The work here reported on *Peromyscus* seems to be the first instance of a diurnal rhythm under controlled conditions where definite amounts of modification of the rhythm are shown to result from definite intensities of an environmental factor. Neither the mechanism by which a diurnal rhythm is maintained under constant conditions, nor the mechanism by which the rhythm is modified in continuous light is adequately understood.[39]

Unfortunately Johnson's accurate observations and lucid writing had little influence on solving the clock dilemma, and circadian biology remained on the plateau

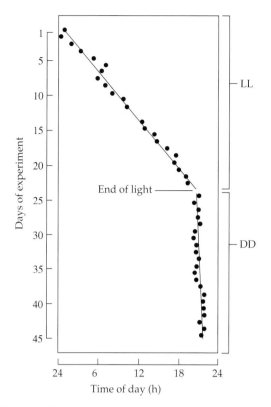

FIGURE 1.10 Early record of a free-running rhythm of locomotor activity in *Peromyscus*. DD, constant dark; LL, constant light. Solid dots indicate weighted midpoint of activity periods. (From Johnson, 1939.)

Often a giant stride is made in the sciences when a new tool such as the microscope is invented or when a new idea is introduced. The coining of a new phrase in 1952 was the spark that ignited the field of chronobiology. In the late 1940s the highly respected German scientist Gustav Kramer was studying time-compensated celestial navigation in bird migration. He had shown that the birds require a sun compass for flying north in spring, and he reasoned that to use such a continuously moving reference point for flying in a fixed direction, the bird must have **eine biologische Uhr**: a biological clock. The term instantly replaced the vague notion of rhythm and suggested a precision instrument in the form of an accurate living physiological entity.

Pittendrigh and Aschoff were founders of modern-day chronobiology[3,4,6,7,17,24,52,53]

Suddenly the whole picture changed. A young insect ecologist in Trinidad took note and decided to study the internal clocks gating pupal hatching rhythms in insects. Simultaneously a young professor in Heidelberg, Germany, turned from studies of human temperature regulation to the daily locomotor activity rhythms of mice. These two, Colin S. Pittendrigh and Jürgen Aschoff, respectively, were to become the driving forces in a new field of science, soon to be called **chronobiology**.

The impact of these charismatic individuals on the course of the field of chronobiology cannot be emphasized strongly enough. Their dynamic personalities, scientific vision, restless energy, and brilliant ideas are some of the highlights of the origins of the field. A memorial note for Pittendrigh reads:

> Biology has lost one of its giants of the 20th Century and it is from his shoulders that we are looking into the 21st. The importance of daily rhythms dawned on Pittendrigh during the war years in Trinidad when he studied mosquito behavior as a malaria biologist. He then decided to study daily organization in *Drosophila*, which remained his organism of choice. In the 1950s, Colin gave the study of biological rhythms its unifying conceptual basis.[18]

Similar words were written for his colleague:

> Aschoff: a life of duty, wit and vision. We commemorate today one of the great men of science of the 20th Century. Jürgen Aschoff and Colin Pittendrigh were the founders of modern biological rhythm research. They developed the unifying concepts and key principles of biological clocks. Through their work, personalities, and influence, the field evolved from scattered studies on exotic curiosities to a branch of science with a clear focus on what turned out to be a core property on a rotating planet. A characteristic which drew so many researchers to [his lab in] Andechs is the exceptionally open mind and broad interest with which Aschoff approached scientists from many different backgrounds.[17]

of the preceding 2,000 years a little while longer. Soon, however, the winds of change would blow.

One of the pioneers was a plant physiologist, the German professor Erwin Bünning. He had started research on circadian rhythms in 1928 and soon proposed the hypothesis that circadian rhythms have adaptive value for organisms. Shortly afterward he published the first evidence for the use of a biological clock for photoperiodic time measurement. Bünning had many insights about internal timing that were far ahead of his time. When he revised his book entitled *The Physiological Clock* in 1973, he wrote poignantly, "As recently as 15–20 years ago, to proclaim the existence of an endogenous diurnal rhythm was regarded, even by some well-known biologists, as subscribing to a mystical or metaphysical notion."[11] Bünning's discoveries are credited today with the beginning of plant photoperiodism research (see Chapter 4). Another brilliant young scientist, James Enright, who worked at the Scripps Institution of Oceanography, was also a pioneer whose chronobiology insights were not fully appreciated in his day. He quipped at a meeting in 1960, "When I studied lunar spawning rhythms in the Pacific grunion fish for my Ph.D. research, my choice was equated with personal lunacy."

Another scientist wrote:

> In 1958 Aschoff had joined the prestigious new Max-Planck Institute for Behavioral Physiology at Erling [Andechs] and Seewiesen, Germany, to collaborate with Gustav Kramer and Konrad Lorenz (later a Nobel Prize Laureate in 1974), in an intellectual arena almost without equal in Europe. Aschoff's great talents as a scholar, teacher, scientist, and mentor were a vital part of the circadian careers of many chronobiologists. Over the half century of his dynamic circadian career he envisioned and explored an extensive array of questions about the fundamental far-reaching web of circadian processes that are vital in the survival of organisms. As we sail forth into untested waters on new circadian adventures, let us remember with gratitude the generous hospitality and mentoring of Professor Aschoff at his remarkable Erling castle.[25] [Figure 1.11]

Interest in biological timing spread rapidly[14]

Enthusiasm for endogenous timing was germinating independently in other laboratories around the world. At Northwestern University a young professor, Woody Hastings, and his collaborator, Beatrice Sweeney at the Scripps Institution of Oceanography, noticed rhythms of bioluminescent flashing and glowing in the unicell *Gonyaulax*. When these rhythms persisted spectacularly in constant conditions, the two scientists realized that even single-celled organisms have the capacity for internal circadian time regulation. Hastings's active laboratory has continued to the present making great strides in understanding *Gonyaulax* rhythms.

At the University of Wisconsin, Kenneth Rawson and his graduate student noted the highly precise circadian rhythms of wheel running in nocturnal deermice and flying squirrels. They used the precise, long-lasting rhythms in continuous darkness to assay light responsiveness to very dim, brief pulses. The resultant phase response curves have now been documented in many other organisms and have been successfully applied to understanding the mechanism of photoentrainment (see Chapter 2).

At the Scripps Institution of Oceanography, Enright documented endogenous rhythms of marine invertebrates for the first time. A young professor at the University of Texas, Michael Menaker, would soon become a leader in the chronobiology field. His initial work on the circadian rhythms of birds was extended to fish and reptiles (see Chapter 5). Many of the early workers in the field nurtured busy laboratories that attracted students and visitors. Isolated sparks of interest and insight continued to ignite in many widely separated laboratories around the world.

An international community of chronobiologists soon flourished[5,14,17,25]

In these early years of the field of biological rhythms, however, the laboratories were relatively isolated from each other and international recognition was lacking. Several factors contributed to the isolation of laboratories. A primary cause was the lack of identification of a specific pacemaker in any organism. The greatest issue at stake in the field during the 1960s, for example, was a controversy over exogenous versus endogenous causation of daily rhythms. The discovery of an insect pacemaker in 1968 was followed in 1972 with localization of the mammalian circadian pacemaker. The recognition of the suprachiasmatic nucleus (SCN) of the hypothalamus of mammals as a primary pacemaker was a major breakthrough. Suddenly a small, discrete circadian pacemaker nucleus was available in the brain for study. Properties of the pacemaker neurons could now be measured directly, since the neurons could be cultured as slices or as dissociated SCN cells.

Neuroscientists sat up and took notice. Using a brain slice technique, they recorded rhythms of neural discharge in the pacemakers (see Chapter 6). The neural afferent and efferent connections were traced to other parts of the brain, or to the appropriate effector organs. Neuro-

FIGURE 1.11 Aschoff's renowned laboratory. "Das Schloss," or The Castle, at Erling-Andechs, Germany. (Courtesy of P. DeCoursey.)

pharmacologists used in vitro and even in vivo microinjection techniques to study SCN function. Molecular geneticists cloned the first clock genes: the *per* gene of *Drosophila* and the *frq* gene of the mold *Neurospora* (see Chapters 7 and 8). Soon the basic principles of circadian biology were cautiously accepted as applicable to human societal and medical concerns (see Chapters 9 and 10).

Interaction between geographically dispersed laboratories was promoted by several crucial events. During the early years of the field, several pivotal symposia were held. The first, at Cold Spring Harbor, New York, in 1960, is often considered the starting point of a unified field of chronobiology.

At this meeting, approximately 150 scientists and students gathered from around the world to discuss biological clocks. Many of them met for the first time and eagerly exchanged ideas with their newfound colleagues. The atmosphere was electrifying. In these pivotal days of the birth of a new scientific field, the most memorable hour was a dramatic lecture by Colin Pittendrigh. The title was prophetic: "Circadian Rhythms and the Circadian Organization of Living Systems." The embryonic field of biological timing was a swirling mist of uncertainty about what clocks were, what significance they had for living organisms, and what directions the field would take in its development. At this stage of doubt and questioning, Pittendrigh's deep booming voice rang out confident and clear. Like some Roman orator of old, he presented a document that has stood as a symbolic charter, a statement of the highest historical importance for the chronobiology field to this day. It recognized the phylogenetic breadth of internal timing in organisms and the ecological significance of living clocks at all levels of physiological organization and presented 16 key properties of circadian rhythms. This document served as the foundation on which the field was built, and it has guided the field through its infancy into maturity. So clear was Pittendrigh's vision that many of these formal defining properties are still valid today. Suddenly, and on a world scale, endogenous timing was seen by scientists and lay people alike as a cardinal feature of living organisms on a rhythmically revolving planet. Biological clocks were core features of life.

So successful was the presentation style that additional symposia were planned at about 5-year intervals. In 1964 the successor was convened in Feldafing, Germany, followed in 1971 by a clocks symposium at Friday Harbor Laboratory in Washington State. Finally, a remarkable NATO-sponsored, month-long workshop at Stanford's Hopkins Marine Station,

California, was staged in 1977. Senior circadian researchers came both to teach and to learn, and they brought droves of enthusiastic students whose interest was fueled by the intellectual sparks of the lectures and workshops. The number of circadian papers published rose precipitously. Engineers, physicists, and mathematicians were attracted to the field. Added to this growing army of new scientists was a substantial number of second-generation scientists who had been rigorously trained in the founding laboratories.

In recent years chronobiology has become one of the most interdisciplinary fields of biology (Figure 1.12). To portray more accurately the relationships of the many overlapping fields that contribute to chronobiology, a multidimensional model would be necessary. Daily or circadian clocks, especially in mammals, have become a favored neurobiology model for decoding the regulatory processes of neurons. Because of the simplicity of the mammalian circadian pacemaker, the suprachiasmatic nucleus (SCN), and the precise behavioral output that the SCN regulates, this pacemaker has been more amenable for study than the nervous system of the simplest worms, which have only a handful of neurons.

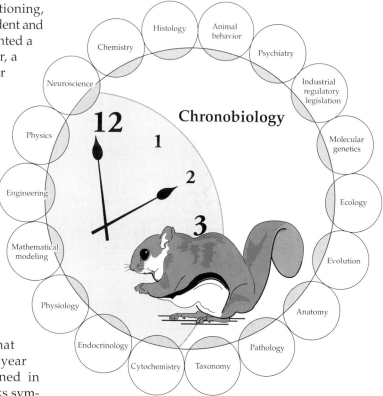

FIGURE 1.12 Interdisciplinary nature of chronobiology. The light shading of the small circles represents a schematic overlap of a field with the central chronobiology circle.

The field has also recently attracted the interest of people in many other walks of life. Among scientists the spectrum of interests runs from neuroscience, cell biology, molecular biology, and molecular genetics to physics and mathematical modeling. Newspapers are peppered daily with reports of the implications of chronobiology for normal function and for chronopathologies. Terms like *shift work dysfunction, jet lag, SAD* (*seasonal affective disorder*), and *insomnia of the aged* are everyday household words. The medical profession has also begun to appreciate the implication of chronobiology for human well-being and health (see Chapter 10).

In the 1980s chronobiology societies proliferated rapidly. The time seemed right to establish societal and publication outlets to foster and encourage research in biological timing. In 1978 a session on chronobiology was added to the roster of the prestigious Gordon Research Conferences. Held every two years, these small, tightly focused meetings bring together the chief scientists of a field for presentations and intense discussions. In 1986 a journal dedicated to chronobiology research was established. The *Journal of Biological Rhythms* has been a towering lighthouse for experienced scientists and newcomers to the field ever since (Figure 1.13). Shortly afterward, in 1988, the Society for Research on Biological Rhythms was chartered in the United States with biannual meetings alternating with the Gordon Research Conferences on chronobiology. In 2002, 14 internationally recognized societies for chrono-

biology are organizing a world chronobiology conference to be held in 2003 in Sapporo, Japan.

Starting Up and Coming to Terms Is the First Step in Any New Scientific Adventure[12]

Every scientific field has its specialty jargon[5]

THE NEED FOR JARGON. Terminology is the nemesis of every nonspecialist in a new field, and chronobiology is no exception. Jargon is the technical terminology of a specialized group. The historical overview that has been presented in this chapter has pointed out that the science of chronobiology started about 1950, and scientists in the field had not at that time located any single living pacemaker. As a result, an approach other than direct anatomical dissection and physiological analysis had to be devised. The rhythmic nature of timing in organisms had all the features of physical oscillators such as pendulums, or chemical oscillations.

The similarities between biological rhythms and oscillators provided a groundwork for studying the unknown properties of clocks. Analysis of rhythms as oscillator-based phenomena was almost inevitable. From the very start, physical oscillator terminology and experimental design quickly became part of the modus operandi of every chronobiologist. Immediately the jargon of the field became bewilderingly arcane for novices. The problem was magnified because of the indirect approach to the unseen oscillator clock that will be described shortly.

Some of the common jargon in chronobiology relates to the lighting schedules used in various protocols. L refers to lights on, D to darkness, and the schedule is indicated by LL for constant light or DD for constant dark. The ratio of L to D on a day–night schedule is referenced as LD 12:12 for a 12 h L to 12 h D schedule. Non-24 h cycles are referred to as T-cycles.

In this text, terms will be clearly explained when first encountered. Selected terms will also be available in the Glossary at the end of the book. An Appendix lists scientific and common names, respectively, for all the organisms mentioned in the book. The chapter material will generally use common English names wherever these are available, in order to simplify reading. Readers who wish to know the scientific names can find them in the Appendix.

THE IMPORTANCE OF DEFINITIONS. Definitions go far beyond jargon terms. They are among the most important conceptual tools of a scientist in any field, and form the intellectual framework for discussions. Definitions prescribe the exact outer limit of a concept or material object and facilitate communication without ambiguity. They convey the essence of a word.

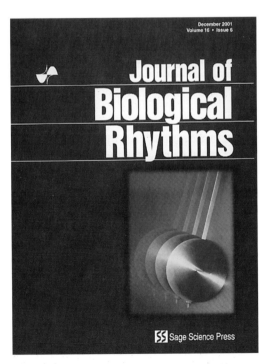

FIGURE 1.13 The *Journal of Biological Rhythms*. Official journal of the Society for Research on Biological Rhythms.

A few technical rhythm terms need to be defined and clarified at the outset. The purpose of the following circadian minilanguage course is to enable the reader to make full use of the diverse information in chapters to come. A **rhythm** is a nonrandom series of events without any statement of causation. A **circadian rhythm** refers to an observed biological activity that oscillates under constant environmental conditions with a period length close but not exactly equal to 24 h. However, not all biological rhythms with period lengths of about a day can accurately be described as circadian rhythms; true circadian rhythms have additional defining characteristics.

First, the rate of a circadian oscillation is internally compensated so that circadian behavior continues to repeat with close the same periodicity even at slightly different ambient temperatures, or in different metabolic states, as long as the environment is unvarying. In this way, for instance, circadian rhythms are distinguishable from cell cycles which can also repeat with a cycle time of about once per day under some conditions but whose period length is quite sensitive to temperature and nutrition. Second, circadian rhythms are acutely sensitive to variations in the constant environment. Such variations can reset a circadian rhythm. A more precise definition of the term circadian rhythm can be found in Chapter 3. The word *circadian* was coined from two Latin words, **circa** ("about") and **dies** ("a day"), reflecting the slight deviation of almost all free-running rhythms from 24 h. For other known or suspected environmentally related pacemakers the equivalent terms are *circatidal* for endogenous tidal rhythms, *circalunar* for endogenous lunar cycles, and *circannual* for endogenous annual rhythms.

Much of the knowledge about circadian clocks is based on observations of overt rhythms, but these rhythms are not the clock. The terms *oscillator, clock,* and *pacemaker* refer to a central timekeeper that controls myriad overt rhythms. Strictly speaking, these three terms have slightly different connotations. An **oscillator** oscillates. A **clock** is an oscillator that tells time, and a **pacemaker** is an oscillator that sets the pace of other oscillators. The term *clockworks* refers to the biochemical mechanism of a clock. Finally, **the clock** is a common figure of speech, but more than one clock may exist in an organism, and circadian clocks are generally systems.

Abundant evidence from multicellular organisms shows that some circadian behavior patterns are composed of the outputs of more than one interacting pacemaker. In fact, multiple circadian clockworks appear to coexist even in some unicellular organisms. Evidence for the multioscillator structure of circadian systems will be discussed in Chapters 5 and 6.

Starting Up Also Involves Learning Basic Concepts and Methodologies[5,14,47]

Rhythm concepts encompass parameters of waves[53]

The idea of rhythmicity merely conveys regular predictable occurrence of an event. No element of causation is implied. Rhythmic events, whether ocean waves moving across the ocean surface or animal locomotor movements throughout the day, can be plotted from a series of regularly recorded descriptors called a time series. Three important parameters include **amplitude**, **period** along with its reciprocal, **frequency**, and **phase** or **phase angle**. The towering, foam-spewing giant in the dramatic Japanese woodcut shown in Figure 1.14A, for example, can be characterized as a very high amplitude wave.

The period could be expressed as the time interval for a reference point in successive waves to pass a fixed point or as distance traveled per unit of time (Figure 1.14B). The crest is the most sharply delineated reference point in each wave cycle and could serve well as the reference point for determining wave period or frequency. A period is normally calculated as the mean ± one standard deviation for at least five waveform cycles. Its reciprocal, frequency, is the number of cycles per unit of time. Finally, the relationship of two mutually synchronized or entrained rhythms is expressed in terms of the phase angle difference. If the beginning of locomotor activity of a nocturnal bat is plotted relative to the solar light–dark cycle, a clear relationship is seen between activity onset and the evening light-to-dark transition. The lag or lead is the **phase angle difference**. In the schematic representation of Figure 1.14C, the phase angle difference is 0. The parameters of ocean waves are directly applicable to circadian rhythms of organisms such as the daily running-wheel cycles of mice.

Biological clock regulation of physiology and behavior is more complex than direct triggering of responses[16,23,45]

In normal studies of physiological functions such as reflexes, a stimulus is often applied and the result directly measured. A simple example is the study of the reflex responses of a frog voluntary muscle. A holder with capacity for measuring changes in pressure or length is attached to a bundle of muscle fibers. When a gentle electric shock from a stimulus generator is applied to a muscle cell, the muscle contracts, or shortens, proportionally to the strength of the signal.

Examples abound. A light directed onto a retinal cone cell generates a neural discharge response, and a spike is carried down the optic nerve to the brain for processing. Exercise elevates body temperature relative to exertion.

(A)

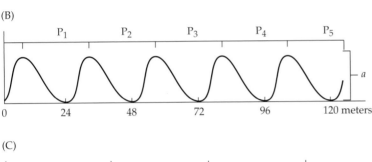

(B)

P_1　　　P_2　　　P_3　　　P_4　　　P_5

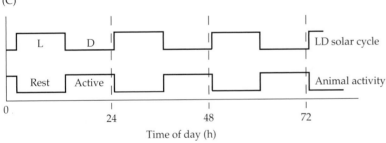

a

0　　24　　48　　72　　96　　120 meters

(C)

L　D　　　　　　　　　　　　　　LD solar cycle

Rest　Active　　　　　　　　　　　Animal activity

0　　　24　　　48　　　72

Time of day (h)

FIGURE 1.14 Parameters of oscillations
(A) Japanese woodcut entitled *Great Wave off Kanagawa* by Katsushika Hokusai, circa 1820. The wave has a high amplitude and a steep waveform front. (B) Five schematic ocean waves (P_1–P_5) with a given amplitude, a, and length in meters. (C) Phase relationships between solar environmental cycles and locomotor activity rhythms of a nocturnal rodent, with both cycles shown as square waves for simplicity. The vertical dashed lines give 24 h reference points. (B and C courtesy of P. DeCoursey.)

By analogy to a house light switch, flipping the switch turns on the light directly by supplying current.

Indirect measures of rhythmicity usually rely on long-term behavioral recording[5,7]

The story is more complicated with animal biological clocks than with muscle cell measurements for two reasons. First, the output of a circadian pacemaker at the cellular level is difficult to measure noninvasively in multicellular organisms (see Plate 3, Photo B). As a result, direct measures of circadian clocks are rarely feasible. The property of circadian rhythmicity in pacemakers lies in the individual cell, and invasion of a cell such as an SCN pacemaker neuron with electrodes usually kills the cell after a few hours (see Chapter 6). Even the use of extracellular electrodes requires the death of the animal

to obtain an SCN slice. Similarly, the cultured SCN slice lasts only a few hours to a few days when the slice is being speared by the electrode every 5 minutes to record from different cells. In light of all the problems with direct measures of most pacemakers, indirect measures are often used instead.

The second reason that direct pacemaker functions in animals are difficult to measure lies in the length of the figurative gear train between a mammalian clock oscillator in SCN cells and the behavioral output as running-wheel activity. Biological clocks such as the bird pineal gland or the mammalian SCN pacemaker have an internal basis and exogenous entrainment elements that fine-tune the endogenous rhythm to local environmental time (Figure 1.15A). The beginning of birdsong at dawn is not merely a passive response to the environment. Instead, a circadian internal clock generates a genetically determined rhythm close to but not exactly 24 h in period.

Because the clock may be either unknown or very inaccessible, an indirect measure of its performance must be used that reflects the clock as accurately as possible. The behavioral outputs, or so-called hands of the clock, are also often called the **behavioral phenotypes** (see Plate 4). The indirect measure should be convenient to automate without invasively disturbing the animal, but it should be as close to the actual clock mechanism as is feasible. Many possible circadian marker processes can serve as clock system indicators. In almost all rodents, for example, voluntary locomotor activity in a running-wheel activity cage is very robust and often highly precise (Figure 1.15B). Rodents actively seek out wheels and turn them for many hours each day. Particularly in nocturnal rodents, the rhythms persist under constant environmental conditions without any environmental clues for months or even years.

The actual rate of a circadian process per unit of time may need to be measured for the eventual analysis. Examples include body temperature rhythms and gnaw-

(A)

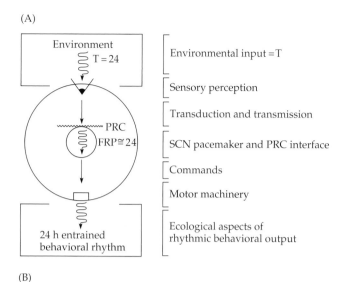

Environment $T = 24$	Environmental input $= T$
	Sensory perception
PRC $FRP \cong 24$	Transduction and transmission
	SCN pacemaker and PRC interface
	Commands
	Motor machinery
24 h entrained behavioral rhythm	Ecological aspects of rhythmic behavioral output

(B)

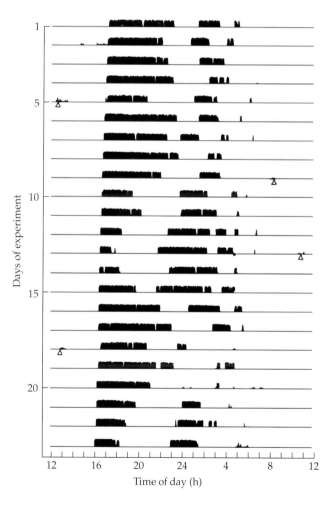

FIGURE 1.15 Behavioral methods in chronobiology
(A) Model of a mammalian circadian clock. (B) Behavioral monitoring showing from left to right a wheel cage with a flying squirrel in the wheel, a full-sized actograph card in the background portraying a free-running flying-squirrel rhythm in DD, an Esterline-Angus recorder with 24 h data for 20 squirrels, and a computer with a free-running rhythm in DD on the monitor. (A and B courtesy of P. DeCoursey.)

ing rate of rodents. Even the rate of locomotor activity may be documented. In the instance illustrated in Figure 1.16, a nocturnal flying squirrel in continuous darkness ran in a running wheel during its subjective night in episodes or bouts of locomotor activity that accelerated from rest to almost 100 revolutions/minute almost instantaneously. In most cases, however, only presence or absence of activity is needed to document activity data (see Figure 1.15B).

In most animals or plants, any convenient behavioral or physiological output can serve. For the single-celled, chlorophyll-bearing protist ***Gonyaulax***, the bioluminescence rhythm can be followed through recording of light arising from the glow rhythm or the flash rhythm (Figure 1.17; see also Figure 2.4). For *Paramecium* (see Figure 3.28) the phototactic rhythm or the mating reactivity rhythm

are excellent choices. The conidiation rhythm of *Neurospora* produces reproductive spores, and these can be documented photographically in special culture tubes.

The importance of developing a noninvasive, reliable, automated, user-friendly methodology can hardly be overemphasized. Historically, the most frequently used behavioral output feature in fruit flies was eclosion, or pupal hatching. The process is gated for a population, and most individuals hatch during a limited time window that is favorable for survival. A simple method of data collection was first devised with clever bang bottles. At regular intervals the incubation bottles were bumped to knock the flies down off the sides of the bottles, and then each bottle was dumped into a collection vial, which was stoppered for later counting of the number of flies hatched per unit of time. Now the circadian rhythm of locomotor activity is measured electronically.

A contrasting hi-tech method involves automation of data collection for the cyanobacteria *Synechococcus*. Originally progress was slow because the population cultures required much hand processing. Advanced

FIGURE 1.16 Properties of actographs Rate of activity of a rodent in a wheel. (Courtesy of P. DeCoursey.)

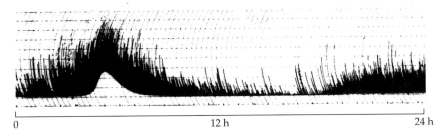

FIGURE 1.17 Luminescent glow rhythm of a vial of *Gonyaulax* in DD for 24 h. The change in baseline is the steady glow, and the vertical lines represent the spontaneous flashing of individual cells. (From Hastings, 1960.)

technology has recently been developed for demonstrating the basic circadian properties in populations of *Synechococcus* for extended periods of time. Automated cell counters have eliminated hand counting. Molecular genetic techniques involve transformation of the cyanobacteria with bacterial luciferase genes that function as reporters of clock-controlled expression of rhythmicity. The visible bioluminescent reporter is easily monitored and recorded by computerized electronic photo detection equipment.

Data processing of circadian behavioral output is often depicted in actograms[7]

The behavioral output of an organism over many days is often best portrayed in a vertical stack of 24 h traces called **actographs** or **actograms** (see Figures 1.15B and 1.16). The daily traces stacked vertically in chronological order give a lucid view of the envelope of activity, the period, and the precision. Because behavioral actograph data collection is noninvasive, it is very often used in protocols. Numerous examples of actograms will appear in succeeding chapters.

Actograms may be single-plotted (see Figure 1.16) as simple 24 h scans. However, sometimes they are clearer as double plots or even triple plots, in which case the data are simply stacked in overlapping 48 h or 72 h traces (Figure 1.18). Each of these rhythms has parameters comparable to the oscillations of a pendulum or other oscillator. The repetition rate or period of the oscillation in constant conditions is called **free-running period (FRP)**, or in some publications, **tau** (τ), and is equal to the reciprocal of frequency such that frequency = 1/FRP. The amplitude can also be measured. The phase is the relationship of a marker point in the organism's rhythm to another oscillation such as the daily environmental cycle.

Circadian time differs from standard time-zone clock time[19,53]

A very important distinction must be made between standard clock time and military time. Standard time runs in every time zone with midnight taken as the middle of the dark period, and noon as the point 12 hours later when the sun is at its zenith. **Military time** is a convention used for most scientific work and in this textbook. Midnight is designated as 00:00, and noon as 12:00. The afternoon hours continue past 12:00 by increments of 1 each hour up to 24:00, or midnight. The addition of "A.M." and "P.M." is therefore not necessary, and data records are much less likely to be confused because the morning and afternoon hours are clearly differentiated. Sunset time on this scale is the time when the sun's disk disappears beneath the horizon at sea level; **civil twilight** is the time when the sun's disk is 6 degrees below the horizon. Civil twilight is important for many nocturnal animals because it is often the time that they start evening activity.

Circadian time (CT) is subjective internal organism time. This distinction will become very clear later in the book. Because an endogenous rhythm is not 24 h in period, a free-running animal in constant conditions will soon lose reference to external environmental time. Most analyses of circadian data are made with reference to circadian time in circadian hours. Each circadian hour for an animal with a free-running rhythm of 23 clock hours is equal to 23/24 h. In the same vein, the light schedule for an organism in the laboratory may have little rele-

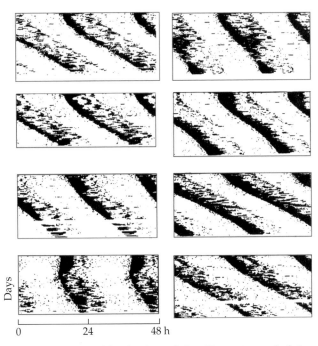

FIGURE 1.18 Double plotting of circadian actograph data. Data here are plotted as 48 h scans for eight hamsters in DD for 5 months. (From Yamazaki et al., 2002.)

vance to standard clock time, and it is usually referred to as ZT, standing for synchronizer time or zeitgeber time. The term *zeitgeber* will be described a little later in the chapter. By convention, the circadian day starts at **subjective dawn**, when the sun would have appeared. This is referred to as CT 0, and **subjective dusk** is CT 12.

A Brief Introduction to Basic Circadian Parameters Will Set the Stage for Discussion of Circadian Behavior and Ecology[31]

Circadian pacemakers are endogenous[7,16,53]

FREE-RUNNING RHYTHMS. A few key concepts will be developed now as background for the discussion of circadian phenotypes and evolution in the next chapter. The text will return again in Chapter 3 to fundamental properties, to develop the ideas in greater depth as needed for the remaining chapters.

When an organism is placed in continuous conditions and the period of its internal clock is measured indirectly, the rhythm is seen to **free-run**. Its period no longer matches the 24 h solar day but runs slightly faster or slightly slower, showing unequivocally that the rhythm is generated within an organism (see Figures 1.16 and 1.19). Because different individuals in an observed group will all free-run with slightly different rates ranging from 23 h to slightly over 24 h in period, no environmental factor could be cuing the rhythms. Endogenous rhythms are innate, which means simply that they have a genetic basis and are not merely acquired by learning. Some of the genes that regulate clocks have now been cloned in the bread mold *Neurospora*, in the fruit fly *Drosophila*, and in the lab mouse (see Chapters 7 and 8).

The pattern of circadian activity varies considerably within an individual and between individuals of a species (Figure 1.19). The daily cycle of organisms is generally divided into a more or less continuously active period called **alpha**, and a quiescent period called **rho**, or rest period. However, the distribution of bouts within the alpha period, the duration of alpha, the presence of bimodal peaks, the regularity of onset and end of activity, the robustness of the alpha period, and the presence of short **ultradian** bouts throughout the day are highly variable. These features can be very important in picking out a model species to test a particular hypothesis. In some hypothesis testing, the highest possible precision may be needed to detect small responses to stimuli (see Chapter 3). Similarly, much diversity is found in interspecific variation of circadian expression. Some species may be very regular, with strikingly precise activity (see Figure 1.19).

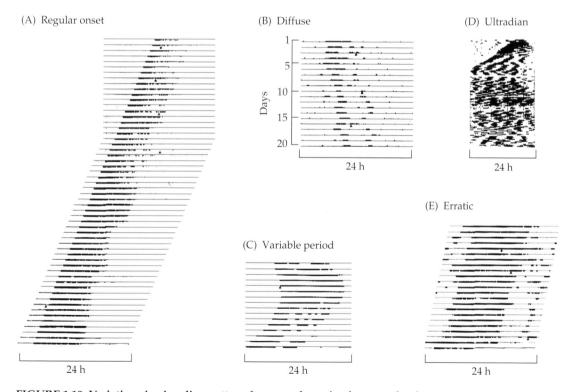

FIGURE 1.19 Variations in circadian pattern for several species free-running in DD. (A) Regular onset in a flying squirrel. (B) Diffuse pattern in another flying squirrel. (C) Variable period in a European dormouse. (D) Ultradian patterns in a redbacked vole. (E) Erratic pattern in another flying squirrel. (A–C from DeCoursey, 1983; D–E courtesy of P. DeCoursey.)

Endogenous pacemakers are widely distributed in cyanobacteria, fungi, plants, and animals. Indeed, endogenous rhythms are often said to be ubiquitous, present in all species. Although that statement may be true, it hasn't been adequately tested. Only a relatively small number of species have been rigorously studied under constant conditions.

Some circadian rhythms are highly precise. In terms of precision of most biological processes, this amounts to a remarkable few tenths of a percent. In some flying squirrels or white-footed mice, the rhythm may be repeated with a standard error as little as ±1 minute per day. In many rodents the start or onset of activity each day is highly accurate and serves well as a convenient marker point for circadian research analysis.

Entrainment requires an environmental time cue or synchronizer for locking on to local time[31,34,35,36,48,49,54,55,56,57,59]

PHOTIC ENTRAINMENT. Normally organisms do not live in constant conditions, but are exposed to the rhythmic changes of the solar day or lunar-related tides. The internal rhythms reflect this influence by **entraining**, or locking on to, the driving oscillation of the environment in what is called **photic entrainment** or **photoentrainment** (Figure 1.20). This critical relation of two coupled oscillators has another big dose of physics jargon attached to it. **Driver and slave oscillators** are self-explanatory. Less obvious is the phase relationship of the two oscillators, the reference of a marker point in the organism's cycle to a point in the environmental cycle. Often the start of wheel turning for a nocturnal rodent is related to the start of darkness in the LD cycle. For *Drosophila* the peak of the emergence rhythm may be related to the time of lights on. The difference in phase of the two rhythms is called the phase angle difference and is usually expressed in time units such as minutes.

One example of a well-designed experiment for testing photoentrainment uses an initial DD free-running rhythm to show the presence of the endogenous component (see Figure 1.20). Then an entraining light schedule is provided to test for capturing the free-running rhythm by the LD cycle. Finally, the animal is returned to DD to check for any possible masking effects (masking will be described shortly). In this example, the free-running rhythm is evident on days in darkness and continuing through the first few days of the LD schedule. Finally, on day 15 it has entrained and remains in a stable phase relationship until the end of the schedule. Immediately following termination of the light cycle, the free-running rhythm again appears.

The effective agent in bringing about the phase control is called a synchronizer, a cuing agent, or an entraining agent. In the German literature, and to some

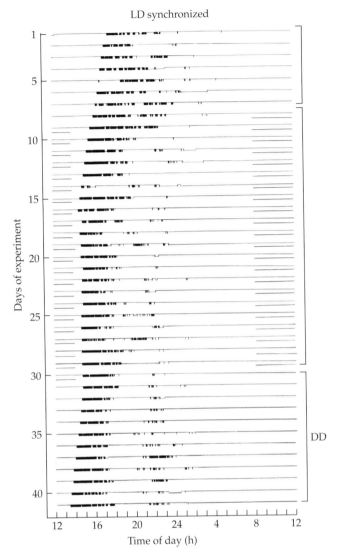

FIGURE 1.20 Photoentrainment of a free-running rhythm in a flying squirrel by a 6-h photoperiod. Days 1–9 in DD; days 10–30 in LD 6:18; and days 31–41 in DD. The light schedule is indicated by underlining. (From DeCoursey, 1961.)

extent in the English literature, another widely used term is **zeitgeber**. Literally translated, the term means "time giver." This textbook will not use the terms *zeitgeber* or *time giver* because the synchronizer does not really give the time, but only couples the two oscillators. Another reason relates to the difficulty of spelling the word. Behind the misspelling lies a humorous tale. One professor, alleged to be Colin Pittendrigh, assigned a term paper on circadian rhythms and was appalled by the various misspellings of the word. The resulting list has become a classic in chronobiology (Figure 1.21). This text will use standard English terms for the synchronization process.

The most common synchronizer for all terrestrial organisms is the day–night solar cycle under natural con-

Zeitgeber

ZEITBURGER!

Zeitgebers: thirty-three alternative spellings used by students in a chronobiology class assignment

Zeigers	Zeitgeibers	Zetgerbers
Zeightgebers	Zeitgeists	Zetgiebers
Zeitagibers	Zeitgerbers	Zetibergs
Zeitberge	Zeitgerers	Zetibers
Zeitebergers	Zeitgibers	Zetiebers
Zeitegebers	Zeitglibers	Zetigebegers
Zeitengebers	Zeitibergs	Zetigebers
Zeiterbars	Zeitigebers	Zietbergurs
Zeiterberg	Zeitinburgers	Zietgeibers
Zeiterbers	Zetebergers	Zietgieters
Zeiterburgs	Zetegerbers	Zitegers

FIGURE 1.21 Thirty-three ways to spell zitterburgers, or zeitgebers for hamburgers.

ditions of the LD cycle in the laboratory, known as photic entrainment (see Figure 1.20). Note that the free-running rhythm often requires a few days to lock on to local time. The mechanism will be explored in Chapter 3.

NONPHOTIC ENTRAINMENT. For a relatively small number of animals, **nonphotic** entraining agents are known (see Plate 3, Photos C, D, and G). In some birds, for example, cycles of species-specific song can act as entraining agents. For some cyanobacteria, fungi, plants, and cold-blooded animals, temperature cycles are effective entraining agents. A limited amount of data are available for effects of social entrainment in animals at least as a reinforcing part of entrainment. Novel wheel turning and access to exercise also are capable of entraining hamsters. Much less is known about nonphotic entraining agents and their mechanisms of action than about photic synchronizers.

MASKING. Still another problem and set of terms arise from the use of indicator processes. Note that the system must have an oscillator as actual pacemaker. It must also have an input of environmental information through the sensory system, and an output of rhythmic physiological and behavioral endpoints (see Figure

1.15A). These are often called **overt rhythms** in contrast to the actual pacemaker states.

In an indirect measurement, anything that interferes with either the input or the output train may **mask** the true state of the oscillator. Therein lies the danger and great pitfall of indirect measurements. A few examples will illustrate the point. An experiment can easily be designed to test the effectiveness of temperature cycles as entraining agents of circadian rhythms. At the start of one such experiment a squirrel was kept in a constant environment at 15°C, and a clear example of a free-running rhythm was observed (Figure 1.22).

After 13 days a temperature cycle was started, with temperature oscillating between 15°C and 25°C. The animal appeared to entrain to the cool part of the cycle. When the animal was returned to the constant temperature cycle, however, the data indicated that the phase of the activity rhythm had not locked on to the temperature cycle but was only masked. The warm temperature prohibited activity because the animal became over-

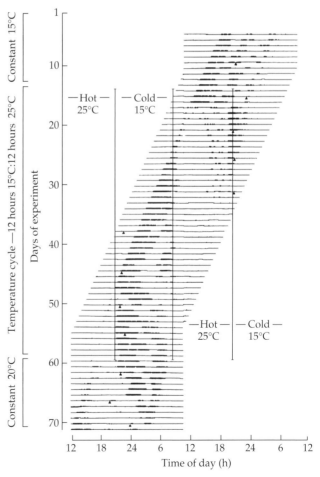

FIGURE 1.22 Masking displayed in the testing of ambient temperature cycles as entraining agents in a flying squirrel. For further explanation, see the text. (Courtesy of P. DeCoursey.)

heated, but the underlying clock was not affected. Metaphorically the **hands of the clock** were broken off and the overt rhythm could not be expressed, but the clock itself was unaffected.

Masking is often a serious problem in pharmacological studies aimed at questions about the pacemaker but measured indirectly at the level of the overt rhythm output. The chemical may not stop the clock, but only break the gear train in the **output pathway**. Masking is also prevalent in human rhythms. With conscious self-awareness, humans are often able to override clock signals. Furthermore, one of the chief overt rhythms used as a circadian marker in humans is the temperature cycle. The range of the temperature cycle is only about 1 to 2°C, and it is affected strongly by heat generation through muscular activity or even body position related to the sleep–wake cycle. Masking in human studies is a major concern of all data collection protocols. Chapter 9 will look intensively at ways to remove artifacts from human data to reveal the true state of the underlying oscillator.

The Rationale Chosen for the Text Is Simple and Logical

The result of the meteoric growth of the field of chronobiology is the interdisciplinary nature of the field (see Figure 1.12). Because of this diversity, construction of a textbook for university-level students is a daunting challenge. No reader, experienced researcher, or student will have sufficient background in all aspects of the field. A carefully designed framework for presenting the material is mandatory. Because of the number of species studied across the entire spectrum of the bacterial, fungal, plant, and animal kingdoms, central unified themes of structure and function are often obscured by minute details differing from species to species, making a possibly confusing story for a beginning student.

The first half of Chapter 1 presented a historic overview and introduction to rhythm terms. The second half has covered an introduction to some of the initial terminology, some beginning rhythm concepts, and a few key methodologies. Sufficient introductory material is included in this chapter that the reader should be able to proceed easily to Chapter 2, where the behavioral expression of rhythmicity in organisms is covered in terms of the enchanting behavioral phenomenology and evolutionary aspects.

Chapter 3 proceeds with an in-depth coverage of the fundamental properties of living clocks in the context of oscillators. The principles are relatively few in number and prevail for all clocks. Clocks are clocks, not thermometers or weather vanes or any other kind of instrument. To function as clocks and pacemakers, they must have certain properties, even if many different kinds of

anatomical structures can provide these properties. Much like the diverse sundials, clockworks, pendulum clocks, quartz crystal watches, and atomic clocks, the cellular components may differ greatly, but all of them measure and display the passage of time. Oscillation theory has been a very useful tool in indirect analysis of rhythms, especially before specific circadian pacemakers were localized.

Chapter 3 sets the stage for completion in Chapter 4 of the behavioral survey covering the longer-period circannual rhythms and photoperiodism. The interjection of oscillation theory between the two behavior chapters is necessary for discussion of the mechanism of photoperiodism.

A sequence of chapters then continues to look at different levels of organization: the structural and physiological basis of circadian timing at the organ level in Chapter 5, the cellular level in Chapter 6, and the molecular level in Chapter 7. Chapter 8 deals with the molecular basis for control of output in circadian systems. The use of selected model species was chosen for these four chapters instead of comprehensive coverage of all known details. The aim is to provide a graphic picture of selected extensively studied cases. Chapters 9 and 10 single out the one highly different species, the human, with respect to both normal circadian function and societal relevance. Chapter 11 closes the book by relating some advances and predicting new wonders that the field of chronobiology will bring in the future.

CONTRIBUTOR
Chapter 1 was written by Patricia J. DeCoursey.

STUDY QUESTIONS AND EXERCISES
1. Jargon and definitions are very important in any scientific field. Review the terms below and then try the following.

   a. Define in 15 words or less:

   Circadian
   Free-running rhythm
   Entrainment
   Masking
   Endogenous rhythm

   b. Construct five rhythm-related sentences using a different example of jargon in each.

2. Design an experiment using an indirect measurement protocol to test whether a sparrow has a circadian pacemaker. (*Hint*: Perching birds such as sparrows actively hop on and off a movable perch during daylight hours.) The moving perch could activate some kind of mechanical, mercury, or photic switch. Collect some imaginary data for a bird in constant dim light. (*Hint*: Most birds do not

feed well in constant darkness and soon decline from starvation.) Plot the data on graph paper, first as a single-plot actograph and then as a double-plot actograph.

3. Now try an entrainment experiment with a mammal such as a nocturnal deer mouse, *Peromyscus*. Collect data and plot an actograph. Why is it important to start first with a DD schedule, then change to LD, and finally end with a terminal DD schedule? Will the new synchronization occur immediately? If not, why not? How can potential masking be avoided?

4. Try visiting the library to locate the *Journal of Biological Rhythms*. Pick out one article to read, and make a list of the jargon terms that appear in it.

5. Scan enough of the journal exemplar from question 4 to get a feel for the organization of a scientific journal article. What standard sections do all data papers contain? Try making an outline of the basic parts. Later assignments in this textbook will ask the reader to write a brief scientific paper based on some of the data presented.

6. Set up a brief laboratory or class demonstration for illustrating simple handmade clocks and calendars. Sundial, obelisk, and water thief clocks are all simple to construct. Include, if desired, PowerPoint demonstrations of some of the monumental clocks and calendars of the world, such as Stonehenge or Big Ben.

7. Organize a minisymposium on entrainment by photic and nonphotic synchronizing agents. Why are nonphotic synchronizers much less effective in most organisms?

REFERENCES

1. Andrewes, W. J. H. (Ed.) 1996. The Quest for Longitude: The Proceedings of the Longitude Symposium, Harvard University, Cambridge, MA, November 4–6, 1993. Collection of Historical Scientific Instruments, Harvard University, Cambridge, MA.

2. Andrewes, W. J. H. 2002. A chronicle of timekeeping. Sci. Am. 287 (Sept): 76–85.

3. Aschoff, J. 1960. Exogenous and endogenous components in circadian rhythms. Cold Spring Harb. Symp. Quant. Biol. 25: 11–28.

4. Aschoff, J. 1963. Comparative physiology: Diurnal rhythms. Annu. Rev. Physiol. 25: 581–600.

5. Aschoff, J. (Ed.) 1965a. Circadian Clocks. North-Holland, Amsterdam.

6. Aschoff, J. 1965b. Circadian rhythms in man. A self-sustained oscillator with an inherent frequency underlies human 24-hour periodicity. Science 184: 1427–1432.

7. Aschoff, J. (Ed.) 1981. Handbook of Behavioral Neurobiology 4: Biological Rhythms. Plenum, New York.

8. Bradbury, J. W. and S. L. Vehrencamp. 1998. Principles of Animal Communication. Sinauer, Sunderland, MA.

9. Brearley, H. C. 1919. Time Telling through the Ages. Ingersoll & Bros., New York.

10. Bruton, E. 1979. The History of Clocks and Watches. Rizzoli International, New York.

11. Bünning, E. 1973. The Physiological Clock. Springer, New York.

12. Carpenter, G. A. 1985. Some Mathematical Questions: Circadian Rhythms. American Mathematical Society, Providence, RI.

13. Castleden, R. 1987. The Stonehenge People: An Exploration of Life in Neolithic Britain 4700–2000 BC. Routledge & Kegan Paul, New York.

14. Chauvnick, A. (Ed.) 1960. Biological Clocks. Cold Spring Harb. Symp. Quant. Biol. Volume 25.

15. Cunliffe, B. W. 1997. The Ancient Celts. Oxford University Press, Oxford, UK.

16. Daan, S. 2000. Colin Pittendrigh, Jürgen Aschoff, and the natural entrainment of circadian systems. J. Biol. Rhythms 15: 195–207.

17. Daan, S. 2001. Jürgen Aschoff 1913–1996: A life of duty, wit and vision. In: Zeitgebers, Entrainment, and Masking of the Circadian System, K. Honma and S. Honma (eds.), pp. 17–47. University of Hokkaido Press, Sapporo, Japan.

18. Daan, S., G. Block, and J. Aschoff. 1996. In memoriam to Dr. Colin Stephenson Pittendrigh. J. Biol. Rhythms 11: 91–92.

19. Daan, S., M. Merrow, and T. Rönneberg. 2002. External time-internal time. J. Biol. Rhythms 17: 107–109.

20. DeCoursey, P. J. 1960. Phase control of activity in a rodent. Cold Spring Harb. Symp. Quant. Biol. 25: 49–55.

21. DeCoursey, P. J. 1961. Effect of light on the circadian activity rhythm of the flying squirrel, *Glaucomys volans*. Z. vergleichende Physiol. 44: 331–354.

22. DeCoursey, P. J. 1983. Biological timekeeping. In: The Biology of Crustacea Volume 7: Behavior and Ecology, F. J. Vernberg and W. B. Vernberg (eds.), pp. 107–162. Academic Press, New York.

23. DeCoursey, P. J. 1989. Photoentrainment of circadian rhythms: An ecologist's viewpoint. In: Circadian Clocks and Ecology, T. Hiroshigi and K-I. Honma (eds.), pp. 187–206. University of Hokkaido Press, Sapporo, Japan.

24. DeCoursey, P. J. 1990. Circadian photoentrainment in nocturnal mammals: Ecological overtones. Biol. Behav. 15: 213–238.

25. DeCoursey, P. J. 2001. Early research highlights at the Max-Planck Institute for Behavioral Physiology, Erling-Andechs and their influence on chronobiology. In: Zeitgebers, Entrainment, and Masking of the Circadian System, K. Honma and S. Honma (eds.), pp. 55–74. University of Hokkaido Press, Sapporo, Japan.

26. DeMairan, J. J. 1729. Observation botanique. L'Histoire de l'Academie Royale Scientifique, pp. 47–48.

27. Ellis, P. B. 1990. The Celtic Empire: The First Millennium of Celtic History, c. 1000 BC–51 AD. Carolina Academic Press, Durham, NC.

28. Ezzell, C. 2002. Clocking cultures. Sci. Am. 287 (Sept): 74–75.

29. Gibbs, W. W. 2002. Ultimate clocks. Sci. Am. 287 (Sept): 86–93.

30. Hastings, J. W. 1960. Biochemical aspects of rhythms: Phase shifting by chemicals. Cold Spring Harb. Symp. Quant. Biol. 25: 131–143.

31. Hastings, J. W., B. Rusak, and Z. Boulos. 1991. Circadian rhythms: The physiology of biological timing. In: Integrative Animal Physiology, C. L. Prosser (ed.), pp. 435–546. Wiley, New York.

32. Hawking, S. 2001. The Universe in a Nutshell. Bantam, New York.

33. Hawkins, G. S. 1965. Stonehenge Decoded. Doubleday, Garden City, NY.

34. Honrado, G. I., and N. Mrosovsky. 1991. Interaction between periodic socio-sexual cues and the light-dark cycle in controlling the phasing of activity rhythms in golden hamsters. Ethol. Ecol. Evol. 3: 221–231.

35. Hut, R. A., N. Mrosovsky, and S. Daan. 1999. Nonphotic entrainment in a diurnal mammal, the European ground squirrel (*Spermophilus citellus*). J. Biol. Rhythms 14: 409–419.

36. Janik, D., and N. Mrosovsky 1993. Nonphotically induced phase shifts of circadian rhythms in the golden hamster: Activity-response curves at different ambient temperatures. Physiol. Behav. 53: 431–436.

37. Jespersen, J., and J. Fitz-Randolph. 1977. From Sundials to Atomic Clocks: Understanding Time and Frequency. National Bureau of Standards, U.S. Department of Commerce, Washington, DC.

38. Johnson, M. S. 1926. Activity and distribution of certain wild mice in relation to biotic communities. J. Mammal. 7: 254–277.

39. Johnson, M. S. 1939. Effect of continuous light on periodic spontaneous activity of white-footed mice (*Peromyscus*). J. Exp. Zool. 82: 315–328.

40. Kern, H. 2000. Through the Labyrinth: Designs and Meanings over 5,000 Years. Prestel, Munich, Germany.

41. Kramer, G. 1952. Experiments on bird orientation. Ibis 94: 265–285.

42. Landes, D. S. 2000. Revolution in Time: Clocks and the Making of the Modern World. Harvard University Press, Cambridge, MA.

43. Lloyd, H. A. 1981. Some Outstanding Clocks over Seven Hundred Years, 1250–1950. Baron, Woodbridge, Suffolk, UK.

44. Matthews, W. H. 1970. Mazes and Labyrinths: Their History and Development. Dover, New York.

45. Menaker, M. 1993. Special topic: Circadian rhythms. Annu. Rev. Physiol. 55: 657–659.

46. Menaker, M., and G. Tosini. 1995. The evolution of vertebrate circadian systems. In: Circadian Organization and Oscillatory Coupling, K-I. Honma and S. Honma (eds.), pp. 39–52. University of Hokkaido Press, Sapporo, Japan.

47. Moore-Ede, M., F. M. Sulzman, and C. A. Fuller. 1982. The Clocks That Time Us. Harvard University Press, Cambridge, MA.

48. Mrosovsky, N. 1995. A non-photic gateway to the circadian clock of hamsters. In: Circadian Clocks and Their Adjustments (Ciba Foundation Symposium 183), pp. 154–177. Wiley, Chichester, UK.

49. Mrosovsky, N. 1996. Locomotor activity and non-photic influences on circadian clocks. Biol. Rev. 71: 343–372.

50. Needham, J., W. Ling, and D. J. D. Price. 1986. Heavenly Clockwork: The Great Astronomical Clocks of Medieval China. Cambridge University Press, Cambridge, UK.

51. Palmer, J. D. 2002. The Living Clock: The Orchestrator of Biological Rhythms. Oxford University Press, New York.

52. Pittendrigh, C. S. 1993. Temporal organization: Reflections of a Darwinian clock-watcher. Annu. Rev. Physiol. 55: 17–54.

53. Pittendrigh, C. S., and S. Daan. 1976. A functional analysis of circadian pacemakers in nocturnal rodents. I. The stability and liability of spontaneous frequency. J. Comp. Physiol. 106: 223–252.

54. Pohl, H. 1998. Temperature cycles as zeitgeber for the circadian clock of two burrowing rodents, the normothermic antelope ground squirrel and the heterothermic Syrian hamster. Biol. Rhythm Res. 29: 311–325.

55. Rajaratnam, S. M. W., and J. H. R. Redman. 1998. Entrainment of activity rhythms to temperature cycles in diurnal palm squirrels. Physiol. Behav. 63: 271–277.

56. Reebs, S. G. 1989. Acoustical entrainment of circadian activity rhythms in house sparrows: Constant light is not necessary. Ethology 80: 172–181.

57. Rietveld, W. J., D. S. Minors, and J. M. Waterhouse. 1993. Circadian rhythms and masking: An overview. Chronobiol. Int. 10: 306–312.

58. Sobel, D. 1996. Longitude. Penguin, New York.

59. Stephan, F. K. 2002. The "other" circadian system: Food as zeitgeber. J. Biol. Rhythms 17: 284–292.

60. Symons, S. 2000. Accuracy issues in ancient Egyptian stellar timekeeping. In: Current Research in Egyptology, A. McDonald and C. Riggs (eds.), pp. 111–114. Archaeopress, Oxford, UK.

61. Waugh, A. 2000. Time: Its Origin, Its Enigma, Its History. Darroll & Graf, New York.

62. Wright, K. 2002. Times of our lives. Sci. Am. 287 (Sept): 58–65.

63. Yamazaki, S., V. Alones, and M. Menaker. 2002. Interaction of the retina with suprachiasmatic pacemakers in the control of circadian pacemakers. J. Biol. Rhythms 17: 315–329.

Summer

Winter

After an introduction in Chapter 1 to the history, terminology, and beginning principles of chronobiology, this second chapter turns to the actors in the drama of life on the revolving stage of planet Earth. Rhythmic behavioral phenomenology of organisms in an immediate sense, and the evolutionary processes that ultimately caused cyclic temporal organization, are explored. The primordial cyanobacteria added a biological pacemaker system to their physiological repertoire as early as 3.5 billion years ago, and most present-day organisms from protists to mammals have multiple circadian rhythms within each single cell. The single-celled *Gonyaulax* has several different daily rhythms, including a spectacular rhythm of bioluminescent flashing.

At the other end of the complexity scale, hamsters and squirrels turn running wheels in a 24 h day–night rhythm, and humans experience daily circadian body temperature cycles. On a longer time scale, hamsters progress through the stages of the 4-day estrous reproductive cycle. Many plants and animals reproduce or show annual hibernation cycles, as in golden-mantled squirrels (pictured above).

Many overt rhythms are governed by underlying high-precision biological pacemakers. These living clocks play a vital role in the adaptation of organisms to environmental cycles. Chapter 2 will systematically cover the types of short-period clock-regulated behavior patterns and explore the selective factors involved in their evolutionary development.

2

The Behavioral Ecology and Evolution of Biological Timing Systems

Time and tide waiteth for no one.
—English proverb

Introduction: Rhythmic Environmental Features Have Shaped the Temporal Pattern of Behavior in Plants and Animals[7,12,19,34,55,57]

Chronobiology deals with both proximate and ultimate questions[1,60]

OVERVIEW OF BEHAVIOR. The scientific study of behavior encompasses changes in the motion or state of organisms, as well as the underlying physiological and evolutionary mechanisms. The threads of the vast interdisciplinary discipline of behavioral research extend into the fabric of practically every field of biology today. In spite of the scope, however, the kinds of questions raised can be readily condensed into two categories: How are the behavior patterns accomplished, and how have they evolved over eons of time in terms of ecological adaptiveness and overall fitness? Behaviorists call the immediate, physiological *how* questions the proximate issues. The long-term ecological and evolutionary *why* concerns are referred to as ultimate questions.

The physiological questions can be measured in the laboratory with various experimental paradigms. In contrast, the ultimate questions of ecological adaptiveness and evolution are difficult to answer and are normally addressed through indirect measures of the fitness imparted by a particular behavioral feature. Inquiries about ultimate causes are best dealt with by attempts to understand the behavioral ecology of organisms and the probable course of evolution of the feature. Neither physiological nor evolutionary studies of biological timing functions are sufficient in themselves, and a complete understanding will certainly rest on a synergistic interaction of the two approaches.

Konrad Lorenz, the distinguished Nobel Prize–winning animal behaviorist, summed up the challenges when he wrote his autobiography, entitled *King Solomon's*

Ring. The wise king was said to possess a gold ring that allowed him to talk to all creatures in their own languages. In this way he could learn from each species the details of its behavioral mechanisms, both present and past. Lacking such a ring, present-day chronobiologists must use many clever strategies to learn about the physiology, ecology, and evolution of biological timing. Because these methods are especially difficult for wild species, a special discussion has been provided later in the chapter.

PROXIMATE AND ULTIMATE CAUSES. The dichotomy of physiological versus ecological/evolutionary approaches is particularly clear in the field of chronobiology. Chapters 5 through 10 of this text are predominantly physiological in character because the detailed structure of living clocks and their workings are a vital part of the story of biological timing. Those chapters deal primarily with the different levels of pacemaker systems in members of various phyla. They explore the physiological and molecular workings of the oscillatory mechanisms, the entraining input signals, and the genetic as well as biochemical output messages to the behavioral machinery. Chapters 2 through 4 cover almost exclusively the behavioral, ecological, and evolutionary aspects of biological timing systems.

The first nonscientific observations of animal rhythmicity date back thousands of years, but actual systematic research on living clocks started less than 100 years ago. The imperative need for a noninvasive assay of biological clock functions in rhythm research has dictated the continued use of behavioral measures of clocks even in hi-tech molecular chronobiology research. As a result, the indirect measure of function through behavioral output still remains very important today. Wheel running in a flying squirrel, for example, reflects essential properties such as the free-running rhythm, the entrainment process, and the phase response curve system of the squirrel's circadian pacemaker in the suprachiasmatic nucleus, or SCN (see Figure 1.15B in Chapter 1).

Whereas circadian ecology involves the temporal relationship of organisms with their environment, evolution delves into the accumulation of heritable changes in populations over extended time periods. The biological clock systems in a particular animal are the clocks of its ancestors, modified by natural selection acting on genetic variation that has been created by mutation in the genes coding for clock components.

Organisms are highly specialized for the diverse niches of the specific ecosystems in which they carry out their life cycles[19,37,60]

ECOSYSTEMS. Living organisms are found in all latitudes from the North Pole to the South Pole and in terms of elevation from the depths of abyssal ocean trenches to the summit of the highest mountains. Organisms flourish along a moisture gradient from starkly dry deserts to rain-drenched forests and along a salinity gradient from limpid freshwater to the highly saline soda lakes of Africa's Rift Valley. Furthermore, organisms abound along a temperature gradient from steamy tropical jungles and hot geyser pools to frigid arctic tundra or tops of the tallest mountains.

Biogeography, the study of floral and faunal distributions, relates spatial patterns of communities to climatic factors. The geographic position of a potential habitat site largely determines its climatic conditions. Particularly important is the distribution of land and water masses of Earth in determining the major ecosystems of the world. Ocean masses cover almost three-quarters of Earth's surface. A significant portion of the remaining quarter of terrestrial habitat is locked up in the arctic/antarctic polar ice caps in a frigid sheet too cold for survival of any life except primitive bacteria or algae. Ecosystems of the world are finely tuned webs of bacterial, fungal, plant, and animal species that are highly adapted physiologically and behaviorally for the specific environmental conditions of a site. Examples of ecosystems include salt marshes, tropical rain forests, prairies, rocky coasts, and mountains (see Plate 1).

NICHES. An ecosystem in turn contains many compartments, or niches, for living things to inhabit. A niche is not merely a topographical place with food and shelter. It is the sum total of all limited resources required by an organism. The niche for a marine fish such as a giant tuna includes food, trace minerals, oxygen supply, and a mate. Some of the obvious structural spatial parameters of a niche are food, shelter, moisture, mate availability, suitable lighting conditions, and appropriate temperature. The temporal factors of a niche are equally vital. In some cases a niche includes a long-distance migration to escape winter's rigors, and the navigational demands require a continuously consulted circadian clock for sun-compass orientation.

Appropriate entraining agents must also be available. In the majority of circadian terrestrial rhythms, the solar day–night cycle is the most important feature, resulting in two predominant categories of terrestrial circadian animals: nocturnal and diurnal (see Plate 3, Photos C, D, E, and G). However, no two niches of an ecosystem are identical, and the parameters of the solar light cycle to which an animal is exposed depend on its life cycle. The daily light patterns in a wooded location on a single day depend on position in the forest from ground level to treetop and on degree of cloudiness (Figure 2.1).

Organisms in natural environments encounter many temporal opportunities as well as challenges[5,13,14,59,65]

TEMPORAL ASPECTS OF GENETIC FITNESS. Every organism requires food resources to survive and to reproduce before death. Whereas feeding and reproduction are primary

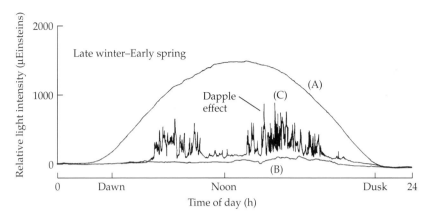

FIGURE 2.1 Light profile of a temperate pine and hardwood forest in spring. (A) Clear sky at treetop level. (B) Ground level on an overcast day. (C) Ground level on a windy, clear day with the dapple effect of sunlight penetrating the moving leaves. (From DeCoursey, 1989.)

behavioral functions, activities such as locomotion, social interaction, grooming, courting, or migration are chiefly auxiliary activities that provide support for feeding and reproduction. One of the key behavioral limitations for reproduction is acquisition of nutrient energy sources for sustaining growth, for maintaining metabolic activities, and for producing young. In an inexorable track, each organism that bridges present and future must complete its developmental stages, grow to adult size, and reproduce in timely sequence. The ability to place viable offspring in future generations, referred to as genetic fitness, depends on both spatial and temporal factors.

Organisms complete the events of their life cycles in a continually changing environment that varies in rhythmic fashion related to the geophysical properties of the solar system. Orbital characteristics of Earth and its moon result in dramatic and predictable changes in the environments of animals. The rotation of Earth on its axis creates the daily solar cycle, and Earth's orbit around the sun causes the annual cycle of the changing seasons. The rotation of the moon around Earth creates the short-term tidal cycles of ebb and flood tides or the longer-term spring and neap tides of differing tidal heights. Every living creature enhances its chances for survival in its niche by maximizing its efficiency. Paramount to survival is the element of the right time. Consequently, most organisms follow environmental changes in a rhythmic fashion.

EXOGENOUS VERSUS ENDOGENOUS TIMING. Rhythmic behavior of a particular animal could be driven exogenously as a direct response to environmental cycles. Alternatively, it could depend on the composite interaction of an internal pacemaker and a corrective rhythmic environmental signal. It might seem simpler for animals to start their activities at the time when a favorable environmental window has opened. Purely exogenous rhythmic responses, however, are often insufficient for optimal

adaptation of organisms. In contrast, internal biological timers provide a combination of precise time structuring and options for behavioral flexibility.

In all of the basic types of endogenous circa-rhythms, a physiological pacemaker generates a free-running rhythmicity that is corrected in phase and frequency to local time by environmental synchronizers or entraining agents. Living clocks include those with circatidal, circadian, circasemilunar, circalunar, and circannual periodicities (Figure 2.2). Internal clocks provide three immense advantages over exogenous timing responses: anticipatory preparation in advance of need, triggering of behavioral events in environments such as dark caves where no environmental cues are available, and continuous time consultation capacity for events such as time-compensated celestial navigation in migration.

Endogenous timing is based on cellular pacemaker systems[44,50,54,69,70]

A clear distinction should be made from the start between rhythms and circa-rhythms. All nonrandom events are, by definition, rhythmic phenomena. They occur at regular intervals with predictable waveform, amplitude, and period without causation necessarily being known. On the other hand, circadian or circannual rhythms are exclusively those that match an environmental periodicity but have been shown to be generated by an underlying endogenous pacemaker system.

All environmentally related pacemaker systems in organisms have three components that are generally neural: the oscillator or generator, the sensory systems for environmental input for entrainment, and the output messengers (see Plates 3 and 4). The details of these systems will be presented in Chapters 3 through 9. In many behavioral studies the arduous work to collect information on even basic aspects of circadian regulation of a rhythmic function has not been completed. A concerted effort has been made to restrict all examples in this chapter to known endogenously controlled rhythmic systems.

Circadian Timing Facilitates a Large Variety of Behavioral Functions[4,10,27,67]

A big challenge for beginners in behavioral chronobiology is the design of experiments for organisms in both the laboratory and the field[1,12,20]

EXPERIMENTAL DESIGN. Several aspects of circadian rhythm research may puzzle the beginner who is just

(A) Circatidal

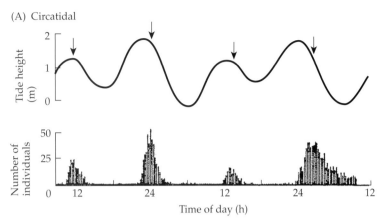

(B) Circadian

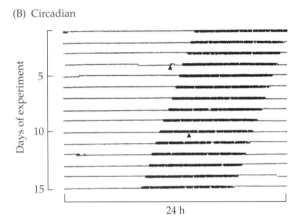

(C) Circasemilunar/circalunar

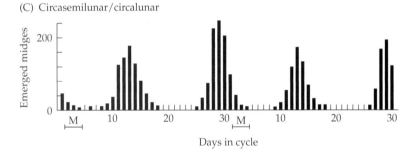

(D) Circannual

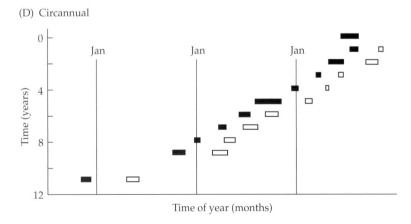

FIGURE 2.2 Circa-rhythms. (A) Free-running circatidal rhythm of swimming activity in a group of freshly collected beach hoppers recorded for two days in constant laboratory conditions. The top trace depicts semidiurnal, unequal tide heights of the southern California collection site; the histograms below indicate the number of individuals swimming in the water column in constant laboratory conditions, based on time-lapse photographs. (B) Free-running rhythm of wheel activity of one flying squirrel in DD. The black horizontal bars on the baseline indicate wheel running. (C) Semilunar emergence of the marine insect *Clunio* sp. from a mass culture in the laboratory during two successive 30-day cycles of LD 12:12 with moonlight entrainment stimulus on days 1–4. "M" indicates moonlight stimulus. (D) Circannual rhythms of summer molt (black bars) and winter molt (white bars) in a garden warbler kept for 10 years in the laboratory. (A from Enright, 1975; B from DeCoursey, 2001; C from Neumann, 1989; D from Gwinner, 1981.)

starting to study animal behavior. The intimidating issues may include the formulation of an important hypothesis, the choice of an animal for testing the questions, or the ways to trap and maintain wild species. Other issues may be how to design the behavioral monitoring systems and data collection equipment, or how to analyze and present the data.

In one sense these are the demands of the scientific method, and they are the same for any biological research. For many chronobiology questions, however, the artifacts introduced by experimental conditions may seriously compromise behavioral data. Wild species of higher vertebrates, for example, are very secretive, easily stressed by changes in conditions such as laboratory caging, and the resultant indirect behavioral markers may then fail to reflect the true underlying state of the clock.

Faced by all of these problems at once, a student may complain, "How do I find a magic ring like King Solomon's?" The following discussion will try to answer some of these questions for a challenging group, the nocturnal insect-eating bats.

BATS AS A CIRCADIAN MODEL. Several intriguing questions about living clocks are particularly relevant to bats, especially the cave-dwelling, insectivorous bats of the Temperate Zone. All species are strictly nocturnal, and many species roost in summer in the remote recesses of cool caves during the day (Figure 2.3A). Most are heterothermic and raise their body temperature to the normal 37°C euthermic level during activity at night, but cool to the temperature of their roosting site during the day. In this way they conserve a large amount of energy. How does an internal, presumably metabolic clock react to these drastic temperature changes and still remain an accurate chronometer? (See Chapter 3 for a discussion of temperature compensation mechanisms as a basic property of circadian clocks.)

A circadian clock awakens the bats several hours before sunset so that they can become warm, alert, and active by twilight. After awakening, they swirl around inside their cave to check the light intensity at the cave entrance; then they emerge at dusk in an impressive stream before fanning out across the countryside. Although they are almost blind, they fly swiftly above ponds and lakes and hunt on the wing in darkness. Their highly developed ultrasonic pulsed calls are used to detect airborne insects and to aid the bats in scooping up half their body weight in flying insects each night.

The famed 8 million bats formerly inhabiting Carlsbad Caverns, New Mexico, darkened the sky as they plumed out of the cavern. A compelling mystery was the way in which they entrained their circadian clocks to local time. In spite of considerable interest in bat circadian rhythms, little progress in solving that mystery was made for several decades because of difficulties in maintaining the bats under laboratory conditions. The story behind a successful research project to answer these questions is a fascinating one that gives insight into behavioral approaches to studying clock questions.

Initial steps involved documenting the nocturnal behavior patterns of free-living bats during the non-hibernating summer months and discovering how they determined twilight departure time. The scientists first located a summer breeding colony of bats in the cavernous attic rooms of an ancient German church. The dark, freely interconnected rooms led gradually to a dimly lit front chamber. A small window in the room provided an exit point for the departure and return of the bats to the colony (Figure 2.3B). Lying on the floor beneath the window for many nights, the scientists took turns counting the number of bat exits. They documented the precise departure time of the bats at a specific light intensity (Figure 2.3C).

The observers found out that members of the bat colony exited during a period of about one half-hour. They also discovered how the bats tested the external light intensity. Several hours before twilight, awakened presumably by their internal clocks, the bats began to warm up gradually. First they shivered, and then they made short test flights up to the exit window to sample the light. At the appropriate twilight intensity each night, they poured quickly out of the exit window and flew off to feed.

The next step of the investigation was even more demanding. Two bats were brought into the laboratory, and each was acclimated to life in its own recording room. The most promising species was the horseshoe-nosed bat of deep Bavarian caves, but insectivorous bats rarely survive in the laboratory at room temperature more than a few days because they require a constant supply of live flying insects for food. In this case, a trainer hand-fed the bats vitamin-fortified mealworms. He gradually conditioned them over a period of months to fly to his hand (Figure 2.3D) and finally to fly to a feeding stand in an activity recording room where they could freely eat mealworms from a dish.

Next a noninvasive recording device for circadian activity was designed. Horseshoe-nosed bats are extraordinary flyers and can hover in a very small space in spite of a large wingspan (see Figure 2.3D). Normally they hang solitarily from exposed stalactites in caves (see Figure 2.3A), but each of the bats studied in the laboratory readily accepted the flight detection device, a simple horizontal piece of wire attached to a microswitch, as a perch. Each bat flew around its small living chamber in short trips all night. It landed every few minutes to hang from the wire perch. This setup allowed a long-term record of circadian activity in a constant environment to be documented with minimal disturbance to the bats (Figure 2.3E).

Some of the data were surprising. The highly precise endogenous rhythm of each bat in its constant isolation room free-ran for 15 days in constant DD with a period close to but not exactly 24 h. The rhythms readily entrained to the light cycle for 31 days and again free-ran in DD afterward (Figure 2.3E and F). The pattern of entrainment was dependent on the time of light interruption of the active flight time. Furthermore, the exact value of the free run after entrainment was considerably shorter than the initial value (see Figure 2.3F). These two important features of all circadian clocks will be discussed in Chapter 3 when phase response curves and aftereffects of entrainment are considered in detail.

Finally the scientists decided to explore light sampling in more detail in the laboratory. Using sheets of plywood, they built a very large simulated cave in a laboratory room. Infrared light beams were installed as motion detectors. One was situated just beyond the roosting perch in the cave and another at the cave entrance. In this way the wake-up and departure from the perch, as well as the light-sampling behavior, could be detected. Over a 28-day period, the activity of three horseshoe-nosed bats living in the cave was recorded automatically, without the bats being disturbed. The summary actograph, especially the expanded portion, impressively documents the early wake-up and warm-up period, and the intense light sampling in the 45 minutes before lights off (Figure 2.3G).

(A)

(B)

(C)

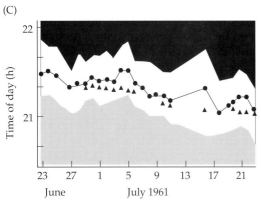

(D)

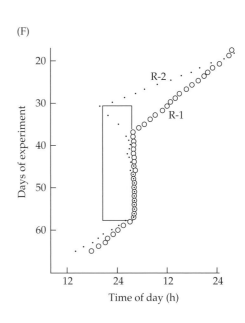

(E)

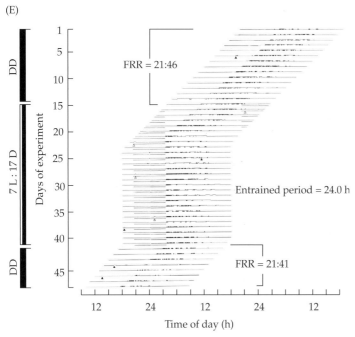

(F)

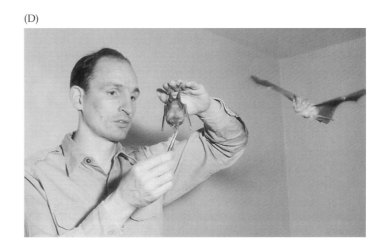

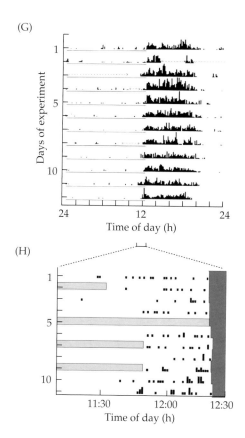

(G)

(H)

FIGURE 2.3 Testing circadian behavioral hypotheses with bats in the field and laboratory. (A) Cave-perching habits of a horseshoe-nosed bat in a natural cave. (B) Schematic drawing of the emergence of the bats at the Wendelsheim Church colony in Germany. The roof cut-aways and numbers 1 to 4 indicate the interior room order and flight path. (C) Summary of the onset of evening activity at the Wendelsheim roost relative to twilight intensity. The white central zone indicates the exit time span. Circles represent mean exit times; triangles, outdoor light intensity of 0.08 lux. (D) Training horseshoe-nosed bats to feed freely in the laboratory. (E) Representative actograph of entrainment of the free-running rhythm by an LD light cycle. The light regime is shown on the left. (F) A summary of entrainment of two bats, R1 and R2, shows the onset of flight of each in its individual constant-environment room. The box indicates the light period. (G) Light-sampling flight activity of three bats living in a simulated cave, showing summed activity as photocell counts/8 minutes, with constant temperature and DD in the cave, and LD 12:12 in the outer room. Underlining indicates the light period. (H) Activity during the bracketed time of day, recorded as flights per minute up to the exit or out of the tunnel. Dark hatching indicates darkness and light hatching indicates no data. (A and D courtesy of P. DeCoursey; B, C, E, G, and H from DeCoursey and DeCoursey, 1964; F from DeCoursey, 1990.)

POSTLUDE. Incidentally, these uniquely trained horseshoe-nosed bats lived for 2 years at the University of Tübingen during the circadian experiments, then traveled across the Atlantic Ocean in a storm-tossed ship for 10 days and continued their journey to Harvard University. Scientists there were able to study their highly developed echolocation sonar pulses. Because the frequency of 100 kilohertz lies far above the 17-kilohertz upper limit of human hearing, the task required precisely focused ultra high-speed photography and ultrasonic microphones. The bats were trained to chase mealworms tossed into the air by a solenoid catapult. Mealworms were aimed at a specific spot for the bats to catch, right in front of the camera and microphone recording equipment.

At the end of the experiments, the bats were maintained for the rest of their long lives at a university and taught many students the marvels of bat flight and echolocation. Surely these three must rank as the most widely traveled and best-educated bats in history. Regrettably, in the four decades since the research was carried out, horseshoe-nosed bats have become endangered because of habitat deterioration in southern Germany.

Circadian rhythms occur in both prokaryotic and eukaryotic unicells[40,41,49,53]

PROKARYOTIC UNICELLS. Single-celled organisms such as bacteria that lack nuclei or membrane-enclosed organelles constitute the prokaryotes. They differ greatly from eukaryotic unicells by having a circular chromosome free in the cytoplasm, with the structural genes grouped into operons. They also lack a cytoskeleton and divide by fission rather than by mitosis. Many are parasitic, and they may be extremely pathogenic. Some are remarkably adapted for existence in the most extreme habitats capable of supporting life, such as extreme salinity sites, near-boiling thermal pools, or deep marine niches in sulfurous underwater volcanic vents.

Only recently has good evidence been collected for true circadian rhythmicity in prokaryotes. In the cyanobacterium *Synechococcus*, regulation of the photosynthetic machinery shows circadian properties. *Synechococcus* colonies free-run in constant conditions, and the rhythms entrain and display temperature compensation (see Figure 5.1).

EUKARYOTIC UNICELLS. In terms of general ecology and circadian rhythmicity, *Gonyaulax* has been the most thoroughly studied of all eukaryotic unicells. It is widespread in ocean waters and often reaches such large numbers that brilliant reddish or orange-gold patches reach out in plumes across the ocean surface. *Gonyaulax* has a highly developed capacity for bioluminescence and is one of the main species contributing to nocturnal luminescence of the sea. The nighttime glowing of a ship wake is the result of the flashes of light induced by mechanical disturbance of myriad *Gonyaulax* and related dinoflagellates. This primitive eukaryote exhibits at least four different circadian rhythms (Figure 2.4A; see also Figure 1.17). The flash rhythm entrains to LD schedules with the free-running period equal to 24.0 h (Figure 2.4B, top) and free-runs in dim LL with a period of 24.4 h (Figure 2.4B, bottom).

(A)

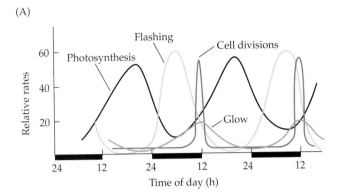

(B)

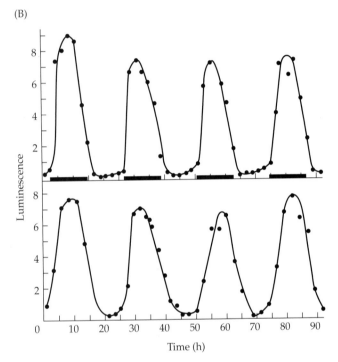

Time (h)

FIGURE 2.4 Circadian rhythms in *Gonyaulax*. (A) Four simultaneous rhythms having different phase relationships. (B) Rhythm of flash bioluminescence of a *Gonyaulax* culture in LD (top) and dim LL of 120 foot-candles (bottom). (A from Hastings and Keynan, 1965; B from Hastings and Sweeney, 1959.)

lowed by observation of the percentage in conjugation. The rhythm is circadian and persists in both LL and DD. The period of the circadian rhythm is light-intensity dependent. The rhythm entrains to an LD schedule and can be phase-shifted by pulses of light.

All of the preceding work was carried out with population cultures containing hundreds or thousands of individuals. In three cases, however, it has been shown that the rhythmicity of unicells is detectable in single organism units, not just in populations. *Acetabularia* produces oxygen during photosynthesis. The rhythmicity persists for many cycles and can be documented in single isolated cells. Similarly, in *Paramecium* the mating reactivity rhythm has been investigated in single isolated cells in culture. A very convincing demonstration comes from the mating reactivity rhythm for 15 randomly isolated *Paramecium* cells taken from an arrhythmic culture. Every cell manifested a clear circadian rhythm out of phase with other individuals (see Figure 3.28). In a third case, *Gonyaulax* exhibits photosynthetic rhythms in single cells.

Daily activity patterns are very widespread in animals[4,12,19,34,61,65,67]

DIURNALITY AND NOCTURNALITY. Diurnal animals are well adapted to the good lighting conditions that prevail during daytime; nocturnal species are specialized for the darkness of night, and a few crepuscular species limit their activity to the dim twilight period (see Figure 2.6). Many diurnal species, for example, have acute visual resolution and color vision, as well as group antipredator tactics. In contrast, nocturnal species have evolved sensory systems and communication strategies such as highly developed olfactory or acoustic senses that do not rely heavily on light, and they use concealing antipredator strategies more often than social antipredator tactics. These behavioral differences depend on basic underlying physiological rhythms. As a result, evolutionary transitions between nocturnal and diurnal lifestyles have been relatively infrequent because they would involve a coordinated suite of

Other protists have pronounced phototactic circadian rhythms. If a brief test light is presented at regular intervals throughout the day in several species of protists, the speed of swimming or the proportion of individuals swimming toward a test light varies with the circadian time of day. One of these protists is *Euglena* (Figure 2.5). Another is the common ciliate *Paramecium*, and still another is *Chlamydomonas*.

Paramecium is highly interesting in terms of its mating reactivity rhythm. Two different mating types occur. The reactivity can be assayed by addition of an aliquot of a population of one type to an aliquot of the other, fol-

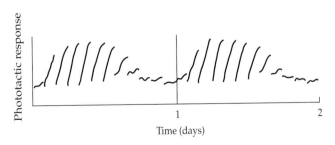

FIGURE 2.5 Circadian phototactic rhythm of a protist, *Euglena*, free-running in constant DD conditions. (From Pittendrigh and Bruce, 1959.)

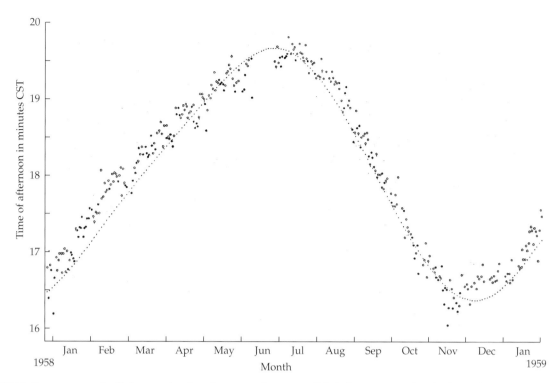

FIGURE 2.6 Emergence of a flying squirrel under natural daylight conditions relative to civil twilight time throughout the year. Open circle, onset of activity on clear days; solid circle, onset on cloudy days; dotted curve, civil twilight. (From DeCoursey, 1989.)

changes in the temporal organization of a wide range of behavior patterns and physiological processes.

One great advantage of a biological timing system is the capacity of an organism to anticipate the arrival of day or night and to prepare in advance for its periods of activity. A circadian clock thus becomes an alarm clock, alerting animals before the time of need. Birds can leave feeding grounds and fly to distant roosting sites in time to arrive before darkness falls. Flying squirrels often build dens deep in hollow trees for protection during sleep. Each squirrel's circadian clock awakens it in advance of darkness and allows it to synchronize the start of its activity with a specific light intensity shortly after sunset throughout the year (Figure 2.6).

Another example comes from a day-active fish, the convict cichlid. A female protects her free-swimming young from predators by retrieving them in late afternoon from the water column and carrying them in her mouth to a pit dug in the sand. She then guards the young through the night. The fact that the females retrieve the young before darkness, even under abrupt LD cycles without a twilight transition in the laboratory, shows that the retrieval is mediated by an endogenous clock.

Nocturnality and diurnality are generally shared by members of relatively broad taxonomic groups. Most birds, for example, are highly diurnal, with the exception of the very specialized owls, whippoorwills, and nighthawks. The first archaic mammals were nocturnal, and a night-active pattern remains the most common one in modern mammals. The majority of mice and bats initiate activity abruptly at twilight. In contrast, the majority of the sciurid family, including ground squirrels, marmots, chipmunks, and tree squirrels, are day-active, and only the specialized flying squirrels are nocturnal.

ACTIVITY RHYTHMS. The solar cycle causes extreme changes of light intensity and temperature in the environment from day to night, and many organisms have become specialized for locomotor activity at a particular phase of the 24 h day. In laboratory conditions the locomotor activity is often the most regular and precise overt rhythm reflecting the underlying internal clock and its entrainment phenomena.

Particularly in rodents, the dramatic wheel spinning exhibits an extremely robust rhythm, and the sharp onset of wheel turning has been very widely used in circadian recording. The entrainment of the free-running rhythm of a nocturnal jumping mouse is very clear-cut and concise (Figure 2.7A). The activity of a diurnal chipmunk (Figure 2.7B) is 180 degrees out of phase with the rhythm of the mouse. The end of activity may often be fairly irregular, changeable, and relatively unreliable. Reasons for the attraction of rodents

(A)

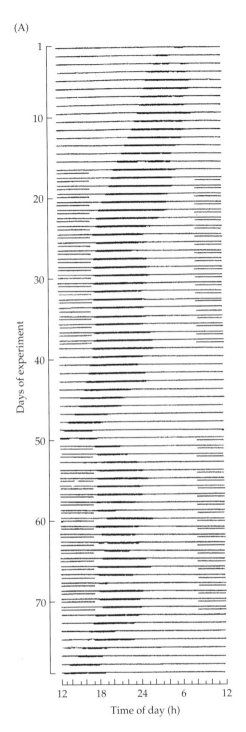

(B)

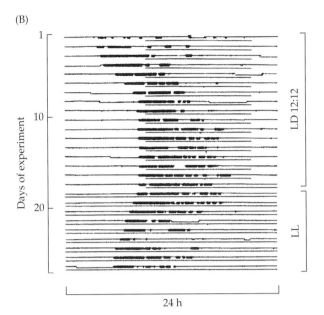

24 h

FIGURE 2.7 Photoentrainment of circadian activity rhythms in artificial LD schedules for a nocturnal and a diurnal species. (A) Wheel-running activity of a nocturnal jumping mouse in two photoentrainment schedules with DD on days 1–15, LD on days 16–37, DD on days 38–48, LD on days 49–70, and DD on days 71–78. Hours of light are indicated by underlining. (B) Wheel-running locomotor activity of a day-active chipmunk reentraining to a new LD schedule on days 1–7, in stable entrainment on days 8–17, and in an LL schedule on days 18–26. Hours of light are indicated by underlining. (From DeCoursey, 1990.)

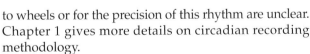

to wheels or for the precision of this rhythm are unclear. Chapter 1 gives more details on circadian recording methodology.

SONG AND OTHER COMMUNICATION RHYTHMS. Other circadian activities may also be very clear and precise. Birdsong chorusing at dawn is a well-known event. Early Greek naturalists noted its regularity over 2,000 years ago. Many insects express circadian activity in a variety of functions. Fireflies flash at twilight in a spec-

tacular display when males begin to court females with a species-specific pattern. The color of the bioluminescent light, the flash pattern, and the flashing rate identify a species. A female ready to mate responds with a flash to identify her position and receptivity.

Other well-known insect activity rhythms include chorusing during the reproductive season. Thousands of katydids drone their monotonous "katydid-katydidn't-katydid-katydidn't" from dusk until dawn on warm summer evenings in favorable habitat. Seventeen-year locusts, on the other hand, rasp their buzzy song on hot days the entire day long. The locust produces the sound by rasping a special comb of spines on its legs against the stiff, hardened edge of a wing. Crickets also use stridulation noises in a circadian mating call. The calls are highly species specific, and in areas where five or more species are sympatric, the calls are important for species identification for the mating males and females. An excellent documentation of the circadian nature of the call is available for the cricket *Teleogryllus* (Figure 2.8). Frogs are also famous for their twilight and nocturnal chorusing using species-specific calls.

SLEEP–WAKE RHYTHMS. Another aspect of daily scheduling of activity time involves the sleep–wake rhythm. During their rest phase, animals conserve energy and remain inactive. In higher vertebrates, inactivity is often

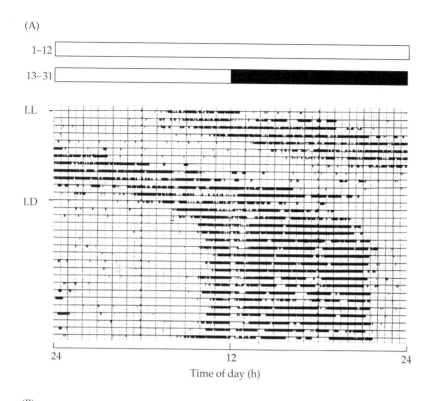

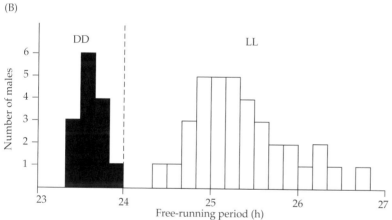

FIGURE 2.8 Mating call of crickets. (A) Single-plot actogram of the stridulation mating call rhythm in a male cricket *Teleogryllus*, free-running in LL on days 1–12, entraining to an LD light schedule on days 13–15, with stable entrainment on days 16–31. Light bars give LD schedules. (B) Histograms of mean free-running stridulation rhythms of 14 male crickets in DD and 36 male crickets in LL. (From Loher, 1972.)

rhythms of humans are covered in Chapters 9 and 10.

Daily torpor and hibernation bouts reduce metabolic demands in some mammals[47,64]

Related to daily sleep–wake rhythms is the periodic arousal from a state of lowered body temperature of several species of rodents and bats at a specific circadian phase (see Plate 2, Photos B and D). Daily torpor may occur in small normally euthermic rodents such as pocket mice or hamsters even during warm months of the year. Such an adaptation conserves energy during an animal's sleep phase because metabolic rate is proportional to temperature. In general, each lowering of the reaction temperature lowers the metabolic rate two- or threefold. As a result, considerable energy is saved even from brief periods of daily torpor. The pronounced daily reduction of body temperature in Temperate Zone bats is even more striking. The horseshoe-nosed bat hangs upside down during daytime hours of summer in cool caves, removed from any clues about the solar day–night progression. It remains torpid until just before twilight. Then its circadian timer starts a process of thermogenesis by muscular shivering, ensuring warm-up to the active level just before dusk. The activity pattern, including warm-up, is controlled by the bat's circadian clock.

In profound hibernators such as the golden-mantled squirrel, summertime temperatures are euthermic and oscillate in circadian fashion from day to night over a range from about 35 to 39°C. In autumn, as the mean daily temperature begins to drop and day length to shorten, the squirrel enters a torpid state of hibernation lasting up to 7 months. Body temperature adjusts to the ground temperature as a squirrel lies curled up in its nest deep underground, as shown in the figure at the beginning of this chapter. However, it cannot survive the 7-month chilling to a few degrees above freezing without periodic warm-up bouts to the euthermic level. It must remain warm but torpid for at least 12 hours every week or two in order to sustain brain and kidney function through the long winter (Figure 2.9).

The wake-up bouts are regulated partly by a circadian timer. Although cyclic bouts persist after removal

enforced by sleep. One possible role of a circadian clock is to ensure that sleep coincides with the time of day for which the animal is ill adapted. Rats and hamsters, for example, are nocturnal and sleep mostly during the day. SCN-lesioned rats, however, sleep in short bouts distributed equally between day and night. In nature, such time-impaired rats would needlessly expose themselves to diurnal predators by being active during the day and would also lose foraging opportunities by falling asleep at night. Important aspects of the well-known sleep

(A) The calendar of the golden-mantled squirrel

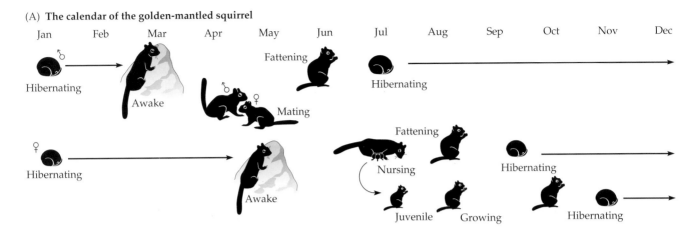

(B) Body temperature September – May

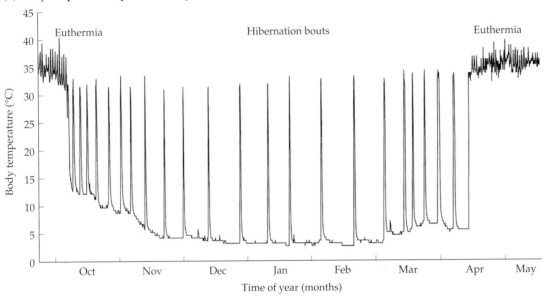

FIGURE 2.9 Hibernation performance for wild, free-living golden-mantled squirrels in a wilderness habitat. (A) Annual reproductive and torpor timetables of adults and juveniles. (B) Body temperature profile for one free-living adult female from September 2000 to June 2001 measured by an implanted telemetric data logger. Warm-up bouts during the hibernation period are shown, flanked by the summer-time euthermic period. (Courtesy of P. DeCoursey.)

of the SCN pacemaker, at least under laboratory conditions, they appear much more irregular and infrequent. Because warm-up bouts by means of muscular shivering are very expensive metabolically, the regularity may be essential in rationing out a squirrel's limited fat supply to last the entire winter. In the extreme case of the Arctic ground squirrel, hibernation may continue for 9 months of the year.

Time sense, time–place learning, and celestial navigation are functions that require continuous consultation of a circadian clock[3,6,23,38,46]

TIME SENSE, OR ZEITGEDÄCHTNIS. Animals that possess a well-developed time sense can use their circadian clocks to arrive at a feeding site at the time of day when food is most abundant. One of the best examples comes from honeybees and is closely linked to the ecology of the flowers from which bees gather nectar and pollen. Cross-pollination is advantageous for most flowers and is enhanced when pollinators visit many flowers of the same species in succession. Consequently, some species of flowers make nectar or pollen accessible for only a short time each day. Honeybees are able to maximize foraging efficiency by appearing daily at the food source only at the time of food availability. They come even on days when no food is available, showing that they are responding to a learned time sense or time memory (see Plate 4, Photo D). The German name, Zeitgedächtnis, is often used to describe time sense phenomena. See also the section on circadian rhythms in plants, later in this chapter.

Time sense in bees is an endogenous phenomenon. It is clear that bees do not use external cues such as position of the sun to remember time, for they can be successfully trained to come to test dishes at specific times even in the depths of salt mines, where daily cycles of light, temperature, and humidity are absent. In translocation experiments, bees in Paris were trained to come to a feeder at 09:00; when moved by airplane to New York, they still fed at 09:00 Paris time (03:00 New York time), at least for the first few days. These bees disregarded external cues, which are very different at 03:00 and at 09:00, and followed their endogenous clock.

Time sense of bees has circadian properties. Time of arrival at a test dish free-runs in LL with a periodicity of 23.4 to 23.8 h. The endogenous rhythm can continue for several days without reinforcement. Such persistence is a definite ecological advantage because bees must stay inside the hive all day during bad weather. After several days, however, the response is extinguished. Short-term persistence is also advantageous because flowers produce nectar for only a few days, and nectar sources may change fairly rapidly from one week to the next. Another circadian property is entrainment of the rhythm after changing of the light schedule; the shifts take place only gradually. A similar ability to anticipate mealtime every day has been demonstrated for several species of fish, birds, and mammals, but links between time memory and the ecology of these animals have not been documented.

TIME–PLACE LEARNING. Even more complex than time memory is time–place learning, for it involves association of food at multiple sites at specific times. As in Zeitgedächtnis, a circadian clock is used to distinguish among the various times of day. Garden warblers and starlings were kept in a laboratory setting consisting of a central area and four adjacent rooms. Each room was equipped with a feeder that delivered food for 3 h per day, with each room's feeder available at a different time of day. Although this complex task involved four different time–place associations, the birds were able to make correct choices at least 75% of the time after 3 to 11 days of conditioning.

On a test day when all feeders were continuously filled with food, these two species still showed a preference for the learned time and place, indicating that their behavior depended on an endogenous mechanism, not on the availability of food. The time–place associations had circadian properties. In constant dim light and constant food availability, the period free-ran with a value slightly less than 24 h. When the LD cycle was advanced by 6 hours, the timing of room visits shifted gradually.

Evidence for time–place learning has been obtained in a few additional species of birds, as well as in some insects, fish, and mammals. For some cases, a clear eco-

logical relevance was evident. The time–place learning ability of seed-eating weaver finches is not as robust as that of insect-eating weaver finches, possibly because seed availability does not vary spatiotemporally as much as insect availability does.

Golden shiner minnows can learn to forage in one part of an aquarium in the morning, in another part at midday, and back at the first area in late afternoon. They have difficulty, however, learning to forage in three different places at three different times. In natural lake habitats, these fish feed in open waters at dawn, move to the littoral zone during the day for shelter, and come back to open waters near dusk. Thus a parallel is seen between the laboratory situation and natural habitat.

Still another example involves sparrow hawks, which feed on mice and insects. Their daily hunting routine often occurs in a predictable pattern that could be attributed to time–place learning related to prey availability.

TIME-COMPENSATED CELESTIAL NAVIGATION. Some animals can use their clocks to compensate for the changing position of the sun on long-distance journeys to a precise end point. Examples include birds migrating thousands of miles south or north with the changing seasons, bees flying several kilometers from a hive to a known food source, or accidentally displaced crustacean inhabitants of a shoreline returning to shelter. The azimuth position of the sun can be used as a reference compass in navigation, provided that the animal can compensate for the daily movement of the sun across the sky by means of an endogenous circadian clock.

Some of the most convincing work on celestial orientation has involved studies of birds kept in small orientation cages during the period of migratory restlessness (Figure 2.10A). If housed in a room with four windows that allow views of the sun throughout the day, a bird will hop in a fixed direction on its perch. The orientation can be modified by deflection of the sun's rays with mirrors. The results indicate that the cue is the sun, not room details. Similarly, if the bird is always trained to seek food in one compass direction, it will fly in this direction even when tested at a time of day that differs from the training time. Once again, the fact that the direction will change if mirrors are used to deflect the sun's rays shows that room cues are not involved. When birds use the sun as a cue for orientation, they must have an endogenous clock to compensate for the daily changes in the sun's position. Sun-compass orientation in migratory birds has circadian properties. A phase shift of the light schedule by 6 hours results in a gradual shift of the directional choice by 90 degrees (Figure 2.10B). In constant light, the choice of direction will free-run together with the bird's activity rhythm.

Similar results have been obtained in experiments with insects, spiders, crustaceans, fish, amphibians, rep-

(A) Activity onset

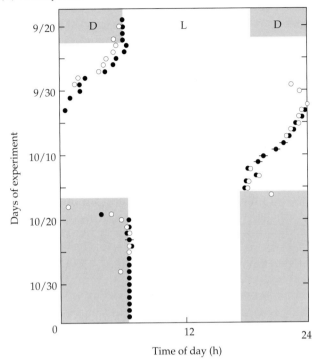

(B) Orientation direction

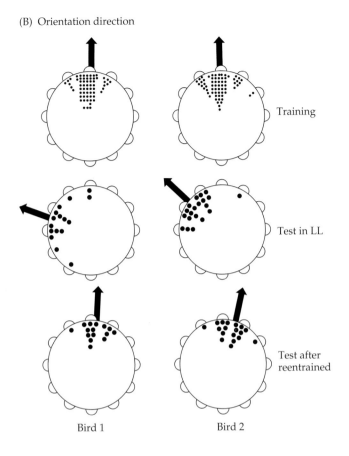

FIGURE 2.10 Time-compensated celestial orientation. (A) Onset of activity for two starlings, represented by open and solid circles, respectively, in constant conditions. Shaded areas indicate darkness. (B) Vector diagrams of orientation choices during training, in LL, and after reentrainment. (From Hoffmann, 1960.)

tiles, and mammals. Honeybees, for example, use sun-compass orientation during foraging. After finding a distant food source, a bee can indicate the flower's location to other bees by performing an elaborate waggle dance on the vertical honeycombs inside its hive (see Figure 1.9 in Chapter 1). The angle of the dance with respect to the top of the hive indicates the horizontal flight angle relative to the position of the sun.

After finding an important food source, some bees stay in the hive and dance for many hours. During this time the dance pattern changes comparable to the sun's movement in the sky. The simplest explanation is that the dancing bees use their circadian clocks to correct for the moving sun compass and consequently are able to keep giving accurate information to other worker bees.

Gated rhythms in development are one-time occurrences in a population[57]

Animals can use their biological clocks to ensure that critical developmental events take place at appropriate times of day. The birth of a mammal, the hatching of a bird or insect egg, and the emergence, or eclosion, of an insect from its pupal case are all examples of events that happen only once in the lifetime of an individual organ-

ism. These single events are not rhythms in the usual sense, but they may involve population synchrony as well as phase control. Although the animal may be ready to hatch or molt at any time of day, it waits for the permissive circadian time.

The classic example of such a gated rhythm is the eclosion of new adult fruit flies from their pupal cases (Figure 2.11). In nature, under a photoperiod of about LD 12:12 eclosion always takes place near dawn. In the laboratory, populations of pupae transferred from LD to DD will show a rhythm of eclosion that free-runs with a period very close to 24 h and persists for several days until all pupae in the population have emerged. A biological pacemaker guarantees that the pupae, developing in the darkness of a soil substrate, emerge during the time of high humidity at dawn. The fruit fly gating clock has circadian properties: entrainment by daily light or temperature cycles, phase shifting by light pulses, and temperature compensation.

Theories about the ecological significance of circadian control over developmental stages abound, but experimental data are rare. Freshly hatched or newborn animals may survive better under conditions of temperature and humidity found only at specific times of day. The argument is particularly good for the pupal eclosion of fruit flies because the middle of the night may be too cold and the daytime too dry for optimal

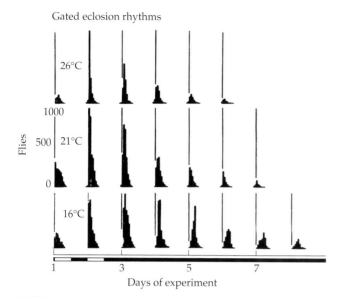

FIGURE 2.11 Fruit fly pupal eclosion rhythms at three temperatures under LD 12:12 for 3 days followed by constant darkness, as indicated in the light bar below. The histograms show the number of flies hatched per hour. Vertical reference lines indicate 24 h periods. (From Pittendrigh, 1954.)

survival. In other cases, young animals may be vulnerable to predation and may benefit by appearing at daily times when predators are absent or satiated with other prey. In some cases, mass eclosion or birth may swamp the predator and provide individual benefits from a group dilution effect. For insects that mate immediately after emergence, gated appearances may synchronize the activity of males and females. The actual determination of the fitness value of gated rhythms for a population phenomenon is difficult because the selective pressures are probably exerted over many generations.

Intra- and interindividual coordination is important for all rhythmic species[39]

Circadian timekeeping systems play a role in the internal temporal coordination of physiological and behavioral processes of organisms. Providing this internal temporal coordination may indeed be one of the most important functions of the circadian system. For example, when animals such as rodents become active after a long period of rest, the energy required for activity must be supplied from storage sites within the body. Prior to the active period, adrenal secretion of glucocorticoid hormones rises, resulting in the release of energy from various storage sites. In this way the circadian system facilitates behavioral activity by appropriately timing the release of necessary energy reserves.

The estrous cycle of female golden hamsters provides another example of adaptive internal coordination by the circadian system. For the 3 days preceding the

day of estrus, ovarian follicles grow and estrogen levels rise. On the day of pro-estrus, when estradiol levels are high, a circadian signal triggers a surge in secretion of luteinizing hormone (LH) from the anterior pituitary gland approximately 4 h before the daily onset of activity. The LH surge acts on the ovary, causing mature follicles to rupture and eggs to be released about 10 h later, at about midnight. The LH surge also triggers a burst of progesterone, and this hormone causes the female to enter a period of sexual receptivity that begins in the evening and lasts until the following morning, encompassing the time of ovulation. The coordinate circadian processes open a favorable window for mating during the middle of the nocturnal activity period.

Examples of coordinated circadian activity in groups of organisms are also known. A female rabbit hides her pups in a nest and returns briefly for feeding only once each day. The young pups nurse voraciously during each daily visit and obtain enough milk to survive for the next 24 h interval. Limited nursing presumably minimizes the chances that a predator will detect the helpless young by sound or odor. The pups anticipate the daily visit of the mother by opening the nest in order to facilitate their access to her when she arrives. The timing of anticipatory activity must be under circadian control because the young rabbits continue to open the nest at the usual time even when the mother repeatedly fails to appear.

Social synchrony is particularly important for animals living in groups. The advantages of group living include better detection or deterrence of predators, improved discovery and harvesting of food, and improved thermoregulation. Endogenous timing promotes group cohesiveness by mutual entrainment of group members. Ecologically relevant examples of social entrainment include the free-running but synchronous activity of beavers under the ice in winter (see Figure 3.3A), entraining by female mice of suckling and related activities of their litters, and the occupation of different temporal niches by rodents or lizards depending on their dominance status.

Circadian Rhythms Are Also Widespread in Higher Plants[9,66]

For this brief survey, higher plants are defined as multicellular, aerobic, photosynthetic, autotrophic organisms. Adaptive timing in the prokaryotic cyanobacteria and in green microalgae was covered already in the discussion of unicells. Almost all species of higher plants possess circadian rhythms of photosynthesis. The concentration of enzymes for photosynthesis is low at night but increases before dawn in anticipation of high activity starting with first light.

In addition, many plants exhibit endogenous daily leaf movements. DeMairan's demonstration in 1729 of rhythmic movements of the sensitive plant under constant conditions, which was described in Chapter 1, has excited plant physiologists over the ensuing centuries. Additional plants that exhibit leaf movement are the *Mimosa* tree and the bean plant *Phaseolus*. Possibly leaf closure is important for reduction of water loss by transpiration or for increased cold tolerance.

Another area of concentrated research on circadian rhythms of higher plants concerns the blooming of flowers, both daily and seasonal. Plants are categorized in major subdivisions. Relatively primitive spore-bearing species are separate from seed-bearing plants. The seed bearers are divided into the cone bearers, or gymnosperms, whose seeds are not encased by parental tissue and the more advanced flower bearers, the angiosperms, whose seeds are encased in an ovary or fruit tissue. Interest in daily blossom time may have originated in the so-called flower clock of the renowned botanist, Linnaeus. Still popular in botanical gardens today, colorful flower clocks proclaim the time of day by the arrangement of blooming plants in clocklike segments relative to the time of opening or closing of their flowers (Figure 2.12).

The flowering seed bearers also attracted attention from chronobiologists because of their intricate pollination symbioses with insects that are dependent on biological timing. Flowering plants coevolved with insects, and many plant species depend on bees for cross-fertilization. As a result, plants have evolved several circadian timing strategies to assist in maximizing pollination by insects at the optimal time of day. Numerous species close their petals at night to keep their pollen dry, thus facilitating packaging by insects prior to pollination of other plants. Many species open their petals for only a few hours each day to ensure the collection of a single species of pollen by bees and the delivery of that pollen to a flower of the same species. Morning glories, for example, open in early morning, but the related four-o'clocks open late in the afternoon. As a result, the chance that an insect will carry pollen between a morning glory and a four-o'clock is very low. One unusual plant, the night-blooming cereus, opens its spectacular flower for only a single night, and it depends exclusively on night-flying moths for pollination. Other flowers dip their heads at night and raise them after dawn to protect the pollen from dew. This motion is also under circadian control.

Many species of flowering plants, such as snapdragons and numerous orchid species, release their bee-attracting fragrances on a daily pattern. Most well-

FIGURE 2.12 A Linnaean flower clock showing the hours in terms of flower opening.

known orchids, such as the cattleyas used in corsages or wedding bouquets, release their perfume during daylight hours (Figure 2.13A). In contrast, the remarkable Darwin's orchid gives off its fragrance strictly at night. Its long spiral petals reach a length of almost half a meter, and its nectary is almost as long (Figure 2.13B). Although the pollinator of this orchid was unknown during Darwin's lifetime, he predicted that a night-flying moth with an extensible proboscis or nectar-siphoning mouthpart would eventually be discovered. Nearly 40 years later, just such an insect was finally seen pollinating this orchid. Presumably such relationships between flowers and their pollinators are based on endogenous circadian timers, but few have been tested under constant laboratory conditions.

Numerous species of plants carry out their life functions on a strict seasonal timetable by means of photoperiodic responses to day length. Long-day or short-day behavior is particularly common among temperate and subpolar plant species. In these regions, most plants grow new leaves and stems, flower, and set seed in the warm or wet times of year. In preparation for the harsh, cold months of winter or for the dry season, the leaves of these

(A)

(B)

FIGURE 2.13 Rhythmic secretion of insect-attracting perfume by orchids. (A) *Cattleya* orchids like this hybrid release their perfume during the day. (B) The moth-pollinated Darwin's orchid releases its perfume at night. (A courtesy of P. DeCoursey; B courtesy of R. Raguso.)

plants drop, their seeds mature, and they enter dormancy. Winterizing requires considerable structural and physiological change that must begin long in advance of the arrival of cold weather. As the day length increases in spring, photoperiodism also initiates the new cycle of budding in deciduous trees and the development of reproductive structures. Only extreme cold can delay the spring behavioral changes of plants brought about by changes in day length. In all these ways, the major events of a nontropical plant's annual cycle are regulated by circadian photoperiodic responses. Photoperiodic responses are considered in detail in Chapter 4.

The Behavior of Many Marine Organisms Reflects Tidal and Lunar Environmental Cycles [18,26,51,52,55,74]

Rhythmic complexity is prevalent in ocean niches [18,26,55]

TIDAL AND LUNAR ENVIRONMENTAL RHYTHMS. Oceans cover 71% of Earth's surface. Because of its fluid nature, this great body of water within confluent basins is easily deformed by the lunar gravitational pull. The dramatic tidal, semilunar, and lunar cyclic changes in water level that result from the moon's influence profoundly affect the lives of many marine organisms. Lunar influences are especially strong for intertidal, estuarine, and nearshore species. In addition, the regular daily solar cycle and moonlight cycles are important rhythmic elements in many marine habitats. However, the lunar-related environmental signals are not entirely regular in frequency because local weather factors may play a role

in their expression. Wind, for example, may greatly affect the amplitude of the tides, and cloud cover may obscure moonlight.

All of these factors result in extremely complex environmental periodicities in marine environments, particularly in shallow coastal waters and estuaries. The lack of predictability of tidal patterns, as well as the uncertainty of expression of moonlight patterns, may explain the relative paucity of well-developed lunar-related rhythms in marine organisms. Elements of tidal, semilunar, lunar, and occasionally diurnal rhythmicity may appear in the behavior patterns of a single coastal species. For this reason the following discussion is organized by behavioral function rather than by periodicity.

LOCOMOTOR RHYTHMS. Activity rhythms are pronounced in some marine organisms. Fiddler crabs of the genus *Uca* leave their burrows and forage in large aggregations on exposed mudflats during the daytime low tide. When fiddler crabs are housed in actograph recording cages in the laboratory under constant conditions, locomotor activity continues for several days with elements of tidal, daily, and semilunar rhythmicity. The rhythms, however, are often quite irregular and usually become arrhythmic, thus casting some doubt on their endogenicity.

Several other marine species in coastal waters move rhythmically in synchrony with the environmental cycles of their habitat. *Emerita*, *Synchelidium*, and *Excirolana*, for example, are three sandy-beach crustaceans that live in the surf zone and move vertically in the water column with great regularity. When freshly collected specimens are held in laboratory seawater tanks under constant conditions, they continue to swim up into the water column in a pattern that closely parallels the timing and amplitude of concurrent tide heights on the beach of origin (see Figure 2.2A). Tide-related movements up the beachfront have also been documented for several species of fish, crabs, and shrimp, as well as the bivalve mollusc *Donax*. These species migrate up and down the beachfront in synchrony with the ebbing and flooding tides. They rise into the surf zone with each breaking incoming wave and burrow as each wave recedes. An endogenous basis for the rhythmicity, however, has not been shown for these latter species.

In deep ocean waters, a stratified layer called the deep scattering layer contains huge numbers of plankton, chiefly larval shrimp and fish. The impressive phenomenon of these organisms rising and falling hundreds of meters in the water column on a predictable daily schedule is called vertical migration. The deep scattering layers were discovered in about 1940, when U.S. ships began using newly developed sonar scanning devices for depth sounding to locate enemy submarines. At first the crews feared that the constantly moving layers of unidentified objects were fleets of submarines. Later exploration with submersibles located and identified the marine organisms involved. Little is known about the regulation of vertical migration.

Relatively few marine vertebrate organisms use lunar-related rhythms. An exception is the Galápagos iguanid lizard. These large lizards graze on algae in the intertidal zone of several islands during low tides. They walk from distant resting areas in anticipation of the low tide, apparently in response to an internal timer.

REPRODUCTIVE RHYTHMS. Lunar-related rhythms of reproduction are well known for a variety of marine organisms. One of the earliest descriptions of lunar rhythms appears in the sailing logbook of Christopher Columbus during the turbulent days of early October 1492. Threatened by a mutiny of his rebellious crew, Columbus peered intently toward the west one evening and saw a faint light on the horizon, glowing like a flickering candle. Assuming that the light emanated from human-built bonfires, he landed at San Salvador on October 11, 1492. The actual origin of the light remained a mystery for almost 500 years, until L. R. Crawshay of Plymouth Laboratory, England, studied the brilliant bioluminescent display and breeding of the Atlantic fireworm, *Odontosyllis*. The bioluminescent glow peaked during an 18 h period each summer month starting one hour before moonrise of a specific moon phase. Crawshay documented that the night of October 10, 1492, was a swarm night for the fireworm. In this unexpected way, lunar rhythms played a role in the momentous discovery of America by European explorers.

Another fascinating example of lunar reproductive rhythms involves a marine polychaete, the palolo worm. For centuries, the natives of Fiji and Samoa have eagerly collected the gametes of these highly solitary benthic worms for food. At the time of breeding the worms fragment their terminal body segments into gametic capsules called epitokes. The capsules float to the ocean surface and rupture, allowing synchronous release of gametes. Fertilization follows immediately. The major rising of the epitokes takes place at dawn of the full moon in November. A lesser breeding event precedes the primary event by one semilunar cycle.

Reproductive behavior of fiddler crabs also reflects the 14 day semilunar cycle of an estuary. The highest amplitude ebb–flood tides of the cycle occur at the new or full moon throughout the year and are called "spring tides." The ebb–flood patterns are modulated during the ensuing days to the lowest amplitude of the cycle, called the "neap tides," which occur at the time of the first and last moon quarters (see Figure 2.15E). An adult fiddler

(A)  (B) 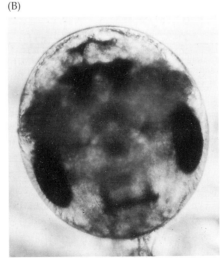 (C)

FIGURE 2.14 Egg release and hatching in the fiddler crab. (A) An egg-bearing female fiddler crab emerges from her burrow after incubation of her egg sponge, which is seen ventrally below the front claws. She is about to step into high-tide water (foreground) to release her larvae, which are called zoeae. (B) An attached egg moments before release. Prominent eyespots are visible. (C) A zoea larva in the first moments after hatching. (Courtesy of P. DeCoursey.)

crab female mates in summer at the time of a spring tide. She extrudes about 100,000 fertilized eggs onto sticky hairs of her abdomen, then retreats to her burrow for the next 14 days. As the flooding waters of the next spring high tide reach her burrow, she steps out onto the mudflat surface (Figure 2.14A) carrying her fully developed eggs (Figure 2.14B) and runs into the water. She quickly ruptures the membranes and releases the mass of newly hatched larvae (Figure 2.14C). The tide soon ebbs and transports her larvae downstream to feeding areas. After a month of development, the larvae metamorphose and migrate back upstream for adult life.

Analysis of field events shows high precision and synchrony of fiddler crab egg hatching throughout an estuary. Larvae are released at the headwater creeks at the new or the full moon, primarily during the hour of the highest spring tide each semilunar cycle (Figure 2.15A). A census was carried out each day to count the number of egg-bearing females on the surface substrate at dusk. The stages of the egg masses were easily deter-

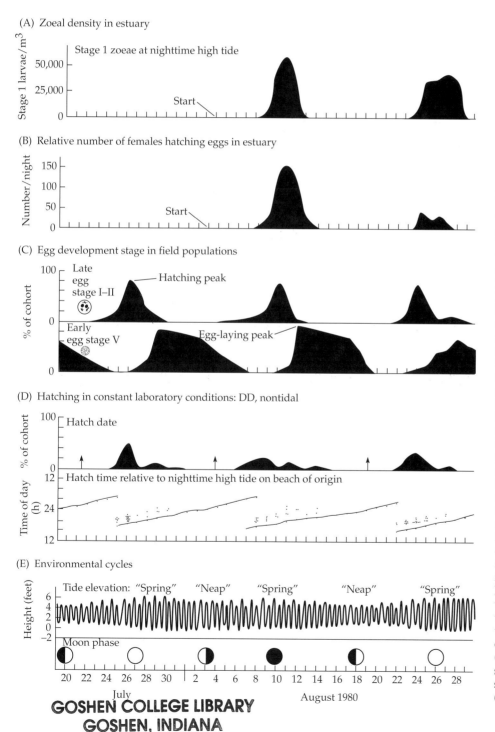

(A) Zoeal density in estuary

(B) Relative number of females hatching eggs in estuary

(C) Egg development stage in field populations

(D) Hatching in constant laboratory conditions: DD, nontidal

(E) Environmental cycles

FIGURE 2.15 Reproduction in the sand fiddler crab. (A) The number of newly hatched fiddler crab larvae per cubic meter pumped each evening during the 90 minutes following nighttime high tide at a shallow estuarine creek station. (B) The number of egg-bearing females walking down to the water's edge to release larvae at evening low tide along a 50 m census line. (C) Egg development stage based on standardized samples of 50–100 individuals collected daily. The stages were determined immediately after collection under a dissecting scope through the use of morphological characters. Late egg stages (top) are within 24 hours of hatching; early egg stages are shown in the bottom trace. (D) Persistent hatching rhythm of three consecutive cohorts of 32 egg-bearing crabs each, collected 7 days before full or new moon as indicated by the arrows, then maintained individually and monitored continuously for peak larval hatch time under constant conditions. Hatch summaries are shown in upper trace; each dot in the lower graph indicates one hatch, and the line shows the time of nighttime high tide on the beach of origin. (E) Predicted tidal elevations on the beach of origin (top), and moon phases (bottom). Open circle, full moon; solid circle, new moon. Date scale applies to all graphs (A–E). (From DeCoursey, 1983.)

mined based on the large amount of orange yolk in the freshly laid stages and the presence of translucent larvae with large, black paired eyespots in the late stages (see Figure 2.14B). Females with late-stage egg masses predominated during the time of high spring tides (Figure 2.15B). A sample of 50 to 100 egg-laden females was also dug from the mudflats each day. Data from females with fully developed egg masses (Figure 2.15C, top) coincided with the behavioral date of egg hatching (see Figures 2.15A,B). Several days later, the samples yielded only females with freshly laid egg masses (Figure 2.15C, bottom).

An egg-bearing female survives well in constant laboratory conditions in a small chamber with a trap for collecting hatching larvae. The hatching event of each female is precise relative to the time of the nighttime high tide on the beach of origin, but the hatches of an entire group of females are spread over about six days (Figure 2.15D). The phases of reproduction of fiddler crabs favor survival in the surging tides of an estuarine and coastal niche.

Another remarkable lunar-correlated reproductive cycle is seen in the midge *Clunio*, the only known marine insect (see Figure 2.2C). All species of *Clunio* are superbly adapted to a specialized sublittoral niche. The short-lived adult lays its eggs on algal mats that are briefly exposed only at the daytime low-water mark of spring tides. Within an hour the eggs are immersed again. Larvae develop underwater on the algae and pupate, still covered with seawater. When the pupae are exposed to air at the next spring ebb tide, new adults emerge. These adults immediately mate and lay their eggs, thus starting a new generation.

Dramatic lunar-related breeding behavior is also seen in a small marine fish, the Pacific grunion (Figure 2.16). Mating and egg deposition are carried out at semi-lunar intervals during the spring and summer months on the night following the highest spring tides. Just after high water, the females ride the waves to shore. They quickly burrow into the sand and deposit their eggs a few inches below the surface. A male arches around each female and fertilizes her eggs. Both fish then follow the receding waves back to the ocean.

The mating is timed for the greatest possible survival of the developing young. The eggs remain buried for 14 days in moist sand above the high-water mark while the embryos develop to hatching stage. The swirling waves of the next high spring tide then wash the newly hatched grunion safely out to sea. In the La Jolla, California, area of the United States, large numbers of tourists often gather on the beaches at the time of the grunion runs to watch the exciting spectacle. These selected descriptive examples of cyclic reproduction in marine organisms include only a small part of a rich tapestry of phenomena spanning phylogenetic diversity from unicells to vertebrates.

Pacemaker systems have not been localized for tidal or lunar-related rhythms[18,55]

Relatively little is known about pacemaker function or entrainment of marine rhythms. The complex rhythmic variables of a marine ecosystem are difficult to simulate in the laboratory, and many marine organisms are too fragile to hold for extended periods of time under laboratory conditions. Furthermore, many of the rhythmic marine invertebrates are extremely small or even single celled in structure. As a result, persistent rhythms have been demonstrated under constant laboratory conditions for only a few species. In most cases neither a pacemaker nor potential entraining agents have been identified yet. Chapter 5 discusses the physiological mechanism of the best-studied pacemaker in marine invertebrates: the circadian eye oscillator of the marine molluscs *Bulla* and *Aplysia*.

Biological Timing Systems Are Involved in Seasonal Behavior of Some Animals and Plants[28,32,75]

Three basic strategies are used in seasonal rhythms of organisms[28]

If animals and plants totally lacked seasonal physiological and behavioral adaptations such as winter hibernation or long-distance migration, much of Earth would be relatively uninhabitable. The annual extremes of temperature, precipitation, and hours of sunlight per day in many habitats would severely stress resident species. Many organisms, however, survive in environments with surprisingly large annual extremes by restricting reproduction to the

FIGURE 2.16 The Pacific grunion. (Courtesy of W. Hootkins.)

optimal time of year for survival of the young and by developing adaptations to protect the adults from severe seasonal conditions. Some of these strategies require a long preparatory time, as well as a mechanism for proper annual timing. A ewe breeds in autumn, for example, to allow for the long gestation and consequent birth of lambs in spring. Three general strategies are used for seasonal adaptation: leaving the area, staying in place with periods of dormancy, or staying in place with physiological adaptations other than dormancy.

The first strategy involves migration to a better location for the duration of the harsh season. Creatures as diverse as monarch butterflies, sea turtles, and whales use this strategy, but the most common long-distance migrants are birds (see Plate 4, Photos F and H). Many bird species from temperate regions carry out mass movements away from the poles as environmental conditions become harsher and return as conditions become favorable. Mass migration is much more conspicuous in the Northern Hemisphere. By moving, for example, from Northern Hemisphere feeding and nesting territories in summer to Southern Hemisphere wintering grounds, many bird migrants tap a perpetually mild climate to avoid the cold of the northern winter. The record migration for insects is the 2,000-mile flight of adult monarch butterflies from central Canada to central Mexico each fall (see Figure 5.5). Banded Arctic terns hold the bird record, with a round-trip migration route over 20,000 miles in length. Many large mammals, such as wildebeest, caribou, and reindeer, also make yearly migrations to escape drought or winter conditions. Numerous species of fish and sea mammals, as well as Antarctic penguins, migrate to take advantage of improved feeding in distant areas. Long-distance migration involves many more risks and much more energy expenditure than dormancy does.

The second and third strategies both involve staying in place rather than migrating. The two strategies have considerable overlap, but an arbitrary division is made for convenience. One group uses a physiological strategy of quiescence to extend energy reserves during times of food scarcity. The other group stays active but alters body structure or behavior during times of extended environmental stress.

The quiescent strategy almost always involves a change in physiological parameters: lowering body temperature, lowering metabolism, and becoming dormant or torpid. Ectothermic organisms such as lizards, crabs, and insects rely on external sources of heat. Because they conform in body temperature to the ambient environmental temperature, they become dormant as the ground cools in winter. Endothermic organisms can mobilize substantial metabolic resources for heat production and can easily maintain a relatively constant elevated temperature during favorable times of year.

Some of these animals are able to enter a prolonged period of hibernation using programmed low temperature just above freezing for up to 9 months of the year. The list includes ground squirrels, chipmunks, marmots, jumping mice, and some bears. Hibernating over the winter allows animals to cease eating for extended periods by gradually using body fat reserves built up during the previous summer. A comparable way of escaping extended heat and drought in summer is to enter estivation by seeking a cool sheltered area where body temperature can be lowered and both activity and metabolic rate decreased. In the insect world, many species use diapause to escape inclement weather. In some species the egg is dormant, and in other species development of the pupal stage is arrested during the harsh months of the year. Diapause can be broken by a critical photoperiod, by appropriate temperature signals, or by moisture stimuli. Overwintering of plant seeds and fungal spores is an equivalent way to survive cold seasons.

The third strategy involves staying in place, alert and active, but changing behavior or structure to increase chances for survival. Examples of effectively coping with extreme cold include molting into thicker fur or feathers; storing layers of blubber fat, as whales do for passive insulation against the cold; or huddling in groups in sheltered areas, as bluebirds do in winter. As part of this strategy, production of young is usually delayed until milder conditions return in spring.

Various combinations of these three strategies are used in different species. At one end of the spectrum, a species may emphasize only one type. Arctic ground squirrels may hibernate as much as 9 months of the year to avoid the temperature extremes of the long arctic winter. In contrast, the ruby-throated hummingbird uses multiple adaptations. These tiny birds have extreme anatomical specializations for nectar feeding. Very small size, hovering flight, and tubelike bills allow them to tap high-energy, cellulose-free carbohydrate food with great trophic efficiency, but they must depart northern temperate realms before the advent of flower-killing frosts. Many hummingbirds migrate in late summer from the eastern United States to the Yucatán Peninsula in Mexico for winter and return northward in spring, back to nesting grounds (Figure 2.17). Presumably they use a circadian chronometer for celestial navigation on their 500-mile, nonstop migration flight across the Caribbean Sea.

Because of their extremely high metabolic rate and great surface loss of heat, these birds have also evolved a daily circadian strategy of heterothermy. To conserve energy, they drop body temperature to ambient during their nocturnal rest period. In addition, they use molting and other structural adaptations to regulate insulation on a seasonal basis. This highly specialized nectar feeder takes advantage of all three strategies for seasonal adaptation.

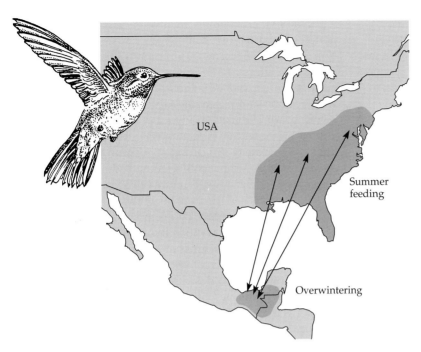

FIGURE 2.17 The ruby-throated hummingbird and its migration route. Part of the ruby-throated hummingbird's seasonal migration trip entails a nonstop 500-mile crossing of trackless water from the southern United States to the Yucatán Peninsula of Mexico.

Timing mechanisms differ among the three strategies[28]

Adaptive strategies for survival in environments with annual climatic extremes involve different mechanisms to guarantee appropriately timed responses. The specific means vary considerably among species. The simplest means of timing is a direct response to a changing and generally unpredictable environment. Some species enter estivation or diapause during extremely dry spells and emerge only when rain falls again. The remaining two mechanisms for programming annual changes in the behavior and physiology of organisms require an endogenous biological timer for efficient time structuring in predictable environments. One evolutionary choice is photoperiodism. By using a circadian clock to measure the length of the daily light period, photoperiodic species can initiate specific functions at the optimal time of year. Finally, timing annual adaptive changes involves rhythms based on a circannual pacemaker. Programming of annual events is common in hibernation of ground squirrels, in reproduction of some mammals such as sheep and deer, and in long-distance migration of many birds. Under constant conditions these rhythms free-run, usually with a period from 10 to 11 months in length but in some species as short as 7 months. In most Temperate Zone species, circannual rhythms are entrained to the 12-month year by photoperiod. Details of circannual rhythms and photoperiodism are considered in Chapter 4.

Evolutionary Theory Suggests the Derivation of Biological Timing[33,63]

Darwin's theory of evolution by natural selection predicts the major steps in development of new traits in organisms[15,17,30,73]

EVOLUTIONARY ADAPTATION. Evolution is the heritable change in populations over time, through the winnowing forces of natural selection on adaptive genes. The term *adaptation* is used in biology in two ways. *Evolutionary adaptation* refers to inherited features that enhance survival or reproduction of an organism. In contrast, *physiological adaptation* refers to the ability of individual animals to adjust or acclimate to an environmental change. The capacity of an animal to entrain its activity to an environmental signal is an evolutionary adaptation, the result of heritable traits. In contrast, the process of entraining to a new light–dark cycle by an organism is primarily a physiological adaptation, for it involves a simple change in an individual's physiological state. In this chapter the term *adaptation* is used to describe an evolutionary feature that promotes survival or reproduction of an organism.

DARWIN'S THEORY OF ORGANIC EVOLUTION. The monumental book *On the Origin of Species by Natural Selection* by Charles Darwin (Figure 2.18) was based largely on observations of the radiation of 14 species of small finches in the Galápagos Islands. Publication of his theory of evolution in 1859 revolutionized the biological sciences. About one century later, the great geneticist Theodosius Dobzhansky would note that nothing in biology can be explained except in the context of evolution.

Darwin outlined the steps by which he deduced that natural selection could lead to evolutionary development of new adaptive features in animals. Because all species produce more offspring than can survive on the limited resources available, intraspecific competition arises. Organisms also exhibit her-

FIGURE 2.18 A lithograph of Charles Darwin from 1859.

itable interindividual variation. In the so-called struggle for existence that results from competition, the fittest variant will produce more surviving offspring and transmit adaptive heritable traits to the young. Organisms with features of structure, physiology, and behavior that make them more likely to survive and reproduce will contribute a disproportionate number of offspring to the next generation. Although Darwin was unable to explain heritability during his lifetime, later research demonstrated the transmission of traits by genetic mechanisms. The fundamental insight of Darwin's theory is that heritable traits that increase survival and reproduction become more abundant within a population with time.

Some caveats are necessary[25,29,63]

COMMON PITFALLS. Oversimplification is a common intellectual trap in evolutionary research. Care is necessary to avoid the numerous pitfalls. Selection acts on the whole individual, not on a particular allele. The persistence of any mutation depends only on the fitness of a phenotype. Because organisms live in extremely complex environments, they are seldom exposed to a single selective pressure but respond to many stimuli in successive life stages. Conflicting demands of environmental stresses may prevent the perfect biological clock from appearing. Furthermore, alleles or genotypes are favored only if they are good for the individual, never because they are good for the species.

One of Darwin's many contributions was to focus attention on the actual individuals constituting a population rather than on the abstract average member of the population. In fact, natural selection will favor alleles that cause population decline if those alleles confer increased individual fitness. A male lion that has just acquired a new harem will immediately kill the young of the pride. Females quickly come into estrus again, giving the male an immediate mating opportunity. Infanticide would thus favor survival of that male's own progeny even though decreased population size might endanger the species.

ADAPTIVE STORYTELLING. A great danger lies in confusing undocumented "just-so" stories about adaptiveness with a rigorous demonstration of adaptive evolutionary change. Because of the powerful logic of Darwin's theory of evolution, biologists have tended to speculate wildly on the adaptive significance of phenomena they study. As a result, adaptive storytelling has become very common in the field of biology. A whole series of reasonable explanations about the adaptiveness of any clock characteristic can generally be made. In his novel *Candide*, Voltaire portrayed a character named Dr. Pangloss who could concoct a justification for any event. No matter how devastating the experience, he always

found a way to describe it as the best of all possible worlds. As a result, the tendency for unbridled speculation about adaptiveness has become known as the Panglossian Paradigm.

Guesswork is easy, but critical testing of hypotheses is often difficult. Contriving scenarios about adaptiveness is productive as long as the explanations are recognized as brainstorming and not rigorous scientific testing. Proper use of methods of modern evolutionary biology encourages valid tests for adaptiveness of observed traits while avoiding the pitfalls.

OTHER CONSIDERATIONS. Not all clock characteristics are adaptations. Particular clock features may merely be vestigial remnants from past selective pressures no longer operating. Alternatively, features may exist as a constraint, simply a necessary consequence of another clock feature. Frequencies of alleles can also be influenced by chance events unrelated to fitness. Resultant genetic drift may be important in small populations. Several different populations of a species should therefore be examined whenever possible in tests of adaptive significance. In addition, the perfect set of traits may never appear because of cost/gain tradeoffs.

Some alleles potentially favorable for a better biological clock may have detrimental interactions with other, nonclock genes. Natural selection generally favors adaptations built onto an existing biological clock rather than novel adaptations that could be the first small steps in the evolution of a new and better pacemaker. The biological clocks of organisms are not the best possible biological clocks but rather the clocks of their ancestors with the best modifications available.

GENES AND OUTPUT FUNCTIONS. Another important fact is that selection acts indirectly on the genes for biological clocks by way of their output functions. Synchronization of any rhythm with an environmental cycle results from the composite interactions of an input system, a pacemaker system, and an output system. The actual component, however, that contacts the environment and is subject to selection is the output rhythm. Questions about the adaptive significance of each feature of the system can be formulated to ask how changes in that characteristic would alter the behavior of the output rhythm and whether the change in output rhythm would convey any selective advantage to the organism.

Demonstration of adaptive evolutionary change for any behavioral feature requires stringent criteria[8,25,33,63]

CRITERIA. Evolution of any trait begins with mutations at genetic loci that control the trait. If those mutations produce favorable functional changes in overt behavior, they can lead to adaptation. Neutral mutations that produce no important functional changes in the clock can result in evolution of clock genes but cannot produce adaptive changes in biological timers. Convincing evidence of simple evolutionary change merely requires a demonstration that allele frequencies have increased or decreased in a population over time. Claims of adaptive evolutionary change, however, must demonstrate the operation of natural selection on adaptive genes. Although some data on adaptive evolutionary change through natural selection has been collected from laboratory populations of animals, few unambiguous data for adaptive evolutionary change in nature are available.

The necessary data should meet very stringent standards. The genetic basis of the trait must first be understood and the competing phenotypes identified. Changes in allele frequencies over extended periods of time must be documented, and the changes must be correlated with relative mortality, as well as the reproductive success of competing alleles. The selective pressures acting on each phenotype must then be identified. These requirements are so demanding that only a few evolutionary studies are widely regarded as having unequivocally supported the theory of adaptive evolution by natural selection in the wild. Difficulties are encountered in all facets of the issue. Demonstration of the genetic basis and the action of natural selection on adaptive traits in wild, free-living species is especially hard. Adhering to the stated criteria is important, however, in the evaluation of unsupported inferences that adaptive evolution has occurred.

GENETIC EVIDENCE. One major problem lies in showing the genetic basis of the behavioral trait under consideration: Genes alone don't make traits. They code for proteins that in turn regulate cell metabolism. Most behavior patterns are very complex and depend on the interaction of many genes. In addition, the eventual phenotypic expression of structure and function reflects both the genetic blueprint and the modifying influence of the environment. The task of tracing the links of a developmental chain from gene to behavioral function is very difficult for even the simplest organism. One indirect type of information involves crossbreeding of two phenotypes or species to demonstrate eventually heritable traits from both parents in the offspring.

A related type of evidence for the genetic basis of behavior involves selection of domesticated and laboratory animals to obtain breeds or strains with particular behavioral traits. Still another line of indirect evidence comes from the observation that phenotypic expression is the product of a genetic template plus its interaction with environmental modifiers. If a highly homozygous strain of organism is observed under different environmental conditions, the genetic components can usually be distinguished from the environmental contributions.

When identical twins are separated at birth and reared under contrasting circumstances, they usually show great similarities in behavior and personality as adults. Such resemblance reflects their identical genes.

The most direct and convincing evidence for the genetic basis of a behavioral trait is a single dominant gene mutant for a behavioral trait. If a single gene mutation appears and simultaneously causes a behavior change, logically a causal relationship probably exists between gene and behavior. Specific biological clock examples are considered later in the chapter.

NATURAL SELECTION APPROACHES. A second major difficulty of research on evolution of behavioral traits lies in clearly demonstrating the role of natural selection in adaptive evolution of populations of free-living, wild organisms. Endler comments in his book, *Natural Selection in the Wild*, that "natural selection is a major part of the theory of evolution yet there is much argument and confusion as to what it is, what it is not, and even whether it exists."[26] Because Darwin's theory is a cornerstone of modern biology, the issue of whether natural selection actually operates in derivation of new traits is not a trivial question. In light of the importance of Darwinian theory, the paucity of data on natural selection is all the more surprising. The lack suggests that data are extremely difficult to obtain.

The requirement at first glance seems simple enough: a demonstration that identifiable environmental agents can bring about changes in relative abundance of two or more polymorphic alleles of a gene in a free-living population of organisms. In reality, the problem has several facets. The finding of subtle, heritable traits based on polymorphic genes in wild species is not easy. In fact, differences adequate for the operation of natural selection in the field may be too small for biological detection or measurement. In addition, identification of the specific selective agent or agents in the complexity of a natural environment is often a very demanding task. Finally, measuring reproductive success and mortality is one of the most difficult assignments an ecologist faces in working out life history tables.

Several highly divergent paradigms have been devised to deal with studies of these facets of natural selection in the field. One way is to change part of an organism's environment slightly and then watch over many generations for any spontaneously occurring mutant that has greater fitness than the original morph. An example is industrial melanism of moths, a pigmentation mutation that appeared in the English midlands. Bird predators picked off more of the contrasting light-colored forms on sooty trees than dark-colored individuals. The data on actual predation rate was obtained by patient direct observation. Adequate data for confirming the true selective agent is still not fully available, and the genetic issue is not well studied.

In another study, two parallel creeks inhabited by a species of small minnow were monitored for many years. A predatory fish was introduced into one creek but not into the other. Over a large number of fish generations, new behavior patterns and life history traits appeared in the predator-filled creek. The fish predators were the most likely natural selection agent, but observation of the actual predator–prey interaction was difficult. Furthermore, little is known of the genetic traits involved.

Waiting around for spontaneous mutations to appear and come under natural selection pressure is rarely practical within the framework of a human lifetime. An alternative track is to fortuitously find or directly engineer polymorphic gene differences and then measure the relative mortality of the two morphs. This approach is also fraught with severe problems, such as identifying the selective agent and characterizing the polymorphic gene involved.

Information about the Evolution of Biological Timing Systems Comes from Several Sources[42,48,68]

Evidence for the genetic basis of circadian clocks is substantial[24,43,71]

The most dramatic evidence for the genetic basis of circadian biological rhythms is very direct. Single-gene clock mutations have been identified in several species. They include *per* gene alleles in the eclosion of *Drosophila*, *frq* gene mutants in the sporulation rhythm of the bread mold *Neurospora* (see Plate 2, Photo G), the *tau* mutant in the locomotor activity of the golden hamster, and the *Clock* mutant in locomotor activity of the house mouse (Figure 2.19). Single-gene clock mutations are considered in detail in Chapters 7 and 8. The mutations are not entirely realistic models for evolution in free-living organisms, for they simply demonstrate that the substrate on which selection might operate is present. Almost no data are available about the genetic basis for other circa-pacemakers with noncircadian periods.

Evidence for adaptive evolution of biological timing by natural selection is quite limited[11,20,25,30,53,62,72,73]

TYPES OF DATA AVAILABLE. The fact is inescapable that biological clocks are found in nearly all known living organisms and are used in a wide variety of ways. The temptation is great to state flatly that living clocks are therefore adaptive and have developed by the action of natural selective forces. In reality, the data are very fragmentary. Several types of evidence are available. Small amounts of indirect laboratory data are available. Observations on regressive evolution in cave-dwelling animals and other species of animals living in constant environments for long periods of geological time are

(A)

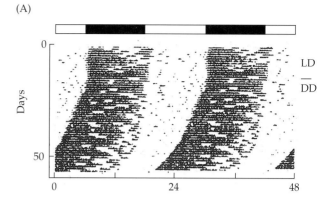

(B)

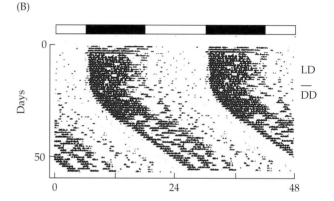

(C)

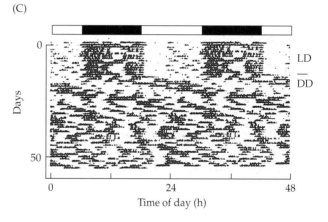

FIGURE 2.19 Locomotor activity of house mice. (A) Wild type. (B) Heterozygous *Clock* mutant. (C) Homozygous *Clock* mutant. Double-plotted actographs of wheel-running activity indicate period changes or arrhythmia; lighting is shown by the light bars above each actograph. (From Vitaterna et al., 1994.)

also useful. Data from studies of geographic clines of circadian parameters suggest adaptiveness. Correlation of essential circadian features with specific niche parameters can be used cautiously. Very recently a small amount of direct evidence has been published.

LABORATORY DATA. The dilemma for the evolutionary biologist interested in circadian clocks is that the natural world is not about to become arrhythmic or change to a periodicity different from 24 h. The laboratory, however, provides an opportunity for subjecting some species to an aperiodic or abnormal temporal environment. Although it is not easy to generalize from laboratory results to an organism's real world, the research permits testing of particular selective agents of known severity.

Laboratory environments with reduced or controlled selection can cause regressive changes in clock characteristics. Many strains of the fruit fly, *Drosophila melanogaster*, exhibit a circadian rhythm in pupal eclosion that is probably an adaptation to daily cycles of humidity. In the lab, flies are cultured in sealed bottles where humidity levels are constant. These flies have not yet lost their circadian rhythmicity, but they show greater variability in period length than wild-caught individuals. Therefore, artificial selection for early or late eclosion times is easier in lab cultures than in wild-caught populations. The results suggest that selective pressures in the wild prune rhythmicity to a narrow circadian range. Another example concerns house mice. Some strains of inbred laboratory house mice have lost the ability to produce melatonin rhythmically, implying that with reduced environmental selection pressure, the loss is no longer a great liability.

A related type of laboratory study has applied known selective forces to populations of animals and documented changes in the resultant phenotypes. One extensive study of genetic polymorphism for reproductive photoresponsiveness was carried out with the white-footed mouse, *Peromyscus*. In this species, reproduction is usually, but not always, photoperiodic, and a circadian timing system measures the critical photoperiod for gonadal growth. For a matrix of groups, selection was applied both for and against the photoperiodicity trait over four generations. In one group, a first-generation strain was obtained in which 80% produced litters with the photoperiodic trait. At the other extreme was a strain in which only 20% had the trait. In the succeeding three generations, the percentage remained nearly the same. The phenomenon has also been demonstrated in the laboratory for other small mammals, such as voles, Siberian hamsters, and deer mice. The development of these microevolutionary changes in flies, house mice, hamsters, voles, deer mice, and white-footed mice in the course of a few laboratory generations by artificial selection supports the hypothesis that early stages of regressive evolution may occur quite rapidly.

One of the best laboratory circadian examples comes from competition studies of fitness for several period mutants of the cyanobacterium *Synechococcus* (see Figure 5.1). Period mutants were first engineered by the use of a mutagenic agent. One strain, called PCC 7942, was particularly suitable for the study because three of its mutants had equivalent growth rates, and all had stable free-running periods. Cultures were obtained with free-running periods of 23 h, 25 h (wild type), and 30 h

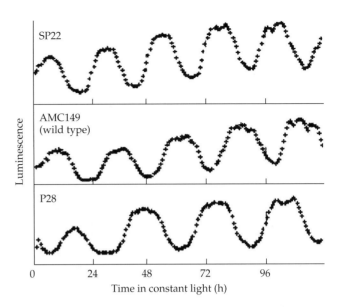

FIGURE 2.20 Circadian phenotypes of three cyanobacteria mutants for competition studies. For further information, see the text. (From Ouyang et al., 1998.)

(Figure 2.20). Pairs of the three cultures with different period lengths were then mixed in equal concentrations in batch cultures, and an aliquot was plated to form single colonies for determining the initial composition of the culture. The strains were allowed to grow for 27 days in either a short day of LD 11:11 or a long day of LD 15:15. Another aliquot of the mixed culture was then used to determine the composition of each culture.

The experiment was repeated three times with essentially the same outcomes. In all cases, one member of the pair dominated or completely took over the culture, and this mutation was invariably the culture whose free-running period was closest to the period of the light cycle (Figure 2.21). For a mixture of 23 h and 25 h colonies, the 23 h colony dominated in LD 11:11, but the wild type with a 25 h free-running period dominated in LD 15:15. Similarly, when a 25 h wild type was mixed with a 30 h mutant, the wild type prevailed in the LD 11:11 condition, and the long-period 30 h mutant prospered in the LD 15:15 condition. Finally, in the third pairing of the 23 h mutant with the 30 h mutant, the short-period mutant entirely took over in the short-day experiment, and the long-period mutant outcompeted in the long-day experiment. The authors concluded that the circadian pacemaker in cyanobacteria confers a significant competitive advantage if the free-running period of the organism's clock matches the environmental cycle and if an optimal phase relationship between the LD cycle and the internal pacemaker is achieved.

At present no field data are available for circadian period mutants in competition with each other. In fact, very few data are on hand for any phenotype in the wild except for surgically induced SCN-lesioned

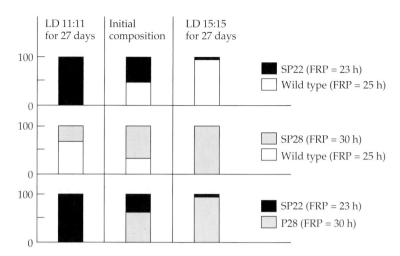

FIGURE 2.21 Growth of various cyanobacterial strains in competition with each other in batch cultures in the laboratory. FRP, free-running period. For further explanation, see the text. (From Ouyang et al., 1998.)

arrhythmic animals, as will be discussed a little later in the chapter. A major deterrent to such studies is the enormous commitment of time, labor, and funds required for obtaining the necessary data. In the next section of this chapter, the field data supporting the genetic basis of clocks and their probable derivation by natural selection will be considered.

REGRESSION OF CIRCADIAN RHYTHMS IN CONSTANT ENVIRONMENTS IN NATURE. An interesting parallel exists between regressive evolution in laboratory situations and regression in the few constant environment niches that exist in nature. Fish, salamanders, crayfish, and insects living in deep caves provide a natural laboratory for the study of regressive evolution. Large caverns generally contain a gradient of temperature and light, with three easily demarcated zones: the outer twilight zone, the middle zone of constant darkness but variable temperature, and the inner zone of permanent darkness and constant temperature.

Species of the outer two cavern zones may commute frequently to the surface for food, but the obligate fauna of the deepest cave regions have been isolated from daily environmental changes for long periods of geological time. These species have lost their eyes and skin pigmentation because those features are no longer needed. The cave crayfish has been isolated from daily cycles for over 25,000 generations and is blind. When timing of its oxygen consumption and locomotor activity were measured, correlated bursts were seen, but the data give very little evidence of precise circadian activity in continuous darkness or of the ability to entrain to an LD 12:12 light schedule.

Similar examples of regressive evolution come from fossorial species that live in dark environments other than a deep cave. Naked mole-rats are burrowing mammals that spend their entire lives in extensive underground tunnels and make only rare trips to the surface. In a sense, naked mole-rats are counterparts to termites. Each colony contains one breeding female, one to three breeding males, and a large number of sterile workers. In this colony setting, most members sleep in bouts randomly spaced throughout the day and night or with a slight preference for daytime (Figure 2.22A, left), but the male dispersers are nocturnal in their activity pattern (Figure 2.22A, right). These rhythms free-run when animals are transferred to DD. In the wild, the large disperser males appear to leave their natal colonies occasionally, perhaps to find breeding opportunities with other colonies. A few marked dispersers were captured aboveground more than 2 km from the burrows of their natal colony, but only at night. Nocturnal dispersal would probably reduce the risks of predation or desiccation in the semiarid, warm environments of this species. The limited data suggest that naked mole-rats have functional circadian systems that are used only for very limited functions and may not be used at all in most individuals.

GEOGRAPHIC CLINES OF CIRCADIAN FUNCTIONS. One important type of correlative data for natural selection comes from the clinal mapping of a circadian feature along a latitudinal gradient from north to south. Climatic factors such as daily photoperiod or mean monthly temperature follow a regular pattern along a latitudinal gradient and could serve as predictable selective agents for many behavioral output functions. The annual cycle of day length of a given location, for instance, is very regular and almost noise free, and it has a rate of change that is easily detectable by many organisms. Photoperiodic cuing is highly adaptive in climates with predictable food resources and weather patterns, especially those with harsh drought or winter seasons. The cuing can serve advantageously to trigger reproduction, metabolic preparation for hibernation, molting of feathers or fur, or migration. The use of opportunistic breeding or torpor patterns can be expected in unpredictable climates with variable rainfall or varying food availability patterns.

One study recorded the incidence in deer mice of photoperiodic responders along a clinal gradient from southern Canada to Texas. Almost all of the mice responded at the northern extreme, some did not respond in South Dakota, and none responded south of 30° latitude in Texas. Presumably winter conditions at the northern extremes of the cline were effective selective agents capable of pruning out those individuals that failed to rely on photoperiodic signals for reproduction. Mice at the southern extreme in Texas could successfully rear young with an opportunistic strategy of breeding whenever ambient temperature and food were favorable.

(A)

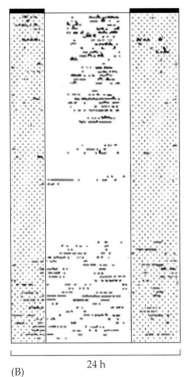
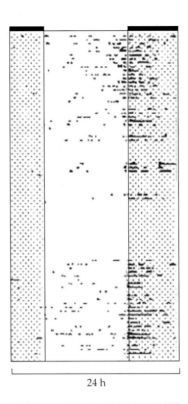

24 h   24 h

(B)

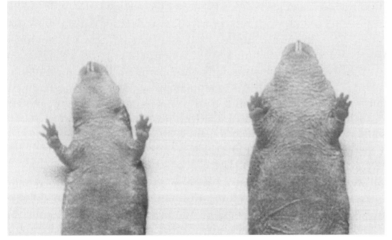

FIGURE 2.22 Circadian activity in naked mole-rats. (A) Wheel-running activity on an LD cycle for sterile workers (left) and disperser males (right). Hatching indicates the time of darkness. The corresponding rats themselves are pictured in part B. For further explanation, see the text. (From Riccio and Goldman, 2000; photo courtesy of B. Goldman.)

series. Experiments with locomotor rhythms in constant darkness showed that the shorter the T–G pair lengths, the less the flies are able to temperature-compensate between 18°C and 29°C. The most common allele, with 17 T–G pairs, shows a significant difference in period at these two temperatures, whereas the 20 T–G allele is able to maintain its period consistently at the two temperatures. Because the T–G region is involved in thermostability of the circadian system, the correlation of decreasing segment length with increasing ambient temperature in the south suggests possible adaptiveness of this feature. The polymorphism is apparently maintained by temperature selection.

CORRELATION OF ESSENTIAL CIRCADIAN FEATURES WITH SPECIFIC NICHE PARAMETERS. A last type of data can be used very cautiously in support of the hypothesis of natural selection in circadian evolution of wild species. Several examples suggest a matching in the fine details of circadian features and ecological niche. The correspondence is logically appealing but has the limitations of all correlative methodologies. Circadian free-running periods range for almost all species from about 22 to 25 h, suggesting that natural selection deletes alleles for periods outside this narrow range. Furthermore, the entrainment of circadian free-running rhythms to the 24 h solar cycle is brought about by photic phase response systems (see Chapter 3). Stable entrainment results when a light stimulus produces a daily phase shift of sufficient magnitude to correct the free-running period to the 24 h period of the entraining light signal. A survey of the known free-running periods and phase response systems of animals shows a fair degree of correlation between the DD free-running period and the shape of the phase response curve (Figure 2.24A). Species with a mean free-running period shorter than 24 h tend to have a large delay segment of the PRC. Those with free-running periods longer than 24 h generally have shallow delay segments of the PRC and large advance segments (see Figure 2.24A). The matching may be the result of natural selection.

Clinal circadian data suggesting adaptive evolution also come from studies of fruit flies. The *per* gene of *Drosophila* controls circadian period length of daily eclosion and daily activity rhythms. This highly polymorphic clock gene encodes a protein with a series of alternating threonine–glycine (T–G) pairs. The number of pairs may vary from 13 to 20 in natural *Drosophila* populations, but the most abundant alleles by far contain 17 T–G pairs and 20 T–G pairs, respectively. A pronounced gradient is observed on a north–south axis from Europe to North Africa, with a higher proportion of 20 T–G alleles in the northern part of the range and increasing representation of 17 T–G alleles in the south (Figure 2.23).

Temperature compensation based on activity rhythms is greater in long T–G repeats than in short

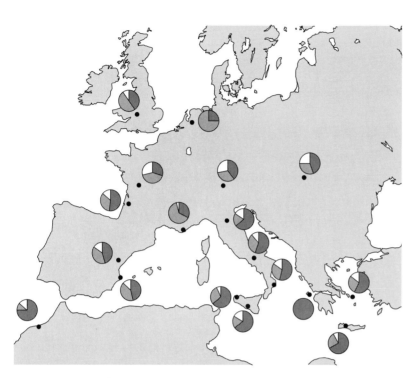

FIGURE 2.23 Geographic clines of *per* gene polymorphism in 18 European and North African populations of a fruit fly species. Each circle represents one population with the proportion of alleles of 20 threonine–glycine (T–G) pairs shown in dark shading, the proportion of alleles of 17 T–G pairs in hatched shading, and the proportion of other alleles in unshaded areas. (From Costa et al., 1992.)

Another indication of possible shaping of circadian features by natural selection is the lability of free-running rhythms. Frequency modulation of free-running rhythms has been demonstrated for several species. The clearest examples come from studies of history dependence of free-running rhythms, using measurements of the effect of prior photoentrainment on the subsequent free-running period (Figure 2.24C). For flying squirrels, the greater the deviation of the free-running period from 24 h before entrainment to a 24 h photoperiod, the more the free-running period was modulated toward a 24 h period. Although such precise relationships may be the result of chance alone, the matching could be due to natural selection.

A remarkable circadian feature of small nocturnal cave-dwelling or den-inhabiting mammals is light sampling for photoentrainment. Flying squirrels, deer mice, and many bats sleep during daylight hours in a dark recess. There they are safe from predators but also out of touch with light conditions of the outside world. They need only a glimpse of light at dusk for entraining (Figure 2.24B). The amount of the daily light schedule seen and the pattern of testing the environmental light conditions were measured quantitatively for flying squirrels in simulated natural dens (Figure 2.24D). Each squirrel spent almost the entire daytime period resting in total darkness in its nest box. It sampled the light level only at subjective dusk after being awakened by its cir-

cadian clock. The average amount of light seen per day on an LD 12:12 schedule was several minutes or less per day. The results are consistent with adaptiveness of a brief light-sampling procedure as an antipredator measure, but the conjecture is untested. With due regard to the Panglossian Paradigm and to the pitfalls of adaptive storytelling, these correlative demonstrations may provide useful insights until definitive direct tests are completed.

Direct testing of circadian adaptiveness in the field has recently been carried out[16,21,22]

Logistics for carrying out a convincing direct demonstration of circadian adaptiveness in the field are daunting. Clearly the only option for addressing circadian evolution in nature lies in engineering a population to a noncircadian periodicity, not in changing the periodicity of the solar cycle. Serious difficulties are encountered in obtaining time-disadvantaged cohorts, but electrolytic lesioning of the SCN pacemaker in mammals and pinealectomy in birds are feasible phenotypic avenues. Alternatively, mutant arrhythmic strains of hamsters or mice could be employed. The available mutant strains are problematic, however, because they are essentially domesticated and not adapted to living in the wild. The accurate measurement of fitness and mortality is also a considerable obstacle, but state-of-the-art radio telemetry and data loggers, combined with diligent field observations of individuals, provide a possible method. A high-density population of small, conspicuous, day-active rodents can be color-marked and fitted with radio collars for observation. The SCN in some individuals can be lesioned to induce arrhythmia and the other individuals left intact or given sham SCN lesions as controls. After reintroduction of the animals into their home territories, predation, feeding success, and reproduction can be painstakingly measured for several seasons.

Until 5 years ago only speculation and imaginary results were available for wild species living in natural habitat. Hypotheses like the following could be generated:

- More SCN-lesioned animals died, picked off by predators, if they were active at an unaccustomed time.

- Lack of synchrony between the sexes delayed their mating season, which shortened the summer foraging season of the young. The juveniles were then less fit when winter arrived.

(A) Free-running period in DD relative
to phase response curve shape

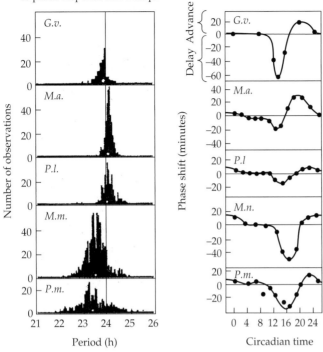

(B)

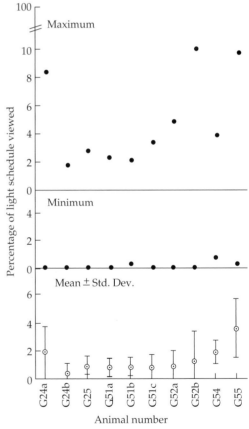

(C) Frequency modulation of free-running period

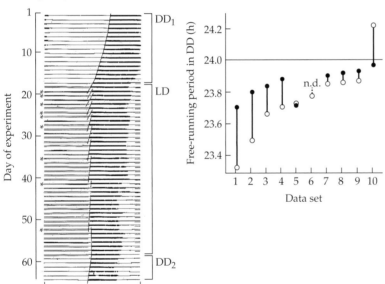

(D) Light sampling

- Fewer offspring were produced in the arrhythmic individual, and these did not survive as well because the mothers could not synchronize her own foraging with suckling needs of the young.

- Control animals huddled together more often during winter than SCN-lesioned individuals did, and consequently they survived the winter better than lesioned animals.

Although this intellectual flight of fancy is simplistic and overlooks many logistical problems, it nevertheless suggests the kind of evidence needed for showing conclu-

◄ **FIGURE 2.24 Circadian properties relative to environmental features.** (A) Distribution of values of free-running period in DD (left) for five species relative to phase response curve shape (right). *G.v.*, flying squirrel; *M.a.*, golden hamster; *M.m.*, house mouse; *P.l.*, white-footed mouse; *P.m.*, deer mouse. Phase response curve values (right) are plotted as phase shift in minutes for *G.v.* and as degrees for all others. (B) Flying squirrel sampling the light level at its den exit at evening twilight. (C) Frequency modulation of the circadian pacemaker by prior light exposures. Wheel-running activity of one flying squirrel in a simulated den cage (left) on a DD_1/LD $12:12/DD_2$ schedule as indicated. Eye-fitted lines connect the slope of onsets, and the asterisks indicate days on which a light exposure occurred at onset time with subsequent phase delay and frequency modulation. DD_1 and DD_2 indicate the free-running periods before and after photoentrainment, respectively. Summary correlation (right) of the degree of frequency modulation for 10 flying squirrels indicated for free-running period in DD_1 (open circles) and for free-running period in DD_2 (solid black circles). (D) Time spent exposed to light for flying squirrels entrained for at least 28 days to a normal light schedule. Maximum per day is shown at the top, minimum per day in the middle, and mean per day at the bottom. Standard deviation indicates the high precision of the response. (A from Pittendrigh and Daan, 1976, and DeCoursey, 1990; C from DeCoursey, 1989; D from DeCoursey, 1990.)

sively that circadian clocks have ecologically adaptive values.

Ideas for experiments of this kind have been circulating for many years, but the demanding field and laboratory work necessary to complete such an ambitious undertaking has long deterred the project. Published data for serious experimental tests have become available only very recently. The first report appeared in 1997 for the highly day-active antelope ground squirrel. A group of SCN-lesioned and intact antelope squirrels were maintained free-living in natural habitat in an outdoor enclosure in the desert with telemetric equipment to monitor the activity patterns and survival of each individual. The SCN-lesioned (SCN-X) animals were surprisingly active above ground at night (Figure 2.25). A feral cat unexpectedly entered the compound one night and removed 60% of the lesioned animals but only 29% of the intact animals. The infrared videotapes showed that the motionless cat waited until an antelope squirrel wandered out from its burrow onto the surface sand, then pounced on the victim and

FIGURE 2.25 Effect of SCN lesioning on antelope ground squirrels in an outdoor natural enclosure. Double-plotted actograph of activity of intact and SCN-lesioned squirrels recorded by a microchip transponder scanner during a 12-day period. Vertical data lines represent the number of bouts of activity per 10-minute intervals. Light conditions are indicated by the schedule bars. (A) Activity of five intact control squirrels and (B) activity for the two surviving SCN-lesioned animals. (From DeCoursey et al., 1997.)

(A) Intact animals

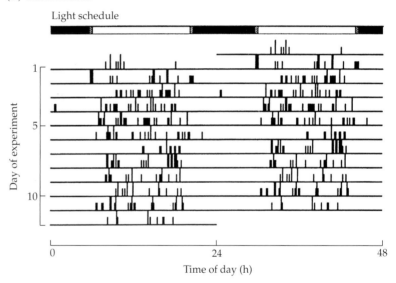

(B) wSCN-lesioned animals

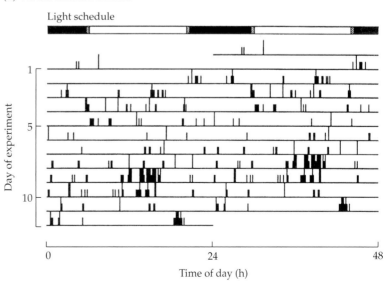

killed it. The data are consistent with the hypothesis that removal of the SCN circadian pacemaker reduced the normal nocturnal quiescence of an antelope squirrel in its burrow at the time of greatest potential predation danger (Figure 2.26).

A second, more comprehensive study was carried out with eastern chipmunks completely free-living in a near-wilderness habitat in the Allegheny Mountains of Virginia. In a pilot study, during a year of medium chipmunk density and few predators, little difference in survival rate between SCN-lesioned and intact control chipmunks was seen during the first year (Figure 2.27). In a large-scale project 2 years later, conditions had changed considerably. Following 2 optimal years of acorn crops, the chipmunk density had increased to the carrying capacity, and predators had multiplied as well. After lesioning and release of a group of experimental chipmunks, weasel predation started immediately. The predator killed about one chipmunk per day. By the end of summer, a significantly higher proportion of SCN-lesioned chipmunks than intact controls or surgical controls had been killed by the weasel predators.

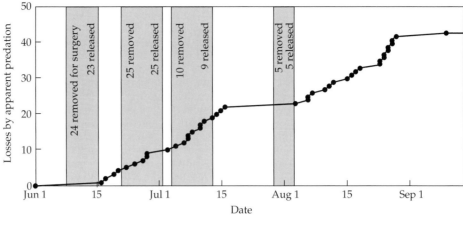

FIGURE 2.27 Predation of wild, free-living chipmunks in a pristine forest by weasel. All animals were followed two to three times daily by radio tracking symbols. Each open circle represents a chipmunk removed by the predator. Shaded boxes indicate the time periods in which the specified numbers of chipmunks were removed from the population to the laboratory for brief recording of circadian activity in wheel cages in dim LL, surgery and recovery, then repatriation of each chipmunk at its den site. (From DeCoursey et al., 2000.)

Telemetric recording of the activity of each chipmunk in its den at night indicated that the clockless, arrhythmic animals did not venture out of their burrows at night, but were more restless within their dens than controls were. Perhaps weasels were able to detect and locate the arrhythmic chipmunks by acoustic or olfactory clues more easily than the immobile sleeping controls. Again, one important value of an SCN pacemaker in small mammals may be to enforce quiescence and sleep at times of greatest predation danger.

A Phylogenetic Survey of Plants and Animals Suggests Trends in the Evolutionary History of Biological Timing Systems[48]

The field of systematics offers tools for studying the evolutionary history of traits[2,33]

Tracing the evolutionary origins of biological rhythmicity is not an easy task. The tool kits of many scientific disciplines have been brought to bear on the issues. Geologists and astronomers have constructed a timetable for the probable origin of Earth and its subsequent changes (Figure 2.28A). Chemists have calculated the condition of Earth's early atmosphere. Systematists have named and classified species of organisms, and have studied the evolutionary relationships between taxonomic groups. Recent advances in identification of molecular components of oscillators will allow sequence comparison of common components between species in the future.

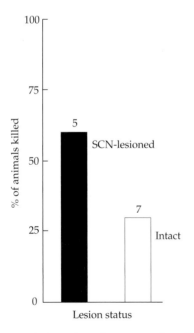

FIGURE 2.26 Predation of antelope ground squirrels by a feral cat for SCN-lesioned (SCN-X) and intact squirrels monitored by infrared video camera. The number above each histogram bar indicates the group size. (From DeCoursey et al., 1997.)

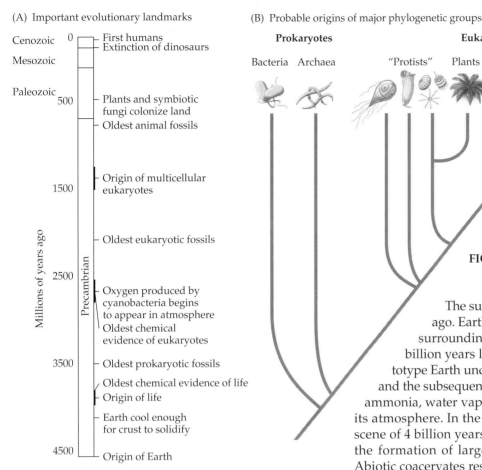

(A) Important evolutionary landmarks

(B) Probable origins of major phylogenetic groups

FIGURE 2.28 The history of Earth.

The sun originated about 5 billion years ago. Earth coalesced from gaseous clouds surrounding the primitive sun about half a billion years later. Shortly afterward, the prototype Earth underwent intense volcanic activity and the subsequent degassing of its interior ejected ammonia, water vapor, methane, and hydrogen into its atmosphere. In the intensely hot, harsh, anaerobic scene of 4 billion years ago, chemical evolution led to the formation of large, complex organic molecules. Abiotic coacervates resembling simple bacteria probably evolved by clumping of these organic molecules within a membranelike envelope.

From the beginning of the systematic study of chronobiology in the early 1950s, scientists recognized the adaptiveness of biological timing and speculated on its evolutionary origins. Circadian timing was documented from the simplest unicellular eukaryotes to humans. Chronobiologists first speculated that circadian timing arose with the development of a eukaryotic lifestyle and its concomitant requirement of protecting fragile DNA during the sensitive mitosis processes from damaging daytime photo-oxidation.

Circadian timing would be one way to segregate daytime from nighttime processes. Light–dark cycles may have acted as the selective force. The idea of possible damage during replication of DNA was later broadened to include other damaging daytime effects. Increasing levels of oxygen-free radicals could have been a decisive factor in delegating some metabolic processes to nighttime. An objection was then raised that photo-oxidative concerns were not the relevant issue. Perhaps the real driving force was the symbiotic fusion of a primitive bacterium with a more complex bacterium host cell to form the first eukaryote. This development would have required the metabolic coordination and synchronization of the cell cycles of both

Evolutionary studies of a trait such as biological timing require first a phylogenetic tree constructed from features unrelated to the trait under consideration. Criteria may include features such as selected anatomical traits, overall similarity, or supposed evolutionary history. The widely accepted five-kingdom tree based on morphology illustrates the kingdoms and major phyla (Figure 2.28B). A character description of clocks must then be constructed for all species available. Possible descriptors include morphological measurements, biochemical features such as DNA sequences, or behavioral traits. The systematist can next map pacemaker characteristics on the tree and look for evolutionary trends in the clock trait. If a trait is widespread in a taxon, the characteristic is assumed to have developed early in the group's evolutionary history. However, a common problem is convergent evolution because similar selective pressures frequently mold look-alike features in unrelated species.

The chronology of Earth's early history gives insight into the origin of biological timing[42]

A series of geological events in Earth's primordial abiotic stages paved the way for eventual living organisms.

partners. Daily synchrony could have been achieved best by evolution of an internal pacemaker. In both these earlier arguments, circadian clocks were postulated to have originated about a billion years ago.

Some prokaryotes have well-developed circadian timing[40]

Recently discovered Australian fossils, 3.5 billion years old, are thought to be primitive blue-green bacteria in which anaerobic photosynthesis first evolved. Among them was a group of cyanobacteria that released oxygen into the atmosphere. Gradually, over the next billion and a half years, some cyanobacteria developed the capacity for aerobic photosynthesis.

About 1 billion years ago a dramatic change in cell structure took place. Symbiotic relationships between several prokaryote groups gave rise to eukaryotic protists such as protozoans. These complex unicells now contained a nucleus enclosing chromosomes and many internal organelles with limiting internal membranes. In addition, they reproduced by mitotic division rather than by the binary fission of prokaryotes. Multicellular organisms soon evolved, and a remarkable radiation took place, with the result that many diverse plant and animal species quickly populated the planet. Single-celled eukaryotic protists gave rise to present-day fungi and metazoan animals. Other protists were invaded by photosynthetic cyanobacteria, and probably evolved into the higher plants.

It seems unlikely that a circadian clock mechanism was transmitted directly from prokaryotes to eukaryotes. The possible independent evolution of circadian clocks in higher phyla is also hard to evaluate. Remarkable similarities are seen between formal properties of the rhythms of prokaryotes and eukaryotes, but the similarity may merely reflect convergent evolution. The serial endosymbiosis theory of evolution provides a plausible explanation for the origin of higher plants from the invasion of a primordial protist by cyanobacteria. The circadian rhythmicity in plants could have been inherited directly from cyanobacteria, but no comparable explanation is available for eukaryotic animals.

Almost all eukaryotic organisms possess biological clocks with similar timing properties but great structural diversity[48,68]

The information currently available on evolution of the structural components of all noncircadian timing systems is too sketchy for a systematic phylogenetic survey or attempted synthesis. The hallmark of circadian clock evolution in terms of structure is diversity in the components that provide essential clock properties. All circadian timers need an oscillatory element for generating rhythmicity, a sensory receptor input of environmental information, and a coordinating messenger to carry timing commands to the output system. The possibility that similar clock properties of circadian systems in organisms from prokaryotes to humans can be generated by many different types of structures is strong.

Such a finding is not surprising in light of the evolutionary opportunism seen in all heritable traits. Evolution has rarely progressed in a tidy, straight line. A chance cataclysmic event can wipe out evolutionary developments of potentially superior features, giving the eventual outcome a strong historical slant. By evolutionary trial and error, new mutant adaptations are also added onto old working parts to form a remarkable patchwork for any particular feature. As a result, a linear tracing of the evolution of circadian structures for the five kingdoms is difficult. Sometimes a thread of phylogenetic continuity is visible, but at other places deletions plus the overlay of new elements on old features make the picture a web of confusing interlocking strands. Any theory generated is at best highly speculative.

An unraveling of evolutionary history is particularly challenging for circadian rhythms. The most essential clocklike properties of circadian systems include a free-running period in constant conditions, temperature compensation of the period, and entrainment by means of phase response systems. These properties are consistently preserved from the oldest prokaryotes to the highest primates. In contrast, features related to the ecology of a particular species are usually liable to change by natural selection.

The Known, the Unknown, and the Wildly Speculative

Circadian clocks appear important to organisms because virtually all organisms have them. Clockless individuals might not rear as many surviving offspring as those with clocks, but that is not known. Circadian scientists imply that most animals could not survive in natural habitats without a circadian clock, but the truth of that statement has just begun to be evaluated. Whether animals in aperiodic environments can survive without a clock, or whether they only lose some of the commonly measured outputs, such as activity, is also untested. Noncircadian clocks may be vital to some animals, but the answer is not available.

Speculation is usually directed at the significance of particular clock features. How important is it for daily timers to free-run within an hour or two of 24 h? The answer is unclear. How critical is circadian clock control of the shedding of retinal discs? It may not matter in the least, for the discs may simply be an insignificant constraint or historical feature that has no survival value at all. Amazingly, our extensive and substantial data do not conclusively reveal the absolute need for clocks. Clocks are not so deeply woven into animal physiology that they are absolutely essential to life in the laboratory, for animals survive quite well when their clocks are removed. Predictions still rely chiefly on indirect evidence and hunches. More careful field and laboratory experiments are required.

The way clocks evolve and the rate of evolution need more systematic consideration. Clocks certainly have a genetic basis because many clock mutations have been discovered or created. These mutations produce different phenotypes that can be subjected to artificial selection. The fact that none of these mutant phenotypes have been found in the wild implies that the features of clocks are subject to selection. In the wild, all currently known induced clock mutations would undoubtedly be eliminated by natural selection, just as Darwin proposed. If, however, the rotational period of Earth should change to 28 hours, a mouse with an induced mutant 28 h clock might prosper and its allele might spread rapidly through a population. To go beyond these simple deductions is only speculation.

Variations in clock properties and outputs clearly exist both within a species and between species. Circadian scientists infer that variation between species has been, at least in part, the product of evolution by natural selection. Both the means by which more complex features of clocks might evolve and the potential rate of evolution are pure guesswork. More knowledge about clock genetics will help the formulation of better hypotheses about clock evolution. Combinations of field and experimental studies will hopefully lead eventually to a clear understanding of the evolution of biological clock systems.

SUMMARY

This chapter has surveyed the major types of rhythmic behavior throughout the animal and plant kingdoms in terms of their probable ecological adaptiveness and their possible course of evolution. Rhythmic environmental variables of an organism's niche are predominantly solar factors related to the 24 h day–night cycle and the annual day length cycle, or tidal/lunar factors related to the gravitational pull of the moon on Earth's oceans. The evolution of internal physiological timers that are locked on to local time by entraining agents presumably imparts fitness to animals by allowing them to anticipate events, to respond rhythmically in the absence of a rhythmic environmental cue, or to consult an internal clock continuously in celestial navigation.

Circadian timing enhances the efficiency of organisms in numerous behavioral functions, including daily activity rhythms, arousal from torpor, time sense, time–place learning, time-compensated celestial navigation, gated events, and intra- and interindividual coordination. Plants rely on circadian timing for regulating photosynthesis and daily blooming time. Many marine organisms coordinate rhythms of locomotor activity and reproduction with favorable phases of tidal or lunar environmental cycles. Photoperiodic responses and circannual rhythms are important in the adaptation of organisms to harsh seasonal changes.

Darwin's theory presents the steps in evolution of new traits that could eventually lead to new behavior patterns or species. The key element is natural selection working on adaptive genes of each individual organism. The demonstration of a genetic basis of any behavioral trait is difficult because of the long pathway between genes and behavior and the intervention of environmental components. The operation of natural selective agents on alleles has also been hard to demonstrate, particularly for free-living animals. Rigorous standards should be applied to experimental design and data interpretation to avoid pitfalls of adaptive storytelling.

Single-gene clock mutants in bread mold, fruit flies, golden hamsters, and house mice are convincing indicators of the genetic basis of circadian timing systems. Most of the data currently available for showing the action of natural selection on adaptive clock genes in field populations are indirect. The first convincing direct field data for free-living small mammals has just become available for rodents, demonstrating that SCN-lesioned, time-disadvantaged animals do not survive as well as normal controls.

CONTRIBUTORS

Patricia J. DeCoursey wishes to thank Paul Heideman, Teresa Horton, Theresa Lee, Stéphan Reebs, and Laura Smale for their material contributions to Chapter 2.

STUDY QUESTIONS AND EXERCISES

The following exercises will give some hands-on experience in studying pacemakers at the behavioral level, both in the laboratory and in the field.

1. Model species:

Some of the most widely used model animals are pictured at right (see also Plate 2). Used in this sense, models refer to representative species that have potential for yielding unambiguous answers to a scientist's hypotheses.

Some factors involved in judging the merits of a potential model species include

- Commercial availability

- Feasibility of capture in the field when needed

- Potential for breeding of wild species in the Plaboratory

- Ease and expense of maintenance

- Nocturnal or diurnal activity habits

- Presence of a good indirect measure of rhythmicity

- Expense of automated recording of the rhythmic output by telemetry

- Clarity of the rhythmic behavioral output

- Published background information on care and surgical procedures for the species

- Feasibility of field studies

- Cost feasibility of establishing controlled laboratory conditions for experiments

- Period of the rhythm to be studied: tidal, daily, annual

    a. For each of the model species shown, list advantages and disadvantages in table form.

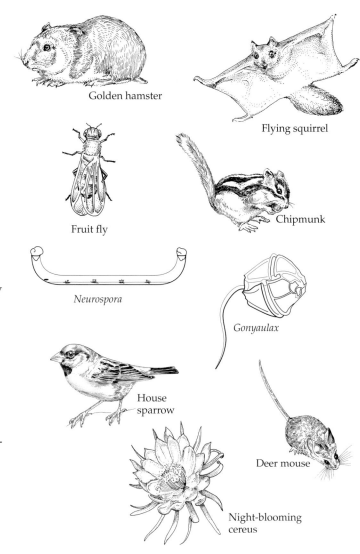

Golden hamster

Flying squirrel

Fruit fly

Chipmunk

Neurospora

Gonyaulax

House sparrow

Deer mouse

Night-blooming cereus

Model species	Advantages	Disadvantages
1. Golden hamster or rat		
2. Flying squirrel		
3. Chipmunk		
4 Fruit fly		
5. *Gonyaulax*		
6. Deer mouse		
7. *Neurospora*		
8. House sparrow		
9. Night-blooming cereus		

b. Is it surprising that over 90% of circadian research is carried out on human or laboratory rats, mice, and hamsters? Why or why not? What bias could this predominance in models introduce into the prevailing view of chronobiology mechanisms?

c. Try looking at ecological chronobiology questions and design a work site for a specific species. List some of the logistics of determining the adaptive value of a circadian pacemaker for one or more wild species under natural conditions.

d. Test your sleuthing skills at the library and try to find out what pivotal discoveries were made in experiments on each of the models above. (*Hint*: The journal of the premier U.S. chronobiology society is the highly respected *Journal of Biological Rhythms*, published monthly.)

2. Temporal and spatial aspects of ecosystems:

Familiarity with the multiple features of ecosystems is essential for understanding ecological and evolutionary aspects of organisms in natural habitats. Choose one familiar ecosystem, such as a sandy desert, a tropical rain forest, an estuary, or a large lake. Describe three possible niches for this ecosystem. For example, three niches in a sandy desert might be sand surface, a cavity in a large saguaro cactus, or a crevice in a rock. Characterize your three niches both spatially and temporally in terms of substrate, light availability, and temperature ranges, as well as availability of water and food. Choose a species for each niche and predict the kinds of rhythms expected. In what ways could these rhythms be adaptive? What selective forces might have shaped these rhythmic adaptations? Answer this question in either chart form, as shown below, or essay form.

REFERENCES

1. Alcock, J. 2001. Animal Behavior: An Evolutionary Approach, 7th Edition. Sinauer, Sunderland, MA.

2. Appenzeller, T. 1999. Test tube evolution catches time in a bottle. Science 284: 2108–2110.

3. Aragona, B. J., J. T. Curtis, A. J. Davidson, Z. Wang, and F. K. Stephan. 2002. Behavioral and neurochemical investigation of circadian time-place learning in the rat. J. Biol. Rhythms 17: 330–344.

4. Aschoff, J. (Ed.) 1981a. A Handbook of Behavioral Neurobiology Volume 4: Biological Rhythms. Plenum, New York.

5. Aschoff, J. 1981b. A survey on biological rhythms. In: A Handbook of Behavioral Neurobiology Volume 4: Biological Rhythms, J. Aschoff (ed.), pp. 3–10. Plenum, New York.

6. Biebach, H., H. Falk, and J. R. Krebs. 1991. The effect of constant light and phase shifts on a learned time-place association in garden warblers (*Sylvia borin*): Hourglass or circadian clock? J. Biol. Rhythms 6: 353–365.

7. Binkley, S. 1990. The Clockwork Sparrow: Time, Clocks, and Calendars in Biological Organisms. Prentice Hall, Englewood Cliffs, NJ.

8. Brooks, D. R., and D. A. McLennan. 1991. Phylogeny, Ecology, and Behavior: A Research Program in Comparative Biology. University of Chicago Press, Chicago.

9. Brown, K. 2002. Something to sniff at: Unbottling floral scent. Science 296: 2327–2329.

10. Chadwick, D. J., and K. Ackrill. (Eds.) 1995. Circadian Clocks and Their Adjustment. Wiley, Chichester, UK.

11. Costa, R., A. A. Peixota, G. Bargujani, and C. P. Kyriacou. 1992. A latitudinal cline in a *Drosophila* clock gene. Proc. R. Soc. Lond. B Biol. Sci. 250: 43–49.

12. Daan, S. 1981. Adaptive daily strategies in behavior. In: A Handbook of Behavioral Neurobiology Volume 4: Biological Rhythms, J. Aschoff (ed.), pp. 275–298. Plenum, New York.

13. Daan, S. 2000. Colin Pittendrigh, Jürgen Aschoff, and the natural entrainment of circadian systems. J. Biol. Rhythms 15: 195–207.

14. Daan, S., and J. Aschoff. 2001. The entrainment of circadian rhythm. In: Handbook of Behavioral Neurobiology Volume 12: Circadian Clocks, J. S. Takahashi, F. W. Turek, and R. Y. Moore (eds.), pp. 7–43. Kluwer Academic/Plenum, New York.

15. Darwin, C. 1859. On the Origin of Species by Natural Selection. Murray, London.

16. DeCoursey, G., and P. J. DeCoursey. 1964. Adaptive aspects of activity rhythms in bats. Biol. Bull. 126: 14–27.

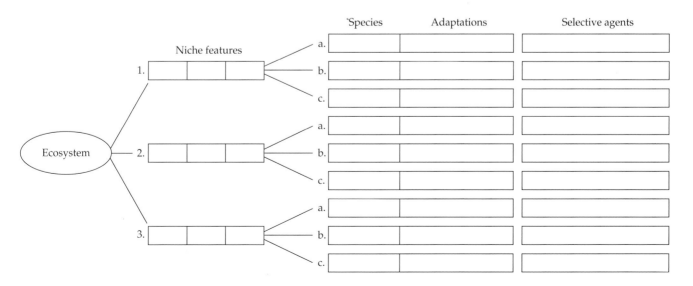

17. DeCoursey, P. J. 1983. Biological timing. In: The Biology of Crustacea Volume 7: Behavior and Ecology, F. J. Vernberg and W. B. Vernberg (eds.), pp. 107–162. Academic Press, New York.

18. DeCoursey, P. J. 1989. Photoentrainment of circadian rhythms: An ecologist's viewpoint. In: Circadian Clocks and Ecology, T. Hiroshigi and K-I. Honma (eds.), pp. 187–206. University of Hokkaido Press, Sapporo, Japan.

19. DeCoursey, P. J. 1990. Circadian photoentrainment in nocturnal mammals: Ecological overtones. Biol. Behav. 15: 213–238.

20. DeCoursey, P. J. 2001. Early research highlights at the Max-Planck Institute for Behavioral Physiology, Erling-Andechs and their influence on chronobiology. In: Zeitgebers, Entrainment, and Masking of the Circadian System, K-I. Honma and S. Honma (eds.), pp. 55–74. University of Hokkaido Press, Sapporo, Japan.

21. DeCoursey, P. J., J. Krulas, G. Mele, and D. Holley. 1997. Circadian performance of suprachiasmatic nuclei (SCN)-lesioned antelope ground squirrels in a desert enclosure. Physiol. Behav. 62: 1099–1108.

22. DeCoursey, P. J., J. K. Walker, and S. A. Smith. 2000. A circadian pacemaker in free-living chipmunks: Essential for survival? J. Comp. Physiol. [A] 186: 169–180.

23. Dittman, A. H., and T. P. Quinn. 1996. Homing in Pacific salmon: Mechanisms and ecological basis. J. Exp. Zool. 199: 83–91.

24. Dunlap, J. C. 1999. Molecular bases for circadian clocks. Cell 96: 271–290.

25. Endler, S. A. 1986. Natural Selection in the Wild. Princeton University Press, Princeton, NJ.

26. Enright, J. T. 1975. Orientation in time: Endogenous clocks. In: Marine Ecology, O. Kinne (ed.), pp. 917–944. Wiley, New York.

27. Fleissner, G., and G. Fleissner. 2002. Retinal circadian rhythms. In: Biological Rhythms, V. Kumar (ed.), pp. 71–82. Narosa, New Delhi, India.

28. Goldman, B. D. 2001. Mammalian photoperiodic systems: Formal properties and neuroendocrine mechanisms of photoperiodic time measurement. J. Biol. Rhythms 16: 283–301.

29. Gould, S. J., and R. C. Lewontin. 1979. The spandrels of San Marco and the Panglossian paradigm: A critique of the adaptationist programme. Proc. R. Soc. Lond. [Biol] 205: 547–565.

30. Grant, B. R. T., and P. R. Grant. 1989. Evolutionary Dynamics of a Natural Population: The Large Cactus Finch of the Galápagos. University of Chicago Press, Chicago.

31. Gwinner, E. 1981. Circannual Systems. In: Handbook of Behavioral Neurobiology Volume 4: Biological Rhythms, J. Aschoff (ed.), pp. 391–410. Plenum, New York.

32. Gwinner, E. 1986. Circannual Rhythms. Springer, Berlin.

33. Harvey, P. H., and M. D. Pagel. 1991. The Comparative Method in Evolutionary Biology. Oxford University Press, New York.

34. Hastings, J. S., B. Rusak, and Z. Boulos. 1991. Circadian rhythms: The physiology of biological timing. In: Integrative Animal Physiology, C. L. Prosser (ed.), pp. 435–546. Wiley, New York.

35. Hastings, J. W., and A. Keynan. 1965. Molecular aspects of circadian systems. In: Circadian Clocks, J Aschoff (ed.), pp. 167–182. North-Holland, Amsterdam.

36. Hastings, J. W., and B. M. Sweeney. 1959. The Gonyaulax clock. In: Photoperiodism and Related Phenomena in Plants and Animals, B. Withrow (ed.), pp. 567–584. American Association for the Advancement of Science, Washington, DC.

37. Hochachka, P. W., and G. N. Somero. 2002. Biochemical Adaptation: Mechanism and Process in Physiological Evolution. Oxford University Press, New York.

38. Hoffmann, K. 1960. Experimental manipulation of the orientational clock in birds. Cold Spring Harb. Symp. Quant. Biol. 25: 379–387.

39. Jilge, B. 1993. The ontogeny of circadian rhythms in the rabbit. J. Biol. Rhythms 8: 247–260.

40. Johnson, C. H., and T. Kondo. 2001. Circadian rhythms in unicellular organisms. In: Handbook of Behavioral Neurobiology Volume 12: Circadian Clocks, J. S. Takahashi, F. W. Turek, and R. Y. Moore (eds.), pp. 61–77. Kluwer Academic/Plenum, New York.

41. Johnson, C. H., S. S. Golden, M. Ishiura, and T. Kondo. 1996. Circadian clocks in prokaryotes. Mol. Microbiol. 21: 5–11.

42. Kerr, R. A. 1999. Early life thrived despite earthly travails. Science 284: 2110–2113.

43. King, D. P., and J. S. Takahashi. 2000. Molecular genetics of circadian rhythms in mammals. Annu. Rev. Neurosci. 23: 713–742.

44. Klein, D. C., R. Y. Moore, and S. M. Reppert (Eds.) 1991. Suprachiasmatic Nucleus: The Mind's Clock. Oxford University Press, New York.

45. Loher, W. 1972. Circadian control of stridulation in the cricket Teleleogryllus commodus Walker. J. Comp. Physiol. 79: 173–190.

46. Lohmann, K. J., and C. M. F. Lohmann. 1996. Orientation and open-sea navigation in sea turtles. J. Exp. Zool. 199: 73–81.

47. Lyman, C. P., J. S. Willis, A. Malan, and L. C. H. Wang. 1982. Hibernation and Torpor in Mammals and Birds. New York, Academic Press.

48. Menaker, M., L. F. Moreira, and G. Tosini. 1997. Evolution of circadian organization in vertebrates. Braz. J. Med. Biol. Res. 30: 305–313.

49. Miwa, I., H. Nagatoshi, and T. Horie. 1987. Circadian rhythmicity within single cells of Paramecium bursarius. J. Biol. Rhythms 1: 57–64.

50. Moore, R. Y., and R. K. Leak. 2001. Suprachiasmatic nucleus. In: Handbook of Behavioral Neurobiology Volume 12: Circadian Clocks, J. S. Takahashi, F. W. Turek, and R. Y. Moore (eds.), pp. 141–179. Kluwer Academic/Plenum, New York.

51. Morgan, S. G., and J. H. Christy. 1995. Adaptive significance of the timing of larval release by crabs. Amer. Nat. 45: 457–479.

52. Neumann, D. 1989. Circadian components of semilunar and lunar timing mechanisms. In: Biological Clocks and Environmental Time, S. Daan and E. Gwinner (eds.), pp. 173–182. Guilford, New York.

53. Ouyang, Y., C. R. Andersson, T. Kondo, S. S. Golden, and C. H. Johnson. 1998. Resonating circadian clocks enhance fitness in cyanobacteria. Proc. Natl. Acad. Sci. USA 95: 8660–8664.

54. Page, T. L. 2001. Circadian systems of invertebrates. In: Handbook of Behavioral Neurobiology Volume 12: Circadian Clocks, J. S. Takahashi, F. W. Turek, and R. Y. Moore (eds.), pp. 79–110. Kluwer Academic/Plenum, New York.

55. Palmer, J. D. 2002. The Living Clock: The Orchestrator of Biological Rhythms. Oxford University Press, New York.

56. Pittendrigh, C. S. 1954. On temperature independence in the clock system controlling emergence time in Drosophila. Proc. Natl. Acad. Sci. USA 40: 1018–1029.

57. Pittendrigh, C. S. 1981. Circadian systems: General perspective. In: A Handbook of Behavioral Neurobiology Volume 4: Biological Rhythms, J. Aschoff (ed.), pp. 57–80. Plenum, New York.

58. Pittendrigh, C. S., and V. G. Bruce. 1959. Daily rhythms as coupled oscillator systems and their relation to thermoperiodism and photoperiodism. In: Photoperiodism and Related Phenomena in Plants and Animals, A. R. Withrow and R. Withrow (eds.), pp. 475–505. AAAS, Washington, DC.

59. Pittendrigh, C. S., and S. Daan. 1976. A functional analysis of circadian pacemakers in nocturnal rodents. I. The stability and lability of spontaneous frequency. J. Comp. Physiol. 106: 223–252.

60. Purves, W. K., D. Sadava, G. H. Orians, and H. C. Heller. 2001. Life: The Science of Biology, 6th Edition, Chapters 52–57. Sinauer, Sunderland, MA.

61. Reebs, S. G. 1996. Time-place learning in golden shiners (Pisces: Cyprinidae). Behav. Proc. 36: 253–262.

62. Riccio, A. P., and B. D. Goldman. 2000. Circadian rhythms of locomotor activity in naked mole-rats (Heterocephalus glaber). Physiol. Behav. 71: 1–13.

63. Ridley, M. (Ed.) 1997. Evolution. Oxford University Press, New York.

64. Ruby, N. F., J. Dark, H. C. Heller, and I. Zucker. 1998. Suprachiasmatic nucleus: Role in circannual body mass and hibernation rhythms of ground squirrels. Brain Res. 782: 63–72.

65. Rusak, B. 1981. Vertebrate behavioral rhythms. In: A Handbook of Behavioral Neurobiology Volume 4: Biological Rhythms, J. Aschoff (ed.), pp. 183–213. Plenum, New York.

66. Sweeney, B. M. 1987. Rhythmic Phenomena in Plants, 2nd Edition. Academic Press, New York.

67. Takahashi, J. S., F. W. Turek, and R. Y. Moore (Eds.) 2001. Handbook of Behavioral Neurobiology Volume 12: Circadian Clocks. Kluwer Academic/Plenum, New York.

68. Tosini, G., and M. Menaker. 1995. The evolution of vertebrate circadian systems. In: Circadian Organization and Oscillatory Coupling, K-I. Honma and S. Honma (eds.), pp. 39–52. University of Hokkaido Press, Sapporo, Japan.

69. Underwood, H. 1992. Endogenous rhythms. In: Biology of the Reptilia, C. Gans and D. Crews (eds.), pp. 229–297. Chicago University Press, Chicago.

70. Underwood, H. 2001. Circadian organization in nonmammalian vertebrates. In: Handbook of Behavioral Neurobiology Volume 12: Circadian Clocks, J. S. Takahashi, F. W. Turek, and R. Y. Moore (eds.), pp. 111–140. Kluwer Academic/Plenum, New York.

71. Vitaterna, M. H., D. P. King, A-M. Chang, J. M. Kornhauser, P. L. Lowrey, et al. 1994. Mutagenesis and mapping of a mouse gene, Clock, essential for circadian behavior. Science 264: 719–725.

72. Walton-Davis, J., and P. W. Sherman. 1994. Sleep arrhythmia in the eusocial naked mole rat. Naturwissenschaften 81: 272–275.

73. Weiner, J. 1995. The Beak of the Finch. Vintage/Random House, New York.

74. Wikelski, M., and M. Hau. 1995. Is there an endogenous tidal foraging rhythm in marine iguanas? J. Biol. Rhythms 10: 335–350.

75. Zucker, I., T. M. Lee, and J. Dark. 1991. The suprachiasmatic nucleus and annual rhythms of mammals. In: Suprachiasmatic Nucleus: The Minds' Clock, D. C. Klein, R. Y. Moore, and S. M. Reppert (eds.), pp. 246–259. Oxford University Press, New York.

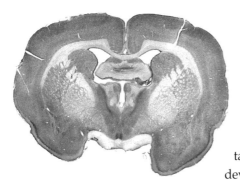

Chapter 1 began the task of presenting chronobiology vocabulary and principles. A vital new idea in the origins of the field of chronobiology was the concept of biological clocks as self-sustained physical oscillators. This chapter will expand the view of circadian pacemaker theory and demonstrate the intellectual power of the self-sustained oscillator model. The ideas portrayed will be developed in the remaining chapters of this book.

A very important function of rhythms is to provide an internal estimate of the external local time. Because most research on plant and animal rhythms has concerned daily timing, Chapter 3 will deal exclusively with circadian properties. Just as a wristwatch tells the time of day, a biological clock allows an organism to program its activities at an appropriate phase relationship, the phase angle, with respect to the daily environmental cycle. Conserving phase angle under varying environments is one of the key function of clocks. By performing this and many other functions, living pacemakers add invaluable survival capacity to plants and animals.

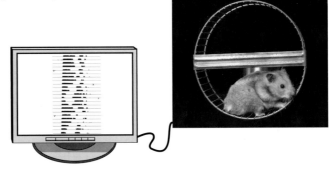

3

Fundamental Properties of Circadian Rhythms

*Your hearts know in silence the secrets of the days and the nights.
But your ears thirst for the sound of your heart's knowledge. You
would know in words that which you have always known in
thought.*

— Khalil Gibran, *The Prophet*

Introduction: Circadian Clocks Are Important for Organisms[2,27,45]

A biological timer can act as an alarm clock to wake up an organism or to initiate a physiological process at an appropriate phase of the daily environmental cycle. It can also help an organism prepare in anticipation of actual need. Another important function in some organisms is the accurate measurement of ongoing time throughout the daily cycle. Such a continuously consulted timer is particularly relevant for time-compensated sun-compass migration. A third function of a circadian clock is to act as a measuring stick for estimating the day length or night length. By this means, seasonal phenomena requiring photoperiodic time measurement of day length can be regulated appropriately.

Understanding the fundamental physical properties of rhythms is a major step toward becoming a chronobiologist. Knowledge about the physical properties of circadian rhythms gives the reader a language to describe circadian phenomena and provides a basis for experimental design and the interpretation of results. The vocabulary of chronobiology draws heavily on the lexicon of physics because the properties of circadian oscillators resemble physical and chemical oscillators. This similarity has allowed exploration of a complex biological system that would otherwise have remained obscure.

Major Defining Features Give Great Insight into the Clock Mechanism[44,53,65]

Although recent breakthroughs in the chronobiology field have identified many proteins that appear to act as biological clock components, scientists have only just begun to understand how the components interact functionally to generate circadian oscillations. The genes encoding clock components clearly encode different proteins in different organisms, but current clues suggest that the ways in which these proteins interact to form circadian feedback loops are similar for many species (see Chapter 7).

Perhaps the biochemical mechanisms of circadian clocks have evolved along a common path for diverse organisms. On the other hand, natural selection may have chosen significantly different mechanisms to accomplish circadian timing in different organisms. Nevertheless, circadian rhythms are defined by major, observable, and well-established criteria, not by a molecular mechanism. Many criteria were postulated by Pittendrigh, Bünning, Aschoff, and others almost from the beginning of circadian research as necessary, formal properties of living oscillator clocks. Three key characteristics will be emphasized in this chapter.

The first of these defining criteria is the persistence of an overt circadian rhythm in constant temperature and constant light or constant dark conditions with a period of approximately 24 h. To demonstrate persistence, the rhythm must be assayed for several consecutive cycles, preferably five or more.

Temperature compensation is a second criterion. This characteristic refers to the observation that free-running circadian period lengths in an organism are very similar when measured at different ambient temperatures. Temperature compensation is often considered to be just one aspect of a general compensation mechanism that keeps period lengths similar despite differences in factors affecting metabolism, such as temperature or nutrition.

Third, endogenous rhythms of approximately 24 h can be entrained by certain 24 h environmental cues, such as light–dark cycles, temperature cycles, or other stimuli. The entrainment section will consider phase response curves (PRCs), as well as the discrete and continuous models of entrainment. The chapter will conclude with a description of the insights into circadian properties that come from the realization that circadian pacemakers are limit-cycle oscillators.

Circadian biologists often study organisms under free-running conditions, and they are apt to forget that the function of the circadian clock is not to provide a precise timekeeper that persists in constant conditions. With the possible exception of organisms dwelling in caves or in deep oceans, all organisms live in a rhythmic environment, and the adaptiveness of circadian systems in a real environment should be kept clearly in mind. Consequently, the three properties of circadian rhythms will also be discussed in terms of their adaptive value under daily environmental cycles (see Chapter 2).

The Free-Running Period Is Approximately 24 Hours Long[1,4,9,16,22,35,39,46,53,60]

Under constant environmental conditions the inherent period of a circadian oscillator is expressed in overt output rhythms. These rhythms are generally highly precise, and the period is called the free-running period (FRP). A synonym in some of the older literature is *tau* (*t*), but the standard English term *free-running period* will be used in this text. The period is dependent on a variety of factors, including species specificity, illumination, ambient temperature, developmental factors, and prior history.

Most individuals and most species have free-run values in DD over a span from about 23 to 25 h. A nocturnal flying squirrel in constant darkness, for example, during one assay episode might be active with a free-running period of 23 h and 31 minutes (Figure 3.1A). It

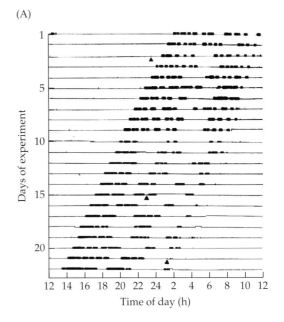

(A)

(B)

Test period	Days	
1	20	23 h: 52 min ± 4 min
2	25	23 h: 48 min ± 9 min
3	15	23 h: 50 min ± 8 min
4	31	23 h: 55 min ± 5 min
5	51	23 h: 55 min ± 6 min
6	16	23 h: 48 min ± 9 min
7	11	23 h: 51 min ± 8 min
8	81	23 h: 31 min ± 7 min

FIGURE 3.1 Free-running locomotor activity of flying squirrels in constant darkness (DD). (A) Segment of wheel-running activity of one squirrel for 81 days, with 23 days shown from test period 8. (B) Summary of eight FRPs in DD for this individual. (A courtesy of P. DeCoursey; B from DeCoursey, 1961.)

would begin its activity 29 minutes earlier each day. Some individuals can show exquisite day-to-day precision, exhibiting a standard deviation for the free-run value of as little as a few minutes per day (Figure 3.1B).

Not all individuals, however, are highly precise. Day-active species tend to be much less precise than nocturnal species. Some rodents, particularly the prairie voles that are grass eaters, display well-developed ultradian rhythms. They feed in irregular 4 h bouts that can obscure the underlying 24 h rhythmicity. A summary of representative actographs gives a glimpse of the variability of form and precision for both intraspecific and interspecific examples (see Figure 1.19 in Chapter 1).

Illumination conditions affect both the period and the amplitude of circadian rhythms. For example, high-amplitude rhythms of most photosynthetic organisms are optimally expressed in LL, but in most nocturnal mammals LL tends to depress the amplitude of circadian rhythms. Dr. Jürgen Aschoff collected large amounts of data demonstrating that the free-running period in LL is a function of the intensity of illumination. He observed that increasing the light intensity tends to shorten the period of rhythms in day-active organisms, while increasing the period of night-active organisms. This intriguing correlation has come to be known as Aschoff's rule (Figure 3.2).

Compared with most other kinds of biological oscillations, the period of circadian rhythms has a surprisingly long duration, approximately 24 h. The free-running rhythms of circadian clocks deviate very little from 24 h, presumably for several reasons. Entrainment between two oscillators will generally be optimal if the inherent frequencies are nearly the same. In the case of circadian oscillators, the driving oscillation is the 24 h cycle resulting from Earth's rotation. Therefore, to entrain stably to the 24 h cycle, the driven circadian oscillator needs an intrinsic frequency approximating Earth's daily rotation period of 24 h.

The long period length has been a fascinating puzzle to circadian scientists. The type of biochemical machinery that could maintain approximately 24 h oscillations of such precision has challenged many minds (see Chapter 7). Some investigators were incredulous that a series of biochemical reactions could underlie such an accurate timekeeper, and they postulated the alternative idea of exogenous timing. According to this theory, organisms could sense and be driven behaviorally by some sort of subtle geophysical cue of external timing in the 24 h environmental cycle. The unknown cue was referred to at the time by the mysterious name *Factor X*. However, the non-24 h nature of a free-running period in constant conditions remains very persuasive evidence against Factor X and the exogenous timing hypothesis. A probable reason explaining why

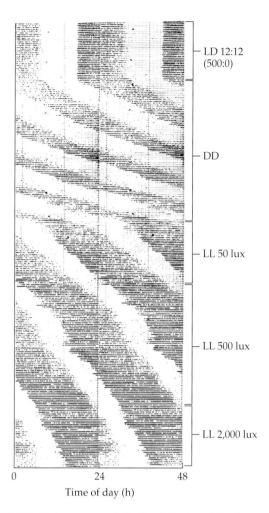

FIGURE 3.2 Maintenance of rhythmicity by a blind sparrow in LL illustrates Aschoff's rule. Data are all double-plotted. At the top the bird is entrained to LD 12:12 with 500 lux in the light phase. The animal is first placed in darkness, for observation of the free-running periodicity, and then transferred sequentially to environments with increasing levels of ambient light. Note changes in both the free-running period length and the ratio of activity to rest time. (From Menaker, 1971.)

the period of circadian systems ranges from 23 to 25 h is discussed later, in the section on entrainment.

Although new molecular investigations are yielding clues to the nature of the feedback loop of the circadian clockwork, an adequate explanation is not yet available for the biochemical means by which the clockwork is able to keep time precisely with a time constant as long as 24 h (see Chapter 7). To explain the remarkable day-to-day precision of the free-running period, some researchers have suggested that the interaction of multiple, imprecise clocks could lead to emergent clock behavior that is precise. These multiple clocks might involve coupling of intercellular or intracellular oscillators.

Although free-running periods can be extremely precise from day to day, they are not rigid and invariable. Most organisms exhibit a range of allowed values of the free run, depending on the organism's prior history. Within this range, period can be modulated, particularly in response to environmental conditions. For example, the free run can be altered by prior entrainment to an LD cycle (see Figures 1.20, 2.3G, and 2.24B).

In the actograms, the free-running rhythm can be visualized by the slope of a line that passes through the activity onset points for each day in DD (see Figures 1.16, 1.18, 1.19, 1.22). Prior to entrainment, a slope is seen reflecting a period less than 24 h. During entrainment to a 24 h LD cycle, the slope is vertical. Following entrainment and release into constant darkness, the slope through activity onsets is more vertical than before entrainment, reflecting a free-running period closer to 24 h in length than in the initial free run (see Figures 1.20, and 2.3F,G). In a history-dependent way, the LD entrainment has modified the innate free run within the allowable range for that individual. Such history-dependent influences on the free-running period, known as aftereffects, can be caused by both 24 h and non-24 h lighting schedules. Daily cycles with different proportions of light and dark periods are called LD ratio schedules. Those with different environmental periods are called T-cycles and can be presented in a range from 18 to 26 h. All of these types of lighting regimes can have profound effects on the subsequent free-running period in constant conditions. Aftereffects usually decay gradually after transfer of the organism to constant conditions.

Circadian rhythms are controlled by oscillators, not by hourglass timers. The persistence of rhythmic behavior in continuous conditions over many days or even years leads to the conclusion that daily biological rhythms are based on a bona fide self-sustaining oscillator. By definition, nonoscillatory hourglass timers can measure time for only 24 h or less and must be triggered anew each day by a defined environmental stimulus. The reason a 24 h oscillator is more valuable as a timekeeper than a non-sustaining hourglass timer is purely guesswork. Evolutionary processes depend on a series of events and selective pressures whose history is still largely unknown. Sometimes, however, intuitive guesses as to function are helpful. Endogenous, self-sustained oscillation may allow an organism to anticipate periodic events in the environment, such as dawn. However, a temperature-compensated hourglass could also allow an organism to anticipate a particular local time in a consistent environ-

ment because dawn could be predicted by timing 12 h after dusk. Perhaps an oscillator allows a much wider range of capabilities than an hourglass. A simple hourglass, for example, cannot always specify a given local time in a fluctuating environment. In particular, the duration of night and day is a function of the time of year except at the equator. Therefore, an hourglass that always measures a certain duration of night will be a precise estimator of the time of dawn on only two days out of the year. On the other hand, an oscillator with appropriate characteristics can do an excellent job of estimating a given phase in the light–dark cycle at all seasons of the year, as will be described later in the chapter. Another advantage of oscillators over hourglass timers is the ability to keep track of time for many cycles if exposure to an environmental time cue is interrupted. Hibernating animals such as bats and ground squirrels become torpid in winter in dark caves or dens that lack daily environmental cues, but many of their circadian physiological functions continue for weeks or months in a predictable way.

Free-Running Periods of Circadian Clocks Are Temperature Compensated[5,8,38,52,57,58,66,73]

Most biochemical processes are very sensitive to temperature in a predictable way[73]

The reaction rates of most biochemical reactions double or triple with every 10°C increase in temperature. For such a process, its Q_{10} value is stated as 2 or 3, respectively. Q_{10} is a gauge of the temperature dependency of a process, and it is usually measured as the ratio of the rate of a process at a higher temperature divided by the rate at a temperature 10°C lower. For a process with a $Q_{10} = 1$, the rate is independent of temperature change.

Surprisingly, the Q_{10} values for the free runs of circadian rhythms range from 0.8 to 1.4 (Table 3.1) when measured within the physiological range for growth of

TABLE 3.1	Q_{10} values for representative species, indicating temperature compensation of free-running rhythms

Species	Phenomenon measured	Q_{10}
Euglena (protist)	Phototaxis	1.01–1.1
Gonyaulax (dinoflagellate)	Bioluminescence	0.85
Neurospora (fungus)	Conidiation	1.03
Drosophila (insect)	Eclosion	1.1–1.25
Lacerta (lizard)	Locomotor activity	1.02
Myotis (bat)	Locomotor activity	1.4
Peromyscus (mouse)	Locomotor activity	1.1–1.4

Source: Sweeney and Hastings, 1960.

the organism. Consequently, these pacemakers proceed with almost the same FRP regardless of the ambient temperature in constant conditions, showing that circadian clocks, although not at all temperature independent, are strongly temperature compensated. Such a property is unusual in biological phenomena, although not totally unique. Other examples of temperature-compensated biological processes include sensory feedback systems in stretch and pressure receptors.

Temperature-compensated biological processes are unexpected, and at the beginning of circadian research, they led to some surprising experimental results. In the 1930s, the well-known scientist Dr. Hans Kalmus tested for the endogenous nature of biological rhythms. He showed that the daily hatching rhythm of the fruit fly *Drosophila* was inherent and persistent. It could also be initiated and entrained by light, and it proceeded with a consistent period length at different constant temperatures. The length of the first period after a temperature shift, however, displayed a Q_{10} of 3.0, leading Kalmus to believe that aspects of the rhythmic process were temperature sensitive. Pittendrigh later pointed out that a temperature-sensitive process would be useless as a clock. He showed that the apparent temperature sensitivity was actually due to transient or temporary behavior of the hatching rather than the clock itself, before the internal rhythmicity had stabilized. Pittendrigh's studies highlighted the importance of temperature compensation for circadian rhythms. His results generated great interest in biological clocks and initiated an avalanche of circadian investigations.

Temperature compensation of circadian clocks means that the period of the clock is approximately the same under different constant ambient conditions [9,24,53,67,73]

Saying that the free-running period of a circadian clock is temperature compensated is not the same as saying that circadian pacemakers are completely insensitive to temperature. Circadian biologists are fond of saying that the biological clock is temperature compensated because if it were not, then the clock would be good only as a thermometer, not as a timekeeper. The statement should be clarified, though, because temperature pulses and cycles can entrain the circadian clock. In fact, the daily temperature transitions at dusk and dawn provide important temporal cues to entrain the internal clock to the external day–night cycle. Thus, temperature compensation of the free-running period exists, but the phase of the clock can be modified by temperature transitions.

Circadian organisms thus have a seemingly paradoxical response to temperature, simultaneously responsive in terms of phase and nonresponsive in terms of period. The resolution of the paradox probably lies in

the need to buffer the biochemistry of the clock and prevent it from running faster or slower when the temperature changes. In this way, an appropriate phase angle is maintained between the internal clock and the external time of day. Without temperature compensation, problems would arise if the organism's clock were entrained by a warm day followed by a cool day. Because the warming sun would appear and disappear at nearly the same times on both days, the phase of the temperature cycle would be the same on both days. Therefore, there should be little difference in the entraining input of the temperature cycle, even though the average temperature each day was different. If the clock's period were not compensated, it would run fast on the first (warm) day and run slower on the second (cool) day. The phase of the circadian pacemaker at dusk on day 1 versus day 2 would be different, merely because the average temperature was different. If the period of the oscillator were not temperature compensated, the phase angle would not be conserved and a key function of circadian timing would be lost.

Circadian clocks used in sun-compass orientation require temperature compensation throughout their cycle [27,67]

The biological uses of circadian clocks also place constraints on the mechanism of temperature compensation. For instance, a clock in which one part of the circadian cycle is accelerated by a temperature increase while another part is slowed by warmer temperature could have temperature compensation of the overall free-running period. However, such a counterbalancing mechanism would be useless for a clock that must be continuously consulted throughout the day for accurate estimation of local time. The temperature compensation of circadian timekeepers apparently has clear adaptive significance for plants and poikilotherms, such as insects, organisms that cannot control their body temperature. However, even homeothermic animals have temperature-compensated clocks.

Some of these organisms hibernate or exhibit daily bouts of torpor that result in dramatic changes in core body temperature. In these cases, the value of a temperature-compensated clock is clear. Most homeotherms maintain a relatively constant body temperature except when suffering from a fever. The evidence suggests that their circadian mechanism retains temperature compensation. Perhaps the compensation mechanism became an integral component of the circadian clockwork early in its evolution and was conserved even beyond its need. Alternatively, the potential disturbance of a temperature-dependent clockwork by the low-amplitude body temperature rhythm of homeotherms may justify maintenance of the temperature compensation mechanism.

Some circadian clocks have a Q_{10} value less than 1[24,73]

Table 3.1 includes one example of an organism whose free-running period displays a Q_{10} value of less than 1. Such oscillators run more slowly at higher temperatures and could thus be described as overcompensated. Temperature compensation is indeed a result of an endogenous biochemical mechanism that can be slightly inaccurate in either direction around a Q_{10} value of 1. Little is known about the mechanism of temperature compensation by circadian pacemakers. Nevertheless, fascination continues for a clockwork composed of biochemical reactions that oscillate at approximately the same rate over a large range of temperatures. Mutants have been isolated whose temperature compensation is impaired, but the mechanism by which these genetic lesions affect clock compensation is also unknown at present. Temperature compensation acts constantly throughout the circadian cycle without speeding up at one part of the cycle and slowing down at another.

Several models for the compensation mechanism have been proposed. Sweeney and Hastings suggested in 1960 that the pacemaker mechanism might include two counterpoised biochemical reactions, the individual rates of which both increase with temperature:

Reaction 1: $X \rightarrow Y$ (mediated by enzyme 1)
Reaction 2: $A \rightarrow B$ (mediated by enzyme 2)

A situation can be postulated in which B, the product of reaction 2, inhibits enzyme 1. As temperature increased, the rate of reaction 1 would begin to increase. However, the rates of reactions 1 and 2 would similarly increase, and more B would be produced at the higher temperature. The increased level of B would cause the rate of reaction 1 to slow in a compensating manner. In such a system, in which the two reactions occur together, the temperature dependencies of the two reactions could be designed such that the rate of reaction 1 was essentially temperature independent. If reaction 1 were the primary determinant of period length in a circadian system, the free-running period would be temperature compensated.

Rhythmic Environmental Stimuli Entrain Circadian Clocks[1,16,42,70]

Entrainment of the circadian clock provides an internal estimate of external local time[1,8,10,56,57,58]

A fundamental realization from chronobiology is that the physiological time of day actually experienced by an organism directly reflects its internal clock but only indirectly reflects the immediate external conditions of light and dark. When a man wakes up at night, walks into the hall, and turns on the lights, he experiences a feeling of nighttime despite the lighting. Similarly, if a woman walks into a planetarium for an afternoon show, she experiences the alertness associated with daytime, despite the darkness. To retain this essential estimate of local time, the internal timekeeper must be entrained to the 24 h environmental cycle.

The consequences of entrainment are that the period of the biological rhythm becomes equal on average to that of the entraining stimuli, with a stable phase relationship or phase angle between the entraining and entrained oscillations. An environmental stimulus that can synchronize circadian clocks is called an entraining agent or circadian time giver. Such stimuli are also referred to in the chronobiology literature as zeitgebers, which is from the German term for "time givers." This text will adopt standard English terms for the entrainment process.

To establish that a time cue has truly entrained a free-running rhythm, two criteria must be demonstrated. First, the period of the overt rhythm must equal the period of the entraining cycle with a stable phase angle. The establishment of stable phase has been observed both in natural and in laboratory conditions (Figure 3.3). Second, after return of the organism to constant conditions, the free-running rhythm must continue with a phase determined by the entraining cycle. A rhythm may superficially appear entrained by the periodic stimuli in some cases, but after removal of the stimuli the free-running rhythm starts up with a phase differing from that of the entraining stimuli (see Figure 1.22 in Chapter 1). The stimuli appear to have forced expression of the overt behavior without actually entraining the central pacemaker. This phenomenon is called masking and can present a serious artifact for investigations into entrainment.

Light and dark signals are the most important environmental entraining agents[1,4,15,40,56,59,72]

In nature, multiple environmental factors oscillate over the daily cycle. Some variables include light, temperature, humidity, food availability, and even social cues in a few species (see Plate 3, Photos C, D, E, and G). Only a few factors can function as entraining cues in most species. The most important cue is generally the 24 h solar cycle of light and darkness. Almost all circadian rhythms can be entrained to light–dark cycles (see Figure 3.3).

Temperature cycles or food availability can function as circadian entrainment cues in some organisms, but most often they play a supporting role to LD cycles. For example, entrainment studies using conflicting phasing of LD versus low-amplitude temperature cycles show that the LD cycle determines the entrained phase angle in a variety of poikilothermic organisms including lizards and fungi. However, high-amplitude 10°C temperature cycles can win out over LD. Because of the

(A)

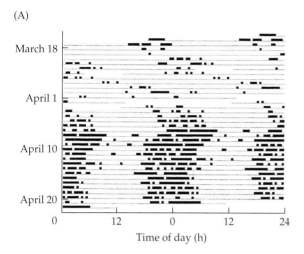

(B)

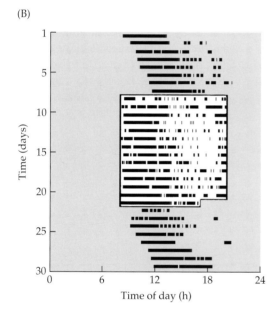

FIGURE 3.3 Entrainment in nature and in the laboratory. (A) Beaver activity was recorded through a microphone placed in a beaver lodge in northern Alberta. At the beginning of the experiment the lodge was under ice and snow, and the animals experienced more or less constant temperature and darkness; beavers express a long free-running period. In the middle of the experiment the spring thaw occurred; recording was temporarily lost in the flood, but then restored. Now exposed to a full day–night cycle, the beavers entrained to the 24 h day after a few days of transients. (B) Entrainment of the activity rhythm of a chaffinch, *Fringilla coelebs*. The record shows a free run in 0.5 lux for 8 days. In this very dim light the bird's clock displays a period of 24.5 h. On transfer to a bright light–dark cycle, the bird is entrained to LD 12:12. It is day-active. On release it reverts to its free-running periodicity. (A from Bovet and Oertli, 1974; B from Pittendrigh, 1981.)

importance of light–dark cycles and the ease with which they can be manipulated, most entrainment research has used LD cycles and light pulses.

Two different models have been proposed to explain circadian entrainment[1,10,14,48,54,61,62]

As mentioned already, entrainment to a time cue such as a light–dark cycle means that the period of the biological rhythm becomes equal to that of the LD cycle, and the phase relationship between the LD cycle and the biological rhythm is stable. Several possible characteristics of an LD cycle in nature could theoretically be responsible for entraining the biological rhythm. Possibilities include the dawn and dusk transitions, the increase and decrease of light intensity during the daytime, the changes in spectral quality or color of light, or the continuous presence of light during the daytime.

Two major classes of models have been proposed to explain the mechanisms by which circadian clocks are entrained to environmental cycles. One kind is the continuous model, also called parametric or tonic, which focuses on the importance of gradual changes in the environment. The second type is called the discrete model, also known as the nonparametric or phasic model. The discrete model focuses on the effects of environmental transitions such as dawn and dusk.

THE CONTINUOUS MODEL. The continuous entrainment model in simplest form states that the angular velocity or rate of motion of the clock mechanism changes proportionally to the intensity of light present. One type of evidence quoted has been the observation that the free-running rhythm depends on light intensity in accordance with Aschoff's rule (see Figure 3.2). However, an explanation based on phase response curves could explain this result equally well (see the next section). Another type of evidence suggests that light has a continuous action on the clock during entrainment to the light–dark cycle, causing it to run faster in light and slower in darkness, for example. The modulation of angular velocity under the light and dark portions of a photoschedule could allow the circadian pacemaker to continuously adjust its cycle length to that of the environment.

In different LD cycles, differing amounts of the daily activity cycle must presumably be exposed to light to achieve the overall angular velocity for a 24 h entrained period. In other words, in different LD ratios the light transition would occur at different phase angles of activity. Entrainment achieved by modulation of the free-running period has sometimes been called *parametric entrainment*, but this term is mathematically inappropriate and the more general term *continuous entrainment* will be used. Aschoff was a strong advocate for the contribution of continuous effects to the entrainment of biological oscillators.

One simple way to test the hypothesis of continuous entrainment is to provide an entraining signal consisting of pure transitions but little or no duration. If a pure-

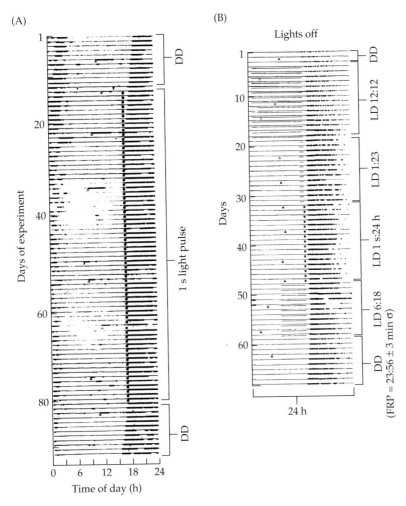

FIGURE 3.4 Testing the continuous-light-action model of photoentrainment in flying squirrels. (A) Entraining a free-running rhythm with a 1 s dim light pulse per day. Dots denote the time of the pulse. (B) Measuring phase angle difference values for four LD photoperiods. Horizontal lines or dots denote the time of lights-on. For further explanation, see the text. (A from DeCoursey, 1989; B from DeCoursey, 1972.)

ly continuous mechanism were responsible, entrainment would fail and the animal would free-run. Just such an experiment was done with flying squirrels (see Figure 3.4A). Initially the squirrels were allowed to free-run in DD for 12 days to demonstrate a free-running rhythm. Then they were provided with an entraining pulse signal of 1 s of very dim light per day (0.5 foot-candles) for 68 days, and finally returned to DD for 12 days to be certain that masking was not occurring. An instantaneous strobe flash for entrainment would have been an even better signal but was not logistically feasible.

In the entrainment part of the experiment, the squirrels continued free-running until their activity onset encountered the light pulse; thereafter they remained synchronized to the daily pulse of light. Finally, in the terminal DD period they free-ran as expected, with onset determined by the prior light pulse schedule

(Figure 3.4A). The investigators tested the second expectation, about a fixed phase angle difference, by providing a squirrel with a progression of LD entraining schedules ranging from 1 s of light per day to LD 12:12. Measurement of the phase angle difference indicated no significant differences among the different LD schedules (Figure 3.4B).

The two kinds of data suggested that continuous effects of a photoschedule had little or no effect, at least in this species. However, nocturnal den-dwelling or cave-dwelling species may be highly specialized for entrainment to the twilight dusk transition at sunset. When they wake up, they test the prevailing light intensity in an almost instantaneous light-sampling behavior, as Figure 2.3E in Chapter 2 showed with bats (see also Plate 2, Photo D). In other species, especially birds, the phase angle difference is highly dependent on LD ratio and here, undoubtedly, light action is continuous.

THE DISCRETE MODEL. Pittendrigh was a champion of the discrete model and considerable data such as the 1 second pulse synchronization experiment support that hypothesis. The greatest success in predicting entrainment of a large number of species comes from the discrete model. Because the model has influenced many chronobiologists, the theory will be described in detail with the data from *Drosophila* and nocturnal rodents from which the theory was originally derived.

In the basic premise of the model, an entrained circadian pacemaker is in equilibrium with an LD cycle when light perceived each day brings about a phase shift sufficient to advance or delay the free-running period to match the external light cycle period. At equilibrium, each light pulse of the LD schedule will fall at the same circadian phase of the organism. The light pulse must cause a shift equal to the difference between the endogenous free-running period and the period of the entraining cycle.

In nature, the entraining cues are the dawn and dusk transitions. These are usually mimicked in the laboratory by square-wave light pulses because the effective action of light is presumably caused by the discrete dawn and dusk cues in nature. For this reason, the model has been called the discrete or nonparametric model. The simplicity of the model lies in its excellent predictive properties, which are based on only two pieces of information: the free-running period and the map of phase-dependent resetting called the phase response curve (PRC).

Phase response curves map the phase-dependent responses of circadian clocks to entraining agents and form the basis of discrete models of entrainment[6,7,29,31,43,49,55,56,68,77,78]

CONSTRUCTION OF PHASE RESPONSE CURVES. For photic entrainment to occur, the circadian oscillator must respond differently to light at different phases of its cycle. The object of entrainment is for the time cue to modify the phase of the internal clock until the internal subjective day corresponds to the external day phase. Several obvious corollaries can be stated: Light seen during the circadian subjective day should have little effect on phase, light seen early at night should move the phase back to the previous day, and light seen late at night should move the phase forward into the next day.

Phase response curves are useful descriptions of this phase-dependent response. A phase response curve is a plot of phase shifts of a circadian rhythm as a function of the circadian phase that a stimulus is given. Most phase response curves are deduced from many individuals or many populations. A typical experiment for nocturnal animals starts with an LD entrainment period. In the case shown in Figure 3.5A, for the first 4 days,

individuals A through E were exposed to an LD 12:12 light schedule. The rhythmic activity was confined to the dark portion of the 24 h LD cycle, and activity started at approximately the same time each day. As a result, onsets are aligned and the activity rhythm appears entrained to the LD cycle.

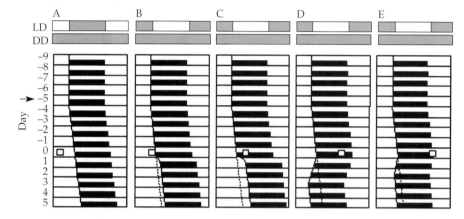

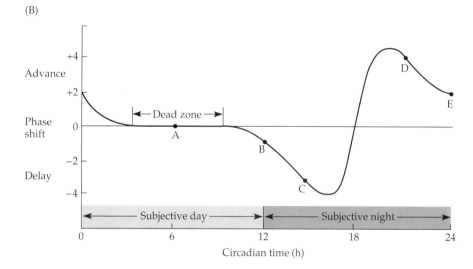

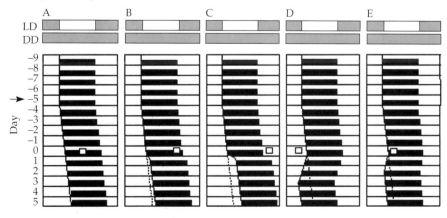

FIGURE 3.5 Phase response curves and their derivation. (A) Applying light pulses to free-running nocturnal animals at different circadian phase points to derive a phase response curve. (B) Phase response curve for the animals in A. (C) Application of light pulses to diurnal animals at different circadian phase points to derive a phase response curve. Activity is shown as dark bars in the activity panels; the light schedule is shown by the bars above panels; arrows in (A) and (C) indicate the start of DD; the small white boxes indicate light pulses. For further explanation, see the text. (Modified from Moore-Ede et al., 1982.)

On the day indicated by the arrow, about day 5, the animals were transferred into DD for the remainder of the experiment. The activity rhythm persisted in DD as a free-running rhythm. Activity started later each subsequent day because the period of the free-run in these individuals is slightly longer than 24 h. If the free-run were shorter than 24 h, as often occurs in night-active organisms, the activity onsets would occur earlier each day.

After 5 days in DD, each individual A through E was exposed to a single 1 h light pulse. For each individual the light pulse was presented at different circadian phases relative to the activity rhythm. By convention, the onset of activity for nocturnal animals is designated as circadian time 12, or CT 12. For individual A, the light pulse was given at CT 6, in the middle of the subjective day. The light pulse at this phase had little or no effect on the onset phase of the activity rhythm on subsequent days. A characteristic feature of phase response curves is that light has little or no phase-resetting effect during most of the subjective day. The subjective-day portion of the phase response curve is often referred to as the dead zone.

In contrast, light presented during the organism's subjective night will usually phase-shift the free-running rhythm. During the first half of the subjective night, light pulses phase-delay the activity rhythm so that activity starts later (Figure 3.5, individuals B and C). By current convention, phase delays are plotted as negative values. During the second half of the subjective night, light pulses will phase-advance the activity rhythm, and activity starts earlier (Figure 3.5, individuals D and E). Phase advances are plotted as positive values. In a phase response curve, the phase advances and delays are plotted against the circadian time of light pulse administration (Figure 3.5B).

To obtain an accurate waveform for the phase response curve, data points on the curve must be closely spaced. The precise waveform of the phase response curve can vary greatly depending on the strength of the stimulus, time in constant LL or DD conditions, and previous photoperiodic history. Generally, however, these factors have a relatively small influence. The phase shift may also show transient cycles before reaching a new stable phase. These transients are usually more common following phase advances (see Figure 3.5). Phase-shift values used in constructing a phase response curve should be calculated after a stable phase has been obtained and transients have subsided. In this way investigators can be certain that the phase response curve represents steady-state phase shifts.

Phase response curves can also be constructed for a day-active organism (Figure 3.5C). In this case the activity is confined to the light portion of the 24 h LD cycle. Light pulse phase response curves usually have characteristics similar to those seen for nocturnal animals. Phase delays occur in the early subjective night, phase advances occur in the late subjective night, and little phase shifting occurs during the subjective-daytime dead zone. The similarity reflects the fact that similar circadian clocks underlie the behavioral rhythms, even though the phase of the activity is reversed in diurnal versus nocturnal animals. These features hold true whether the overt rhythm peaks in the night, during the day, or at twilight. The phase response curves of nocturnal and diurnal organisms are similarly phased to the light–dark cycle, even though their overt rhythms are not.

The two earliest-published phase response curves described the response of the clock to light in the protist *Gonyaulax* and the flying squirrel *Glaucomys* and were derived completely independently. The first was done at Harvard, and the second at the University of Wisconsin for two species at opposite poles of the phylogenetic tree, but they bear a remarkable resemblance to each other. Since that time, phase response curves have been derived from hundreds of organisms, and all show distinct similarities for a collection of organisms (Figure 3.6).

Stimuli that can phase-shift circadian clocks include light pulses, temperature pulses, pulses of drugs or chemicals, food availability, and induced activity. Because light and dark are probably the most important entraining agents in nature, phase response curves for light stimuli have special interest and have been studied most extensively. A precise definition of the term *light pulse* is important. In this text, light pulses will be considered an exposure to light whose intensity exceeds a species-specific threshold. Initially the phase-shift response increases quite rapidly with increases of pulse duration, but at durations typically on the order of a few minutes, pseudosaturation sets in. Beyond this point, significant additional phase shifts are achieved only for much longer durations, of several hours. If light intensity is raised, the time to achieve pseudosaturation is shorter but the size of the response at saturation is essentially independent of intensity. For the filamentous fungus *Neurospora*, phase shifts of 8 h are achieved in response to 2-minute pulses, and in nocturnal rodents phase shifts of 1 to 3 h appear in response to pulses of 15 minutes or less.

In the natural environment, dawn and dusk are characterized by gradual changes in light intensity. Most laboratory studies have used light stimuli with abrupt transitions between light and dark. In general, however, light stimuli with abrupt transitions appear to mimic entrainment under natural conditions. Evidence from studies of entrainment in hamsters, however, suggests that the range of periods to which a hamster can entrain is larger if the dawn and dusk transitions are gradual, than if they are abrupt. This enhancement of entrainment with gradual transitions suggests participation of continuous entrainment in nature.

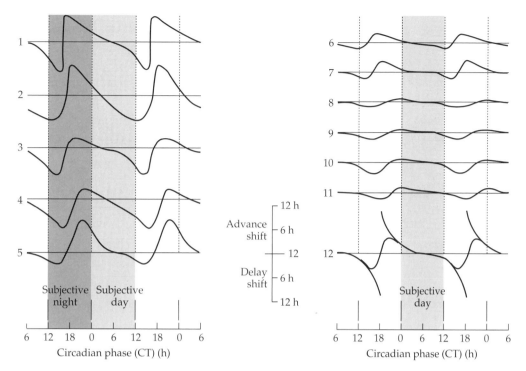

FIGURE 3.6 Phase response curves for light pulses from plant, invertebrate, vertebrate, and unicellular organisms. (1) *Sarcophaga* (flesh fly), 3 h of light at 100 lux; (2) *Coleus* (green plant), 4 h of light at 13,000 lux; (3) *Leucophaea* (cockroach), 6 h at 50,000 lux; (4) *Euglena* (unicellular protist), 4 h at 1,000 lux; (5) *Gonyaulax* (dinoflagellate), 3 h "bright" light; (6) *Anopheles* (mosquito), 1 h at 70 lux; (7) *Mesocricetus* (hamster),

0.25 h at 100 lux; (8) *Peromyscus leucopus* (white-footed mouse), 0.25 h at 100 lux; (9) *Peromyscus maniculatus* (deer mouse), 0.25 h at 100 lux; (10) *Mus musculus* (house mouse), 0.25 h at 100 lux; (11) *Taphozous* (bat), 0.25 h at 100 lux; (12) *Drosophila pseudoobscura* (fruit fly), 0.25 h "bright light" (strong resetting, Type 0 PRC; see text) or 1 ms "bright light" (weak resetting, Type 1 PRC; see text). (From Pittendrigh, 1988.)

TYPE 0 AND TYPE 1 PHASE RESPONSE CURVES. A phase response curve is a graph of phase shifts on the ordinate plotted against the original circadian phase at which a phase-resetting stimulus is administered on the abscissa. However, the same data may be plotted as a phase transition curve (PTC). In this format the new circadian phase to which a pacemaker is reset is plotted on the ordinate, while the initial circadian phase at which the stimulus was administered is plotted on the abscissa. For example, a light pulse given at an initial circadian phase of CT 6 produces no phase shift (Figure 3.7A, upper left plot). In the corresponding PTC (Figure 3.7A, lower left plot), the new circadian phase is identical to the initial circadian phase for stimuli presented at CT 6. A light pulse given at CT 12, however, elicits a phase delay of approximately 1 h (Figure 3.7A, upper left plot) and resets the pacemaker to a new circadian phase of CT 11 because 12 – 1 = 11. Thus, at the initial circadian phase of 12 (ordinate) the new circadian phase is plotted corresponding to 11 on the abscissa (Figure 3.7A, lower left plot).

In this type of plotting, a stimulus causing no phase shifts would result in a diagonal line with a slope of 1. Alternatively, a very strong stimulus that always drives

the clock to the same phase, regardless of delivery phase, would yield a PTC with a horizontal line having a slope of 0. Whereas weak phase response curves (Type 1) yield a PTC curve with an average slope of 1 (Figure 3.7A, right), strong phase response curves (Type 0) yield an average slope of 0 on a PTC (Figure 3.7A, left). In practice, no curves have monotonic slopes of 1 or 0. Type 1 resetting curves display relatively small responses amounting usually to phase shifts of less than 6 h, and they display a continuous transition between delays and advances.

If the phase shifts of a strong Type 0 phase response curve are plotted as advances and delays, a discontinuity or break point often appears at the transition between delay and advance phase shifts (Figure 3.7B, left). The break point discontinuity of phase response curves is in some cases merely a plotting convention of arbitrarily assigning phase shifts in one half-cycle of 12 h as delays and the other half-cycle as advances. To avoid these arbitrary distinctions, sometimes Type 0 phase response curves are plotted monotonically, with all phase shifts plotted as delays from 0 to 24 h (Figure 3.7B, right). For example, in a monotonic plot a phase advance of 10 h would be plotted as a phase delay of 14 h.

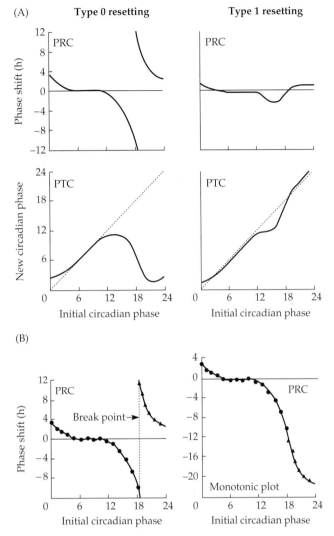

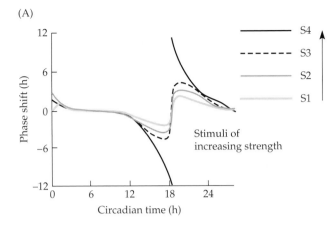

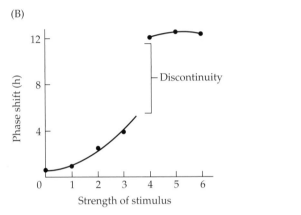

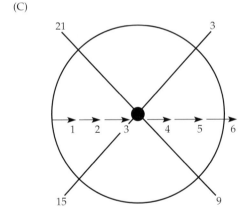

FIGURE 3.7 Types of phase response curves. (A) Type 0 and Type 1 phase response curves (PRCs) and their associated phase transition curves (PTCs). (B) Nonmonotonic and monotonic plotting of Type 0 phase response curves. For further explanation, see the text. (A modified from Pittendrigh, 1981; B from Johnson, 1999.)

FIGURE 3.8 Effect of increasing stimulus strength on phase response curve type. (A) Hypothetical family of phase response curves with increasing stimulus strengths. (B) Dose response curve for stimuli near the middle of the subjective night, at CT 19.5. (C) Limit-cycle model. For further explanation, see the text.

Whether Type 1 or Type 0 resetting is exhibited often depends on the strength of the stimulus. In the flies *Drosophila* and *Sarcophaga*, and in the mosquito *Culex*, for example, increasing the light dose of the stimulus converts Type 1 to Type 0 resetting (Figure 3.8A) and leads to a discontinuity in the dose response curve (Figure 3.8B). A similar transition from weak to strong resetting—that is, from Type 1 to Type 0 resetting—is seen in panel 12 of Figure 3.6 for pulses of different duration given to the fruit fly *Drosophila pseudoobscura*. The unexpected discontinuity in the dose response curve provided scientists with a strong clue concerning the limit-cycle nature of the underlying oscillator (Figure 3.8C), as will be discussed later in the chapter.

TRANSIENTS. A light-induced phase response curve is a map of the phase-dependent responsiveness of circadian pacemakers. The map resets rapidly within one cycle in response to light stimuli, whereas overt rhythms often require many transient cycles to attain a steady-state phase shift. Examples of transients seen on exposure to entraining stimuli in two organisms are shown in Figure 3.9. These transient cycles are therefore thought to reflect a disequilibrium or altered phase angle between the overt rhythm and the pacemaker in response to a phase shift. In some systems, however, the overt rhythm resets rapidly, and therefore phase-shifting transients may more directly reflect the motion of the pacemaker.

Transients are more prevalent in advance shifts than in delays, and also more common in complex circadian systems such as those of animals than in the cellular systems of microbes. Jet lag represents the occurrence of transients in resetting human clocks. Because of transients, phase response curves measured on the first day after a stimulus can misrepresent the effect of the treatment on the circadian system. Thus, predictions of the discrete entrainment model are typically based on phase shifts measured after transients have subsided and the pacemaker has reached a steady state.

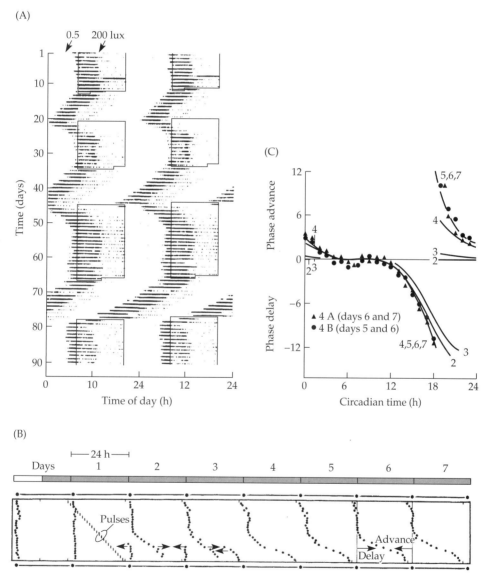

FIGURE 3.9 Transients. (A) Experimental demonstration of entrainment in a chaffinch, *Fringilla coelebs*, kept in an alternating bright/dim light schedule (LD 12:12) followed by constant dim light (DD), then back again to the LD cycle. Note that after the entrainment signal appears, several days of transients are seen prior to stable entrainment. (B) Derivation of the light PRC in *Drosophila pseudoobscura* shows transients. Each of a series of 24 populations of pupae was given a 15-minute, 100-lux light pulse on successive hours (short vertical black bars), and one control was not light treated. Pulses scan the whole day under free-running conditions. The time course of the rhythm in each experimental population (and one control) is then given as a series of points, each marking the midpoint of an eclosion peak. (C) The PRC plots the phase shift as a function of the phase at which the pulse was given. Different curves were derived for the shift as measured on successive days after the pulse. The advance transients go more slowly than the delays, but all are at steady state by day 5. (A from Daan and Aschoff, 2001; B and C from Pittendrigh, 1981.)

Entrainment behavior of many organisms can be predicted by the discrete entrainment model[1,4,25,42,56,61,62,67]

The discrete entrainment model uses the light phase response curve as a tool to predict entrainment behavior. Phase response curves can help investigators envision pacemaker responses to phase-shifting stimuli (see Figure 3.10). A simplified hypothetical case involves the circadian oscillator of a day-active organism that free-runs in DD with a period of 24 h (Figure 3.10A). Initially the phase response curve and the associated activity rhythm free-run in DD. At hour 45.5, which is equivalent to CT 21.5, a single light pulse is presented to the organism. The light pulse elicits a phase advance of about 6 h in the phase response curve and the overt rhythm. A control phase response curve and rhythm that did not experience the light pulse are shown for comparison. The phase shift is called an advance because the phase of the curve and rhythm occurs earlier than in the control. When a light pulse is administered at hour 38, or approximately CT 15, the phase is delayed about 6 h (Figure 3.10B).

In the discussion of the discrete model that follows, the predictions of entrainment will be limited to steady-state entrainment observed after transients have disappeared. Two key assumptions of the model should be mentioned.

First, the free-running period measured in constant conditions is assumed to be an accurate reflection of the circadian period functioning under entrainment conditions. Second, the stimuli used in entrainment and to derive the PRC are assumed to be effectively the same.

For one light pulse per cycle, the discrete entrainment model proposes that the circadian pacemaker is entrained solely by light pulses falling at a specific phase of the phase response curve. The phase shift evoked must equal the difference between the free-running period and the period of the entraining cycle. Stated in elegant simplicity:

$$\text{Phase shift} = \text{FRP} - T$$

where FRP is the free-running period, and T is the length of the entraining cycle.

If the free-running period is 23 h, the pacemaker must experience a net delay phase shift of 1 hour in order to entrain to a 24 h LD cycle. For a free-running period of 21 h, the phase shift in steady state must be a delay of 3 h. A light pulse striking the PRC in the early subjective night could produce such an effect (Figure 3.11B). Conversely, for a free-running period of 27 h, the steady-state phase shift must be an advance of 3 h, and the light pulse will fall in the late subjective night. The light pulse must thus fall at a different phase of the pacemaker for a free run of 21 h versus a free run of 27 h to achieve steady-state entrainment to an LD cycle of 24 h. The relationship between circadian phase and entraining-cue phase, called the phase angle, will be different for different free-running periods (see Figure 3.11).

These predictions have been upheld by studies in which the FRP has been changed by mutation. For example, the *tau* mutation of hamsters changes the free-running period from a wild-type value of 24 h to a mutant value of 20 h in homozygotes. Although not all the short-period *tau* mutant hamsters are able to entrain, those that can entrain to LD cycles do so as predicted, with a significantly earlier phase angle than that of wild-type hamsters (Figure 3.12). Equivalent results have been obtained with clock mutant *Drosophila melanogaster* fruit flies. In LD 12:12, evening activity peaks occur earlier in short-period flies and later in

(A) Effect of phase advance on PRC

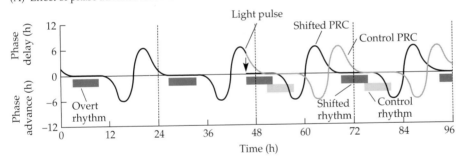

(B) Effect of phase delay on PRC

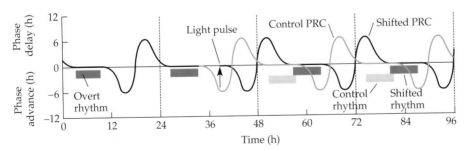

FIGURE 3.10 Impact of phase-resetting light pulses on the time course of circadian phase response curves and overt activity rhythms. (A) Phase advance. (B) Phase delay. Solid black lines are phase response curves of experimental animals; gray lines are for controls. The dark boxes indicate activity of experimental animals; the light boxes, activity of controls. Transients are ignored. For further explanation, see the text.

(A) Free run (FRP = 21 h)

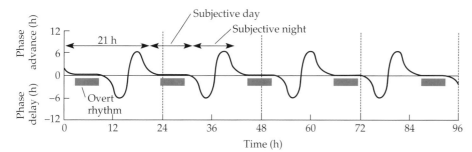

(B) One-pulse entrainment (FRP = 21 h, *T* = 24 h)

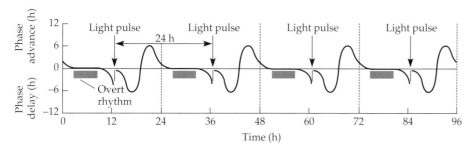

FIGURE 3.11 Phase response curves tracking a short-period circadian oscillator under free-running and entrained conditions. The PRC and overt rhythm are plotted as in Figure 3.4. (A) Free-running of PRC and overt rhythm with 21 h FRP. Notice that the PRC and rhythm recur earlier each day. (B) One-pulse entrainment. A discrete light pulse is administered to the same organism (FRP = 21 h) once every 24 h. After entrainment has stabilized, the light pulse will occur at a circadian phase such that the steady-state phase shift will be a 3 h delay in every cycle. In the case shown here, that light pulse strikes the PRC at approximately CT 13. Notice that under entrainment, the phase of the PRC and rhythm is the same each day relative to the abscissa.

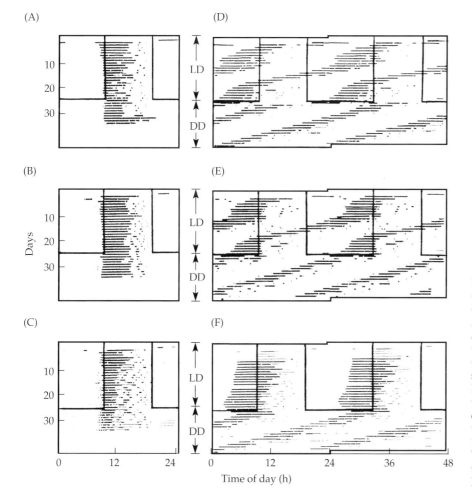

FIGURE 3.12 The phase angle of entrainment in *tau* mutant hamsters is earlier than in wild-type hamsters. Activity records from six litter mates are shown, three (A–C) having wild-type alleles at the *tau* locus, and three (D–F) having a mutation that confers short circadian period length. Panels D–F are double-plotted as shown on the time axis. Boxes superimposed on each panel enclose the light portion of the light–dark cycle, which was LD 14:10 with 100 lux in the light phase. Wheel-running activity is marked by the dark bands on each day. (From Ralph and Menaker, 1988.)

long-period flies than in wild-type flies. Wild-type and period length mutant strains of *Neurospora*, and even lowly pond scum, the cyanobacteria, show similar entrainment characteristics, reinforcing the universal strength of the discrete model to predict circadian entrainment. The same effect, in which entrainment of an inherently short-period endogenous clock to a 24 h synchronizer results in an early phase angle, may be the basis of the sleeping patterns of human families that exhibit familial advanced sleep phase syndrome (ASPS; see Chapter 10). Individuals with this syndrome have a genetic variant of the human clock gene *hPer2* apparently resulting in an inherently short-period clock. Affected individuals tend to go to sleep at about 7:30 P.M. and wake up about 3:00 or 4:00 A.M., as entrainment theory predicts.

Because the phase angle of clock to entrainment cycle depends on a free-running period that is consistent from cycle to cycle, circadian theory predicts that the free run must be kept relatively stable in order to preserve phase angle. In general, that prediction is upheld. However, several factors can modulate the FRP within a narrow range and thereby affect the phase angle of clock to entraining agent. To some extent, the phase response curve acts as an internal compensator for small day-to-day variability in free-running period length. On a day in which the FRP is a little longer or shorter than usual, the entraining cue will strike the phase response curve at a phase that will elicit a slightly larger or smaller phase shift and will thereby counterbalance the effect of the period change.

The phase angle between the driving light cycle and driven circadian clock can also be modulated easily in the laboratory simply by a change to the period of the light cycle, *T* (Figure 3.13). Although these exotic T-cycles, as they are known, have no counterpart in the natural envi-

(A) FRP = 27 h; *T* = 24 h; day-active rhythm

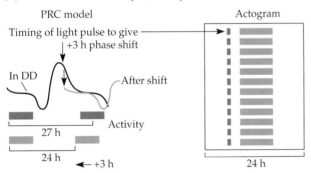

(B) FRP = 21 h; *T* = 24 h; day-active rhythm

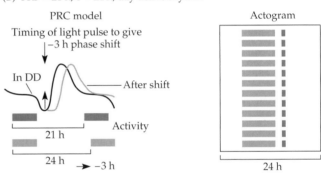

(C) FRP = 25 h; *T* = 22 h; day-active rhythm

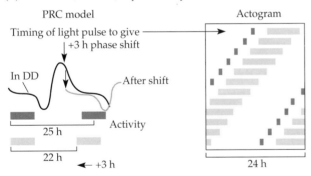

(D) FRP = 23 h; *T* = 26 h; day-active rhythm

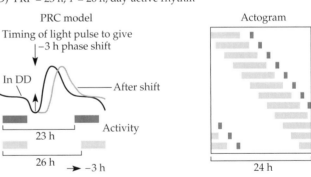

FIGURE 3.13 Predicting the phase angle of a circadian rhythm under entrainment. The phase relationship (phase angle) of a circadian rhythm under entrainment to a particular light–dark cycle can be predicted from the PRC, *T*, and FRP. This figure illustrates how the phase shift needed for stable entrainment (as predicted by the following equation: phase shift = FRP – *T*) varies as a function of FRP (panels A and B) or *T* (panels C and D). In each panel, the left side depicts a PRC explanation of entrainment. In free-running conditions (in this case DD), the PRC is the solid line, and the dark shaded bars underneath represent the timing of overt activity (for a day-active organism) under free-running conditions. Under entrainment, the action of light at the indicated phase angle is represented by the lightly shaded PRC, which shows the phase shift, and by the light shaded bars representing activity.

On the right side of each panel, a simulated activity actogram represents 12 successive days of recorded activity (long bars), and the timing of the brief light pulses (short bars) is presented cyclically with period *T*. (A) Entrainment of a circadian rhythm with a 27 h FRP to a light pulse cycle with a 24 h period (*T*). Stable entrainment in this case requires that the light pulse occur in the late subjective night, where it elicits a phase advance of +3 h. The activity bout occurs after the light pulse. (B) For stable entrainment of a rhythm with a 21 h FRP to a light pulse cycle with a 24 h period (*T*), the light pulse will fall in the early subjective night to elicit a phase delay of –3 h. The activity bout occurs before the light pulse. (C, D) The same interdependency of phase angle on both FRP and *T* holds for entrainment to light pulse cycles with *T* either shorter (C) or longer (D) than 24 h (see the text).

ronment, they have proven to be excellent research tools. A clock having a 25 h free-running period that is entrained by a 22 h light–dark cycle must phase-shift by advancing 3 h in each cycle. The light pulse should strike the late-subjective-night region of the phase response curve in order to achieve a 3 h phase advance. For a given free run and phase response curve, the phase angle between the light–dark cycle and oscillator will vary as a function of T (see Figure 3.13C and D).

Stable entrainment occurs within certain limits[3,35,56,60,61,64,75]

The range or limit of T-cycles that permit stable entrainment is largely determined by the free-running rhythm and the phase response curve (Figure 3.14). Because aftereffects can sometimes modulate the free run, the prior entrainment history can determine whether a circadian pacemaker will entrain to a given T-cycle. Phase response curves with large shifts can permit entrainment to T-cycles of a broader range than low-amplitude PRCs can. Stable, steady-state entrainment is achieved when the light signal engages the phase response curve where the phase shift equals FRP – T, and where the slope of the phase response curve lies between 0 and –2. When the entraining agent cycle fails to meet these two conditions, either frequency demultiplication or relative coordination occurs. Frequency demultiplication results when the entraining cue repetitively falls at the same phase of the phase response curve every second cycle, third, or higher harmonic (Figure 3.15). In these instances, clock period and entraining-cycle period are not equal.

Another possibility, called relative coordination, takes place when the phase shift evoked by the time giver is not large enough to equal FRP – T. Under these conditions, the rhythm looks somewhat like a free-running rhythm, but with a scalloped pattern caused by phase-shifting effects of the repetitive stimuli as the pacemaker attempts to entrain. As seen already with the hamsters on LD cycles that had gradual rather than abrupt transitions, the nature of the entraining cue can also influence the limits of entrainment.

Phase angles are stabilized by free runs that are not exactly 24 hours[21,26,61,69,71]

Circadian periods are almost never exactly or nearly exactly 24 h. Two exceptions are shown by *Drosophila* and *Sarcophaga*, whose pupae have free runs that are nearly equal to 24 h. For these species, cellular biochemistry can be adapted to accomplish precise, nearly 24 h periodicity. In addition, for many coupled oscillator systems the phase relationship between oscillators is most stable if their periods are equal or nearly equal, for both physical and biological oscillators. To achieve synchrony, some kinds of noncircadian biological oscillators modulate period over a rather large range to match the frequency of the driving oscillation.

Because the circadian clockwork can oscillate with an almost exactly 24 h period, and because synchronization of most coupled oscillator systems is most stable with equal periods, it seems as if natural selection should have chosen only biological clocks with period lengths very close to 24 h. Yet, as we have seen and as is implied by the very name, circadian rhythms in many organisms display period lengths that differ from 24 h by several hours. The question underlying this paradox really lies in the target of selection.

The answer apparently is in the difference between an oscillator system that strives for synchrony among the component oscillations and one that is designed for entrainment. As long as they can be reset every day, circadian clocks do not need to achieve synchrony with the environmental cycle. However, they do need to entrain to the environmental cycle with a stable phase angle to ensure that clock-regulated activities always happen at appropriate times of day.

The coupling of circadian oscillators to environmental cycles is described by phase response curves that often have a significant dead zone, the part of the PRC, typically during the day, where light has no substantive phase-shifting effect. In this particular case, stability of phase angle is maximized in one-pulse entrainment by a free-running period length close to but significantly different

Time (h)

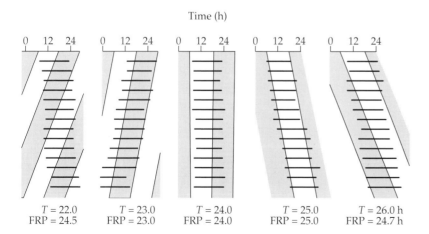

| $T = 22.0$ | $T = 23.0$ | $T = 24.0$ | $T = 25.0$ | $T = 26.0$ h |
| FRP = 24.5 | FRP = 23.0 | FRP = 24.0 | FRP = 25.0 | FRP = 24.7 h |

FIGURE 3.14 Limits of entrainment and phase angle differences seen during entrainment of chaffinches, *Fringilla coelebs*, to exotic T-cycles. Schematic activity records of chaffinches exposed to five different LD cycles, each with a 1:1 ratio of light to dark, but with different cycle lengths. Note the absence of entrainment to T = 22 and 26, and the phase angle changes with other cycle lengths. (Modified from Daan and Aschoff, 2001.)

(A)

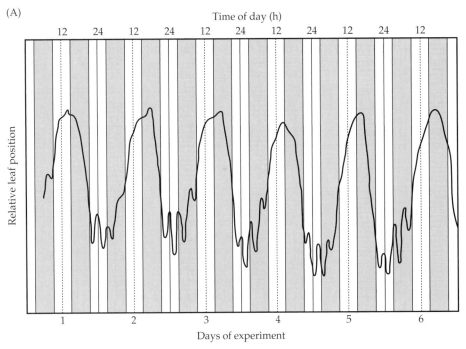

Time of day (h)

Days of experiment

(B)

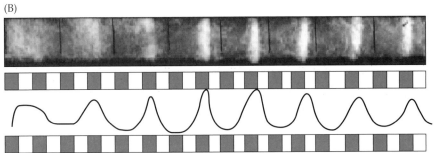

FIGURE 3.15 Frequency demultiplication. (A) The leaf movement rhythm was measured in the plant *Canavalia ensiformis* growing under an LD 6:6 cycle. A stylus attached to the leaves moves with the leaves such that a downward trace corresponds to upward leaf movement. (B) The circadian rhythm in conidiation in the *bd* strain of *Neurospora crassa* was measured under a 6:6 warm–cold cycle of 27°C (white bars) alternating with 22°C (gray bars). Conidiation, which normally peaks near the end of the dark period on a 24 h LD cycle, is seen peaking here near the end of every other cool period. The culture tube is shown above; a densitometric trace of the culture, below. (A from Kleinhoonte, 1928; B courtesy of A. Pregueiro, J. Dunlap, and J. Loros.)

from 24 h. If the FRP and *T* are both equal to 24 h, the phase shift per cycle is zero. The light period's striking at any of many phases throughout the dead zone can cause such a phase shift (Figure 3.16). A zero phase shift would be maladaptive in terms of preserving phase angle because many possible phase angles would be allowed. On the other hand, if the FRP and entraining agent cycle are not equal, there will be only one time of day when an entraining agent will strike the PRC and result in just the right phase shift, equaling FRP – *T*, and where the slope of the phase response curve will be between 0 and –2. Consequently, only when FRP and *T* are unequal will a unique phase angle exist under steady-state entrainment (see Figure 3.16). Thus to preserve phase angle, FRP should not be too close to *T*, meaning that in the real world, for stable entrainment circadian period lengths should not be too close to 24 h.

In this context, the cases of the flies *Drosophila pseudoobscura* and *Sarcophaga* are intriguing. In *Drosophila* pupae, the FRP of the eclosion rhythm is 24 h, and the phase response curve is Type 0, even to weak light stimuli. After eclosion, however, the FRP for adult locomo-

tor activity takes on a distinctly non-24 h value (22.5–23 h), and the phase response curve becomes Type 1. Similarly, in the flesh fly *Sarcophaga*, the FRP for pupal eclosion is 24 h, but for larval wandering the FRP is shorter than 24 h. Clearly the FRP and PRC characteristics of clocks in adult *Drosophila* and wandering *Sarcophaga* larvae match the predictions for optimal stabilization of phase angle, but the properties exhibited by the pupal clocks mock those predictions.

The contradictions are puzzling at first. Possibly the answer lies in the ecology of the pupae. In *Sarcophaga* and possibly also in *Drosophila*, pupae complete development in soil or under bark, where, absent light, they may be effectively free-running in DD. To eclose at an appropriate local time, their clocks need to time the passage of each day accurately for the 1 to 3 weeks of developmental metamorphosis. Hence, a 24 h free-running period length is required. However, in developmental stages that are exposed to the natural LD cycle, such as the *Drosophila* adult and *Sarcophaga* larvae, the timekeeping strategy conforms to the non-24 h FRP tactic to optimize phase-angle stability.

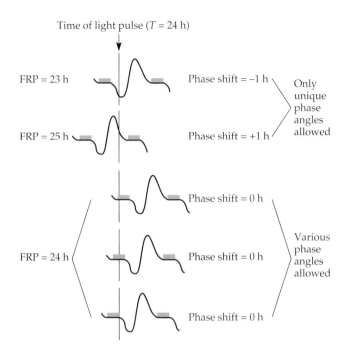

Time of light pulse (*T* = 24 h)

FRP = 23 h — Phase shift = −1 h

FRP = 25 h — Phase shift = +1 h

Only unique phase angles allowed

Phase shift = 0 h

FRP = 24 h — Phase shift = 0 h — Various phase angles allowed

Phase shift = 0 h

FIGURE 3.16 Stabilization of the phase angle of entrainment. The phase angle of entrainment to a short photoperiod at *T* = 24 h is stabilized by an FRP that is different from 24 h. For a phase response curve that has a significant dead zone, a non-24 h FRP results in a unique stable phase relationship, as predicted from the PRC and the equation: phase shift = FRP − *T*. Examples include the top two cases shown here, where FRP is 23 h (first example) or 25 h (second example). In contrast, if FRP = *T* = 24 h, a zero phase shift may be obtained at many different points throughout the dead zone, leading to a phase angle of entrainment that is not stable or predictable.

Circadian pacemakers can entrain to skeleton photoperiods[54,56,61,62]

The one-pulse version of the discrete entrainment model described in this chapter leads to many insights into circadian entrainment. Most organisms, however, live in the real world with full day–night cycles rather than the experimental laboratory environment with brief daily pulses of light. In fact, the only organisms whose light environment approximates that of the lab are nocturnal organisms that normally live underground but come to the surface near dusk to sample the ambient light intensity.

Given the relative unreality of the brief light pulses used in the development of the discrete model, it is all the more surprising that the discrete model has proven to be accurate in reflecting many organisms' entrainment to complete photoperiods (that is, an LD 12:12 cycle). The model succeeds merely by making the assumption that the relevant parts of the photoperiod for entrainment, the time cues, are the discrete lights-on and lights-off transitions. The corollary assumption is that these lights-on and lights-off transitions can apparently be mimicked with a light–dark cycle of, for example, 15-minute light pulses every 12 h. Such periodic pulses provide a "skeleton" of the real photoperiod.

The fact that skeleton photoperiods consisting of light pulses can mimic complete photoperiods is counterintuitive because a light pulse includes both lights-on and lights-off transitions, so it was unexpected that a pulse at dawn would mimic the single lights-on transition of a real dawn, and that a pulse at dusk would mimic the single lights-off transition of a real dusk. Nevertheless, skeleton photoperiods do mimic complete photoperiods reasonably well in some organisms. This finding has provided a useful tool to chronobiologists.

Skeleton photoperiods provide an entraining agent of two pulses per circadian cycle, a two-pulse entrainment. According to the discrete model, photoentrainment is determined by the net effect of phase-shifting by both pulses:

First phase shift + second phase shift = FRP − *T*

A schematic diagram can help clarify differences in entrainment to a skeleton photoperiod (Figure 3.17A) compared with entrainment to a standard LD 12:12 light schedule (Figure 3.17B), as predicted by the discrete model.

In the case of the *Drosophila* pupal eclosion rhythm, entrainment to skeleton photoperiods is similar to that for complete photoperiods within the photoperiod range between 1 h and 13 to 14 h (Figure 3.18). Eclosion of *Drosophila* pupae on LD 12:12 occurs near dawn. For skeleton photoperiods longer than 13 to 14 h, the eclosion rhythm undergoes a phase-angle jump. When fruit fly pupae are entrained to a long complete photoperiod, eclosion also peaks near dawn. After the pupae are transferred to a skeleton photoperiod longer than 13 to 14 h, the rhythm reentrains to the skeleton photoperiod such that the timing of the eclosion peak jumps away from the original subjective dawn of pulse 1 to be close to pulse 2 (see Figure 3.18B). Therefore, under steady-state entrainment to skeleton photoperiods, the subjective-day portion of the PRC shifts so that it coincides with the shorter of the two dark intervals. This phase-angle jump lies in the shape of the PRC and the amount of shift required daily to achieve stable entrainment. The jump can be predicted accurately by simple pencil-and-paper calculations from the discrete model and the *Drosophila* pupal phase response curve with an FRP of 24 h.

In general, most circadian oscillators entrain to skeleton photoperiods in a manner similar to *Drosophila*, such

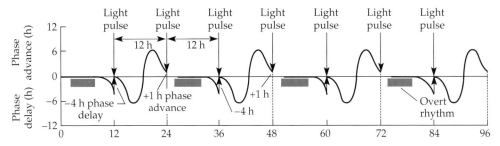

(A) Two-pulse entrainment (FRP = 21h, *T* = 24 h)

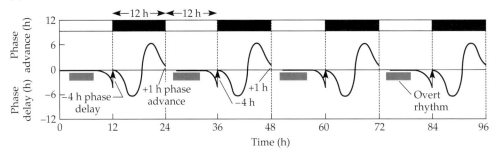

(B) Entrainment to LD 12:12 (FRP = 21 h, *T* = 24 h)

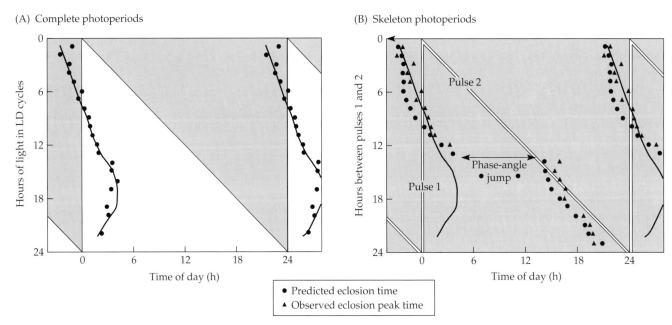

(A) Complete photoperiods

(B) Skeleton photoperiods

- Predicted eclosion time
- Observed eclosion peak time

FIGURE 3.18 The phase of the *Drosophila pseudoobscura* eclosion rhythm as a function of complete versus skeleton photoperiods. (A) The phase of the eclosion rhythm in complete photoperiods. LD cycles range from 1:23 to 22:2. The plotted points are the medians of eclosion distributions of pupae. The solid line is a curve eye-fitted to the points. (B) The phases of the eclosion rhythm in skeleton photoperiods. *Drosophila* pupae are initially entrained to complete photoperiods as in panel A, then transferred to the corresponding skeleton photoperiods using light pulses lasting 15 minutes each. The triangles are the experimentally derived medians

of pupal eclosion in steady-state entrainment; the solid circles are predictions computed from the discrete entrainment model by a PRC to 15 minute light pulses. The black line is the curve fitted to the points from panel A for eclosion medians in complete photoperiods. Both experimental data and modeling indicate that the phase angle of the eclosion rhythm "jumps" when exposed to skeleton photoperiods longer than 13 to 14 h. White areas indicate illuminated intervals; shaded areas, intervals of darkness. (Modified from Pittendrigh and Minis, 1964.)

◀ **FIGURE 3.17 Comparison of entrainment to a skeleton photoperiod with entrainment to a complete photoperiod.** The characteristics of a circadian oscillator with a 21 h FRP entrained to a skeleton photoperiod (two-pulse entrainment), are compared to the predictions of its characteristics under entrainment by a complete photoperiod. The phase response curve and overt rhythm are depicted as in Figures 3.10 and 3.11. (A) Two-pulse entrainment by a skeleton photoperiod of LD 12:12. A discrete light pulse is administered once every 12 h to the same organism, as shown in Figure 3.11 (FRP = 21 h). After entrainment has stabilized, the light pulses will occur at circadian phases such that the net steady-state phase shift will be a 3 h delay in every cycle. In the case shown here, the

net phase shift is achieved by a 4 h delay (by one light pulse striking the PRC in the early subjective night) combined with a 1 h advance (by the other light pulse striking the PRC at the beginning of the subjective day) in each cycle. The phase of the PRC and rhythm is the same each day relative to the abscissa. (B) Entrainment of the same organism to a complete photoperiod (LD 12:12). In this case, the dawn transition is postulated to cause a discrete 1 h phase advance, and the dusk transition is thought to elicit a discrete 4 h delay. Again, the net phase shift is a 3 h delay. The bars across the top of the panel illustrate the light–dark cycle: White bars represent the 12 h of light; black bars, the 12 h of darkness.

that the shorter dark interval is selected as subjective day. However, the rhythm of some organisms can entrain stably to either of the two dark intervals over a narrow range of skeleton photoperiods. In the case of *Drosophila*, this range lies between 10.3 and 13.7 h. Which dark interval is selected as subjective day depends on the phase of the pacemaker that experiences the first pulse of the skeleton photoperiod. This phenomenon, in which the pacemaker can adopt either of two possible phase angles, is called bistability. The discrete model predicts well the experimental responses of *Drosophila* pupae to skeleton photoperiods, including those that allow bistability.

Another interesting insight provided by simulations of two-pulse entrainment is that the phase angle of a circadian pacemaker is much more stable under two-pulse entrainment than under one-pulse entrainment. Assuming some imprecision in actual FRPs measured from day to day, the phase angle is optimally stabilized under two-pulse entrainment to a skeleton photoperiod of sufficient duration to elicit both phase advances and phase delays in each cycle. In so doing, the phase shifts act as checks and balances that prevent the entraining cue from striking near the dead zone of the phase response curve, where phase-angle instability is most likely.

Another example of greater stability from two-pulse entrainment is evident in the comparison of long versus short complete photoperiods. In complete photoperiods greater than 12 h, a balance of delays and advances is elicited by dusk and dawn transitions. The resulting entrainment will be more stable than entrainment to a short photoperiod that

entails only a single phase delay or advance per day. The limit-cycle interpretations discussed later in this chapter will help clarify the situation.

The discrete entrainment model may account for the pacemaker's tracking of the seasons[13,15,16,29,31,58,60,61]

Annual changes in day length are an important challenge for the circadian pacemaker in maintaining an ecologically appropriate phase angle, which, as mentioned already, is a key function of rhythms. Over the year, environmental day–night cycles are constant only at the equator. In other locations, daily environmental cycles are a function of season (Figure 3.19). Preserving phase angle under different seasonal cycles poses a real challenge for organismal clocks. If the clock did not adjust to annual changes in the photoperiod, the early bird would not be early enough to get the worm during some months of the year, and the bird would starve.

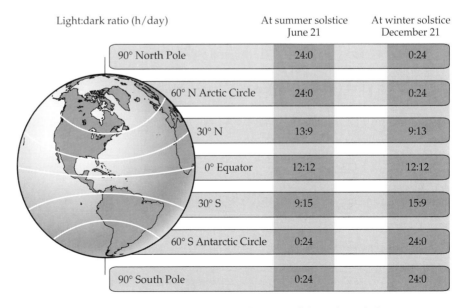

Light:dark ratio (h/day)	At summer solstice June 21	At winter solstice December 21
90° North Pole	24:0	0:24
60° N Arctic Circle	24:0	0:24
30° N	13:9	9:13
0° Equator	12:12	12:12
30° S	9:15	15:9
60° S Antarctic Circle	0:24	24:0
90° South Pole	0:24	24:0

FIGURE 3.19 LD ratios relative to latitude at the time of the solar solstices.

Fortunately, the obstacle to relevant phasing has been overcome in the evolution of circadian clocks. The onset of activity in nocturnal rodents has been known for a long time to track sunset very accurately throughout the year (see Figure 2.6). Likewise, the phases of activity rhythms in day-active and night-active mammals track seasonal changes in photoperiod quite well at moderate latitudes (48 degrees north). Even at the very high latitude of 66 degrees north, the tracking is good except at extremely short and extremely long photoperiods.

The fact that circadian systems can adapt appropriately to annual changes in entraining cycles poses some interesting questions. The basic incarnation of the discrete entrainment model in *Drosophila* postulates symmetrical delay and advance regions of the phase response curve and an FRP near 24 h. The model does not account for seasonal changes. Further, although the model works well for *Drosophila*, it does not work well for all organisms, and many PRCs are not symmetrical as the *Drosophila* PRC is. In addition, many FRPs deviate from 24 h. In fact, even entrainment of the *Drosophila* eclosion rhythm to a wide range of photoperiods shows a rather poor conservation of phase angle to short and long photoperiods (see Figure 3.18A).

The basic model does not fit the data; however, Colin Pittendrigh and Serge Daan had a brilliant idea. They found that through the use of appropriate but realistic combinations of FRPs and phase response curve shapes, the discrete entrainment model could mimic the observed tracking of rhythms to seasonally changing photoperiods. For example, a long FRP (25 h) could be coupled with a phase response curve that has a predominantly advancing shape and a large advance-to-delay ratio. This combination allows an entrainment pattern that will track dawn over a wide range of photoperiods, a day-active strategy. Conversely, a short FRP (23 h) in combination with a predominantly delaying phase response curve will track dusk. The latter combination would be appropriate for a night-active organism that needs to initiate its activity at dusk throughout the year. The satisfying simplicity of these predictions arises from the fact that seasonal tracking is predictable from the discrete entrainment model merely by use of the appropriate FRP value and phase response curve shape.

Pittendrigh later suggested that appropriate FRP values alone can allow seasonal tracking: long FRPs for dawn tracking, short FRPs for dusk tracking. Aschoff's rule also states that brighter light results in shorter period length in day-active animals, whereas in night-active animals the reverse is true (see Figure 3.2). When Pittendrigh's correlation of long versus short FRPs with dawn versus dusk tracking is joined with the hypothesis that light has a continuous action on FRP that can be predicted from PRC shape, the adaptive basis for Aschoff's rule can potentially be explained.

Overall from this discussion, yet another possible justification for a non-24 h FRP emerges: seasonal adaptation of phase angle to photoperiod. Furthermore, recall that entrainment to photoperiods of various durations can result in significant aftereffects on the FRP. For this reason the FRP will also be a function of the photoperiod, and thus it is expected to undergo an annual change that will affect phase angle.

Although at first it may seem that most rhythmic phenomena can be explained or at least rationalized, it becomes clear now that some stimulating hypotheses have emerged that beg for further research. Interspecific comparisons among some nocturnal rodents indicate FRPs and phase response curve shapes that are consistent with the explanation just given. However, comparison of data for all phase response curves does not show a convincing correlation between day and night activity patterns on the one hand, and combinations of FRPs and phase response curves on the other.

The phase response curves of night-active organisms show a definite trend toward small advance-to-delay ratios, but the FRPs are not distributed toward short values. For day-active animals and plants exposed to the complete photoperiod, no significant trend is evident either for phase response curve shape or for the length of free runs. Somewhat mitigating the failure of predictions, though, is the fact that such a comparison is based on studies from many different laboratories whose protocols vary widely. In addition, it is not always clear whether to classify an organism as day-active or night-active. For example, crickets express circadian rhythms of locomotion during the subjective day, but they sing during the subjective night. It is therefore difficult to know which of these rhythms to use for defining day versus night patterns of crickets.

More carefully controlled experiments are needed to test the predictions of Pittendrigh and Daan for seasonal tracking in a variety of organisms. Their hypothesis may prove to be an excellent model in the case of nocturnal animals, whose daily exposure to light in nature is behaviorally modulated to simulate skeleton photoperiods. For organisms exposed to the complete photoperiod each day, however, other factors may need to be taken into account, as suggested in the next section.

Circadian oscillators exposed to complete photoperiods must also experience continuous entrainment[1,12,13,15,18,34,56,58,61,69,72]

Despite the apparent success of the discrete entrainment model, examples that have been noted already highlight some of its shortcomings. The continuous entrainment model of Aschoff was originally based on the observation that the FRP depends on light intensity. It follows that light must exert continuous action on the clock to entrain it to the light–dark cycle. Even in organisms whose entrainment behavior apparently conforms to the discrete entrainment model, at least one important action of continuous light has already been noted: The continuous action of light prevents the phase-angle jump observed in long skeleton photoperiods.

Additional evidence supports the idea that the continuous action of light has an impact. For example, when Syrian hamsters are placed in short photoperiods, the entrainment to skeleton photoperiods does not mimic well the entrainment to natural photoperiods. A dramatic example for which the discrete model is inadequate is that of day-active ground squirrels that entrain to light–dark cycles without ever seeing dawn or dusk. These animals stay in their burrows until several hours after sunrise and return to the burrows well before sunset. Clearly they never see the transitions upon which the discrete model depends, yet they entrain stably in natural conditions.

Continuous entrainment very likely plays a role here, but even where skeleton photoperiods mimic complete photoperiods well, as in *Drosophila, Sarcophaga*, and some nocturnal rodents, doubts remain. Particularly in tests of skeleton photoperiods in these species where $T = 24$ h and the FRP is close to 24 h, the daily net phase shift is very small. The situation is not a definitive condition in which to test whether skeleton photoperiods mimic complete photoperiods well. Instead, skeleton photoperiod experiments with T significantly different from the FRP would be more convincing. In summary, the discrete model does not fully explain circadian pacemaker entrainment.

In the species of nocturnal rodents and fly pupae used for research on discrete entrainment, light exposure in nature mimics skeleton photoperiods. Entrainment properties and light sensitivity in these species may not be the same as for organisms exposed to a daily complete photoperiod. For the majority of organisms that are exposed to complete photoperiods, circadian entrainment will probably be a composite of continuous and discrete mechanisms. Why, then, has so little attention been paid to continuous entrainment by chronobiologists?

One probable reason for the lack of general acceptance of the continuous model of entrainment is the difficulty encountered in quantitative modeling of continuous entrainment. The discrete model allows quantitative predictions to be made with simple pencil-and-paper calculations, and the quality of the predictions ranges for different species from adequate to excellent. For continuous entrainment, no convenient metric exists that is analogous to the discrete phase shift resulting from a single light pulse at a specific time.

Continuous entrainment occurs through environmental modulation of period length[12,72]

One major distinction between the discrete and continuous models of entrainment is the means of phase and period adjustment. In the discrete model, the free-running period is generally taken to be constant, and it adjusts to T solely by phase shifts caused by the abrupt environmental light transitions at dawn and dusk. On the other hand, the continuous model anticipates modulation of the instantaneous free-running clock period continuously throughout the day. It also allows other effects that cannot be inferred strictly from the phase response curve.

The instantaneous acceleration and deceleration of FRP are often referred to as changes in angular velocity. Modulation of the free-running period by daily changes in light intensity could allow the circadian pacemaker to continuously adjust its cycle length to that of the environment. One way to visualize the magnitude of this effect is to transform the shape of the phase response curve into a velocity response curve that translates the observed discrete effects of light on phase into continuous effects of light on period. One observation supporting this interpretation has been the entrainment of some organisms to sinusoidal profiles of light intensity in LL.

Another potential consequence of period changes on entrainment is the possibility for annual adjustment of phase angle by the pacemaker. Because the free-running period is affected by photoperiod, it changes during the year. If alterations in FRP can account for annual tracking of the pacemaker to the annual cycle, then the modulation may be a seasonal adaptation as well. This seasonality may also help explain Aschoff's rule concerning the effect of continuous light on day-active and night-active animals.

Some past discussions of discrete versus continuous entrainment mechanisms have indicated that one or the other mechanism entrains circadian clocks. However, it seems most likely that the majority of organisms use both mechanisms to some degree. Perhaps light has fundamentally equivalent effects on the central clock of all organisms, but there might be differences in photobiological sensitivity or other parameters. Such differences might underlie the data that favored discrete entrainment in the past.

Temperature transitions can also entrain circadian clocks[5,29,37,53,63,73,76,79]

In addition to light and dark, nonphotic environmental factors can entrain circadian rhythms. Among these factors, the most universal and best studied are temperature steps and cycles. In entrainment of all responsive organisms, transitions from cold to warm are interpret-

(A)

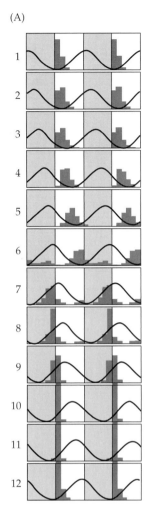

(B)

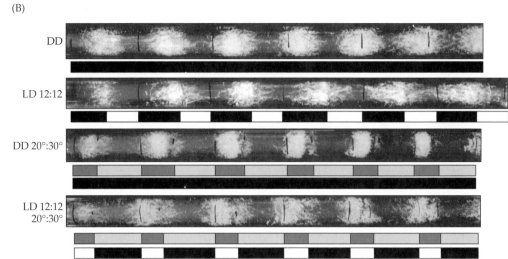

FIGURE 3.20 Temperature or light can be the stronger entrainment factor for a circadian rhythm. (A) Under conflicting light and temperature cycles, the *Drosophila* eclosion rhythm usually follows light. The cultures in this experiment were arranged such that the light–dark cycle (shaded) remained synchronized while the phase of the 10°C/26°C temperature (dark curved line) was offset by 2 h in each culture so as to span the day in all 12 cultures. The number of flies eclosing in 2 h intervals (shaded bar graphs) over two successive days from the 12 cultures are shown. (B) Temperature can be a stronger time giver than light for the circadian rhythm in the development of *Neurospora*. When cultures of *Neurospora* grow over a solid surface, the clock controls a daily developmental cycle so that in the late night to early morning, aerial hyphae and spores (white on the dark background) are made. The top three tubes show the phase of the free-running rhythm and control rhythms entrained either by LD 12:12 cycles (white bars represent the light period; black bars, the dark period) or by warm–cold 12:12 cycles (12 h 20°C followed by 12 h 30°C; dark gray bars represent the cool periods, and light gray bars the warm periods). When the two entraining cycles are placed in opposition (bottom tube), the rhythm follows the temperature cycle. (A from Pittendrigh and Bruce, 1959; B adapted from Liu et al., 1998.)

ed in the same way as lights-on, and steps from warm to cold are interpreted as lights-off. The fact that warm periods correspond to subjective day and cold periods to subjective night follows intuitively from experience because the ambient temperature in nature tends to be higher when the sun is up and it is light outside.

Temperature entrainment probably occurs for all organisms that do not regulate their own body temperature, a group including the fungi, plants, and many animals. In these organisms, this phenomenon may have true adaptive significance for life in the real world. For some nocturnal organisms, such as cockroaches that hide from light during the day, the temperature cycle might be an important environmental entraining agent. Even single pulses of high or low temperature can reset circadian clocks. The phase angle that the pacemaker establishes with an LD cycle can also be a function of the ambient temperature. Even though the free-running period of circadian oscillators is compensated for different ambient temperatures, the phase of circadian clocks can be influenced by temperature.

Although light is generally considered to be the most important time giver, few rigorous studies have been carried out to actually evaluate the relative strength of light and temperature cycles that have comparable amplitudes. Some entrainment studies using conflicting phasing of LD versus low-amplitude temperature cycles show that the LD cycle is the primary determinant of the entrained phase angle in a variety of organisms. Other studies, however, including examination of entrainment in *Anolis* lizards and in *Neurospora*, clearly demonstrate that temperature cycles well within the range typically encountered by organisms in the wild are more influential in determining phase than are opposing full-photoperiod light–dark cycles (Figure 3.20). More investigations of the relative role of light versus temperature cycles in natural environments are needed.

Modeling Scientific Processes Is Useful

Models are ways of organizing data to help explain scientific processes that are only partially understood.

Models can be pictorial, graphical, or verbal, but the most rigorous models usually involve mathematics to define the relative contributions of the different components. In aspects of biology that can be reduced to simple equations like the entrainment equation (phase shift = FRP – T), or that resemble simple physical phenomena such as oscillators, modeling can be an extremely useful and powerful research tool.

Through the use of models, chronobiologists can ask whether their assumptions about a phenomenon are truly necessary or sufficient for the phenomenon. In the earlier section on discrete versus continuous entrainment, the idea of modeling was introduced. The discrete and continuous models were juxtaposed, and their relative virtues in explaining entrainment were compared. More specifically, scientific models are useful because

1. They help organize data into a logical framework.

2. They allow the investigator to focus on the most important parameters of the phenomenon.

3. They suggest new experimental approaches by making predictions.

4. They establish a criterion for how well the phenomenon is understood; models that accurately describe the data (point 1 above) and make accurate predictions (point 3) are generally considered to be most likely to reflect the fundamental nature of the phenomenon.

A famous scientific model comes from astrophysics. It is the sun-centered model of the solar system proposed by Copernicus in the sixteenth century. Although it is obvious to us now that the solar system is heliocentric, most people before Copernicus assumed that the universe was centered around Earth. The movements of the stars and planets had been interpreted by complicated geocentric models prior to Copernicus. By careful measurements and the introduction of a new idea, however, Copernicus was ultimately able to show that the simpler heliocentric model did a better job of explaining the available data, and he was able to make predictions that were later confirmed. We take the heliocentric model for granted now, but most of us would have great difficulty proving the heliocentric model intuitively if we could not resort to photographs from spacecraft. Copernicus did not have photographs from spacecraft to directly prove the nature of the solar system, yet he reorganized the existing celestial data into a new model that could be proved indirectly.

Similar modeling goes on in nearly every scientific discipline. An example from biology that is relevant to our discussion of circadian clock models is a famous case of modeling the dynamics of predator–prey populations by Lotka and Volterra. These scientists noted that predator and prey populations tend to oscillate in time. For example, consider a simple case of rabbits and foxes growing on an island where the rabbits eat grass and rabbits are eaten by foxes. In a constant environment that has no other predator or prey species, the populations of rabbits and foxes will oscillate with rhythms that are out of phase with each other. At initial conditions, the numbers of rabbits and foxes are both low (Figure 3.21A). Rabbits are renowned for their fecundity, however, so their numbers begin to increase. As the rabbit population increases, more food becomes available for the foxes and they gorge and reproduce. As the fox population skyrockets, the rabbits are eaten and their population crashes. After the rabbit population crashes, not enough food is available to support the large fox population and it, too, plummets. When the number of foxes is low, the predation pressure on the rabbits is relaxed, and the rabbit population grows again.

Such a population system can oscillate indefinitely as long as there is grass for the rabbits. The insight is that both dependent variables, the numbers of each animal, are changing as a function of the same independent variable, time. Time can be eliminated and the two dependent variables plotted with respect to each other. In this case the dynamics of these population changes can be visualized differently, where the number of foxes (abscissa of Figure 3.21B) is plotted against the number of rabbits (ordinate of Figure 3.21B). Figure 3.21B depicts the population oscillations in steady state. The size of the circle will depend on the size of the island, the fecun-

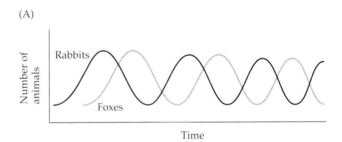

(A)

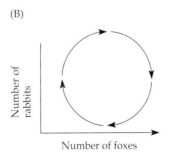

(B)

FIGURE 3.21 A simple limit-cycle oscillator derived from animal population dynamics on an isolated island. (A) Population sizes of predators (foxes) and prey (rabbits) show oppositely phased rhythms. (B) Limit cycle depicting the population oscillations of foxes and rabbits in steady state.

dity of the rabbits and foxes, and several other factors, but the cycle will remain. This type of depiction is called a limit cycle.

Suppose now that a trapper visits the island briefly to capture a few of the foxes before leaving. The short-term effect will be a reduction in fox numbers. Because a lower fox population results in lower predation, the immediate effect will be to allow a longer increase in that cycle of rabbit numbers. Eventually, however, the fox population will rebound, and the two interdependent population cycles will resume. The only difference is that the phase will be altered compared to what it would have been had the trapper never visited. Clearly, if the trapper removes all the foxes, the limit cycle will cease. A similar change in the phase of the rhythm might follow a drought that lasted long enough that the reduced availability of grass limited rabbit growth. A limit cycle that can be perturbed within limits but that later returns by itself to a stable oscillation is referred to as a stable limit cycle.

Circadian Pacemakers Are Limit-Cycle Oscillators[36,49,77]

Experimental studies and mathematical modeling have demonstrated that circadian pacemakers are limit-cycle oscillators. As introduced in the preceding section, stable limit-cycle oscillators are characterized by a standard waveform and amplitude to which they return after relatively small perturbations. Limit-cycle models help explain reactions of circadian pacemakers to stimuli such as entraining agents. The final section of this chapter reviews the empirical findings that support this general model and some of the applications of the model to circadian research.

The fundamental modeling concept is that components underlying rhythm generation change rhythmically in time. Consequently, the rhythmic process can be described by a system of differential equations in which these oscillating components are called state variables. State variables define the state of the oscillation. The equations also contain state parameters that establish the amplitude and period of the oscillations. In a constant environment, the parameters can be constant, but the levels of the state variables must, by definition, oscillate periodically.

External stimuli such as entraining agents may reset a limit-cycle oscillator in several ways. One way is by changing levels of state variables, as seen in the earlier example of the trapper collecting foxes. Another way to reset a limit cycle is by altering the parameters, as in the drought example; this is called parametric excitation. Rhythms can also be reset through more complicated means that can be mathematically represented and modeled. For a simple oscillator consisting of only two

state variables, such as numbers of rabbits and foxes, it is convenient to graphically portray the changes of the state variables in phase space, an abstract space whose coordinates describe the state of the system. Figure 3.21B shows a phase-space diagram.

Most people have an intuitive sense of how pendulum oscillators behave, and this behavior provides a good basis for a thought experiment. Imagine a hypothetical self-sustained, frictionless pendulum that oscillates indefinitely. The displacement of the pendulum bob from vertical is denoted as the variable x, and x changes as a function of time, t. Mathematically, this relationship is written as $x(t)$ and pronounced as "x as a function of t." The pendulum's oscillation can be depicted in phase space as a function of two state variables: position $x(t)$ and rate of change of position $x'(t)$, as illustrated in Figure 3.22. As the pendulum sweeps from left to right, $x(t)$ increases from -1 to $+1$ and $x'(t)$ starts from 0 (at position x_{min}), speeds as it passes the bottommost position, and then slows as it nears the rightmost position (x_{max}). At the rightmost position, $x(t) = 1$ and $x'(t) = 0$. Then the pendulum sweeps back in the opposite direction, with $x(t)$ decreasing from $+1$ to -1 and $x'(t)$ becoming negative until the leftmost position is reached, and so forth. These changes can be depicted as changes of the two state variables in phase space (see Figure 3.22C).

Specifically in this example, each position of the pendulum bob in Figure 3.22B corresponds to a phase in the oscillation described in Figure 3.21A. In addition, each position and phase maps to a specific point on the limit cycle graphed in phase space in Figure 3.22C. Stable limit-cycle oscillators differ from standard pendulum oscillators in two respects: Limit-cycle oscillators do not damp, and the state variables of a limit-cycle oscillator eventually return to the same trajectory in phase space (the limit cycle) after small perturbations away from that trajectory.

As applied to circadian oscillators, the limit cycle is the trajectory in phase space around which the values of the state variables change. Just as with the pendulum, each phase of the pacemaker is defined by a particular point on the limit cycle in phase space. To depict a circadian oscillation on a two-dimensional graph as a function of only two state variables is almost certainly a gross oversimplification. Circadian pacemakers very likely consist of more than two state variables. Nevertheless, simplification is convenient for illustrating the most important points. Indeed, even two-dimensional depictions of limit-cycle behavior have helped explain some of the circadian behaviors already described in this chapter.

As defined earlier, limit-cycle oscillators are those whose state variables oscillate along a trajectory (the "limit cycle") in phase space like horses running around a racetrack. If the oscillator is perturbed such that the state variables are pushed off the limit cycle, they will

(A)

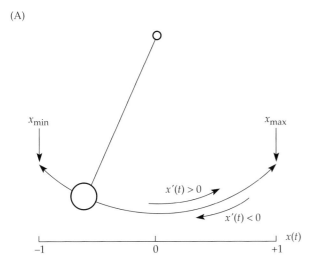

(B)

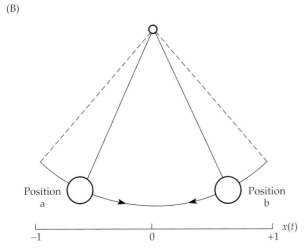

(C)

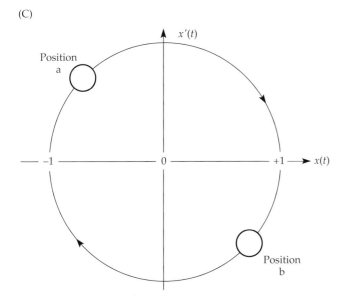

FIGURE 3.22 Graphical portraits of a hypothetical frictionless pendulum oscillator as an example of limit-cycle dynamics. The position [$x(t)$] and rate of change of position [$x'(t)$] are indicated in each panel. (A) As a pendulum sweeps back and forth, its position can be defined as positive values [$x(t)$] when it is to the right of center and as negative values [$-x(t)$] when it is to the left of center. The pendulum's rate of change of position can be defined as positive values [$+x'(t)$] when it is sweeping from left to right, and as negative values [$-x'(t)$] when it is sweeping from right to left. (B) Two points on the pendulum's arc are labeled "position a" (sweeping to the right) and "position b" (sweeping to the left). These positions are plotted in phase space in panel C. (C) Portrayal of this pendulum's motion in phase space, plotted as a function of the two state variables: position [$x(t)$, plotted on the abscissa] and rate of change of position [$x'(t)$, plotted on the ordinate].

return to the limit cycle. Various different disturbances can move the state variables from the limit cycle. The perturbations can occur at different times corresponding to different points around the cycle, but if it is a stable limit cycle, the values of the state variables will eventually return to the cycle. However, depending on the sizes of the perturbations and when in the cycle they occur, the variables can return to the limit cycle at different points. The perturbations can be grouped, though, into sets yielding state variables that return to the limit cycle in phase with the same point, the same phase on the cycle. The set of points in phase space that specify values for the state variables, all of which will return to the limit cycle in phase, are called isochrons. For example, in Figure 3.23 an oscillator at CT 21 (position a) is given a stimulus that moves its state variables to the isochron of CT 3. This means that the state variables of the oscillator will return to the limit cycle in phase with another oscillator that started at CT 3 (position b) and never received a stimulus. A less precise, but more intuitive definition of an isochron is that it maps "those points in phase space that are at the same time" (*iso* means "same"; *chronos* means "time").

Phase response curves can be modeled in terms of limit cycles[77,78]

The limit-cycle interpretation of phase resetting is based on the assumption that entraining agents change the value(s) of one or more state variables. As a result, the oscillator is perturbed from the limit cycle to another position in phase space. If this change moves the state variables from one isochron to another, a steady-state phase shift will be observed. This phase shift occurs because as the state variables move back to the limit cycle, they will return to it at a different phase from the one they would return to if no displacement had occurred.

In Figure 3.23, light is postulated to greatly increase the value of variable X while variable Y increases only

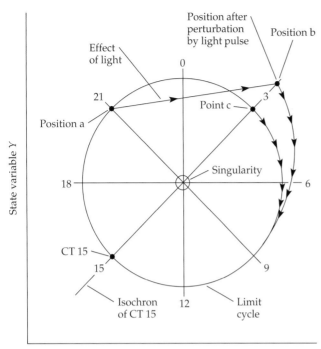

FIGURE 3.23 Limit-cycle depiction of phase resetting as a stimulus causing movement of state variables from one position to another in phase space. The circadian pacemaker proceeds clockwise around the circular limit cycle. The isochrons are shown as straight lines that intersect at the singular point within the limit cycle. A light pulse administered to the oscillator when it is at CT 21 (position a) carries the system to the state indicated by position b on the isochron, corresponding to CT 3. As time proceeds after the stimulus, the perturbed oscillator moves toward the limit cycle so that it becomes asymptotically indistinguishable from a rhythm initiating at point c on the limit cycle, at CT 3. Because CT 3 may also be designated CT 27, the light pulse has advanced the phase of the perturbed oscillator by 27 – 21 = 6 h.

slightly, so a light pulse given at CT 21 moves the state variables to a position on the isochron of CT 3. After the light pulse, the state variables will return to the limit cycle along the trajectory shown by the thin arrowed line. When the oscillator, or rather the state variables describing it, reaches the limit cycle, it will be in phase with another oscillator that had started from CT 3, not from the original CT 21. The oscillator in this example has been phase-advanced by 6 h. When this type of perturbation analysis is applied to many phases around the limit cycle, a "resetting contour" can be derived that shows the positions to which the state variables are moved immediately after identical resetting stimuli are applied at different phases in the oscillation. An example of a resetting contour is shown in Figure 3.24C and F.

One might assume that stimuli presented at phases in the dead zone would not modify the state variables because no phase shift results from stimuli occurring in the dead zone. Although this can be true for some specific models, it is not a necessity of a limit-cycle model. An alternative explanation is that stimuli presented during the dead zone do result in changes of the state variables, but these altered values move the variables approximately along the original isochron. An example is shown by the perturbations delivered between CT 3 and CT 8 in Figure 3.24C and F. All of these perturbations change the values of the state variables, but all the values return to the limit cycle in phase with an oscillation that had never been reset. Therefore, no steady-state phase shift results. Consequently, state variables of the oscillator are not necessarily insensitive to the stimulus during the dead zone. In fact, the stimulus could induce large changes of the state variables, but these changes do not move the oscillator to a different isochron.

This insight gleaned from limit-cycle modeling has important implications for identifying the molecular correlates of state variables that will be discussed in Chapter 7. Particularly in attempts to define criteria for identifying state variables in circadian oscillators, a variable's responsiveness to phase-resetting stimuli does not have to be correlated directly with the magnitude of the phase shift.

Limit-cycle modeling can explain how a "critical" stimulus could evoke arrhythmic and/or unpredictable behavior[28,33,41,50,74,77,78]

Isochrons radiate from a point somewhere inside the limit cycle. This hub, called the singularity, is an essentially phaseless point because isochrons of all phases converge there. Depending on the particular model, the singularity might be a single point, or it might be a significant region of phase space. Also depending on the specific model, the singular point or region could be a relatively stable place such that the state variables could be trapped there indefinitely in a nonoscillating state. Alternatively, the singularity could be unstable such that state variables entering the singular region would squirt back out to the limit cycle. In the latter case, note that because isochrons of all phases diverge from the singularity, an oscillator whose state variables have been reset to this region could return to the limit cycle along any of the isochrons. In other words, in the singular region, small differences in the values of the state variables translate into large differences in phase. Therefore, it would be impossible to predict the final phase of an oscillator whose state variables return to the limit cycle after having been in the singular region.

A limit-cycle model of circadian oscillators would suggest that stimuli presented at just the right phase and just the right strength (a "critical" stimulus) could move the state variables to the singular region. An oscillator moved to the singularity could result in arrhythmia if the state variables became "stuck" in the singularity.

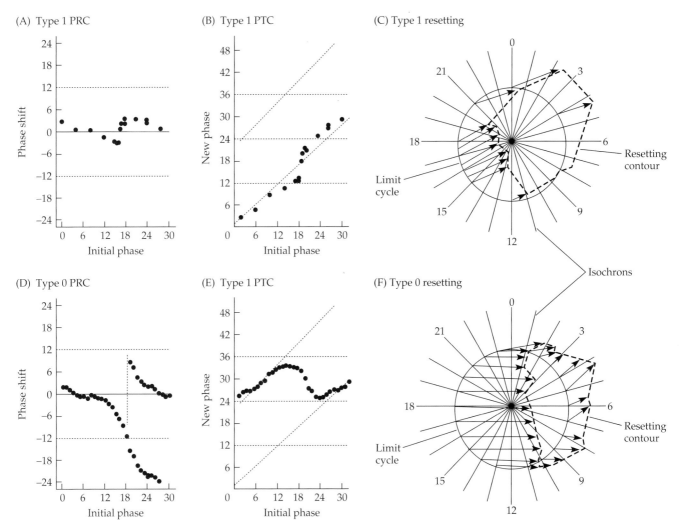

FIGURE 3.24 Phase response curves, phase transition curves, and limit-cycle diagrams for Type 1 versus Type 0 resetting. (A–C) Type 1 resetting of mosquitoes (*Culex quinquefasciatus*) in response to light pulses. (D–F) Type 0 resetting of fruit flies (*Drosophila pseudoobscura*) in response to light pulses. Panels A and D are PRCs, plotted as circadian time of stimulus (initial phase) on the abscissa versus phase shift on the ordinate (advances are positive, delays are negative). Panel D is plotted both in the conventional fashion with a break point, and monotonically. Panels B and E are PTCs, plotted as circadian time of stimulus (initial phase) on the abscissa versus new phase (the phase of the clock after the phase shift) on the ordinate. Panels C and F are limit-cycle diagrams, where the limit cycles are the circles with isochrons radiating from the central "singular" points. The bold dashed lines are the resetting contours, the points on the phase plane to which the state variables are changed by the resetting stimuli. Resetting of the state variables is illustrated by the arrows from points on the limit cycles to points on the resetting contours. (Modified from Johnson, 1999.)

Alternatively, the result could be highly variable and irreproducible resetting behavior if the state variables could return to the limit cycle along any of many different isochrons.

Arthur Winfree showed that the response of the circadian rhythm of *Drosophila* to such critical stimuli behaves in precisely this fashion. He determined resetting contours empirically for this oscillator and then predicted the location of the singularity. From this information he could predict the critical time at which a stimulus needed to be delivered and the critical strength of the stimulus in order to push the oscillator to the sin-

gular state. A light pulse of the correct duration and intensity given at exactly the correct time evoked arrhythmicity in populations of fruit flies. The interpretation of this arrhythmicity was that the oscillators in the population of flies had all been moved to the singularity. This was a strong prediction of the limit-cycle model that was upheld by experimental results.

Later studies have reproduced this phenomenon in other systems. For example, drugs that inhibit protein synthesis will reset the clock of many organisms, including the circadian bioluminescence rhythm of the dinoflagellate *Gonyaulax*. After resetting contours of this rhythm were sufficiently mapped, arrhythmicity could be elicited by critical stimuli. The arrhythmicity induced by the drugs was shown not to be due to irreversible damage because the cells recovered from even higher concentrations of the drug with a large phase shift. More recently, when light pulse stimuli of many different intensities were tested on the alga *Chlamydomonas*, the same phenomenon was discovered. That is, for light pulses given at some phases, the rhythm exhibited small phase shifts after moderate stimuli, arrhythmicity after stronger stimuli (Figure 3.25), and large phase shifts after still stronger stimuli. This general pattern of phase shifting to stimuli of different strengths is depicted schematically in Figure 3.26A.

These data can be interpreted in terms of limit cycles. In the *Chlamydomonas* experiment, all the pulses were delivered at a time known to be the critical time for reaching the singular state. For moderate stimuli, such

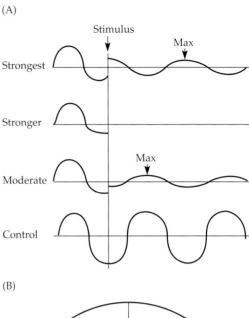

(A)

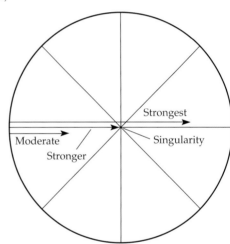

(B)

FIGURE 3.26 A limit-cycle oscillator is subjected to a perturbation at a phase (the critical phase) that drives the state variables directly toward the singularity. (A) Behavior of overt rhythmicity after light pulses of various strengths given at the critical phase. A moderately strong pulse reduces the amplitude of the rhythm, but the phase is little changed from that of the control. A stronger pulse reduces the amplitude of the rhythm essentially to zero; the rhythm appears to stop. The strongest pulse leaves the rhythm at an amplitude less than that of the control, but the phase has been shifted by a large magnitude. (B) Limit-cycle interpretation of the hypothetical data in panel A. A moderate pulse drives the oscillator toward the singularity. After the pulse, the state variables spiral back to the limit cycle on an isochron that is similar to that of the control. A stronger pulse leaves the state variables at or near the singularity; the state variables either remain there or only very slowly return to the limit cycle. The strongest pulse drives the oscillator past the singular region to a point in phase space from which the state variables spiral back to the limit cycle on an isochron whose phase is far removed from that of the original control.

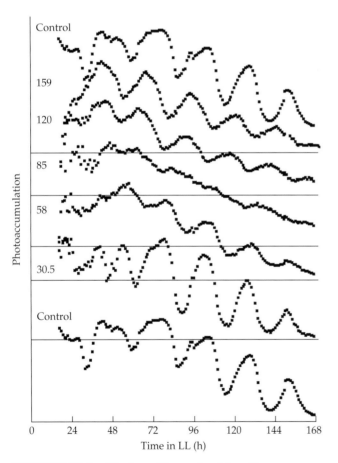

FIGURE 3.25 A critical resetting pulse drives the *Chlamydomonas* clock to a singular point. Changes in rhythm amplitude, shown as a function of fluence of the light pulse for pulses all given at the same circadian time. The control cultures (top and bottom curves) received no light pulse. Each of the other cultures received a light pulse at the fluence rate indicated (in micromoles per square meter per second). (From Johnson and Kondo, 1992.)

as the 30.5 μmol/m²/s fluence example shown in Figure 3.25, the state variables are moved toward the singular region, but after the stimulus the state variables are on an isochron close to their original phase, so only a small phase shift is observed. This phenomenon is shown schematically in Figure 3.26B.

A different stimulus, also presented at the critical phase but having just the right strength, placed the state variables close to the singular region. From this point the state variables in each of the clocks in the population either got stuck if the singularity was a stable region, or moved back to the limit cycle with unpredictable phases if the singularity was unstable. This critical "stronger" stimulus case is illustrated in Figure 3.26; it is also shown by the 85 μmol/m²/s fluence example of Figure 3.25, which resulted in arrhythmicity.

Finally, an even stronger stimulus moved the state variables beyond the singular region to an isochron quite different from the original isochron. The result was a large phase shift, as in the 120 and 159 μmol/m²/s fluence cases of Figure 3.25. The correspondence between these predictions and the experimental data, as seen in the original data from Winfree, is excellent. The success of similar analyses using a variety of phase-resetting agents in plants, algae, fungi, and animals, including humans, serves to emphasize the general significance and utility of a limit-cycle view of resetting.

As implied in the interpretation of the *Chlamydomonas* data in Figure 3.25, several explanations are possible for the observed arrhythmicity, even after the validity of the limit-cycle model has been accepted. The fact that just the right stimulus at just the right time is needed to achieve a singular state implies that the singularity point is likely to be highly unstable. Thus it is unlikely that an individual circadian oscillator would remain stuck at the singularity for a long time. It follows that observation of singularity-induced arrhythmicity in individual oscillators would be highly improbable.

In accordance with this prediction, the investigations that have suggested arrhythmicity or reduction in oscillator amplitude due to critical stimuli have mostly been studies of circadian rhythms in populations of organisms (*Drosophila*, *Culex*, *Gonyaulax*, *Chlamydomonas*), or in individuals having populations of oscillators (humans). In populations, the stimulus might be moving the state variables to a region around the singularity point where the isochrons converge. Because the individuals in the population are almost certainly not perfectly in phase, phase shifting of the population will move the collection of state variables in each oscillator of the population to the area around the singularity point. Over time, the individual oscillators will move from the singular region and re-join the limit cycle, but their initial small phase differences will place them on different isochrons in the closely packed region at the singularity. In that way, the

individually rhythmic organisms will become desynchronized such that the population is arrhythmic (Figure 3.27). An experimental example comes from research on the unicellular protozoan *Paramecium*, in which arrhythmic populations can be composed of rhythmic cells that are out of synch (Figure 3.28).

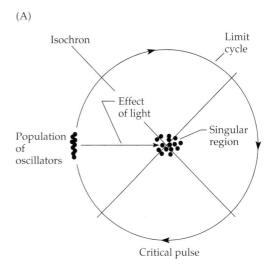

(A)

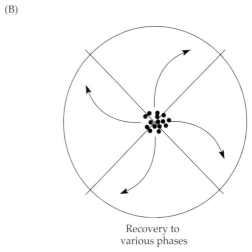

(B)

FIGURE 3.27 Predicted behavior of a population of circadian oscillators subjected to a resetting stimulus of critical strength presented at the critical phase. (A) A population of oscillators, mutually entrained with slightly different phases and amplitudes, is exposed to a light pulse of critical strength and timing. Each dot is an individual oscillator. The oscillators are driven close to the singular region and are slightly dispersed, as a result of small individual variations in responsiveness and initial conditions. (B) As the oscillators individually spiral back out to the limit cycle, their rhythms recover amplitude. In addition, because of their original dispersion around the singularity, they spiral back to the limit cycle on various isochrons. Therefore, the final phases of the individual oscillators will vary, resulting in a population pattern that is apparently arrhythmic (potential entraining interactions between the oscillators in the population are small when the amplitudes are small).

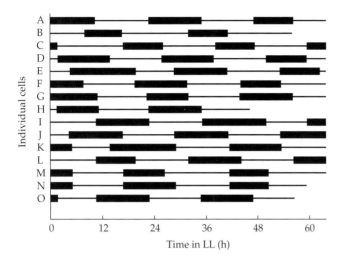

FIGURE 3.28 Mating reactivity rhythm of 15 single *Paramecium bursaria* cells isolated randomly from an arrhythmic population. Prior to analysis, cells that had been in LL for 4 weeks. Each single cell was maintained in LL, and its mating reactivity was tested every 3 h by mixing with about 100 highly reactive cells of the opposite mating type. After a 5-minute testing period, the original cell was reisolated and returned to LL. Each horizontal line represents the data of a single cell; bars mark mating-reactive phases, horizontal lines mark unreactive phases. The letters along the ordinate identify each single cell. Every single cell shows a clear circadian rhythm of mating reactivity, even though the pattern of mating reactivity of the entire population from which these cells were isolated was arrhythmic. (Modified from Miwa et al., 1987.)

Limit-cycle organization can explain Type 1 and Type 0 resetting[20,21,32,33,49,51,68,74,77,78]

When phase response curves were first introduced in this chapter, resetting stimuli were described as exhibiting two patterns: Type 1 and Type 0 (see Figure 3.7A). Type 1 PRCs display a relatively low amplitude and a continuous transition between phase delays and advances; Type 0 PRCs show large phase shifts in which it is difficult to distinguish delays from advances.

The switch between these PRC types appears to depend on the strength of the stimulus used. The strength of the stimulus can be modulated by changes in light intensity or duration, drug dosage, or other such factors. In studies in which the strength has been systematically varied, weak stimuli evoke Type 1 PRCs. As the stimulus strength is gradually increased, the amplitude of the Type 1 PRC increases until there is an abrupt switch to Type 0.

In addition to varying stimulus strength, however, other factors can also cause the conversion from Type 1 to Type 0. These include genetic mutation, background light quality and/or intensity in LL, and developmental stage. This effect is probably not fundamentally different from that of changing stimulus strength because it

is likely that these other factors affect the sensitivity of the clock to the stimulus, and in so doing they affect the perceived stimulus strength.

The limit-cycle interpretation of Type 1 versus Type 0 resetting is illustrated in Figure 3.24. A comparison of Figure 3.24A and C shows that Type 1 resetting is seen if the resetting contour is not moved beyond the singularity. In this case the resetting contour encloses the singularity. On the other hand, a comparison of Figure 3.24D and F shows that Type 0 resetting occurs if the stimulus is strong enough to move the variables beyond the singular region. In this case the resetting contour does not enclose the singularity.

Visualization of phase shifting from the limit-cycle perspective is valuable in that it illustrates how a stimulus that changes the value of a state variable by an equivalent amount at every phase could cause a wide variety of observable phase-shifting behaviors: delays versus advances, break points, and an apparently discontinuous switch between Type 1 and Type 0 resetting. From the PRC visualization alone, it would be easy to conclude falsely that phase delays are mechanistically different from phase advances. The false reasoning behind such a conclusion might be that whereas phase advances result from a pacemaker component being changed in one direction, phase delays change the component in the opposite direction. Limit-cycle models suggest that such mechanistic distinctions are not necessary to explain advances versus delays. Moreover, limit-cycle models interpret the transition from Type 1 to Type 0 resetting to depend merely on whether the magnitude of the stimulus is sufficient to shift the resetting contour beyond the singularity.

An important thing to keep in mind in this discussion is that levels of state variables presumably correspond to real and potentially measurable things. A state variable could be the amount of a clock protein, for instance, and the limit cycle could describe the daily change in the number of protein molecules in the nucleus of a rhythmic cell. A resetting stimulus could be a signal that causes rapid synthesis or destruction of the protein. By maintaining a real-world view of state variables and limit cycles, one can maintain an alternative but parallel explanation for the switch between Type 1 and Type 0 resetting.

This alternative is based on differences in the sizes of limit cycles. In this view the same stimulus that leads to Type 0 resetting for a limit cycle that has a small diameter might elicit only Type 1 resetting for a larger-diameter limit cycle. A stimulus that displaced the state variables of a small-diameter limit-cycle oscillator beyond the singularity to yield Type 0 resetting could fail to move the variables in a large-diameter limit cycle beyond the singular region, thereby yielding Type 1 resetting. The reason would be that the distance is farther in a large-diameter limit cycle, meaning that a greater actual change in the number of protein molecules is necessary in the larger cycle to elicit the same phase shift.

A prediction can be drawn from the limit-cycle interpretation of phase shifting: Oscillator systems that differ only in the amplitude of the oscillating state variables, and therefore in the diameters of the limit cycles, may have different PRCs to the same stimulus. Such an effect has been observed in the case of the *tim^UL* mutant of *Drosophila* (*tim^UL* is a long-period allele of the *tim* gene discussed in Chapter 7). In this mutant the amplitude of the oscillation of two likely molecular correlates of state variables (*per* and *tim* mRNAs) is significantly smaller than in wild-type flies. This observation could mean that the diameter of the limit cycle in *tim^UL* flies is smaller than for wild-type flies. The PRCs for wild-type and *tim^UL* flies to 10-minute light pulses show the predicted result. The wild-type PRC is Type 1; the *tim^UL* PRC is clearly Type 0. This is an excellent example of classic phase-shifting experiments and state-of-the-art molecular approaches converging on explanations based on modeling insights.

There are additional corollaries to the use of limit cycles to visualize the switch from Type 1 to Type 0 resetting. One is that stepwise increases of light intensity or drug dosage presented at some phases can lead to fluence or dose response curves that show a discontinuity (see Figures 3.8B and 3.26). A discontinuous fluence response curve means that a process downstream from the photopigment's absorption of light is converting the initially continuous photochemical response into a discontinuous biological response. In the case of circadian rhythms, the limit-cycle organization of the circadian oscillator is likely to be responsible for converting the initially monotonic response into a discontinuous response as increasing stimuli strengths move the pacemaker past the singular region (see Figure 3.8C). Such discontinuous response curves have been observed for light in the fungus *Neurospora*, the dinoflagellate *Gonyaulax*, and the green alga *Chlamydomonas*; and for drug-induced phase shifting in *Gonyaulax*.

Limit-cycle modeling provides an alternative explanation for cases in which the clock appears to be stopped[23,49,51,54,56]

Some insects, including *Drosophila* pupae, *Sarcophaga* pupae, and *Culex* adults, express no discernible rhythm in LL. Upon transfer to DD, however, a rhythm appears with a phase corresponding to a start at approximately CT 12. If the LL treatment lasts less than 12 h, the phase of the subsequent rhythm appears to be set by the lights-on signal. For longer durations, however, the rhythm appears to occur at a fixed interval of 12 h after lights-off. The most straightforward explanation for this phe-

nomenon has been that the circadian pacemakers in these organisms are stopped at CT 12 by long durations of light, and that upon transfer to DD, the rhythmicity starts up again from CT 12 (Figure 3.29A). Upon closer inspection of the data, however, the phase of the rhythm can be seen to fluctuate about a fixed interval of 12 h after lights-off, with a period of about 24 h (Figure 3.29B). This phase effect was so slight that it was obscured in the biological variability of the phase measurements in early experiments. However, alert researchers noticed the same fluctuating pattern in the phases of the flight activity rhythm of mosquitoes (*Culex*) as in the eclosion rhythm of flesh fly pupae (*Sarcophaga*), which prompted closer examination of the *Drosophila* data.

These data can be interpreted in light of limit cycles as follows: The state variables of the pacemaker oscillate around a different limit cycle in DD than they do in LL. The LL limit cycle is displaced from and centered on the CT 12 isochron of the DD limit cycle (see Figure 3.29B). In real terms, this difference between LL and DD limit cycles could correspond to a higher average daily number of state variable molecules per nucleus in LL than in DD. During exposures to LL longer than 12 h, the state variables continue to oscillate on the LL limit cycle. That cycle is bounded by the isochrons of CT 11 and 13 of the DD limit cycle. Therefore, upon transfer from LL to DD, the state variables find themselves on isochrons of the DD limit cycle somewhere between CT 11 and 13. They then return to phases on the DD limit cycle corresponding to a range between CT 11 and 13. The exact phase of return to the DD limit cycle depends on the phase of the state variables on the LL limit cycle at the time of the transfer from LL to DD. Because the state variables fluctuate rhythmically on the LL limit cycle, the final phase of the rhythm in DD will show a slight rhythmicity (see Figure 3.29A). This model provides an elegant alternative to the original interpretation that LL "stops" the clock. The new interpretation is that the clock is running in LL, but on a different limit cycle.

Limit-cycle modeling can provide an integrated entrainment mechanism[4,37,49,56]

The limit-cycle model has important implications beyond the experiment shown in Figure 3.29A. One potential insight illuminates the issue of discrete versus continuous entrainment discussed previously. The limit-cycle model interprets long-duration light treatments as having a single action on circadian pacemakers: to move the state variables to a different area of phase space, from the DD limit cycle to the LL limit cycle. An interpretation of entrainment based on this limit-cycle model has advantages: it can enable apparently discrete and continuous modes of entrainment behavior, it can

(A)

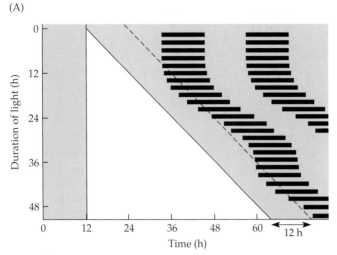

(B)

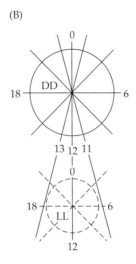

FIGURE 3.29 Limit-cycle interpretation of transitions from light to darkness. Reinterpretation of the idea that constant light "stops" the clock at CT 12. (A) To a group of individuals (or populations) free-running in DD, light pulses of from 0 to 50 h long are presented beginning at the same phase (each horizontal group is an individual or a population that has been exposed to a duration of light that differs from that of each other horizontal group). Light pulses that are longer than 12 h reset the rhythms (bars) to an average phase of CT 12, denoted by the diagonal dashed line that is drawn parallel to, and 12 h after, the end of the light pulses. The example shown is for a day-active organism with onset of behavior at CT 0. However, the rhythmic behavior does not coincide exactly with what would be expected if the underlying pacemaker started from precisely CT 12 at the light-to-dark transition. In particular, a "scalloping" of the rhythmic onsets occurs around the average phase of extrapolated CT 12.

(B) Limit-cycle interpretation of the scalloping illustrated in panel A. The state variables of the pacemaker are postulated to oscillate on two different limit cycles in LL versus DD. The LL limit cycle is centered on the isochron of CT 12 of the DD limit cycle. In this hypothetical example, although the state variables oscillate around the 24 h of phase on the LL limit cycle (that is, the pacemaker is not stopped at CT 12), on the DD limit cycle they oscillate only between CT 11 and 13. Therefore, when the organism is exposed to light, the state variables are attracted away from the DD limit cycle and instead oscillate around the LL limit cycle. When the organism is transferred back to DD, the state variables return to the DD limit cycle from a narrow range of isochrons with an average phase of CT 12. Consequently, the pacemaker is reset by the light-to-dark transition to an average phase of CT 12 that actually shows a small but significant oscillation of phase. (A modified from Peterson and Saunders, 1980.)

explain how complete photoperiods can prevent the phase-angle jump, and it can provide the basis for understanding the bistability seen with entrainment to skeleton photoperiods. The basic premise is that two distinct limit cycles exist: the DD and LL limit cycles. Transitions between light and dark initiate movements of the state variables toward these limit cycles, and the kinetics of these movements are crucial.

For example, if the DD and LL limit cycles were well separated in phase space, as is envisioned to be the case for *Culex*, a complete photoperiod would elicit a larger-amplitude oscillation of the state variables than would occur in either DD or LL (Figure 3.30A). The amplitude of the oscillation is defined by the approximate diameter of the limit cycle. On the other hand, entrainment to skeleton photoperiods is expected to look different. The amplitude of the state variables' oscillation on a skeleton photoperiod would be much smaller and remain close to the DD limit cycle (Figure 3.30B). Therefore, even though the entrainment of a circadian rhythm to skeleton photoperiods might appear to be essentially the same as that to complete photoperiods, the state vari-

ables of the pacemaker could be oscillating in significantly different areas of phase space.

One result of this effect would be to prevent the phase-angle jump phenomenon observed with skeleton photoperiods (see Figure 3.18B). With skeleton photoperiods, the state variables always remain close to the DD limit cycle, and after each lights-off, the state variables are at different isochrons (see Figure 3.30B). This topology in phase space allows the oscillator to interpret the smaller dark interval as subjective day regardless of the prior phase of the oscillator. Therefore, either light pulse could become the sunrise or sunset pulse once steady-state entrainment has been achieved. For example, in a skeleton photoperiod of 16:8, the oscillator is free to select the 8 h interval as subjective day, even though it might take many cycles to attain steady-state entrainment.

With complete photoperiods, on the other hand, the state variables are displaced far off the DD limit cycle (see Figure 3.30A). Therefore, at lights-off the variables must return to the DD limit cycle along a restricted range of paths. For sufficiently long photoperiods, the

(A)

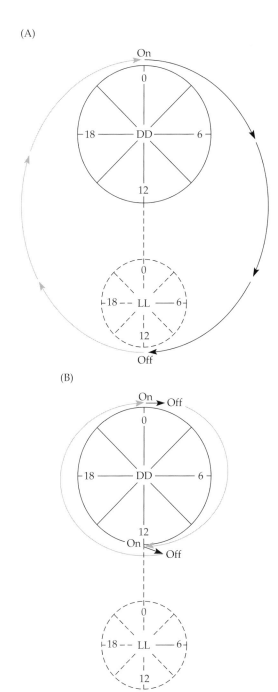

(B)

FIGURE 3.30 Predicted behavior of circadian oscillators under entrainment to complete and skeleton photoperiods of LD 12:12. DD limit cycles are shown at the top of each panel, LL limit cycles at the bottom. The directions of motion of state variables across the phase planes are indicated by arrowed lines: solid black lines for trajectories in light, gray lines for trajectories in darkness. (A) Entrainment to complete photoperiods of LD 12:12. Under stable entrainment, the state variables at lights-on ("On") move from the DD limit cycle near CT 0, initially to the right, then arc toward the LL limit cycle. At lights-off ("Off"), the state variables are attracted back to the DD limit cycle. (B) Entrainment to a skeleton photoperiod of LD 12:12. At lights-on ("On" near CT 0), the state variables move away from the DD limit cycle. When that light pulse is turned off ("Off" near the isochron of CT 1.5), the state variables return to the DD limit cycle until the next light pulse begins (near CT 12), at which time the state variables move again toward the right. When the second light pulse is turned off (near the isochron of CT 11), the state variables again spiral back toward the DD limit cycle. (A modified from Peterson, 1980.)

entrainment to LD thereby provides an explanation for why circadian oscillators do not undergo a phase-angle jump in complete photoperiods.

If the limit-cycle model is generalizable, then similar explanations may also apply to entrainment by other entraining agents. Good examples may be phase shifting and entrainment by temperature cycles. Examination of the levels of the FRQ protein, a likely molecular correlate of a state variable of the *Neurospora* circadian system, have shown that the average level around which FRQ levels oscillate is higher at a higher temperature (see Figure 7.21B). In the limit-cycle interpretation, this implies that the position on the phase plane of the limit cycles at different temperatures is different. Steps from one temperature to another may be interpreted as steps from one limit cycle to another. The interpretation given to temperature steps in *Neurospora* draws on and parallels the limit-cycle interpretation for light in Figures 3.29 and 3.30.

If entrainment of various organisms can be explained by this single solution, why do the entrainment properties of different organisms appear to be different? For example, the pacemakers of *Drosophila* pupae and *Neurospora* hyphae are exquisitely sensitive to light: A 1-minute light pulse of moderate intensity elicits very large phase shifts, and their outputs become arrhythmic in very dim LL. At the other extreme, plants and many birds require long, high-intensity light pulses to achieve relatively small phase shifts. A truly generalizable model must be able to accommodate these differences in sensitivity. One possible explanation is that the clock systems of different organisms have very different sensitivities of their circadian photoreceptors and phototransduction pathways. In a very photosensitive system, light and dark transitions might very rapidly move the state variables between the DD and LL limit cycles. In this case the dim light of dawn and dusk would be sufficient to entrain the circadian rhythm, and the behavior would appear to match the predictions of the

return path is approximately in phase with the isochron of CT 12. Therefore, the oscillator can interpret the photoperiod in only one phase relationship. For example, in a complete photoperiod of LD 16:8, the state variables cannot be interpreted such that the 8 h interval is subjective day. The reason for this limitation is that the 16 h light exposure forces the variables to remain in the region of the CT 12 isochron of the DD limit cycle, oscillating around the LL limit cycle. Consequently, lights-off can be interpreted only as approximately CT 12 in a complete photoperiod. The limit-cycle interpretation of discrete entrainment model. In such a high-sensitivity system, the only remaining function supplied by a complete photoperiod would be to prevent the phase-angle jump, as explained already.

Simply assuming a lower photosensitivity, however, might make the entrainment behavior appear superficially quite different, even though the fundamental characteristics are similar. In this case the state variables might make the transition between the DD and LL limit cycles slowly. The consequence of slow kinetics is that much stronger stimuli (in duration and/or intensity) might be required to elicit significant phase resetting. For stimuli below the strength threshold, the state variables will not complete the journey between the DD and LL limit cycles, a situation that was modeled for *Culex*. Therefore, a decrease in the system's photosensitivity without any change in the fundamental response of the pacemaker to light and dark could alter the overall behavior into one that is not predictable from the discrete entrainment model. Organisms that are routinely exposed to complete photoperiods probably do not need a highly photosensitive circadian system. Therefore it is reasonable to propose that their entrainment mechanism need not have developed extreme photosensitivity during the course of evolution.

Past discussions of entrainment mechanisms have sometimes indicated that either discrete or continuous mechanisms are operative in entraining circadian clocks. Through use of the limit-cycle model, a different, more integrated perspective can be advanced that reconciles these differences.

Limit-cycle modeling establishes criteria for components of the central clockwork[11,30,47,49]

Much of the work described in Chapter 7 is the result of searches for molecules whose biochemistry and regulation correlate with mathematical expectations for state variables and parameters. Limit-cycle concepts establish criteria for deciding whether a candidate molecule is a state variable or not. In the context of entrainment, the limit-cycle model makes the strong prediction that the level of at least one state variable should be induced or repressed by light.

The limit-cycle models such as those suggested for *Culex*, *Drosophila*, and other organisms predict that light will strongly affect the value of one state variable such that it will be moved to a region of the phase plane that is significantly off the DD limit cycle (*Culex*) or to one extreme of the DD limit cycle (*Drosophila*). These are testable predictions that are consistent with experimental data. For example, in *Neurospora* the mRNA levels of the *frq* gene, which encodes the FRQ protein that behaves as a correlate of a state variable of the *Neurospora* clock, are induced to high levels by light

exposure (see Chapter 7). In the corresponding counterexample, levels of the *Drosophila* PER protein, another correlate of a state variable, are rapidly reduced by light (see Chapter 7). Another strong prediction of the limit-cycle model is that the values of the state variables describing the limit cycle under different constant environmental conditions will be distinguishable. Data consistent with this prediction also exist in *Neurospora*. Temperature treatments have a substantial impact on the level of the state variable correlate FRQ (see Chapter 7).

Limit-cycle models of circadian clocks have already made significant contributions toward the main uses of models noted earlier: They have organized data, they have established criteria for determining how well phenomena are understood, and they have made accurate predictions for future investigations. The value of these modeling endeavors is likely to increase as the complexity of circadian phenomena is progressively unveiled.

SUMMARY

This chapter has discussed the fundamental properties of circadian rhythms, paying special attention to the property of entrainment. An essential function of circadian rhythms is to provide an internal estimate of the external local time, thereby allowing the organism to program its activities so that they occur at an appropriate time in the daily environmental cycle. Three salient characteristics of circadian clocks, each important for maintaining an appropriate phase relationship of clock to environment, have been emphasized. First, circadian rhythms are controlled by self-sustained oscillators, not hourglass timers, and therefore they continue to oscillate in constant conditions. Second, these oscillators are temperature compensated so that they run with approximately the same period at different constant ambient temperatures. Third, entrainment is the most important property for determining the phase relationship of the clock.

Various rhythmic environmental stimuli are known to entrain circadian clocks. Light–dark cycles are usually thought to be the most important environmental entraining agents because they are the most consistent from day to day. However, other environmental stimuli, such as temperature, may play important roles for entrainment in nature. Phase response curves are maps of the phase-dependent responses of circadian clocks to entraining agents.

Two different models have been proposed to explain circadian entrainment: the discrete model and the continuous model. Whereas the discrete model predicts entrainment on the basis of phase changes, the continuous model predicts entrainment on the basis of frequency (period) changes. The entrainment behavior of many organisms can be predicted by the discrete model, which uses the phase response curve and the free-running period to estimate entrainment characteristics. However, the discrete model cannot explain all relevant features of circadian entrainment, especially for organisms that are exposed to sunlight throughout the day. Therefore, some aspects of the continuous model will probably ultimately be found to be important.

Modeling scientific processes can be useful. With growing sophistication in chronobiology, mathematical modeling of circadian oscillators is assuming increased importance. Because circadian oscillators behave as limit-cycle oscillators, their behavior can be predicted by limit-cycle modeling. Limit-cycle perspectives can explain several important features of circadian clocks, including phase response curves, arrhythmic behavior, and the criteria for identifying essential components of the central clockwork. This chapter has presented a limit-cycle model for understanding and predicting entrainment.

CONTRIBUTORS

This chapter was written by Carl Hirschie Johnson, Jeffrey Elliott, Russell Foster, Ken-Ichi Honma, and Richard Kronauer.

STUDY QUESTIONS AND EXERCISES

1. Indirect measures of biological timing
   a. **Overview.** Behavioral indicators of biological timing are used in almost every aspect of studies at the organismic level because they preserve the integrity of whole-animal functioning. The hands of a wristwatch allow an observer to read time of day without having any inkling of the mechanical or electrical mechanism in the enclosing case. In the same sense, an internal biological clock in a mammal, the suprachiasmatic nucleus, within the fragile, complex brain, can be understood by observation of its output behavioral "hands," thus avoiding damage to the delicate nerve cells. The formalisms of this chapter concerning free-running period, PRC, and entrainment are almost all based on indirect measures. For the widely used mammalian rodent models, the most common hands are daily locomotor activity cycles, measured in a running wheel and using the onset of running activity as the marker point for data analysis. Review Figure 3.1 and be certain that you understand the equipment, the type of data collected, and the ways this data can be interpreted.
   b. **Data collection for indirect measurement of the oscillator.** Now collect your own data, if you have the proper facilities. You may also use the table of data from running-wheel activity

onset in a flying squirrel in continuous darkness. Plot the data carefully on graph paper or on the computer, assuming an 8 h active period following each activity onset. The result will be a circadian actograph. Calculate the free-running period and standard deviation of the onsets for this data set. Is the value exactly 24 h? Why or why not? Is the standard deviation 0? Again, why or why not?
   c. **Masking.** Look back at the actographs in Chapter 2 and in this chapter and try to interpret them. After a little practice, the actographs will be extremely informative. They constitute a basic item in the tool kit of almost all chronobiologists who study animals. Certain caveats, however, must be emphasized. Many errors of interpretation can be made with the indirect tool kit. Try to think of some disadvantages and sources of potential error. *Hint*: If wheel-running activity rhythms cease in a chipmunk when temperature rises above its normal range, has its clock stopped, or have the hands merely been "broken off"? If a flying squirrel appears to entrain immediately to a 12 h–shifted light schedule, is the data valid, or could the highly nocturnal animal simply be refusing to carry out activity in daylight? These considerations of potential masking are vital issues of interpretation of data by chronobiologists.

2. Direct measures of pacemaker activity

   Most direct measures are technologically very challenging, if not state-of-the-art. The technique of electrophysiological recording of nerve output signals from a pacemaker neuron in the SCN, for example, contrasts greatly from the indirect behavioral approach. Review the section in Chapter 6 concerning techniques for studying cellular mechanisms in pacemakers. Think about the challenges of direct techniques. Set up a hypothesis for the cellular basis of timing in a pacemaker like the SCN, and design a feasible experiment to test an original hypothesis of your choice. The tutorial of Chapter 5 on the mammalian brain and how to answer questions about its functioning will provide some help.

3. Choosing a research model

   Make a list of some of the advantages and disadvantages of direct measures, including such considerations as cost, damage to cells, viability of the pacemaker, and possible length of recording. Brain cells do not normally divide and replace themselves, nor do they usually live more than a few hours outside the brain. Crucial in dealing with some of the questions of chronobiology is the choice of the most appropriate model animal system. For example, whereas the pineal circadian pacemaker of the cold-blooded salmon can be cultured in vitro for many days with the greatest ease if it is simply excised and dropped into a bath of physiological saline, the culturing of SCN neurons is a state-of-the art task mastered by only a few of the most skillful chrononeurobiologists.

REFERENCES

1. Aschoff, J. 1960. Exogenous and endogenous components in circadian rhythms. Cold Spring Harb. Symp. Quant. Biol. 25: 11–28.

2. Aschoff, J. 1965a. The phase-angle difference in circadian periodicity. In: Circadian Clocks, J. Aschoff (ed.), pp. 262–276, North-Holland, Amsterdam.

3. Aschoff, J. 1965b. Response curves in circadian periodicity. In: Circadian Clocks, J. Aschoff (ed.), pp. 95–111. North-Holland, Amsterdam.

4. Aschoff, J. 1981. Free-running and entrained circadian rhythms. In: Handbook of Behavioral Neurobiology Volume 4: Biological Rhythms, J. Aschoff (ed.), pp. 81–93. Plenum, New York.

5. Barrett, R. K., and J. S. Takahashi. 1995. Temperature compensation and temperature entrainment of the chick pineal cell circadian clock. J. Neurosci. 15: 5681–5692.

6. Boivin, D. B., J. F. Duffy, R. E. Kronauer, and C. A. Czeisler. 1994. Sensitivity of the human circadian pacemaker to moderately bright light. J. Biol. Rhythms 9: 315–331.

7. Boulos, Z., M. M. Macchi, and M. Terman. 2002. Twilights widen the range of photic entrainment in hamsters. J. Biol. Rhythms 17: 353–363.

8. Bovet, J., and E. Oertli. 1974. Free-running circadian activity rhythms in free-living beaver (Castor canadensis). J. Comp. Physiol. 92: 1–10.

9. Brown, F. A., Jr., J. W. Hastings, and J. D. Palmer. 1970. The Biological Clock, Two Views. Academic Press, New York.

10. Bruce, V. G. 1960. Environmental entrainment of circadian rhythms. Cold Spring Harb. Symp. Quant. Biol. 25: 29–48.

11. Crosthwaite, S. K., J. J. Loros, and J. C. Dunlap. 1995. Light-induced resetting of a circadian clock is mediated by a rapid increase in frequency transcript. Cell 81: 1003–1012.

12. Daan, S. 2000. Colin Pittendrigh, Jurgen Aschoff, and the natural entrainment of circadian systems. J. Biol. Rhythms 15: 195–207.

13. Daan, S., and J. Aschoff. 1975. Circadian rhythms of locomotor activity in captive birds and mammals: Their variations with season and latitude. Oecologia 18: 269–316.

14. Daan, S., and J. Aschoff. 2001. The entrainment of circadian rhythms. In: Handbook of Behavioral Neurobiology Volume 12: Circadian Clocks, J. Takahashi, F. Turek, and R. Moore (eds.), pp. 7–44. Plenum, New York.

15. Daan, S., and C. S. Pittendrigh. 1976. A functional analysis of circadian pacemakers in nocturnal rodents. III. Heavy water and constant light: Homeostasis of frequency? J. Comp. Physiol. 106: 267–290.

16. DeCoursey, P. J. 1961. Effect of light on the circadian activity rhythm of the flying squirrel, Glaucomys volans. Zeit. vergl. Physiol. 44: 331–354.

17. DeCoursey, P. J. 1972. LD ratios and the entrainment of circadian activity in a nocturnal and a diurnal rodent. J. Comp. Physiol. 78: 221–235.

18. DeCoursey, P. J. 1986. Light-sampling behavior in photoentrainment of a rodent circadian rhythm. J. Comp. Physiol. 159: 161–169.

19. DeCoursey, P. J. 1989. Photoentrainment of circadian rhythms: an ecologist's viewpoint. In: Circadian Clocks and Ecology, T. Hiroshige and K. Honma (eds.), pp. 187–200. Hokkaido University Press, Sapporo, Japan.

20. Dharmananda, S. 1981. Studies of the circadian clock of Neurospora crassa: Light-induced phase shifting. Ph.D. dissertation, University of California at Santa Cruz.

21. Engelmann, W., and J. Mack. 1978. Different oscillators control the circadian rhythm of eclosion and activity in Drosophila. J. Comp. Physiol. 127: 229–237.

22. Eskin, A. 1971. Some properties of the system controlling the circadian activity rhythm of sparrows. In: Biochronometry, M. Menaker (ed.), pp. 55–80. National Academy of Sciences, Washington, DC.

23. Goto, K., and C. H. Johnson. 1995. Is the cell division cycle gated by a circadian clock? The case of Chlamydomonas reinhardtii. J. Cell Biol. 129: 1061–1069.

24. Hall, J. 1997. Circadian pacemakers blowing hot and cold—But they're clocks, not thermometers. Cell 90: 9–12.

25. Hamblen-Coyle, M. J., D. A. Wheeler, J. E. Rutila, M. Rosbash, and J. C. Hall. 1992. Behavior of period-altered circadian rhythm mutants of Drosophila in light:dark cycles (Diptera: Drosophilidae). J. Insect Behav. 5: 417–446.

26. Hanson, F. E. 1978. Comparative studies of firefly pacemakers. Fed. Proc. 37: 2158–2164.

27. Hoffmann, K. 1971. Biological clocks in animal orientation and in other functions. In: Proceedings of the International Symposium on Circadian Rhythmicity, pp. 175–205. North Holland Publishing Co., Wageningen, Netherlands.

28. Jewett, M. E., R. E. Kronauer, and C. A. Czeisler. 1991. Light-induced suppression of endogenous circadian amplitude in humans. Nature 350: 59–62.

29. Johnson, C. H. 1990. PRC Atlas: An atlas of phase response curves for circadian and circatidal rhythms. Department of Biology, Vanderbilt University. Technical bulletin. Also available online at http://johnsonlab.biology.vanderbilt.edu/chj/prcatlas/index.html

30. Johnson, C. H. 1994. Illuminating the clock: Circadian photobiology. Sem. Cell Dev. Biol. 5: 355–362.

31. Johnson, C. H. 1999. Forty years of PRCs—what have we learned? Chronobiol. Int. 16: 711–743.

32. Johnson, C. H., and Hastings, J. W. 1989. Circadian phototransduction: Phase resetting and frequency of the circadian clock of Gonyaulax cells in red light. J. Biol. Rhythms 4: 417–437.

33. Johnson, C. H., and T. Kondo. 1992. Light pulses induce "singular" behavior and shorten the period of the circadian phototaxis rhythm in the CW15 strain of Chlamydomonas. J. Biol. Rhythms 7: 313–327.

34. Johnsson, A., and H. G. Karlsson. 1972. The Drosophila eclosion rhythm, the transformation method, and the fixed point theorem. Department of Electrical Measurements, Lund Institute of Technology, Report No. 2/1972.

35. Kleinhoonte, A. 1928. Dedoor het licht geregelde autonome bewegingen der Canavalia-bladeren. Ph.D. dissertation, Utrecht University, Utrecht, Netherlands.

36. Lakin-Thomas, P. L. 1995. A beginner's guide to limit cycles, their uses and abuses. Biol. Rhythm Res. 26: 216–232.

37. Liu, Y., M. Merrow, J. J. Loros, and J. C. Dunlap. 1998. How temperature changes reset a circadian oscillator. Science 281: 825–829.

38. Menaker, M. 1959. Endogenous rhythms of body temperature in hibernating bats. Nature 184: 1251–1252.

39. Menaker, M. 1971. Synchronization with the photic environment via extraretinal receptors in the avian brain. In Biochronometry, M. Menaker (ed.), pp. 315–332. National Academy of Sciences, Washington, DC.

40. Millar, A. J., M. Straume, J. Chory, N-H. Chua, and S. A. Kay. 1995. The regulation of circadian period by phototransduction pathways in Arabidopsis. Science 267: 1163–1166.

41. Miwa, I., H. Nagatoshi, and T. Horie. 1987. Circadian rhythmicity within single cells of *Paramecium bursaria*. J. Biol. Rhythms 2: 57–64.

42. Moore-Ede, M. C., F. M. Sulzman, and C. A. Fuller. 1982. The Clocks That Time Us. Harvard University Press, Cambridge, MA.

43. Nelson, D. E., and J. S. Takahashi. 1991. Sensitivity and integration in a visual pathway for circadian entrainment in the hamster (*Mesocricetus auratus*). J. Physiol. 439: 115–145.

44. Njus, D., L. McMurry, and J. W. Hastings. 1977. Conditionality of circadian rhythmicity: Synergistic action of light and temperature. J. Comp. Physiol. 117: 335–344.

45. Ouyang, Y., C. R. Andersson, T. Kondo, S. S. Golden, and C. H. Johnson. 1998. Resonating circadian clocks enhance fitness in cyanobacteria. Proc. Natl. Acad. Sci. USA 95: 8660–8664.

46. Page, T. L. 1991. Developmental manipulation of the circadian pacemaker in the cockroach: Relationship between pacemaker period and response to light. Physiol. Entomol. 16: 243–248.

47. Pavlidis, T. 1973. Biological Oscillators: Their Mathematical Analysis. Academic Press, New York.

48. Pavlidis, T. 1981. Mathematical models. In: Handbook of Behavioral Neurobiology Volume 4: Biological Rhythms, J. Aschoff (ed.), pp. 41–54. Plenum, New York.

49. Peterson, E. L. 1980. A limit cycle interpretation of a mosquito circadian oscillator. J. Theor. Biol. 84: 281–310.

50. Peterson, E. L. 1981. Dynamic response of a circadian pacemaker. II. Recovery from light pulse perturbations. Biol. Cybern. 40: 181–194.

51. Peterson, E. L., and D. S. Saunders. 1980. The circadian eclosion rhythm in *Sarcophaga argyrostoma*: A limit cycle representation of the pacemaker. J. Theor. Biol. 86: 265–277.

52. Pittendrigh, C. S. 1954. On temperature independence in the clock system controlling emergence time in *Drosophila*. Proc. Natl. Acad. Sci. USA 40: 1018–1029.

53. Pittendrigh, C. S. 1960. Circadian rhythms and the circadian organization of living systems. Cold Spring Harb. Symp. Quant. Biol. 25: 159–184.

54. Pittendrigh, C. S. 1966. The circadian oscillation in *Drosophila pseudoobscura* pupae: A model for the photoperiodic clock. Z. Pflanzenphysiol. 54: 275–307.

55. Pittendrigh, C. S. 1967. Circadian systems I. The driving oscillation and its assay in *Drosophila pseudoobscura*. Proc. Natl. Acad. Sci. USA 58: 1762–1767.

56. Pittendrigh, C. S. 1981. Circadian systems: Entrainment. In: Handbook of Behavioral Neurobiology Volume 4: Biological Rhythms, J. Aschoff (ed.), pp. 95–124. Plenum, New York.

57. Pittendrigh, C. S. 1988. The photoperiodic phenomenon: Seasonal modulation of the "day within." J. Biol. Rhythms 3: 173–188.

58. Pittendrigh, C. S. 1993. Temporal organization: Reflections of a Darwinian clock-watcher. Annu. Rev. Physiol. 55: 17–54.

59. Pittendrigh, C. S., and V. G. Bruce. 1959. Daily rhythms as coupled oscillator systems and their relation to thermoperiodism and photoperiodism. In: Photoperiodism and Related Phenomena in Plants and Animals, R. B. Withrow (ed.), pp. 475–505. American Association for the Advancement of Science, Washington, DC.

60. Pittendrigh, C. S., and S. Daan. 1976. A functional analysis of circadian pacemakers in nocturnal rodents. I. The stability and lability of spontaneous frequency. J. Comp. Physiol. 106: 223–252.

61. Pittendrigh, C. S., and S. Daan. 1976. A functional analysis of circadian pacemakers in nocturnal rodents. IV. Entrainment: Pacemaker as clock. J. Comp. Physiol. 106: 291–331.

62. Pittendrigh, C. S., and D. H. Minis. 1964. The entrainment of circadian oscillations by light and their role as photoperiodic clocks. Amer. Nat. 98: 261–294.

63. Pittendrigh, C. S., and D. H. Minis. 1971. The photoperiodic time measurement in *Pectinophora gossypiella* and its relation to the circadian system in that species. In: Biochronometry, M. Menaker (ed.), pp. 212–250. National Academy of Sciences, Washington, DC.

64. Ralph, M. R., and M. Menaker. 1988. A mutation of the circadian system in golden hamsters. Science 241: 1225–1227.

65. Roenneberg, T., and D. Morse. 1993. Two circadian oscillators in one cell. Nature 362: 362–364.

66. Ruby, N., D. E. Burns, and H. C. Heller. 1999. Circadian rhythms in the suprachiasmatic nucleus are temperature compensated and phase shifted by heat pulses in vitro. J. Neurosci. 19: 8630–8636.

67. Saunders, D. S. 1977. An Introduction to Biological Rhythms. Blackie, London.

68. Saunders, D. S. 1978. An experimental and theoretical analysis of photoperiodic induction in the flesh-fly, *Sarcophaga argyrostoma*. J. Comp. Physiol. 124: 75–95.

69. Saunders, D. S. 1986. Many circadian oscillators regulate developmental and behavioural events in the flesh-fly, *Sarcophaga argyrostoma*. Chronobiol. Int. 3: 71–83.

70. Smolensky, M., and L. Rensing (Eds.) 1989. Masking in circadian rhythmicity. Chronobiol. Int. 6: 1–187.

71. Strogatz, S. H. 1994. Nonlinear Dynamics and Chaos. Addison-Wesley, Reading, MA.

72. Swade, R. H. 1969. Circadian rhythms in fluctuating light cycles: Toward a new model of entrainment. J. Theor. Biol. 24: 227–239.

73. Sweeney, B. M., and J. W. Hastings. 1960. Effects of temperature upon diurnal rhythms. Cold Spring Harb. Symp. Quant. Biol. 25: 87–104.

74. Taylor, W. R., R. Krasnow, J. C. Dunlap, H. Broda, and J. W. Hastings. 1982. Critical pulses of anisomycin drive the circadian oscillator in *Gonyaulax* towards its singularity. J. Comp. Physiol. 148: 11–25.

75. Toh, K. L., C. Jones, Y. He, E. Eide, W. Hinz, et al. 2001. An *hPer2* phosphorylation site mutation in familial advanced sleep phase syndrome. Science 291: 1040–1043.

76. Underwood, H., and M. Calaban. 1987. Pineal melatonin rhythms in the lizard *Anolis carolinensis*: I. Response to light and temperature cycles. J. Biol. Rhythms 2: 179–193.

77. Winfree, A. T. 1970. Integrated view of resetting a circadian clock. J. Theor. Biol. 28: 327–374.

78. Winfree, A. T. 1971. Corkscrews and singularities in fruit flies: Resetting behavior of the circadian eclosion rhythm. In: Biochronometry, M. Menaker (ed.), pp. 81–109. National Academy of Sciences, Washington, DC.

79. Zimmerman, W. F., C. S. Pittendrigh, and T. Pavlidis. 1968. Temperature compensation of the circadian oscillation in *Drosophila pseudoobscura* and its entrainment by temperature cycles. J. Insect Physiol. 14: 669–684.

Chapter 2 already introduced the topic of seasonal rhythms based on physiological timing mechanisms. The fascinating story of circannual rhythms and closely related photoperiodism in plants and animals is very large in scope, covering the regulation of diverse behavioral rhythms in a broad spectrum of phylogenetic groups. The cameos in this chapter's opening illustration give an inkling of the sagas to follow.

After nesting in Europe, the garden warbler flies to Africa to escape winter. Canada geese undertake long-distance migrations, flying high in wedge formation. Many insects undergo a stage of arrested development such as the pupal cocoon of the giant silkworm moth, *Antheraea polyphemus,* and ladybugs climb under protective logs and leaves to escape winter's rigor. Almost all temperate birds and mammals breed at the time of most favorable temperature and food supply. The cocklebur plant requires short days to flower and set seed. Buffalo young are born in spring when prairie plants begin to grow after the long winter. All of these phenomena require elaborate metabolic preparation in advance. In almost all cases, anticipatory preparation is regulated by circannual and photoperiodic timing mechanisms. Because of the magnitude of the topic, all of Chapter 4 is dedicated to expanding the drama of endogenous and exogenous factors in timing seasonality.

4

Circannual Rhythms and Photoperiodism

For everything there is a season, and a time for every purpose.
—Ecclesiastes 3:1–2

Introduction: The Lives of Most Plants and Animals Are Organized on a Seasonal Schedule[9,20,25,26,28,51,58,67,75]

SEASONAL CYCLES. Most plants and animals are exposed to annual as well as daily variations in their environment. Seasonal influences are particularly marked at extreme latitudes, where variations in external conditions are most pronounced. Plants and animals in these locations have adjusted to the seasonal variations in factors such as ambient temperature, food supply, and predator abundance. Annual biological cycles are widespread even in regions close to the equator, where rainfall and food availability may be highly cyclic even if photoperiod is uniform throughout the year.

Organisms generally concentrate reproductive effort during seasons that favor survival of offspring (see Plates 2 and 4). Other activities, such as growth, molt, migration, and hibernation, are also timed in an adaptive manner, relative to the seasonal environment and relative to each other. For example, the reproductive season of birds varies with latitude. With increasing distance from the equator, breeding tends to become more and more concentrated into spring and early summer months and begins progressively later in the year, reflecting the window of conditions favorable for production of offspring (Figure 4.1). In equatorial regions breeding birds are found year-round, but only for the avian fauna when viewed as a whole. Even in tropical regions, a particular population usually has a seasonal reproductive pattern. Temporal relations between various seasonal activities are also important. For migratory birds, migration must not interfere with reproduction, and molts must not overlap with times of migration.

Ultimate factors shape the evolution of biological seasonality; proximate factors provide the direct stimuli for physiological regulation of seasonality. One example

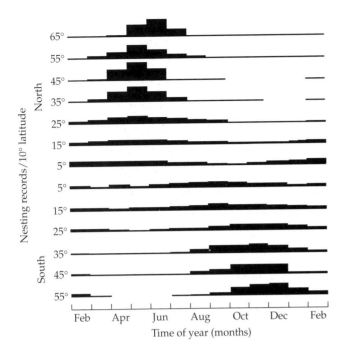

FIGURE 4.1 Distribution of breeding seasons in avian species as a function of latitude, shown as the relative number of times that eggs have been found in each 10° band of latitude. (From Baker, 1938.)

will illustrate this important distinction. The most widely used environmental cue for predicting favorable timing of breeding in temperate and arctic latitudes is photoperiod, or day length. Photoperiod has a direct influence on physiological parameters that influence reproductive status in many organisms and is therefore clearly a proximate cause for seasonal reproduction in these species. Photoperiod is a useful cue only because it correlates perfectly with time of year. It has little direct significance for biological activities and is therefore not an ultimate factor. Ultimate causes are the environmental factors that directly affect reproductive fitness of organisms by optimizing seasonal investments of energy in reproduction and other activities.

In temperate latitudes, proximate causes such as photoperiod frequently differ from ultimate factors. In tropical latitudes photoperiod is relatively constant throughout the year, but rainfall and related food availability are quite variable. These other cyclic environmental factors may supplant photoperiod as proximate cues in the Tropics. For nonmigratory tropical organisms, cyclic rainfall and food availability are often critically important in both a proximate and ultimate sense.

TWO GENERAL TYPES OF ANNUAL BIOLOGICAL RHYTHMS. Some seasonal rhythms are self-sustaining under constant environmental conditions. These have been called true circannual rhythms or Type II seasonal rhythms. They have been documented to persist for as many as

12 consecutive cycles in birds and 7 cycles in mammals, even when all significant environmental conditions are held constant (Figure 4.2A). True circannual rhythms have free-running periods that approximate 12 months but can vary considerably among and within individuals. Long-lived species such as squirrels, sheep, deer, bats, ferrets, starlings, and Old World warblers are good vertebrate examples. Although they can persist in the absence of environmental cues, the use of such cues is still essential for synchronization to local seasonal time. Photoperiod is the most common cue for entrainment of true circannual rhythms.

A more commonly studied type of seasonal rhythm (Type I) has elements of endogenous timekeeping combined with obligatory exogenous components (Figure 4.2B). This type of cycling persists for not more than one cycle, and often for only a fraction of a cycle in constant environmental conditions. Both long-day and short-day cues are required at specific times during the cycle to sustain Type I rhythms and to synchronize the cycle to local seasonal time. Type I rhythms have been documented extensively in relatively short-lived mammals and in many avian and reptilian species. Good examples are voles, mice, hamsters, and New World sparrows. Behavioral and physiological functions expressed in Type I and II annual rhythms include yearly cycles of reproduction, body mass, molt, thermoregulation, migration, and social behavior.

This chapter will first consider properties of true circannual rhythms, as well as analogies and differences between circannual and circadian. The chapter will then proceed to a discussion of photoperiodism covering the mechanism of synchronization of seasonal cycles to local environmental seasonal time.

Many Seasonal Rhythms Are Based on Endogenous Circannual Rhythms[20,26,28,53,75]

Circannual rhythms are widely distributed in plants and animals[22,26,27,43,49,53]

HISTORICAL OVERVIEW. Circannual rhythms were first clearly demonstrated in the late 1950s and early 1960s. Blake worked with the beetle *Anthrenus*. Also at this time, Pengelley and colleagues described circannual hibernation rhythms in the golden-mantled ground squirrel. Individual squirrels that were kept in constant conditions of day length and temperature continued to begin the hibernation season at approximately yearly intervals and to show associated seasonal changes in body weight and food consumption. Soon circannual rhythms encompassing diverse functions were noted in a wide variety of animals and plants.

FOUR PHYLOGENETICALLY REPRESENTATIVE EXAMPLES. Circannual rhythms occur even in primitive plants.

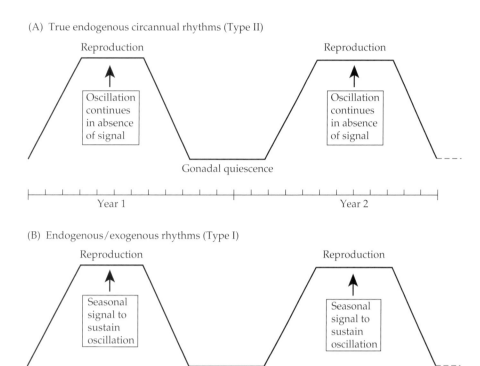

(A) True endogenous circannual rhythms (Type II)

Reproduction   Reproduction

Oscillation
continues
in absence
of signal

Oscillation
continues
in absence
of signal

Gonadal quiescence

Year 1   Year 2

(B) Endogenous/exogenous rhythms (Type I)

Reproduction   Reproduction

Seasonal
signal to
sustain
oscillation

Seasonal
signal to
sustain
oscillation

Gonadal quiescence

FIGURE 4.2 The two main types of annual rhythms in hypothetical animals. Seasonal reproduction is the example used here. (A) True circannual rhythms (Type II) persist indefinitely in constant environmental conditions and need day length information only for entrainment to local seasonal time. (B) Type I annual rhythms contain an endogenous timekeeping component, such as spontaneous gonadal growth in spring, but they require environmental (exogenous) cues both for sustaining the oscillation and for synchronizing to local seasonal time. (Modified from Prendergast et al., 2002.)

Growth rate of a terminal blade of the sporophytes of the marine seaweed *Pterygophora californica*, maintained for 2 years at 5°C on a daily LD 16:8 lighting schedule, changed periodically, showing an average period of about 11 months (Figure 4.3A).

Circannual rhythms have been amply demonstrated in birds for such diverse functions as reproduction, molt, and migration. The many pressing temporal demands of a long-distance migratory lifestyle have resulted in evolution of synchronous, tightly linked rhythms of gonadal growth and regression, molt pattern, courtship and breeding, and migratory restlessness. These patterns serve as good markers of circannual rhythms. Changes in testicular width and in molt occurred in European starlings that were kept for 43 months in an LD 12:12 light schedule at constant temperature (Figure 4.3B). A pattern similar to cycling of free-living conspecifics persisted throughout the experiment, except that the period deviated from 12 months. An African stonechat continued its circannual molt and fattening pattern under constant conditions in the laboratory for 10 years through 12 reproductive cycles (see Figure 4.5).

Circannual rhythms have also been documented in sheep (Figure 4.3C). Ovariectomized ewes were treated with constant-release estradiol implants that provided fixed negative feedback signals to the reproductive neuroendocrine axis. Circannual cycling of plasma concentrations of luteinizing hormone (LH) continued for 5 years in an LD 8:16 light schedule. The mean period of cycles, measured between successive rises in plasma LH concentrations, was about 11 months.

Other good examples are seen in a widely separated mammalian group, the ground squirrels. Hibernation cycles persist in many species for extended periods of time. Twelve individual golden-mantle ground squirrels were maintained for 47 months on an LD 12:12 light schedule at 3°C (Figure 4.3D). The squirrels entered hibernation earlier each year than the previous year, exhibiting a free-running period of about 10 months. Many other instances of hibernation rhythms in rodents are now known, especially for species living in cold climates such as high-elevation montane habitats or high-latitude arctic areas.

DIVERSITY OF GROUPS AND FUNCTIONS. The four examples just discussed are only a small fraction of a long list of well-documented circannual rhythms. Birds and mammals are the best-studied groups, but circannual rhythms are known also for higher plants, dinoflagellates, coelenterates, molluscs, insects, fish, and reptiles. Functions controlled by circannual rhythms range widely, from deposition and metabolism of body fat through hibernation, molt, migration, and reproduction. The presence or absence of circannual rhythms in humans

(A) Seaweed

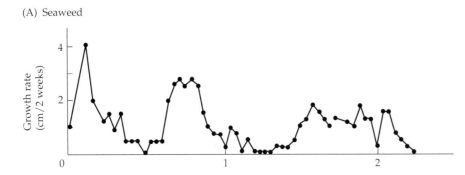

(B) Starling

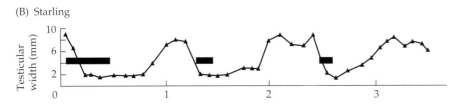

(C) Sheep

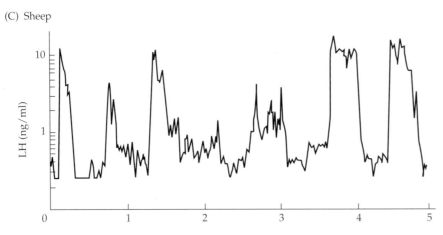

(D) Golden-mantled ground squirrel

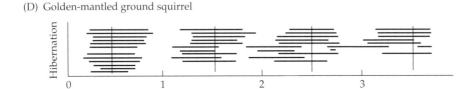

Time (years)

FIGURE 4.3 Examples of circannual rhythms. (A) Changes in growth rate of a seaweed, in LD 12:12 at 5°C. (B) Changes in testis width (triangles) and occurrence of molt (bars) in a starling for 3.5 years in LD 12:12 at 20°C. (C) Changes in plasma LH concentration in a ewe for 5 years in LD 8:16. (D) Occurrence of hibernation (black bars) in 12 golden-mantled ground squirrels for 4 years in LD 12:12 at 30°C. (A from Lüning and Kadel, 1993; B from Gwinner, 1981; C from Karsch et al., 1989; D from Pengelley et al., 1976.)

species. For conclusive demonstration of the existence of a circannual rhythm, organisms must be maintained for more than a year under conditions that do not provide potential seasonal time cues. Seasonally constant conditions are analogous to those of circadian experiments lacking time-of-day cues. In most studies of circannual rhythms, both photoperiod and temperature are held constant.

Particular environmental conditions are necessary in some species for the expression of circannual rhythms. In the sika deer, for instance, a circannual rhythm in antler replacement is observed under daily LD 18:6 and LD 6:18 light schedules but not in LD 12:12. In contrast, European starlings show circannual rhythms in reproduction and molt in LD 12:12 but not in longer or shorter photoschedules. The need for permissive environmental conditions has led to confusion about the endogenous nature of the annual rhythm in these instances. If a circannual rhythm is expressed in an organism kept under constant conditions, then the organism itself must vary periodically with a self-sustaining rhythm of internal change. The requirement for a tightly defined range of constant conditions in certain organisms suggests that rhythm maintenance requires a specific type of environmental input at a particular stage of the circannual cycle.

The need for narrowly permissive photoperiodic conditions does not seem to be a general principle. In golden-mantled ground squirrels, sheep, and many birds, circannual rhythms seem to persist in virtually all unvarying photoperiods. For these species the only function of photoperiod is entrainment. An under-

has not been established. Long-term studies under carefully controlled conditions are required to verify the presence of a circannual rhythm, and such studies are difficult in humans.

Specific environmental conditions are sometimes required for expression of circannual rhythms[11,23]

Though circannual rhythms are found in a variety of species, they are not as widespread as circadian rhythms. Even closely related species may differ in their reliance on circannual rhythms. The rigidity and degree of persistence of circannual rhythms is probably related to the ecology and life history characteristics of a standing of the circannual rhythms is clouded by the lack of localization of any particular anatomical structure or any discrete biochemical substrate as a circannual oscillator. As a result, SCN-style analysis or pineal-style studies have not yet been possible for circannual rhythms.

The annual cycle of photoperiod is the most important synchronizer of circannual rhythms[16,22,34]

In most true circannual species, photoperiod is the most important cue for synchronizing the endogenous cir-

cannual rhythm with the natural year. Starlings are a good example. Sinusoidal photoperiodic cycles mimicking the amplitude and general shape of natural annual cycles at 40 degrees north latitude can synchronize a starling's circannual testicular and molt cycles to a period of 12 months. If the sinusoidal photoperiod cycle is shortened to 8, 6, 4 or 3 months, the bird's rhythms become compressed accordingly. Only under cycles shorter than 3 months is the rhythm irregular (Figure 4.4A). Similarly, the rhythm of antler replacement in sika deer can be synchronized to sinusoidal photoperiodic changes with periods of 12, 6, and 4 months, respec-

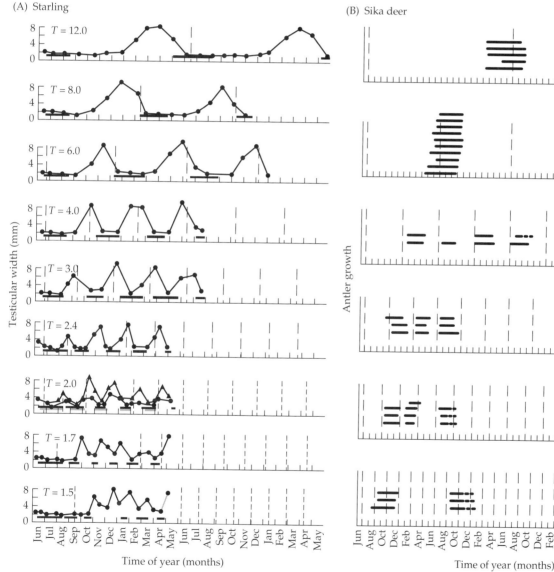

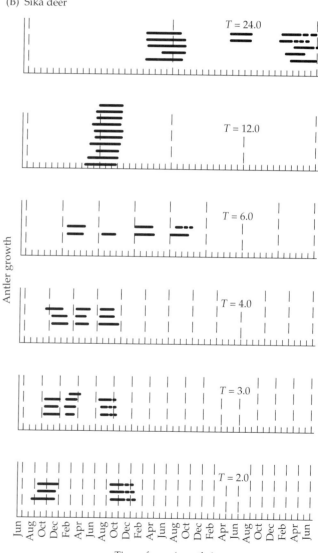

FIGURE 4.4 Responses of two species to sinusoidal light cycles simulating natural conditions. (A) Variation in testis width (curves) and occurrence of molt (bars) in 10 groups of starlings exposed to sinusoidal changes in photoperiod. Duration (*T*) varied from 12 months in the top panel to 1.5 months in the bottom panel. Dashed vertical lines represent times of the longest photoperiods at the summer solstice.

Two replications of *T* = 2.0 are shown: one with triangles and gray bars, the other with circles and black bars. Vertical bars at data points represent standard deviations. (B) Phases of antler growth (bars) in individual sika deer exposed to sinusoidal changes in photoperiod. Light schedules are as in part A, with duration from 24 months in the top panel to 2 months in the bottom panel. (From Gwinner, 1986.)

tively. In shorter periods, one or more cycles are skipped (Figure 4.4B).

Synchronization patterns of circannual rhythms are similar in several respects to circadian rhythms. Most species are able to entrain to cycles close to 12 months, but the limits of yearly entrainment are frequently proportionally much larger than the range for circadian rhythms. Some organisms may also entrain to multiples or fractions of the synchronizing period. When sika deer were exposed to a cycle duration of 24 months, two of five animals replaced antlers twice per photoperiodic cycle, thus synchronizing to a fraction, or submultiple, of the period (see Figure 4.4B). At the other end of the range of entrainment, animals skipped cycles and synchronized with multiples of the short photoperiodic cycles. The ability to entrain to multiples of the environmental cycle is analogous to a similar phenomenon observed for circadian rhythms.

Like circadian rhythms, many circannual rhythms entrain to phase shifts in which a photoperiodic cycle is abruptly advanced or delayed by several months. In studies of these rhythms, the circannual cycle followed these phase shifts. Some transients were usually present. The results attest both to the entraining properties of these photoperiodic cycles and to the endogenous, self-sustained nature of the entrained rhythms.

Temperature has been suggested as a potential circannual synchronizer in ground squirrels and European hamsters, but very little convincing experimental evidence is available. Social stimuli may possibly synchronize circannual rhythms in some species, such as sheep, but the database is small. Seasonal food supplementation has also been tried without success. Compared with photoperiod, the importance of these nonphotic synchronizers appears quite limited.

Circannual rhythms are ecologically adaptive[26,27,30,51]

The resistance of circannual pacemakers to phase shifting may be an important feature of endogenous circannual rhythms. The inertia may help buffer against environmental noise and help organisms achieve precise seasonal timing. A stabilizing mechanism may be particularly important in animals with tight seasonal schedules, such as long-distance migrating birds. In addition to reproducing and molting annually, these birds spend many months each year in migration. Some high-latitude hibernating mammals are under similar time pressure. Their reproductive and molting season is restricted to a few months in summer because they hibernate for as long as 7 to 8 months each year. Circannual rhythms may be more robust and tightly organized in long-distance than in short-distance avian migrants and in mammals with extended hibernation seasons compared to species with short hibernation seasons.

Circannual rhythms may also play important roles in the timing of behavioral and physiological processes in animals that inhabit relatively constant environments devoid of obvious seasonal timing cues during much of the year. Again, examples are found in long-distance migrant birds that spend the winter under constant equatorial photoperiods. In these species the winter events are almost exclusively under the control of a circannual mechanism. Spring migratory behavior and the associated onset of gonadal development occur at the appropriate times even if individuals are isolated throughout winter from seasonal environmental changes. Ground squirrels spend many months each year underground in DD and constant low temperature. Their emergence from hibernation and gonadal growth is triggered by internal circannual timing signals.

Circannual rhythms may play very specific roles in providing adaptive timing. African stonechats from equatorial regions have particularly rigid and long-lasting circannual rhythms. Variations in testicular width, plasma LH concentrations, and molt occurred in a wild-born male held for 10 years under a constant 12.8 h photoperiod and under constant food and temperature conditions (Figure 4.5).

In nature, gonads develop in African stonechats during the dry season from October through March. Gonadal regression and postnuptial molt take place during the main rainy season in April and May. The rhythm does not depend on seasonal environmental changes because it free-runs under constant conditions with a period of about 10 months. More than 95% of African stonechats displayed clear and persistent circannual cycles under constant conditions. In contrast, only about 50% of individuals in a closely related subspecies from temperate zones of Europe manifested clear cycles under the same constant conditions. Rhythmicity was arrested or severely disturbed after one or two cycles in the remaining birds. Circannual cycling may be less prevalent in Temperate Zone animals and plants than in tropical organisms, which reproduce annually in spite of only slight seasonal environmental variations.

A comparison of the molt and reproductive cycles of European birds with their African counterparts revealed other differences. Under fixed conditions in the laboratory, postjuvenile molt started earlier and progressed faster in the European birds than in the African birds, reflecting corresponding differences in free-living birds. The differences are probably related to the fact that European birds are migratory and must finish molt prior to the onset of fall migration. The African conspecifics, in contrast, are permanent residents that can afford to molt slowly.

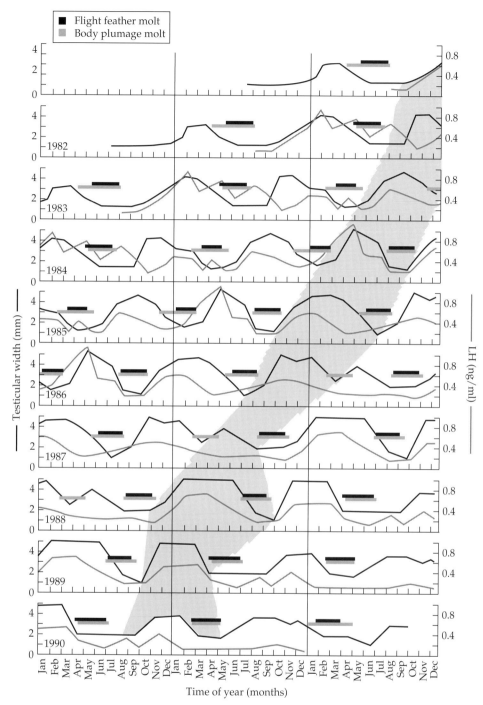

FIGURE 4.5 Circannual rhythms of the East African stonechat. Circannual rhythm of testis width (black line) and plasma LH concentration (gray line) as well as occurrence of molt in a male East African stonechat held 10.25 years in a constant LD 12.8:11.2 light schedule. Data are triple-plotted by year with the preceding 2 years on left. Shaded background marks one series of gonadal cycles. For further explanation, see the text. (From Gwinner, 1996.)

The evidence strongly suggests a genetic basis for the differences because these differences are evident even in birds raised in a constant photoperiod. Such a conclusion is further supported by the intermediate behavior of a group of F$_1$ hybrids (Figure 4.6A). The two subspecies also differ in the length of the breeding season. The reproductive window from gonadal growth to regression was significantly wider in European stonechats than in African stonechats held in the same equatorial photoperiod (Figure 4.6B). The results paral-

lel differences in the duration of the breeding season in free-living birds. European stonechats are double or triple brooded with a longer overall breeding season than the single-brooded African birds.

Endogenously programmed circannual patterns play another important role in some long-distance migratory birds by helping to determine the duration and distance of migration. Warblers of the genus *Sylvia* are active primarily during the day, but they make their migratory flights exclusively at night. When held in

(A)

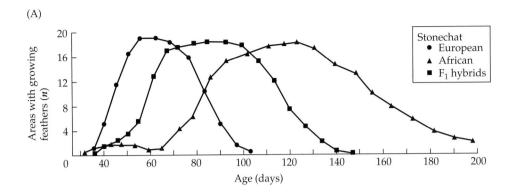

(B)

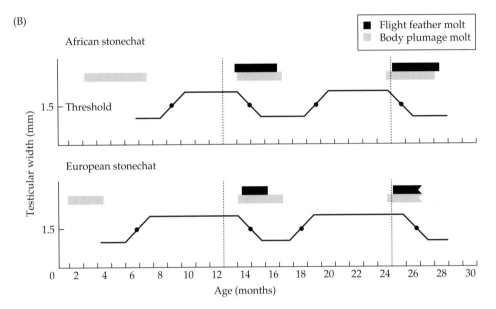

(C)

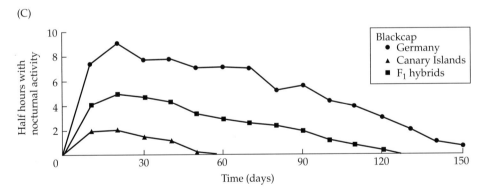

FIGURE 4.6 Examples of genetically determined circannual patterns. (A) Timing of postjuvenile molt in European and East African stonechats and in F₁ hybrids held in constant LD 12.8:11.2 conditions, with approximately weekly recording of molt progress in 19 plumage areas. (B) Testicular and molt cycles of male European and African stonechats for 29 months in constant LD 12.25:11.75. Circles represent mean onsets and ends of gonadal cycles, defined as times when testis widths first reached 1.5 mm in growth or regression. Left and right bar margins represent mean onsets and ends of flight feather molt (black bars) and plumage body molt (hatched bars). (C) Changes in intensity of autumn nocturnal migratory restlessness in first-year blackcaps from Germany and the Canary Islands, plus F₁ hybrids under lighting conditions of the German breeding grounds until day 50, then in LD 12.5:11.5. (A from Gwinner and Neusser, 1985; B from Gwinner 1991; C from Berthold and Querner, 1981.)

cages, these birds develop intense nocturnal activity during the migratory seasons. The fact that the birds continue to show nocturnal activity during their subjective spring and autumn, even when kept in constant environments, shows that migratory behavior is controlled by a circannual clock.

The detailed pattern of migratory behavior developed during the first fall was species specific and even population specific. The patterns of fall migratory activity from groups of German and Canary Island blackcaps illustrated that the duration and overall amount of migratory activity was correlated with migratory distance (Figure 4.6C). Whereas German blackcaps migrate about 1,000 km, the birds of the Canary Islands are nonmigratory or move only short distances. The difference is reflected in the drastic reduction in duration and amount of migratory activity in the latter group. F₁ hybrids from matings of German birds with Canary

Islands birds showed an intermediate pattern. The differentially programmed patterns of fall migratory activity of the various warbler subspecies or populations may be relevant for the orientation of their flights. The data suggest that these programs contain information about migratory distance by determining how long and thereby how far a bird migrates.

The direction of migration also shows a circannual shift. Garden warblers from southern Germany leave their breeding grounds in a southwesterly direction but shift their flight patterns to the southeast upon reaching the Mediterranean area. After wintering in central Africa, they return to their breeding grounds by migrating in a northerly direction. The same directional preferences were observed in a group of caged warblers that had been hand-raised and held for 1 year in a fixed photoperiod (Figure 4.7). The overt circannual rhythm expresses information pertinent to both the distance and direction of seasonal migrations. The information could support a vector navigational system that helps birds to find their specific wintering and breeding grounds.

Attempts have been made to localize a circannual clock[26,56,75]

A dominant circannual pacemaker site for generation of true circannual rhythms has not been identified in any animal. The most concerted attempt to localize such a clock involved a series of studies in golden-mantled ground squirrels. One study tested whether the master circadian pacemaker in the SCN (suprachiasmatic nucleus) was also important for control of circannual rhythms. Ground squirrels were held for several years at 23°C in a constant LD 14:10 schedule and then shifted to 6°C for 2.5 years. A robust circannual rhythm of body mass persisted in the control animals before and after entry into the cold. Discrete hibernation bouts lasted 5 to 6 months during the weight loss phase of each circannual cycle in the cold.

Most SCN-lesioned squirrels showed convincing circannual rhythms of body mass at the higher ambient temperature. Only half of the SCN-lesioned animals experienced severely disrupted circannual body mass and hibernation rhythms at the lower temperature. The

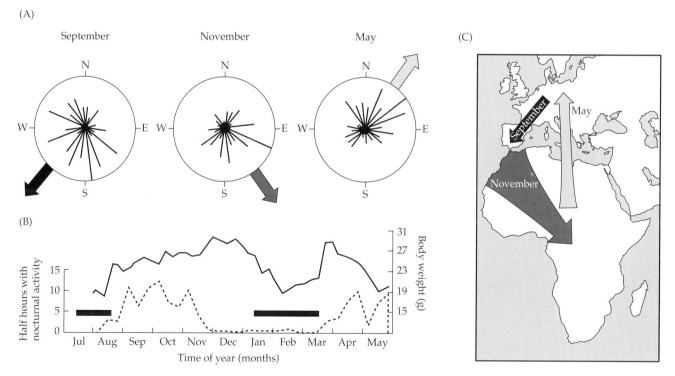

FIGURE 4.7 Migratory behavior of garden warblers, *Sylvia borin.* (A) Spontaneous seasonal changes of directional preferences during nocturnal restlessness for birds kept in a constant LD 12:12 photic regime. Testing was repeated in circular orientation cages with no view of sky but exposed to Earth's natural magnetic field. Circular vector diagrams summarize results in August through September (left), October through December (middle), and April through June (right) of the following year. Data are plotted on a relative scale, with the radius equal to the greatest amount of activity in any one 150-degree sector. Large arrows at the diagram periphery show directions of mean vectors calculated for each test series. (B) Variations in nocturnal activity (dashed line) and body weight (solid line) plus occurrence of molt (black bars) in one representative bird. (C) Approximate annual changes in migratory direction for free-living garden warblers. (From Gwinner and Wiltschko, 1980.)

variable effect of SCN ablation was not attributable to differences in lesion size or location. Because most SCN-lesioned squirrels showed no signs of coherent circadian organization and yet were able to generate Type II annual rhythms, the circannual and circadian mechanisms must be quite separate.

The SCN remains relatively active during deep hibernation. Its neural firing rates are reduced much less than for other subcortical structures. The ablation experiments suggest that the SCN is not a master circannual oscillator but may help synchronize several true circannual rhythms. The strength of coupling among non-SCN oscillators may determine whether or not an SCN-lesioned squirrel continues to show coherent circannual rhythms.

Destruction of several other brain and endocrine structures failed to prevent normal expression of circannual rhythms. Ablation sites in the hypothalamus included the paraventricular, ventromedial, anterior, and lateral hypothalamic nuclei. Lesioning of the medial preoptic nucleus, olfactory bulbs, and septal area had no effect. Deletion of gonads and pineal gland also did not eliminate circannual rhythmicity. Removal of the thyroid gland blocks normal expression of circannual reproductive cycles of sheep and deer. Thyroidectomy probably does not arrest the rhythm-generating mechanism itself, but it may affect coupling between the pacemaker and neuroendocrine enablers that regulate reproduction. Studies involving ablation of brain and endocrine structures in mammals suggest some degree of independence of different rhythms. On the other hand, they do not define the level at which this independence occurs. The separation may be downstream from a common pacemaker.

In the absence of anatomical identification of a discrete circannual pacemaker, the existence of such a structure has been questioned. An alternative view suggests that various tissues interact to form a circannual feedback loop analogous to the estrous cycle mechanism of rodents. The brain, anterior pituitary gland, and ovaries interact via coordinated feedback interplay to produce the sequence of steps that constitutes the ovulatory cycle. In a similar model of the circannual system, some components may be located outside the brain, such as in the thyroid gland, and others may be neural and coupled to functions that exhibit overt rhythmicity. The existence of multiple distinct circannual rhythms in individual animals requires a degree of coordination ordinarily associated with CNS pacemakers, but it does not necessarily contradict the alternative model described here.

Single- or multioscillator circannual pacemakers may exist[26,75]

Successive seasonal events such as reproduction, molt, or migration occur in a regular seasonal progression even under constant conditions for most animals that have endogenous circannual rhythms. The data are compatible with one single basic pacemaking system. However, the possibility of multiple circannual pacemakers is raised by observations that different functions may persist for varying periods of time in individuals held in constant environmental conditions.

In blackcaps held in 10, 12, or 16 h photoperiods, a circannual rhythm of migratory restlessness and molt persisted for up to three cycles, but body weight cycles were absent. Independent control of different circannual functions is even more strongly suggested by some studies in which different circannual functions gradually changed their internal phase relationships to each other. When garden warblers were maintained in constant photoperiods, their cycles of molt drifted completely out of phase with the circannual cycles of gonadal size. Definitive proof for the existence of multiple circannual pacemakers would require a demonstration that different rhythms free-run with different periods for several cycles. Such experiments have not yet been carried out.

Neuroendocrine enablers are important in annual reproductive rhythms[3,7,12,37,47,48]

Several neuroendocrine mechanisms have been implicated in the regulation of seasonal changes in physiology and behavior of animals. Some of these mechanisms are directly involved in annual timekeeping, particularly through the photoperiodic mechanism that will be discussed later in this chapter. Other neuroendocrine pathways are vital for the regulation of particular overt seasonal responses, though they are not directly involved in actual timekeeping. Mechanisms of the latter category are called neuroendocrine enablers. They link the biological pacemakers with the physiological and behavioral functions whose rhythmic output depends on that machinery. Such enablers can be thought of as the gear chain connecting the seasonal clock mechanism to its outputs. Enablers have been studied most intensively in birds and mammals. The enablers that are discussed here are essentially identical for Type I and Type II annual cycles. Therefore, examples from both types are used.

Gonadotropin-releasing hormone (GnRH) is a peptide hormone secreted by the brain. It is required for stimulation of the secretion of pituitary reproductive hormones and subsequent activation of the reproductive system in both sexes. An oscillator of the vertebrate hypothalamus regulates the release of periodic pulses of GnRH into the portal blood shunt from the hypothalamus to the pituitary gland. Discharge frequency can vary from 1 to 48 pulses per day, depending on the species and physiological condition. In sheep, GnRH pulse frequency is closely tied to seasonal reproduction, with high-frequency pulses yielding reproductive induction and low-frequency pulses failing to do so.

Both photoperiodic input and circannual function profoundly influence the activity of the GnRH pulse-generating mechanism. The GnRH pulses stimulate the secretion of follicle-stimulating hormone and luteinizing hormone (LH) from the anterior pituitary gland (see Figure 4.8). These two hormones, collectively designated as the gonadotropins, stimulate the gonads in both sexes, leading to both the production of the gonadal steroid sex hormones and gametogenesis.

Typically, gonadal steroid hormones such as testosterone and estradiol feed back in a negative fashion to reduce the rate of secretion of the pituitary gonadotropins. Many species show a marked seasonal variation in sensitivity to this negative feedback action of gonadal steroids. In the male Syrian hamster and the Ile-de-France ram, for example, the negative feedback effects of testosterone on gonadotropin secretion are greatly intensified during the nonbreeding season.

A striking seasonal variation in GnRH secretion was evident in Suffolk ewes treated with small, constant-release implants containing estradiol (Figure 4.8). GnRH pulse frequency was far lower during the anestrous season than during the breeding season, reflecting enhanced responsiveness to the negative feedback action of the fixed concentration of estradiol during anestrus. The intensified negative feedback action of gonadal steroid hormones during the nonbreeding season leads to reduced secretion of pituitary gonadotropins and the subsequent failure of gametogenesis. In sheep and many other species, gonadectomy at any time of year disrupts this negative feedback loop and leads to increased release of the pituitary gonadotropins.

In some species, secretion of the pituitary gonadotropic hormones is markedly reduced during the nonbreeding season even in gonadectomized individuals. Hence pituitary reproductive hormone activity is seasonally reduced in these animals even in the absence of the typical negative feedback actions of gonadal hormones. The snowshoe hare, pony mare, female golden-mantled ground squirrel, and a variety of birds all show such a steroid-independent, seasonal decrease in gonadotropin secretion.

GnRH also enables seasonal reproductive rhythms in birds. The hypothalamic GnRH content of several avian species changes markedly in association with seasonal transitions between reproductive activity and reproductive quiescence. Whereas nonbreeding birds have low concentrations of GnRH, reproductively active birds exhibit much higher levels. An increase in GnRH content in the cell bodies of the preoptic–septal GnRH system and in the terminals in the median eminence is the first readily identifiable change in the brain associated with the onset of the reproductive season. Similarly, in most birds the termination of breeding is heralded by a decline in the hypothalamic content of GnRH just prior to gonadal regression. In several birds and mammals, the ability of the pituitary gland to respond to

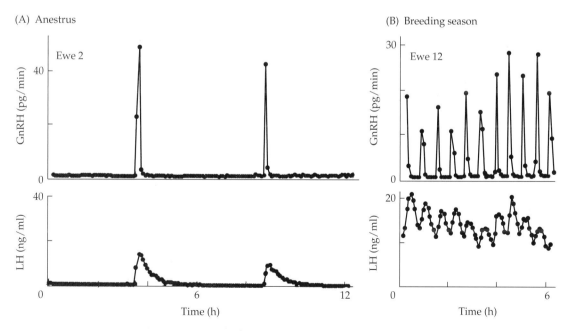

FIGURE 4.8 Hormonal titers from representative ovariectomized ewes with an estradiol implant to maintain physiological concentration of circulating estradiol. (A) Anestrous levels of GnRH in pituitary portal blood (above) and LH in peripheral blood (below). (B) Breeding-season levels of GnRH (above), and LH (below), determined from blood samples obtained at 6-minute intervals for 7 to 12 h. (From Karsch et al., 1993.)

stimulation by GnRH does not change markedly on a seasonal basis. Thus the seasonal variations in activity of the reproductive axis appear to be based primarily on changes in the patterns and/or amounts of GnRH release.

The thyroid gland may serve as a general hormonal enabler for some circannual reproductive rhythms. The requirement of the thyroid for the onset of seasonal gonadal regression was first demonstrated in starlings in the 1940s, and similar findings have subsequently been reported for other bird species. Exogenous thyroxin can block the onset of breeding in starlings held for long term on short days, including the associated changes in hypothalamic GnRH content. Similarly, in reproductively active birds, exogenous thyroxin can induce a decline in GnRH content characteristic of the nonbreeding state.

These observations should be interpreted with caution: most of the data suggesting a role of thyroid hormones in avian seasonality stem from experiments with birds that were thyroidectomized and hence severely hypothyroid or with birds that were possibly hyperthyroid following treatment with thyroid hormone. No data exist to indicate that seasonal changes in thyroid activity are required to drive the annual cycle of reproduction. Rather, low levels of thyroid hormones may simply be required to permit progression of the annual rhythm.

Similar effects of thyroidectomy have been described in various mammals, including mink, Syrian hamsters, sheep, and red deer. In mammals with Type II rhythms, the involvement of thyroid hormones has been best documented in sheep. In both the ewe and the ram, a normal seasonal reproductive cycle is not evident after thyroidectomy. Instead, breeding persists year-round.

Although the mechanism by which thyroid hormones affect expression of the circannual reproductive rhythm is not well understood, thyroid hormones must be present for development of the enhanced response to the negative feedback action of estradiol that leads to anestrus in ewes. The concentrations of circulating thyroid hormones vary during the course of the year, but such changes in themselves are not necessary to terminate the breeding season. Of great importance, however, is the time of year that thyroid hormones act. A limited window of time appears to surround the transition to anestrus, when thyroid hormones exert their action in the ewe.

The study of seasonal enablers sometimes leads to discovery of hormone functions[5,53]

Unsuspected functions of particular hormones have been revealed in some cases by examination of seasonal phenomena. Studies in mammals that undergo seasonal pelage molts in spring and fall suggest prolactin

as an important hormone in determining the type of pelage that grows following a molt. Whereas summer fur is produced under the influence of elevated prolactin concentrations, winter pelage is produced when prolactin is low or absent. This finding is particularly interesting in view of prolactin's well-known role in lactogenesis of mammals. The mammary glands may be related, in an evolutionary and developmental sense, to sweat glands. The role of prolactin in influencing the activity of mammary gland and hair follicles may reflect an ancient function of this hormone in the regulation of integumentary structures.

Entry into torpor in several mammalian hibernators will not occur in the presence of elevated blood testosterone. For these species, testicular regression may be a requirement for expression of the annual cycle of winter torpor. Many mammals that show seasonal reproductive rhythms, particularly those that engage in some form of winter torpor, display marked annual variations in body weight and lipid metabolism. Some of these species are providing excellent models for the investigation of neuroendocrine factors, which are important in the regulation of food intake and lipid metabolism. Recent observations suggesting seasonal variations in responsiveness of the immune system also hold promise for the development of new model systems for studying immune responses.

The Study of Photoperiodic Time Measurement in Plants and Animals Began Early in the Twentieth Century[10,18,55]

The role of photoperiodism in annual cycles of organisms was first noted in plants[8,18]

The seasonal progression of day lengths is the environmental cue most frequently employed to regulate the timing of annual cycles in both true circannual and other annually regulated organisms. Study of the photoperiodic mechanisms of a wide variety of species has revealed a close relation between circadian systems and the ability of both categories of organisms to measure day length accurately. The remainder of this chapter is devoted to mechanisms that underlie photoperiodism and the significance of photoperiodism for both groups.

In 1920 Garner and Allard first noted that the most important environmental factor controlling annual flowering in plants was photoperiod. Similar phenomena were soon described for insects, birds, and mammals. Photoperiodism was thought to be the discrimination between day lengths, or daylight phases of different durations. Photoperiodic time measurement was believed to involve the determination of day length, night length, or a combination of the two. Early workers in the field assumed that photoperiods were measured by some sort of interval timer or hourglass mechanism. Details of a possible mechanism, however, were not clearly addressed. The idea of an hourglass timer survives today. Except for some insects, however, the hourglass concept for photoperiodic time measurement has been largely abandoned in favor of an oscillatory, circadian-based mechanism.

Bünning first used resonance experiments to test a concrete photoperiodic hypothesis[8,32]

The concept of a circadian-based photoperiodic clock has its origins in the writings of the eminent plant physiologist Erwin Bünning. He proposed that the measurement of photoperiods depended on an endogenous diurnal rhythm. His entirely theoretical notion portrayed the free-running circadian period as two half-cycles differing in sensitivity to light. The first 12 h constituted a photophil, or light-requiring half-cycle, and the second 12 h a scotophil, or dark-requiring half-cycle. These two half-cycles were roughly comparable to the current terms *subjective day* and *subjective night*, respectively. Short-day effects might be produced when light was restricted to the photophil and long-day effects when it extended into the scotophil (Figure 4.9). Bünning's hypothesis was based, therefore, on an oscillatory mechanism rather than an hourglass mechanism.

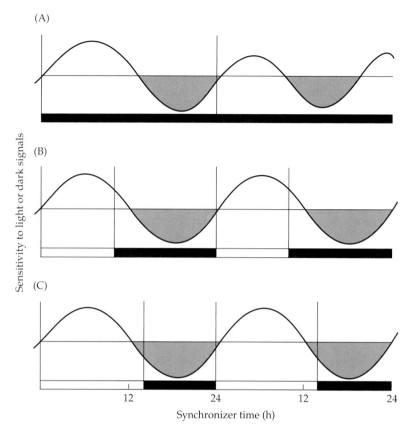

FIGURE 4.9 Bünning's 1936 model for photoperiodic time measurement by circadian oscillation. (A) Free-running oscillation in DD. The unshaded portion of the curve represents the photophil (subjective day), the shaded portion the scotophil (subjective night). (B) Short day with no light in the scotophil phase. (C) Long days with light extending into the scotophil phase. Solid and open bars represent dark and light periods, respectively. (From Bünning, 1960.)

Bünning's theory drew attention to the two probable functions of light in photoperiodic time measurement: entrainment of constituent circadian oscillators and interaction with a specific circadian phase to induce seasonal responses based on day length.

Circadian involvement in photoperiodic time measurement was first revealed by exposure of organisms to non-24 h light–dark cycles. In the 1940s and 1950s, investigators explored the circadian basis of photoperiodism by subjecting plants to light–dark cycles longer than 24 h. In one set of experiments, now called the Bünsow protocol, plants were maintained in 48 h or 72 h cycles with a greatly extended night that was systematically interrupted by a light pulse.

The second type of experiment was introduced by the plant physiologists K. K. Nanda and K. C. Hamner, who named their design resonance experiments. Different experimental subsets were subjected to light–dark cycles that contained the same short light pulse, but with a range of dark intervals to give overall cycle lengths up to 72 h. In both protocols, light illuminates a different circadian phase in each group and there-

by leads to either a short-day or a long-day response (Figure 4.10). These two basic types of experiments were subsequently applied to birds, insects, and mammals with the same general results, indicating that circadian rhythmicity is involved in day length measurement.

Pittendrigh formulated the circadian hypotheses of internal and external coincidence[14,51,58]

At the pivotal Cold Spring Harbor Symposium on Biological Clocks in 1960, Bünning's ideas finally reached the non-German and nonbotanical world. The circadian basis of photoperiodism was then championed by Colin Pittendrigh. In a succession of papers, Pittendrigh and his colleagues developed the external coincidence and internal coincidence models. The external coincidence hypothesis proposed a temporal interaction between a photosensitive circadian phase and an external light stimulus, much as in Bünning's model. Light falling during the photosensitive phase would have a photoinductive action, leading to the expression of seasonal responses normally associated with long days.

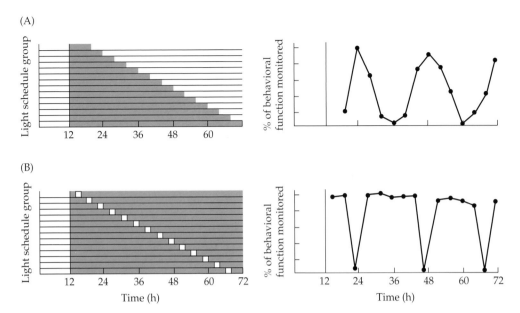

FIGURE 4.10 Schematic representation of two non-24 h light protocols for photoperiodism research. (A) Nanda–Hamner resonance experiment. (Left) Each subset schedule here is exposed to a light cycle such as 12 h L plus a variable time in D. The L+D cycle is repeated throughout the experiment. (Right) Schematic representation of a positive Nanda–Hamner resonance experiment showing peaks of short-day effects at about 24 h intervals as D is extended. (B) Bünsow-style light-break experiment. (Left) Each subset here is exposed to 12 h L followed by 60 h D interrupted by a 1 or 2 h supplementary light pulse that occurs at later and later times in sequential subsets such that, in this aggregate, the light scans all phases of the cycle. (Right) Schematic representation of a positive Bünsow experiment showing periodic long-day effects at about 24 h intervals as dark is systematically scanned by the light pulse. (Courtesy of D. S. Saunders.)

Internal coincidence was a theoretical alternative based on evidence that the circadian systems of higher organisms could be complexes of multiple circadian oscillators, some of which might be coupled to dawn and others to dusk. Consequently, as photoperiod changed with season, the phase relationship between these subsystems would also change. In some photoperiods, for example, biochemically interactive phases might coincide, whereas in others they might not. The internal coincidence model received its name from the suggestion that it involved the internal coincidence of two or more oscillators, rather than just one oscillator with an external signal, namely light. It also differed from external coincidence in that the only role of light in the internal model was photoentrainment.

The external coincidence model was more conservative, in that it postulated the use of only one rather than two oscillators. In contrast, the internal coincidence model was more conservative with respect to the actions of light because it required only the well-established circadian phase-shifting, or entraining, effect but not an additional photoinductive action of light.

Pittendrigh's second major contribution to the study of photoperiodism stemmed from several experiments based on his own studies of the entrainment phenomenon. He showed how phase response curves might be used to predict steady-state phase relationships between a photoinducible phase and the light cycle and to make predictions about the measurement of day length or night length within the framework of the external coincidence hypothesis. In the last 20 years these investigations have been extended to numerous species with diverse and sometimes conflicting results. As a result, several models for photoperiodic timing have been proposed, particularly for insects. In nearly every case, however, Bünning's and Pittendrigh's theoretical and experimental approaches aid analysis. They will undoubtedly continue to bear fruit until the concrete nature of photoperiodism is uncovered.

Plants Use Photoperiodic Timing Extensively[65]

Several photoperiodic responses are known in plants[18,42,64]

In plants, typical responses to the shortening days of autumn include induction of cold hardiness and bud dormancy. The most obvious manifestation of photoperiodism in plants, however, is a seasonal rhythm in reproduction (see Plate 2, Photo A). The induction of flowering provides the best-understood example of photoperiodism in plants. Scientists noted early in the twentieth century that short days and long nights induced flowering in hops and marijuana.

Garner and Allard refined the link between length of the day or night and flowering. They introduced the term *photoperiodism* and showed that flowering depended on photoperiod in some species but not in others. In a short-day plant, flowering is induced by a sufficiently long dark period that exceeds a critical night length. In a long-day plant, flowering is induced only when the day length exceeds a critical day length. By analogy to better-known animal circadian pacemakers, the photoperiodic system presumably involves a clock, a photoreceptor, and an output signal.

The site of photoperiodic perception is in the leaves[64,65]

The bulk of cell proliferation in plants occurs at the apex of the shoot or root in clusters of actively dividing cells called apical meristems. Apical meristem in shoots of vegetatively reproducing daughter cells differentiates into leaves, stems, and secondary vegetative meristems. Flowering requires a change from vegetative to reproductive activity in the shoot apical meristem, with daughter cells differentiating into floral organs or into secondary reproductive meristems. The site of perception of the photoperiodic stimulus was initially assumed to lie within the targeted shoot apical meristem. However, a series of extensive experiments indicated that the leaves were the site of photoperiodic perception.

Exposure of the shoot to an inductive photoperiod and the leaves to a noninductive photoperiod failed to induce flowering. In contrast, exposure of the shoot apical meristem to a noninductive photoperiod and the leaves to an inductive photoperiod was sufficient to induce flowering. Grafting experiments showed that a diffusible signal called florigen was generated in the leaves and then traveled to the shoot apical meristem via phloem vascular connections.

The nature of florigen remains elusive. It may be a complex mixture that includes gibberellins, a hormone class termed cytokinins, sucrose, polyamines, and possibly RNA molecules. Many molecules promote flowering in certain plant species. For example, the plant hormone gibberellic acid brings about flowering of many species, including the widely studied plant photoperiodism model *Arabidopsis thaliana*. However, no single molecule has been shown to support flowering in all plants.

Multiple photoreceptors contribute to photoperiodic light perception[65]

Phytochromes have long been implicated as photoreceptors in plants. This conclusion is based in part on action spectra for plant functions (see Chapter 6), including the promotion of flowering or tuberization in potatoes. Each of these pigments is composed of an apopro-tein that is covalently linked to a tetrapyrrole chromophore. Light absorption results in a conformational change of the pigment molecule to yield an active form. Phytochromes are synthesized in a red-absorbing Pr form, with an absorption maximum near 660 nm. Pr is converted to a biologically active Pfr form by the absorption of red light.

Red light is effective in Bünsow night-break experiments. However, if the red light pulse is quickly followed by a pulse of far-red light, the active Pfr form, with an absorption maximum near 700 nm, is converted back to the inactive Pr form. Thus a second pulse of far-red light will prevent the induction in long-day plants, or the repression in short-day plants, of flowering by the initial red light pulse. This photoreversibility is the hallmark of a phytochrome-mediated response.

Genetic analysis has identified genes important in the photoperiodic induction of flowering[63]

Biologists have identified more than 80 genes that are important in the flowering process. Some of these appear to encode components of the circadian oscillator. A genetic basis for flowering provides additional evidence for a central role of the circadian clock in plant photoperiodism. A second group of genes functions in the light signal transduction pathway leading to the clock, including genes that encode photoreceptor pigments. Particularly interesting is the *CONSTANS* (*CO*) gene, which does not fit in either of the two preceding classes.

The CO protein was found to provide a molecular link between the circadian clock and the photoperiodic induction of flowering in *Arabidopsis*. Mutations that abolish CO function lead to a failure to accelerate flowering in response to long days, and overexpression of CO leads to rapid flowering. Neither type of mutation affects circadian rhythmicity per se. These results suggest that CO does not function either in the circadian oscillator or in the input pathway to the clock. However, expression of *CO* itself is regulated by the circadian clock.

The *CO* gene must lie on an output pathway from the clock. In short days, *CO* mRNA begins to accumulate after dusk and declines before dawn. In contrast, in long days *CO* mRNA accumulates before dusk and persists into the morning light phase (Figure 4.11). Only in long days is *CO* mRNA abundant during part of the light period. Furthermore, the biological activity of the CO protein is increased by exposure to light, possibly as a result of a posttranslational modification such as phosphorylation.

At any rate, CO activity will be elevated in long days but not short days because light coincides with elevated levels of CO only in the former case. As a consequence, preferential induction of the downstream flowering pathway will occur in long days. These observations suggest that CO links the circadian clock with

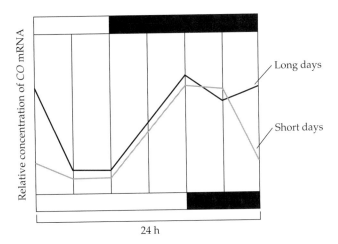

FIGURE 4.11 Circadian oscillation of *CO* mRNA. Light schedule bars at the top show short day (gray line) and at the bottom, long day (black line). Solid and open bars represent dark and light periods, respectively. For further explanation, see the text. (From Suárez-López et al., 2001.)

the photoperiodic induction of flowering, and they also support an internal coincidence model for photoperiodic time measurement in plants.

Photoperiodism Is Widespread in Insects[46,58,69]

Diapause is often timed by photoperiod in insects[46,57,58]

Changes in photoperiod regulate a wide variety of seasonally appropriate behavior patterns in insects. The most obvious is the overwintering cessation of either development or reproduction. The period of dormancy is a phenomenon called diapause. Species at higher latitudes enter diapause in response to short autumn days. The dormant stage restricts active stages of the life cycle to favorable seasons in spring and summer when food is abundant. Dormancy also brings about a mutual developmental synchrony within a population upon reactivation in the spring, and thereby maximizes reproductive success.

Insects have complex life cycles (see Figure 5.5 in Chapter 5). Eggs hatch into feeding larvae that pass through successive stages, called instars, each separated by a molt. In primitive insects such as cockroaches and crickets, metamorphosis is slight, with a gradual change from nymph to adult. In advanced insects such as moths and flies, wormlike larvae undergo an abrupt metamorphosis to the adult stage involving extensive reorganization of the body within a pupal case.

Insects enter diapause at one specific developmental stage of the life cycle. Depending on the species, that stage may occur in the larva, nymph, pupa, or adult, or in an embryo within the egg. Diapause is always preceded by

a stage sensitive to photoperiod. Photosensitivity often occurs at a much earlier stage in development, or even in the maternal generation. Variation in this respect can occur within a single group such as flies. Eggs of the flesh fly are retained within a uterus until birth of the first-stage larvae. Sensitivity to photoperiod reaches a peak in the intrauterine embryos and in the newly deposited larvae, and a subsequent intense diapause occurs in the pupal stage (Figure 4.12A).

In contrast, photoperiodic sensitivity in the blowfly is almost entirely maternal, although in some strains it extends into the eggs and larvae of the next generation (Figure 4.12B). Short days experienced by an adult female blowfly influence the type of eggs produced, which give rise to larvae that exhibit a relatively shallow diapause. On the other hand, eggs laid by a long-day female produce larvae that develop without arrest.

In the fruit fly, *Drosophila melanogaster*, another pattern is seen. When short days and a temperature below about 14°C occur shortly after emergence from the pupa, the adult female enters a shallow reproductive diapause in which the ovaries remain small, with no yolk in the oocytes. Female flies emerging into long days at the same temperature commence a slow cycle of egg maturation (Figure 4.12C).

Insects can be exposed to a range of experimental photoperiods to develop a photoperiodic response curve[58,68,69]

Exposure of a particular species to a range of experimental daily photic schedules results in varying percentages of diapause induction. The determination of the proportions entering diapause provides a photoperiodic response curve (Figure 4.13). The curves show a wide variety of shapes, but in all cases they are dominated by the ecologically important critical photoperiod. In long-day species, the critical photoperiod separates diapause-inducing short days from the long days that allow continuous, nondiapause development. Short-day insects, such as some strains of the commercial silk moth, may show the opposite response, with diapause occurring only in longer days.

Critical day length is generally related to the latitude inhabited by a species and to the range of environmental temperatures encountered during the annual cycle. Even within a single species that occurs over a wide north–south range, the critical day length may vary with latitude. In the knotgrass moth, for example, populations at 60 degrees north in Russia show a critical day length of about 19.5 h, whereas those from the southern part of its range, at 43 degrees north, have a critical day length close to 14.5 h (see Figure 4.13A). These responses are probably ecologically appropriate because insects at more northerly latitudes experience an early onset of harsh winter conditions while day lengths are still relatively long.

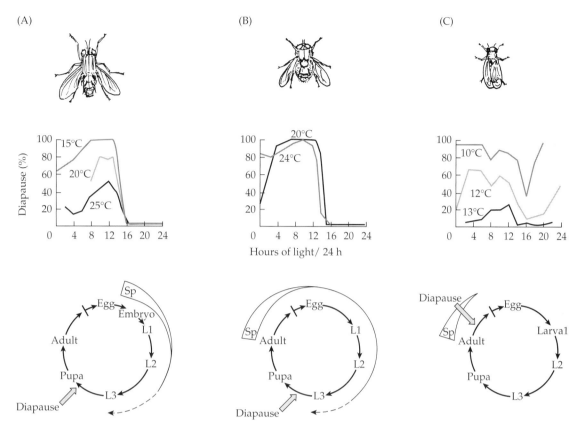

FIGURE 4.12 Photoperiodic responses and sensitive periods in three species of flies. (Top) The effects of temperature on the photoperiodic response curves. (Bottom) Sensitive periods in relation to the life cycle. (A) In flesh flies, photoperiodic sensitivity (Sp) begins in the intrauterine embryo and continues through larval development to regulate an intense pupal diapause. (B) In blowflies, the maternal adults are max-

imally sensitive, although photoperiodic sensitivity extends into the egg and larva; a relatively shallow diapause then occurs in the postfed larva. (C) In *Drosophila melanogaster*, the newly eclosed adult fly is sensitive to photoperiod, and the shallow ovarian diapause occurs in the same instar. (Courtesy of D. S. Saunders.)

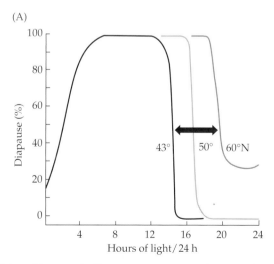

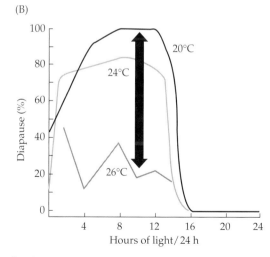

FIGURE 4.13 Photoperiodic response curves, showing the dependence of diapause on latitude (A) and temperature (B). (A) The moth *Acronicta rumicis* shows an increase in the critical day length in more northerly populations. (B) The blow-

fly, showing a reduction of diapause incidence and a shortening of the critical day length at higher temperature. (A from Danilevskii, 1965; B from Kenny and Saunders, 1991.)

Temperature sometimes directly affects the photoperiodic response (see Figure 4.13B). Some species, such as the flesh fly (see Figure 4.12A) show a critical day length that appears to be conserved over a range of temperatures. Others, such as the fruit flies *Drosophila auraria* and *D. melanogaster* (see Figure 4.12C) and the blowfly (see Figure 4.12B) are sensitive to both temperature and photoperiod. In all long-day species, however, diapause incidence under short days is increased at lower temperatures, and very high temperatures frequently override the whole response. In still other cases, a daily thermoperiod may regulate diapause incidence in the complete absence of light. In such cases temperature presumably acts as an entraining agent in its own right.

Insects require exposure to several short or long days to reach a threshold for subsequent diapause[58,69]

Insects need several inductive days before they are programmed to enter diapause. In many species, a temperature-compensated mechanism called the photoperiodic counter counts the environmental cycles. In flesh fly larvae, for example, the number of short-day cycles needed to program the pupae is temperature compensated, whereas the duration of the sensitive period of larval development is not. The interaction between the required day number and the sensitive period accounts for much of the effect of temperature on diapause incidence. For example, at 26°C the larval sensitive period is shorter than the required day number and diapause is not observed; at 18°C, the larval sensitive period is longer than the required day number, and all pupae enter diapause. At intermediate temperatures of 22 to 24°C, the most rapid developers proceed without arrest, but the slowest in the same cohort become dormant.

Populations of the mosquito *Aedes atropalpus* from Ontario, Canada (45 degrees north) needed only 4 short-day cycles, whereas those from Georgia (34 degrees north) and El Salvador (14 degrees north) needed 7 and 9, respectively. Interaction with temperature effects on the sensitive period ensured that larval development lasted less than 9 days in the El Salvador mosquitoes and that no diapause occurred in that population. In contrast, in Canada all the larvae entered diapause. An intermediate situation was found in Georgia, where some larvae entered a diapause stage and others did not. Latitudinal distribution and environmental temperatures, therefore, are related to species differences in the photoperiodic counter as well as the photoperiodic response curve.

Insect photoreceptors and clocks are often located in the brain[58,69]

Organized photoreceptors, such as the compound eyes, may be used in photoperiodism in relatively primitive insects. In contrast, extraretinal photoreception may be the rule in more advanced groups, such as moths and flies, in which tissues are extensively reorganized during metamorphosis. A similar generalization holds for the entrainment of overt circadian rhythmicity.

Several techniques have been used to determine the anatomical location of extraoptic photoreceptors in insects. Working with the green vetch aphid, A. D. Lees used very fine microilluminators attached to various parts of the head and body to localize the site for photoperiodic photoreception. This aphid produces successive generations of viviparous, parthenogenetic daughters called virginoparae during the long days of summer, but it switches to the production of egg-laying, sexual daughters called oviparae as days shorten in the fall.

The insects were held in a short-day photoregime of LD 14:10 and provided daily with a 2 h supplementary light pulse. If the aphids responded to the full 16 h of light per day they produced virginoparous summer progeny. With less light, they produced autumnal oviparae. Long-day illumination delivered to the dorsal midline of the head was the most effective, suggesting direct light perception by the brain. Light directed laterally through the eyes was ineffective.

Brain transplantation experiments have been used to identify both photoreceptor and clock components in the brain of silk moth pupae of *Antheraea pernyi*. Brains disconnected from all nerve connections could still respond to photoperiod. They were capable of regulating the appropriate diapause or nondiapause pathways, whether transplanted to the tip of the abdomen or replaced in the head. Unequivocal evidence that both components lie within the brain and associated retrocerebral complex has been obtained by in vitro experiments.

Brains were first removed from fifth-instar larvae of the tobacco hornworm moth, *Manduca sexta*, and maintained in vitro for 3 days in either a long day of LD 18:6 or an LD 12:12 short-day regime. The brains were then implanted into the abdomens of otherwise diapause-destined larvae. Whereas recipients of the short-day implants molted to diapausing pupae, a significant proportion of those receiving long-day implants proceeded to the adult stage through a nondiapausing pupa. A similar result was subsequently obtained for another species, the silk moth.

Insect photoreceptors are probably based on carotenoids[58,69]

One of the chief methods for determining potential photoreceptors is to measure an action spectrum for the pigment based on behavioral data (see Chapter 6). In green vetch aphids, the effect of photoperiod on production of summer- or autumn-type morphs was used to investigate spectral sensitivity of the photoperiodic system. A night-interrupting monochromatic light pulse was used to convert short-day responses to long-day responses, and the irradiance of the pulse was varied to generate an action spectrum. Maximum sensitivity was in the blue-green region of the spectrum at about 450 to 470 nm. The majority of other species tested showed a similar peak of sensitivity in the blue-green part of the spectrum.

A blue-green sensitivity indicates a pigment that absorbs light preferentially in that part of the spectrum, such as one based on carotenoids. Carotenoid-based photopigments like rhodopsin are widely involved in photoreceptors, including the insect eye. In several insects, dietary carotenoids were shown to be essential for normal photoperiodic adjustments. After rearing on a carotenoid-free diet, photoperiodic responses of *Pieris* and *Apanteles*, for example, were restored following the addition of dietary vitamin A.

A central problem in photoperiodism of insects is the nature of photoperiod time measurement[68,69]

Many experiments have addressed the issue of which parameter a photoperiodic clock measures: day length, night length, or another environmental feature. Another important question is whether time measurement is regulated by a nonoscillatory hourglass or by a circadian system. With the exception of the linden bug and a few others, the photoperiodic clock in insects measures night length rather than day length. The experimental evidence favoring night length measurement comes from investigations in which the light component and the dark component of the daily cycle were varied independently. In the flesh fly, for example, a long night lasting 12 h always induces a high incidence of pupal diapause. It doesn't matter whether the night is coupled to a short day, as in LD 12:12, or to a long day as in LD 18:12. In contrast, a short night has no diapause-inducing effect in either LD 12:6 or 18:6.

With a few possible exceptions, night length measurement is based on circadian rhythmicity. For several decades the literature on insect photoperiodism has discussed whether time measurement is carried out by a nonoscillatory hourglass mechanism or by a circadian-based timer. Experimental evidence that distinguishes between these hypothetical mechanisms has been obtained through use of the Nanda–Hamner and Bünsow protocols. So-called positive responses indicative of circadian involvement are found in many species, including the parasitic wasp *Nasonia vitripennis*, the flesh fly, and the spider mite, as well as various other species of flies, butterflies, and beetles. On the other hand, negative responses suggesting an hourglass involvement have been recorded for several moths and for the green vetch aphid.

The outcome of a Nanda–Hamner experiment, and therefore its interpretation, may be affected by various environmental or developmental factors. For example, in the flesh fly, Nanda–Hamner experiments conducted

at 20 to 22°C gave positive resonance, but at a lower temperature (16°C) the response resembled that of an hourglass. The ground beetle showed positive resonance in a German strain from 51 degrees north but apparently negative, or hourglasslike, responses in a population from Swedish Lappland, located at 64 to 66 degrees north. Experiments with the cabbage butterfly suggested a circadian involvement for diapause induction but not for its termination. With the cabbage butterfly, a change from a natural to an artificial diet apparently changed an oscillator clock to an hourglass. Changes in temperature or diet probably could not have such a fundamental effect on the nature of the photoperiodic time measurement mechanism. More likely the environmental factors affect the rate of oscillator damping, and extremely damped clocks may take on the characteristics of an hourglass mechanism.

In the flesh fly, autumnal long nights give rise to overwintering diapause pupae, but the short nights of summer lead to nondiapause pupae that produce adult flies without delay. The flesh fly has been exposed to a wide range of experimental light cycles in investigations of the circadian basis for photoperiodic time measurement. Nanda–Hamner- and Bünsow-style experiments gave positive responses, suggesting a circadian involvement. "T-experiments" in this species also yielded results in support of circadian involvement in photoperiodic time measurement.

When the free-running period of a circadian oscillator becomes entrained to a light–dark cycle of period T, the light pulse must lie on that part of the phase response curve that generates a phase shift equal to the difference between T and the free-running period. Therefore, when T is less than the free-running period, the light pulse in steady state must fall in the late subjective night to generate a phase advance. When T is greater than the free-running period, the pulse must fall in the early subjective night to generate the necessary phase delay. Systematic alteration of T will therefore cause a short light pulse to fall at different circadian times in various experimental groups. The prediction that a photoinducible phase occurs late in the subjective night can then be tested. The T-cycle paradigm was applied to cultures of flesh fly larvae, and the results were consistent with the external coincidence hypothesis for photoperiodic time measurement.

Photoperiodism Is Becoming More Widely Recognized in the Lower Vertebrates[12,71]

The relation between the circadian system and photoperiodic time measurement has not been studied extensively in cold-blooded vertebrates[66,67]

Following the discovery of photoperiodism in vertebrates in the 1920s, nearly 30 years elapsed before the

Nanda–Hamner protocols originally devised for insect research were used for studies in vertebrates. The results were consistent with the external coincidence model, which proposed that long-day responses such as gonadal growth in long-day breeders occurred when light was coincident with a photoinducible phase of the circadian cycle.

The environmental control of annual cycles in ectothermic (cold-blooded) vertebrates, including fish, amphibians, and reptiles, is a fascinating and complex topic, but these classes have received far less attention than birds and mammals. Metabolic processes in ectothermic vertebrates are coupled to environmental temperature. It is not surprising, therefore, that temperature may interact with photoperiod in the regulation of annual rhythms in these groups. Several selected representative examples will illustrate the relative roles of photoperiod and temperature in the control of annual cycles in ectotherms.

Limited information is available for fish, amphibians, and reptiles[66]

Rainbow trout are fall breeders, in which gonadotropin secretion and spermatogenesis are advanced by exposure to decreasing photoperiods. These responses are much greater at 16°C than at 8°C in some Temperate Zone anuran amphibians. Temperature and rainfall appear to be the principal cues for seasonal timing. In the frog *Rana esculenta*, however, both temperature and photoperiod are important for the stimulation of gonadotropin secretion and testicular activity.

Temperature has been claimed as the single most important factor influencing the timing of breeding for reptiles. In reality, a complex interaction between temperature and photoperiod often occurs. Perhaps the best-known reptile with respect to the environmental control of reproductive cycles is the anole lizard, *Anolis carolinensis*. The testes of the anole regress in late summer. Gonadal growth begins in October, proceeds gradually throughout the winter, and is completed in early spring. Both photoperiod and temperature are involved in the control of this cycle, but the importance of the two stimuli varies with the phase of the cycle. Testicular growth between late autumn and early spring is controlled mainly by temperature, whereas the maintenance and subsequent regression of the testes in late summer are influenced primarily by photoperiod.

Because birds and mammals evolved from reptiles, the study of present-day reptiles may give valuable insights into the evolution of photoperiodic systems in birds and mammals. Hamner and Elliott used Nanda–Hamner lighting protocols to demonstrate that a circadian rhythm is involved in the measurement of day length in house finches and hamsters. In a similar fash-

ion, exposure of male anoles to these protocols also showed that a circadian rhythm is involved in measuring day length. Male anoles were exposed for 21 days to light cycles in which 11 h of light was combined with dark to produce cycles of 24, 36, 60, and 72 h. Only the 36 and 60 h cycles stimulated testicular growth, presumably because on some days the 11 h of light in these schedules coincided with the photoinducible phase of a circadian rhythm of light sensitivity.

A more precise measure of the location and duration of this photoinducible phase was obtained when the male anoles were exposed to light–dark cycles composed of 11 h of light combined with varying durations of dark to produce cycle lengths ranging from 18 to 30 h. The circadian system of the anoles became entrained to these LD cycles, and the phase relation between the circadian system and the light varied systematically as the period of the light cycles lengthened. Light, therefore, scanned different phases of the circadian cycle in the various treatment groups. Testicular growth was stimulated only when the light illuminated the latter half of the subjective night. The data supported the hypothesis that this part of the circadian cycle contains the photoinducible phase of a circadian rhythm of light sensitivity. The results in the anole suggest that the ectothermic ancestors of birds and mammals used a circadian clock to measure day length. This mechanism for photoperiodic time measurement is probably an ancient property of vertebrates.

Some species of ectotherms, including reptiles, show a postbreeding refractoriness to photic or thermal stimulation. The reproductive system regresses under conditions of light or temperature that are normally stimulatory. In this case, refractoriness prevents the animal from producing young at a time of year that may not be conducive to their survival. A period of short days or cold temperatures may be required to break refractoriness and allow the animal to regain sensitivity to stimulatory conditions of light or temperature. In some ectothermic vertebrates that are torpid during the winter, dark and cold conditions may break refractoriness.

In other species, including some Temperate Zone lizards, the gonads may grow spontaneously after a certain number of days under constant conditions, whether in a natural torpor site or in the laboratory, suggesting an endogenous ability to measure time on a seasonal scale. The ability is often incorrectly assumed to represent a component of a circannual rhythm. A true demonstration of circannual rhythmicity, however, requires that endogenous mechanisms both generate and terminate a refractory period at yearly intervals. To obtain the required data, animals would need to be held under constant conditions for several years.

Only a few pertinent studies have been performed in ectotherms. In rainbow trout held under constant photoperiodic conditions, the gonads exhibited cycles of growth and regression with periods ranging from 5 months under continuous light to 15 months under unchanging LD cycles. In reptiles, a circannual rhythm in activity has been seen in the iguanid lizard *Sceloporus virgatus*, and a circannual rhythm in reproduction has been observed in the teiid lizard *Cnemidophorus uniparens*.

The natural history of many ectothermic reptiles suggests that they have circannual rhythms as well as circadian involvement in photoperiodic time measurement. Many mid- and high-latitude reptiles overwinter in dens that provide few or no light and temperature cues, yet the animals must be able to measure the passage of time in order to anticipate and prepare for emergence and postemergence breeding. Selection in reptiles may have favored the development of endogenous circannual rhythmicity, and homeotherms may have retained this mechanism. In this sense, present-day reptiles should offer insights into the evolution of circannual rhythmicity.

Photoperiodism Is Vitally Important in Avian Physiology[3,12,74]

Early observations of vertebrate photoperiodism began with birds[3,55]

In 1925, Rowan showed that spring migratory behavior and the associated fattening in dark-eyed juncos are initiated by increasing day length in spring. His experimental birds were held in outdoor cages in Canada and had regressed gonads in November and December. When the short winter days were lengthened by the addition of several hours of artificial light, the birds' gonads grew in the middle of winter, much earlier than in short-day controls.

Rowan's demonstration that an increase in day length could stimulate premature gonadal development was subsequently confirmed by many other investigators. Although this was a discovery of fundamental importance, the precise physiological mechanisms controlling avian seasonal reproductive development and migratory behavior are still not well understood. The annual cycle consists of a series of events such as growth and regression of the gonads, as well as molt, migratory behavior, and premigratory fattening. The general principle that changes in day length influence the timing of these events applies to most, if not all, Temperate Zone avian species.

Photorefractoriness and photoperiod history effects are important concepts in avian photoperiodism[47,48]

At a certain time of the year the avian neuroendocrine system responds to increasing day lengths with a dra-

matic increase in gonadotropin secretion, gonadal growth, and a range of hormone-dependent reproductive processes. The cascade of events leads to reproductive behavior. The ability to respond to long days is called photosensitivity. Eventually birds become unresponsive to the previously stimulatory photoperiods, and long days no longer support reproductive activity.

The portion of the avian annual cycle associated with regression of the reproductive system during long-day photoperiods is called the photorefractory phase. The state in birds resembles the refractoriness to long photoperiods that develops in some reptiles and mammals with true circannual rhythms during extended exposure to long days, but mammals that exhibit Type I seasonal rhythms most commonly become refractory to short photoperiods (see Figure 4.2B).

The rate of gonadal growth and the onset of refractoriness have been studied in detail in Temperate Zone passerine bird species such as the European starling, the white-crowned sparrow, and the tree sparrow. In long photoperiods, greater than 12 h, these Type I birds show a rapid rate of gonadal growth followed by precipitous gonadal regression. When groups of birds were kept in similar conditions of food and temperature, but in different photoperiods, the rate of development of the reproductive system was proportional to the length of the day. The rate of gonadal growth was faster in photosensitive males placed in a photoperiod of LD 18:6 than in males placed in photoperiods of LD 13:11 or 11:13 or left at LD 8:16. The subsequent development of refractoriness also depends on photoperiod. The longer the day length, the earlier the birds initiate gonadal regression. The threshold for the induction of rapid gonadal growth may be lower than for the initiation of refractoriness. Under natural conditions, breeding in these species is terminated prior to the summer solstice.

Short days are required to terminate photorefractoriness and reinstate photosensitivity. Under natural conditions, the short days characteristic of late autumn and early winter render the reproductive system of birds again sensitive to long days. The breaking of refractoriness is reminiscent of the mammalian phenomenon (see Figure 4.2) in which photorefractory individuals regain photosensitivity when exposed to a contrasting photoperiod. In typical Type I mammals, however, exposure to long days is needed to break the refractoriness that develops during prolonged exposure to a short photoperiod. The breaking of long-day refractoriness in birds by exposure to a short photoperiod occurs gradually.

As with photostimulation, a direct relationship exists between the dissipation of refractoriness and the reinstatement of full photosensitivity. The shorter the day, the more rapid the dissipation of refractoriness. One example, the rate of recovery of photosensitivity in castrated male starlings as measured by LH secretion, will clarify the relationship. If starlings are rendered photorefractory on a long photoperiod of LD 18:6 and then placed in short days between LD 8:16 and 12:12, the rate of recovery of photosensitivity is inversely proportional to the day length.

Some avian species exhibit a complex, graded response to day length as the breeding season progresses. Reproductive development occurs in Japanese quail in early spring, but by late spring relatively longer days are required to maintain reproductive competence. At least some decline in day length is necessary for the gonads to begin premature regression at any point during the breeding season. However, a photic regime such as LD 14:10 that is stimulatory early in the breeding season becomes potentially inhibitory later in the summer, after the birds have been exposed to longer day lengths.

If gonadal regression is induced in quail artificially by transferal to a decreased photoperiod during the summer, the birds are not photorefractory because they retain the capacity to reactivate the reproductive system immediately upon renewed exposure to long days. This phenomenon is quite similar to the photoperiod histo-

ry effects that will be described in more detail later, in the section dealing with mammalian photoperiodism.

Circannual rhythms may result from spontaneous alternation between the photosensitive and photorefractory states permitted under certain photoperiods. Specific constant photoperiods allow for the transition from the photosensitive to the photorefractory state and also for a return to the photosensitive phase. For example, European starlings require at least 12 h light per day for the induction of photorefractoriness and 12 h or less for its termination. Starlings go though circannual cycles only if kept in a 12 h photoperiod. Under photoperiods shorter than 12 h, the system becomes arrested in the photosensitive state, but under photoperiods longer than 12 h, the birds are arrested in the photorefractory state (Figure 4.14A). Rhythmicity that has been completely arrested in the refractory state in an LD 13:11 schedule can be restarted by transfer to an LD 12:12 regime (Figure 4.14B).

In the experiment illustrated in Figure 4.14, after transfer to an LD 12:12 schedule, 5 months elapsed before gonadal growth resumed, irrespective of whether the birds had previously been kept for 10, 16, or 20

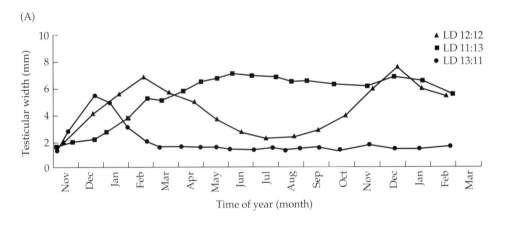

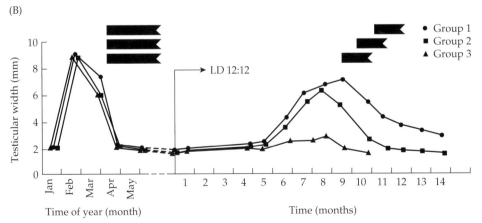

FIGURE 4.14 Photoperiodic regulation of reproduction in male starlings. (A) Changes in testis width in starlings in a constant daily schedule with 11, 12, or 13 h of light for 17 months. (B) Changes in testis width and occurrence of molt (black bars) in three groups of starlings that were initially kept in an LD 13:11 photoperiod. Subsequently they were changed to an LD 12:12 schedule on November 25 of the year in which the experiment was started (Group 1), or on May 17 (Group 2) or September 20 (Group 3) of the following year. (From Hamner, 1971; Schwab, 1971; and Gwinner et al., 1989.)

months in the LD 13:11 light schedule. Only the amplitude of the induced cycle was smaller in the birds previously held for a longer time in the 13 h photoperiod, a yet unexplained phenomenon.

A relationship may exist between the particular photoperiodic conditions restricting the expression of circannual rhythms and the light conditions normally experienced by a particular species. An example is provided by the pied and the collared flycatchers. These two closely related species of songbirds breed in Eurasia. In fall they migrate to Africa to spend the winter either beyond the equator in the Southern Hemisphere (collared flycatcher) or slightly north of it (pied flycatcher).

Consistent with the different photoperiodic environments in winter, these two species have different critical photoperiods for termination of the refractory state. The collared flycatcher is exposed in winter to long Southern Hemisphere photoperiods and is capable of terminating refractoriness under relatively long photoperiods. The pied flycatcher, in contrast, is normally exposed to the much shorter photoperiods of its wintering ground at 10 degrees north and needs short photoperiods to terminate refractoriness. As a result of these differences, the circannual system of the collared flycatcher functions even under photoperiods as long as 14 h. The rhythmicity of the pied flycatcher is arrested in such a photoperiod.

Birds use diverse nonretinal photoreceptors in photoperiodic time measurement [12,50,54,61]

A characteristic feature of photoperiodic time measurement in birds and other nonmammalian vertebrates is that photoreception for mediating the photoperiodic response is accomplished partially by photoreceptors located deep in the septal–hypothalamic region of the brain, rather than exclusively by ocular elements. Studies in several avian species revealed that removal of the eyes does not block the occurrence of photostimulation, the onset of photorefractoriness, or the reinstatement of photosensitivity by short days. However, when the top of the skull is painted black to prevent light from passing through it, the photoperiodic response is blocked.

Presenting focused light to specific areas of the brain with fiber-optic methodology can mimic the effects of external light. Studies in Japanese quail suggest that the photopigment in these encephalic receptors has an action spectrum for response to light resembling that of rhodopsin. The definitive morphological localization of these photoreceptors has not yet been completed, although opsinlike immunoreactivity has been identified in the septum and tuberal hypothalamus of doves.

Recent studies have revealed that birds and all other classes of nonmammalian vertebrates have photoreceptors in the pineal gland, as well as in the brain. These brain and pineal photoreceptors are so-called irradiance detectors rather than image detectors such as the retinal rods and cones. The ventral telencephalon appears to contain photoreceptive cells in a range of vertebrate groups, but other brain sites have been implicated as well. Multiple brain photoreceptor pigments may be used for photoperiodic responses in nonmammalian vertebrates. Recent studies have implicated a pigment called melanopsin in photoreception for the mammalian circadian system. Melanopsin has also been identified in both the SCN (see Chapter 6) and the nucleus preopticus of the clawed frog, *Xenopus*.

The neuroendocrine basis for photoperiodic time measurement in birds is poorly understood [12,35]

The projections of the photoreceptor cells to pacemakers and their relation to the mechanisms mediating the photoperiodic response are not as well understood in birds as they are in mammals. A neural pathway from the retina via the suprachiasmatic nucleus, the paraventricular nucleus, and the superior cervical ganglia to the pineal has been identified in birds. The pathway resembles the anatomically homologous mammalian projection. However, several prominent differences exist between birds and mammals with respect to the function of this neural pathway.

In mammals, the neural input to the pineal is strictly stimulatory to melatonin biosynthesis, whereas in birds it appears to be primarily inhibitory. Mammals rely entirely on neural signals to drive the circadian rhythm of pineal melatonin biosynthesis and secretion, but the avian pineal has a self-sustaining rhythm of melatonin production that is only modified by neural signals. The duration of elevated avian pineal and plasma melatonin varies with day length, but the melatonin rhythm does not mediate photoperiodic responses in birds (see also mammals in the next section). Most attempts to demonstrate an effect of pinealectomy on seasonal variations in reproduction or other functions in birds have failed. The avian retina is an additional minor source of periodic melatonin biosynthesis, but the annual reproductive cycle of tree sparrows was unaffected after combined pinealectomy and optical enucleation.

Many Mammals Show Photoperiodic Responses in Reproduction, Seasonal Fattening, and Other Functions [5,13,73]

Circadian timing is part of mammalian photoperiodism [20,22]

A study of locomotor activity rhythms and testicular responses of Syrian hamsters exposed to light–dark cycles of different resonances provided the first compelling evidence for a circadian basis of photoperiodic time measurement in a mammal (see Plate 3, Photo C). Hamsters that had attained large testes in an LD 14:10 schedule were exposed for 89 days to one of the following LD cycles: 6:18, 6:30, 6:42, or 6:54. For photoschedules that are multiples of 24 h, such as LD 6:18 and LD 6:42, each successive 6 h light pulse falls at the same phase of the entrained animal's circadian cycle. In contrast, for resonance cycles of LD 6:30 and LD 6:54 light falls at different times of the circadian cycle on different days. Only the animals exposed to the LD 6:30 and 6:54 cycles developed enlarged testes, a result indicative of a long-day response (Figure 4.15B, upper panel).

Similarly, when hamsters that initially had regressed testes were subjected for 63 days to the same lighting schedules, only the groups of animals exposed to the LD 6:30 and 6:54 cycles showed gonadal growth. Reproductive activity in both paradigms was sustained by exposure to short, 6 h, light pulses when pulses were presented at 36 h or 60 h intervals. Only in these LD cycles, with period lengths that were not integral multiples of 24 h, did the light pulses occur at different circadian phases on different days, thereby scanning across all phases of the circadian cycle (Figure 4.15A).

These results clearly implicated a circadian mechanism for photoperiodic time measurement and were not compatible with an hourglass-type measurement of the total amount of light or darkness. Several other studies have used a variety of paradigms, including T-cycles, night-break experiments, and SCN ablation. All of these experiments have added to the evidence of circadian involvement in photoperiodic time measurement in many mammals.

Both absolute day length and photoperiod history are important in mammalian photoperiodism[15,24,34,62]

The concept of critical photoperiod has been important to our understanding of photoperiodism in mammals. Early experiments on Syrian hamsters indicated that reproductive function was stimulated by photoperiods with greater than 12.5 hours of light and inhibited by photoperiods with less than 12.5 hours of light. On the basis of these observations, 12.5 hours of light was considered to be the critical day length for maintaining reproductive function. Later studies of other mammals, however, forced reconsideration of the concept of a single species-specific critical photoperiod.

For example, Siberian hamsters that have been transferred from LD 16:8 to LD 14:10 undergo testicular regression but do not molt from summer to winter pelage. Transfer to schedules with less than 13 hours of light per day is sufficient to provoke both inhibition of reproduction and development of winter pelage. Clearly, different traits, even in a single species, can have different critical photoperiods. In a natural setting the existence of different critical photoperiods for different traits might be instrumental in establishing a timed sequence of seasonal physiological changes.

The response to a given day length can also be influenced by photoperiod history. For many organisms, both the absolute day length and the direction of photoperiod change are important. Siberian hamsters, for example, undergo testicular regression when transferred from LD 16:8 to LD 14:10. For this trait and protocol, LD 14:10 is an inhibitory photoperiod. However, when hamsters were

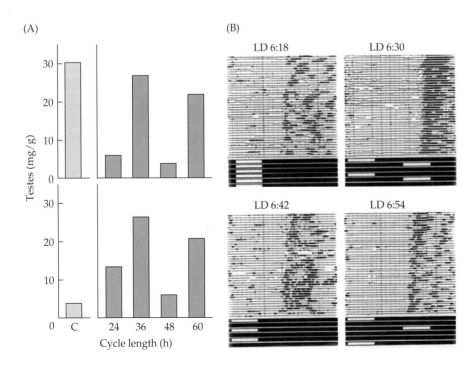

FIGURE 4.15 Circadian involvement in photoperiodic regulation of reproduction in mammals. (A) Testis weights of hamsters exposed to each of the resonance cycles in the upper panel of part B were in LD 14:10 prior to the study, then either exposed for 89 days to the resonance cycles (dark shaded bars) or continued in LD 14:10 as controls ("C," light shaded bars). Hamsters in the lower panel were in LD 6:18 for 10 weeks before the start of the experiment to induce regression, then transferred for 63 days to resonance cycles or continued in LD 6:18 as controls. (B) Wheel-running records of Syrian hamsters that were exposed to light cycles with one of four resonances. LD cycles are indicated above each actogram and shown graphically below with open bars representing 6 h light pulses and black representing darkness. For further explanation, see the text. (From Elliott, 1976.)

first exposed to LD 8:16 to induce testicular regression and were subsequently transferred to LD 14:10, gonadal regrowth was stimulated by the latter photoperiod.

A similar influence of photoperiod history has been demonstrated in birds, sheep, montane voles, Syrian hamsters, and several other species. The particular photoperiod history phenomenon described already for Siberian hamsters has adaptive implications. Hamsters normally begin to breed during the increasing day lengths of spring. Breeding terminates for most hamsters in late summer, at which time reproduction is inhibited by the decreasing photoperiods, even though the absolute day lengths may be the same as at the beginning of the breeding season. Similar history-dependent effects were noted earlier in birds and may be a widespread characteristic of photoperiodic organisms.

Perhaps the most intriguing photoperiod history effect is the transmission of a day-length cue from a mother rodent to her fetuses. In a typical experiment, pregnant Siberian hamsters were exposed to either LD 16:8 or LD 12:12 until the day of parturition. Thereafter, dams and their litters were maintained under LD 14:10. At 30 days of age, testes were threefold larger in juveniles whose mothers had been kept during pregnancy in LD 12:12 compared to the offspring of dams from LD 16:8 (Figure 4.16). The photoperiod history of the dam clearly influences the responsiveness of her offspring to day length during early postnatal life.

Further experiments employed cross-fostering of newborn pups between mothers exposed to different photoperiods during gestation. The results demonstrated conclusively that mother hamsters provide signals to their fetuses before birth. The signal that transmits relevant photoperiod information from the dam to her fetuses during late pregnancy may be the mother's plasma melatonin rhythm because melatonin is known to cross the placenta. Further support comes from a melatonin infusion experiment. Timed infusions for just 2 days during late pregnancy could mimic the effects of different photoperiods in influencing the photoperiod responses of hamster offspring. The rate of testis development in juvenile hamsters in response to a fixed postnatal photoperiod is greater when the photoperiod increases rather than decreases between prenatal and postnatal life.

Similar relations have been established in several other rodents. Photoperiod information transmitted by the mother to her fetuses may facilitate the process by which juvenile mammals determine whether photoperiod is increasing or decreasing, thereby accelerating development of seasonally appropriate responses in spring and autumn. At these times the absolute lengths of intermediate photic schedules, such as LD 14:10, may not provide adequate predictive cues.

Photoperiod is clearly the most important factor in annual reproductive responses; other factors may be modifiers. In Djungarian hamsters, social factors may influence reproductive development and fertility in females. In another species, the golden hamster, social cues attenuate the photoresponsiveness of males.

Photosensitivity and photorefractoriness are important in mammalian photoperiodism[17,70]

Two physiological states are important for regulation of seasonal function in all photoperiodic mammals: photosensitivity and photorefractoriness. The annual reproductive cycle of a long-day breeding rodent, the Syrian hamster, is a good example. During late summer, the photosensitive hamster can breed if exposed to stimulatory long photoperiods, but it will undergo gonadal regression in inhibitory short photoperiods. After gonadal regression has occurred during the decreasing photoperiods of late summer and early autumn, the hamster remains photosensitive for several weeks. Exposure to artificially lengthened photoperiods during this time induces rapid gonadal regrowth.

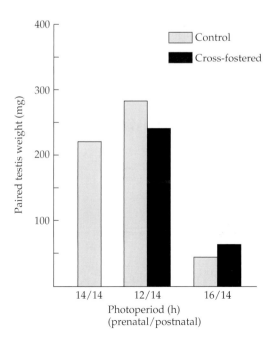

FIGURE 4.16 The effects of prenatal photoperiod on the prepubertal testicular response to LD 14:10. Young gestated in LD 12:12 or LD 16:8 were cross-fostered at birth to mothers from the other photoperiod. On the graph they are identified as 12/14 and 16/14, respectively. Foster mothers and pups were then transferred to LD 14:10. Control litters born under LD 12:12 and LD 16:8 conditions were also transferred to LD 14:10 at birth but not cross-fostered. A third group of control litters (14/14) were gestated in LD 14:10 and remained in LD 14:10 after birth. Testes of male offspring were removed and weighed at 30 days of age. The histogram bars indicate testes weights of 13 to 18 animals per group. (From Elliott and Goldman, 1989.)

If hamsters remain in short days, gonadal growth occurs anyway, but not until 5 to 6 months after initial exposure to the inhibitory photoperiod. Such gonadal regrowth has been termed spontaneous regrowth because it transpires in the absence of long days. In fact, it occurs equally well in hamsters kept in continuous darkness. The onset of spontaneous gonadal growth defines the beginning of the photorefractory phase, in which reproductive function is no longer directly responsive to the ambient photoperiod (see Figure 4.2).

Individual hamsters, all introduced to a short photoperiod at the same time, show remarkable interindividual synchrony both in the time of gonadal regression and in subsequent gonadal growth. The precision is impressive, considering that photoperiod and other environmental cues have been held constant for a 5- to 6-month interval. Also intriguing is the correspondence of the interval from start of short photoperiod to onset of gonadal growth with the length of the winter season. These two observations implicate an endogenous timing mechanism in the induction of photorefractoriness. Photorefractoriness is not always absolute in species such as Siberian hamsters and sheep. In some cases animals that have become refractory to a particular short day length may reinitiate winter-type responses if transferred to a still shorter photoperiod.

Photorefractory hamsters remain reproductively active and unresponsive to short photoperiods indefinitely if maintained exclusively under short days. Nevertheless, when photorefractory animals are transferred to long photoperiods for greater than 10 weeks and then are returned to the original short photoperiod, a second period of gonadal regression ensues. The breaking of photorefractoriness to short days by subsequent exposure to long days establishes that photorefractory hamsters discriminate between long and short days, even though this is not apparent from observation of the immediate responses, such as reproductive state. A major action of long summer days in photoperiodic mammals is the breaking of short-day photorefractoriness (see Figure 4.2).

The concepts of photorefractoriness and photosensitivity, and the relations that have been elaborated here for reproduction, apply equally well to nonreproductive traits in various species. Among such traits are spring and fall molts, seasonal thermoregulatory adaptations, seasonal changes in body weight and lipid metabolism, and changes of several behavior patterns. All show similar changes in the state of responsiveness. The spontaneous return to the spring/summer phenotype associated with refractoriness to short photoperiods probably has adaptive value, particularly in some short-lived rodents.

Gonadal regrowth begins in midwinter, when the photoperiod is still quite short, and it is completed by early spring. A period of approximately 6 weeks is required for a mammalian testis to complete the transition from quiescence to full spermatogenesis. Animals that do not rely on the advent of long days to initiate this process can begin breeding at the very onset of favorable conditions in the early spring. Spontaneous gonadal growth is probably less important for female mammals because ovarian follicular maturation is far more rapid than spermatogenesis.

Type I and Type II annual cycles share mechanistic similarities[20,22,53]

Among the long-day photoperiodic rodent species studied to date are hamsters, mice, and voles. All of these species become refractory with prolonged exposure to short photoperiods and therefore cannot be held indefinitely in a winter state. Commonly, however, these same species sustain the summer state indefinitely when they are held continuously under long photoperiods. Other mammals develop photorefractoriness in long as well as short photoperiods.

Ferrets initiate reproductive activity during the spring, when they become refractory to the previously inhibitory short days of winter. The spring portion of the ferret annual reproductive cycle is similar to the spring pattern for hamsters. Ferrets remain reproductively competent during the long days of summer but, unlike hamsters, do not maintain this state indefinitely, even when treated with artificial long photoperiods. After several months in long days, ferrets spontaneously enter a period of reproductive inhibition. They become refractory to the stimulatory effects of long days.

Sheep also become refractory after prolonged exposure to either long or short photoperiods. The development of refractoriness to both short and long photoperiods implies the capacity for spontaneous changes in both spring and fall, and may form the basis for circannual, Type II rhythms. No mammalian species are known to become refractory exclusively to long days. Mammals become refractory either to short days, or to both short and long days.

The possible relation between Type I and Type II cycles can be viewed from a slightly different perspective, compatible with the earlier discussion. Mammalian Type I cycles require a change in photoperiod to initiate the autumn/winter phase; that is, the decreasing day lengths of late summer/autumn are obligatory for triggering the responses that characterize the winter phase. However, the increasing day lengths of spring are not required for the return to summer condition. Instead an internal timing mechanism can trigger a "spontaneous" switch to the spring/summer state as the animal becomes refractory to short days. The long summer days are required, however, for breaking refractoriness to short days. They permit development of the subsequent winter-phase responses as days shorten (see Figure 4.2).

Type II organisms can be viewed as animals that are capable of a spontaneous, or photoperiod-independent, switch from summer to winter state in addition to the photoperiod-independent switch from winter to summer state that is seen in both Type I and Type II cycles. The ability to accomplish two such switches, even in the absence of environmental change, might lead to the expression of a circannual rhythm under constant environmental conditions.

Output signals include pineal melatonin as the photoperiodic messenger in mammals[5,20,45]

Photoperiodic signals in mammals are relayed by ocular photoreceptors to neuroendocrine regulation centers. Light information in mammals is relayed to the pineal gland via a multisynaptic pathway. Initially light cues that regulate circadian activity of the pineal are transmitted from the retina to the suprachiasmatic nucleus (SCN) by the same direct retinohypothalamic pathway utilized for entrainment of other circadian rhythms. Photic information is then transmitted from the SCN to the paraventricular nucleus of the hypothalamus, the medial forebrain bundle, the spinal cord, the superior cervical ganglia, and finally the postganglionic sympathetic fibers that innervate the pineal by a noradrenergic mechanism (Figure 4.17). Interruption of this pathway at any point deprives the pineal gland of photic information and prevents photoperiodic regulation of several traits by interfering with a mechanism that will be described shortly.

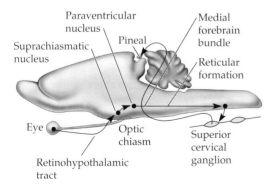

FIGURE 4.17 Pathway from eye to pineal gland in mammals. Environmental light is detected by ocular photoreceptors and is transferred to the SCN primarily via the retinohypothalamic tract. One output system from the SCN includes projections to the paraventricular nuclei, and from there to the medial forebrain bundle, through the reticular formation to the superior cervical ganglion, and finally via postganglionic noradrenergic fibers to the pineal. (From Klein et al., 1983.)

Circadian regulation of pineal enzymatic activity produces a pronounced diurnal variation in melatonin synthesis and secretion, with serum concentrations of melatonin at night exceeding the daytime values by as much as 50-fold in some species. In constant darkness, rhythmic secretion of melatonin persists with a free-running period similar to that of other circadian rhythms, and the light–dark cycle entrains this rhythm. Light determines the duration of elevated melatonin secretion each night both through its action on circadian oscillators and through an acute inhibitory effect on melatonin synthesis. In all mammals studied, the duration of elevated melatonin secretion is directly proportional to the length of the dark phase of the photic schedule.

Regulation of the pattern of melatonin secretion by the light–dark cycle is critically important for the transduction of photoperiodic signals. Melatonin acts as a neuroendocrine messenger for photoperiod and is a vital link in the regulation of seasonal processes of mammals with Type I and II annual rhythms. Studies with Siberian hamsters and sheep employed timed daily infusions of melatonin, thereby permitting precise control over various parameters of experimentally generated melatonin signals.

Infusion experiments in pinealectomized hamsters indicated that exposure to daily melatonin infusion durations of 6 h or less stimulated the reproductive system, whereas daily infusions lasting 8 h or more led to gonadal regression. Conversely, in sheep, reproduction was stimulated by 16 h daily infusions of melatonin but inhibited by 8 h daily melatonin infusions. Thus the duration of the circadian elevation of circulating melatonin appears to be critical for determining the nature of a photoperiodic response in both sheep and hamsters, but the reproductive responses of the two species to long versus short melatonin infusions are opposite (Figure 4.18).

The complementary results in sheep and hamsters showed that sheep are short-day breeders with Type II rhythms and hamsters are long-day breeders with Type I rhythms. These data proved especially informative because they suggested that the involvement of melatonin in photoperiodic time measurement is similar for Type I and Type II mammals. In a given species, long-duration melatonin elevations are associated with responses typical of short days in a given species, and short-duration elevations of melatonin evoke long day–type responses. In this sense, the duration of the circadian elevation of circulating melatonin appears to serve as an internal "code" for day length. Note that melatonin is secreted through virtually the entire night in sheep, but for only part of the dark period in hamsters. Thus the quantitative relation between hours of elevated melatonin and day length varies somewhat among species.

(A)

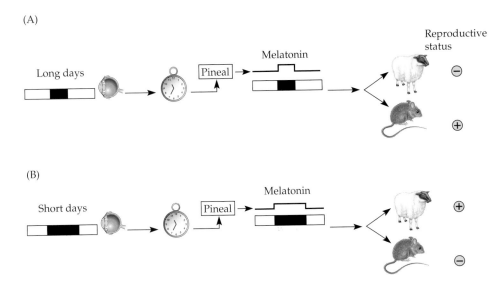

FIGURE 4.18 Schematic representation of the encoding of day length by the pineal melatonin rhythm. (A) Long days. (B) Short days. For explanation, see the text. (From Goldman, 2001.)

(B)

The reproductive responses of Siberian hamsters and sheep to daily melatonin infusions of various durations discredit a popular earlier hypothesis that melatonin is primarily an antireproductive hormone because it is possible to obtain reproductive stimulation as well as inhibition by appropriate melatonin treatments in both species. In addition, melatonin mediates effects of photoperiod on seasonal pelage changes, thermoregulatory adaptations, immune responsiveness, and lipid metabolism, as well as seasonal reproductive rhythms. Evidence supports a human melatonin rhythm patterned according to day length much like that of other mammals, but the possible seasonal effects of melatonin in humans have not been investigated.

A close relation exists between the actions of light and melatonin in mammals[5,20,53]

In both sheep and hamsters, short-duration melatonin infusions elicited long-day responses and long-duration infusions provoked short-day responses, regardless of the time of the day or night the infusions were administered. The typical nocturnal phase of pineal melatonin secretion is not crucial for the hormone's role in evoking photoperiodic responses, nor is the presence of melatonin at a critical phase of the circadian cycle required. The rules by which light affects photoperiodic time measurement are quite different.

The mere presence or absence of light at certain phases of the circadian cycle determines whether long- or short-day responses will be evoked. Very brief pulses of light (even 1 minute or less per day) at the appropriate circadian time can induce long-day responses because they may shorten the duration of elevated melatonin by several hours, and the change in melatonin pattern may persist for several cycles. The actions of light

and melatonin, seemingly so different, are both parts of the same photoperiodic mechanism. Light not only acutely inhibits melatonin secretion but also reprograms circadian oscillators that drive the pineal melatonin rhythm, thereby affecting the phase and duration of nightly melatonin secretion. However, the melatonin target sites that make up part of the photoperiodic mechanism appear to attend only to the duration, and not the phase, of elevated circulating concentrations of melatonin.

When mammals are subjected experimentally to a sudden change in day length, a period of several days or weeks of exposure to the new photoperiod is required before overt responses such as changes in reproductive state or molting of pelage become evident. Likewise, when mammals are treated with melatonin to evoke photoperiodic responses, the treatment must generally be carried out over a period of several days to weeks. No special name has been applied to these types of time lags documented in the mammalian literature, but mammals may well have a parallel to the so-called photoperiodic counter of insects that was discussed earlier in this chapter.

One proposal suggests that these time lags evolved to reduce the likelihood of inappropriate responses in the event that an organism might incorrectly assess the day length on a single day. Thus animals must record several short days in succession before activating the responses that are typically associated with short photoperiod.

Pineal melatonin secretion is regulated by the circadian system in mammals[1,9,14,19,51,56,60]

The circadian system regulates the most critical feature of the mammalian melatonin pattern (Figure 4.19). In

(A)

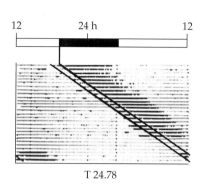
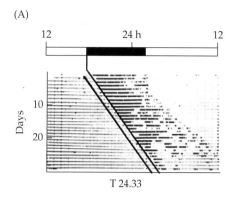

(B)

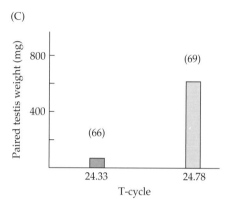
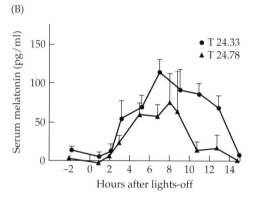

(C)

FIGURE 4.19 T-cycles, melatonin rhythm, and seasonal reproductive response. (A) Wheel-running activity rhythms for individual Siberian hamsters exposed to 1 h light pulses in a 24.33 or 24.78 h T-cycle. Solid and open bars represent prior 16:8 light and dark periods, respectively; the T light pulse is indicated by the oblique parallel lines. (B) Circadian variations in serum melatonin concentration in hamsters entrained to a 24.33 or 24.78 h T-cycle. Time is plotted relative to lights-off on the day of sampling. (C) Paired testis weights of hamsters exposed to a 24.33 or 24.78 h T-cycle; mean is indicated by the height of bars and numbers in parentheses are the group sizes. (From Goldman, 2001.)

one experiment, two non-24 h T-cycles were used, which had different effects on testis growth in hamsters. The data indicated correspondingly different phase angles of entrainment in terms of locomotor activity rhythms, as well as different effects on the duration of elevated serum melatonin concentrations. Light fell during the time of activity onset in the 24.78 h cycle but not the 24.33 h cycle (Figure 4.19A). The two T-cycles also had contrasting effects on testis growth (Figure 4.19C).

Only animals on the 24.78 h cycle showed a typical long-day response of enlarged gonads. The duration of elevated serum melatonin concentrations was shorter in hamsters exposed to the 24.78 h cycle compared to those in the 24.33 h cycle (Figure 4.19B). The data imply that light acts through the circadian system to entrain the locomotor activity rhythm and to determine both the phase and duration of melatonin secretion. The duration of elevated melatonin determines the reproductive response, thus linking the action of the circadian pacemaker to the photoperiodic response.

Extensive experimental results in laboratory rats suggest that two separate circadian oscillators play a role in the control of melatonin secretion. One controls the onset and the other the termination of secretion in each circadian cycle. The phase relation between these oscillators may determine the duration of nocturnal melatonin secretion and ultimately the photoperiodic responses. The hypothesis is compatible with a dual-oscillator model for the circadian system and with an internal coincidence model for photoperiodic time measurement. Similar hypotheses have attempted to explain photoperiodic regulation of the duration of the active phase of locomotor activity using a dual-oscillator model.

The proposal that two separate circadian oscillators act in concert to establish the duration of the nocturnal episode of melatonin secretion offers a tempting format for evaluating the internal coincidence hypothesis. Experiments are needed to demonstrate conclusively that the phase relation between two oscillators determines melatonin duration. It is already well established that melatonin duration determines the photoperiodic response. Obtaining an unequivocal demonstration that two oscillators are involved may be a difficult task. It will require anatomical, genetic, or biochemical "dissection" of the dual-oscillator system, and the two oscillators might well reside within the same brain locus or even within the same cells.

Melatonin physiology is influenced by photoperiod history in mammals[20,52]

The melatonin rhythm unequivocally serves as an endocrine signal that codes for day length. However, photoperiod history in terms of prior melatonin signals can influence how an animal responds to a melatonin signal of a given duration. An example is the case of photorefractoriness. The prolonged exposure to short days and the accompanying long-duration melatonin signals eventually lead to a reversal of the overt photoperiodic responses. The animal finally reverts to the condition that is typically associated with long days. At this stage of the annual cycle, winter-type responses can no longer be obtained either by continued exposure to short days or by treatment with melatonin.

Photoperiod history also may influence the rhythm-generating apparatus by altering the pattern of melatonin secretion. For instance, the light regime for a mother hamster during late pregnancy can influence the melatonin rhythm-generating system of her male offspring.

Melatonin mediates photoperiodic entrainment of mammalian circannual rhythms[4,72]

The daily melatonin secretion pattern varies seasonally with changing photoperiod, and the annual rhythm of melatonin appears fundamental to photoperiodic entrainment in mammalian true circannual rhythms. An experiment with sheep, which show a marked variation of luteinizing hormone (LH) secretion associated with the circannual reproductive rhythm, provides an example. The rhythm of LH secretion was synchronized for six pineal-intact control ewes in an artificial light schedule that simulated the naturally occurring changes in photoperiod (Figure 4.20A). Higher plasma LH concentrations were typical of the breeding season. A second control group of sheep was pinealectomized at the start of the study and did not receive melatonin replacement. In these females the circannual rhythm of LH secretion free-ran with a period different from 365 days (Figure 4.20B).

Entrainment was investigated by the use of physiological doses of melatonin to reinstate circadian patterns of the hormone in pinealectomized ewes. Seasonal blocks of melatonin replacement were provided over the course of 3 to 4 years through stopping and restarting of an infusion apparatus, which provided a circadian melatonin pattern on an annual basis. Initially, 365-day cycles of LH were entrained by exposure to just one block of melatonin each year, in this case 70 days of a long-day melatonin pattern mimicking the pattern secreted around the summer solstice. Both the phase and the period of the circannual rhythm of LH were set by suitably timed blocks of melatonin infusion. The rhythm could be shifted 180 degrees out of phase with the natural reproductive cycle if pinealectomized ewes were infused with the long-day melatonin pattern during early winter.

Subsequent experiments examined whether specific patterns of melatonin secretion at a selected time of the year are preferentially effective in entrainment. Four groups of pinealectomized ewes were treated once a year for 3 years with blocks of melatonin infusions designed to simulate natural melatonin rhythms during one of the four seasons: winter, spring, summer, or fall (Figure 4.20C–F). In each case, the pattern of the melatonin infusions mimicked the normal melatonin pattern during that season.

Infusion of the summertime pattern of melatonin (see Figure 4.20E) entrained the rhythm, with period, phase, and degree of synchrony corresponding to those of pineal-intact ewes. Springtime melatonin was effective in synchronizing the rhythm, but the phase of reproductive activity was advanced (see Figure 4.20D). Autumnal melatonin was less effective, and the phase of reproductive activity was delayed (see Figure 4.20F). Winter melatonin did not entrain the rhythm (see Figure 4.20C). Follow-up studies revealed that the winter melatonin pattern was not effective, whether it was delivered in the winter or the summer stages of the circannual rhythm.

The results of these infusion studies support the hypothesis that melatonin is a neuroendocrine messenger that mediates the photoperiodic entrainment of circannual rhythms in sheep. In addition they provide evidence that a seasonal specificity exists with respect to the circadian melatonin patterns that are capable of entraining the circannual rhythm.

Multiple target sites for the photoperiodic actions of melatonin may exist[17,33,40,41]

The primary target sites for the photoperiodic actions of melatonin are not yet known. Some melatonin effects involve changes in the secretion of hypothalamic releasing factors and anterior pituitary hormones. Changes in the rates of secretion of the pituitary gonadotropic hormones play an important role in mediating seasonal variations in reproductive activity, and variations in pituitary prolactin secretion are important in the regulation of seasonal changes in pelage. These modulations of pituitary hormone secretion are mediated by seasonally changing patterns of melatonin, but they do not appear to be brought about by a direct action of melatonin on hormone-secreting cells in the hypothalamus or anterior pituitary.

Recent studies using radiolabeled melatonin have identified several potential sites of melatonin action in brain and other tissues. Species vary considerably with respect to the various sites of melatonin uptake. In mammals, the pars tuberalis is the most intensely and consistently labeled melatonin uptake tissue. The pars and the anterior pituitary gland both have their embryonic origins from the tissue of Rathke's pouch, though they become anatomically separate structures. The specific function of the pars tuberalis is not known for any vertebrate. Because of its small size and location, it is rel-

(A) Pineal-intact

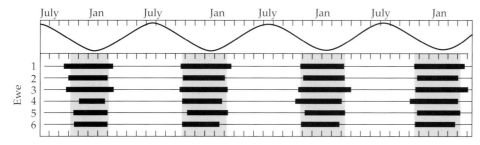

(B) Pinealectomized, no melatonin

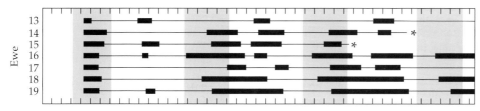

(C) Winter melatonin

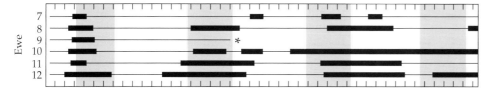

(D) Spring melatonin

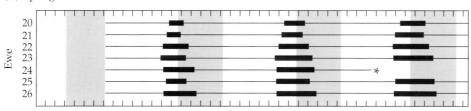

(E) Summer melatonin

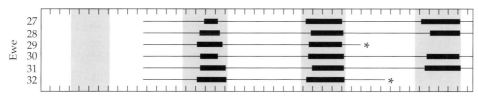

(F) Autumn melatonin

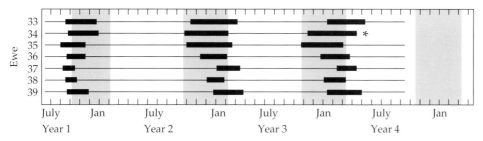

FIGURE 4.20 Circannual LH release cycle in ewes maintained indoors in a simulated sinusoidal natural photoperiod, as indicated at the top. (A) Intact ewes. (B) Pinealectomized ewes not treated with melatonin. (C–F) Individual pinealectomized ewes treated with a winter, spring, summer, or autumn pattern of melatonin, respectively. The horizontal black bars in each graph represent high stages of the LH cycle. Shaded panels span average dates of onset and end of the high LH stage each year in intact controls of panel A. (From Woodfill et al., 1994.)

atively inaccessible and has proved difficult to test directly for a photoperiodic function.

The SCN contains circadian oscillators that generate the pattern of pineal melatonin secretion, and it is also a primary melatonin-binding site in several rodent species. Ablation of the SCN rendered male Siberian hamsters unresponsive to daily systemic infusions of melatonin. The brain-lesioned hamsters failed to undergo testicular regression when subjected to melatonin infusions that inhibited testis function in SCN-intact males. The ability of a brain lesion to prevent the action of a hormone suggests that neurons, glial cells, or fibers of passage disrupted by the ablation may be required for expression of the hormone's action.

Such results do not establish that the hormone acts directly on cells at that site. More direct evidence for an action of melatonin in the SCN was obtained when small amounts of melatonin, administered via microdialysis to the SCN of Siberian hamsters, inhibited reproduction. Similar treatment in adjacent tissue was ineffective. Melatonin uptake also occurs in the paraventricular and reuniens nuclei of the thalamus of this species, and microinfusions of melatonin into either of these nuclei also led to reproductive inhibition. The SCN may be an important target site for melatonin action in Siberian hamsters, but other neural pathways are likely to be involved as well. Neither is the possibility ruled out for an additional role of the pars tuberalis in this species, since this tissue was not studied in the microdialysis experiments.

Further studies in Siberian hamsters revealed that microimplants of melatonin in the suprachiasmatic, paraventricular, or reuniens nuclei led to gonadal regression, as did the daily melatonin infusions in the microdialysis study. After the continuous-release implants had been left in place for more than 5 months, the animals exhibited gonadal regrowth. That is, they became refractory to melatonin in a fashion that parallels photorefractoriness following months of exposure to short days. The hamsters were then administered melatonin systemically to test whether the sites that had not been exposed to melatonin implants were responsive to the hormone. Most of the hamsters underwent a second phase of gonadal regression in response to systemic administration of melatonin. Thus, not only can melatonin at any one of three brain sites lead to reproductive inhibition in Siberian hamsters, but each site appears to become refractory to melatonin independently of the others.

In contrast to the results obtained for Siberian hamsters, destruction of the SCN of Syrian hamsters failed to prevent testicular regression in response to daily injections or infusions of melatonin. However, lesions of either the anterior hypothalamus or the mediobasal hypothalamus prevented this action of melatonin on the reproductive axis. The SCN of Syrian hamsters shows less uptake of melatonin as compared to the Siberian hamster SCN. These apparent differences between the two hamster species are puzzling because their melatonin patterns and photoperiodic responses are similar. Little melatonin uptake has been detected in the SCN of some photoperiodic mammals, such as sheep. None has been observed in the SCN of three species of mustelids—ferrets, mink, and spotted skunks—in which uptake has been demonstrated only in the pars tuberalis.

Do different melatonin target sites regulate distinct seasonal traits such as pelage, thermoregulation, lipid metabolism, or reproduction? Partial evidence from sheep and Syrian hamsters suggests two distinct melatonin target sites for photoperiodic regulation of prolactin and luteinizing hormone, respectively. Alternatively, the photoperiodic system may include some redundancy, with multiple sites of melatonin action activating separate but functionally similar pathways. Species differences in sites of melatonin uptake might then simply reflect taxonomic differences in the relative importance of the various melatonin targets.

Comparative Studies Give Clues to the Evolutionary History of Mammalian Photoperiodism[20,51]

The neuroendocrine basis of photoperiodism shows both similarities and differences among the various vertebrates[3,20]

Pineal melatonin exerts similar photoperiodic actions in many different mammals. Well-known species include representatives of the rodents, mustelids, ungulates, and marsupials. In all cases, pinealectomy either eliminates or severely compromises the animal's ability to develop species-typical seasonal responses to changes in photoperiod. Furthermore, in all species tested, melatonin can elicit appropriate seasonal responses. The universality of duration of the elevated melatonin signal for transducing photoperiodic information remains to be established. The causal relation between long-duration melatonin elevation and short-day winter responses has been extensively tested only in sheep, Siberian hamsters, and Syrian hamsters. Compatible results exist for several other species, including one marsupial.

The similar action of pineal melatonin in various photoperiodic mammals supports the concept of a single ancestral origin for the mammalian photoperiodic mechanism. In birds and lizards, however, the fact that seasonal photoperiodic responses persist even after pinealectomy shows that the pineal gland is not an essential component of the photoperiodic mechanism. The pineal route is apparently not the only means that has evolved for vertebrate photoperiodism. The apparently universal role for the pineal among photoperiodic mammals suggests fixation early in mammalian history or perhaps in a reptilian lineage that gave rise to the mammals.

Photoperiodism has a genetic basis[21,44]

Within some breeding populations of photoperiodic organisms, a subset of individuals may not show species-typical responses to changes in day length. For example, white-footed mice collected in Connecticut are almost always responsive to photoperiod. They cease reproductive activity and grow winter pelage when exposed to short photoperiods. In contrast, white-footed mice from Georgia do not show these responses. Separate laboratory breeding colonies established from each of the two populations continue to produce the phenotypes of the wild populations, indicating a genetic basis for the differences in photoperiodic responsiveness.

Different phenotypes may also exist for photoperiodic responsiveness in a single wild breeding population of deer mice or prairie voles. A small proportion of the population engages in winter breeding in the field, and these animals fail to exhibit gonadal regression when exposed to short photoperiod or when treated with melatonin. Differences in the melatonin response system may be the basis for the interindividual differences in photoperiodic responsiveness in these species. The maintenance of such nonphotoperiodic, year-round breeders suggests that winter reproduction sometimes confers selective advantage. A 3- to 4-month life span is not uncommon in the field, and most adults do not survive the winter. In relatively mild winters, individuals that gamble on reproduction may succeed. They may temporarily contribute disproportionately to the gene pool, compared to their reproductively quiescent conspecifics.

Artificial selection for nonresponsiveness to short photoperiods has been carried out with laboratory populations of Siberian hamsters. For several generations, only hamsters that failed to exhibit the species-typical short-day responses were used for breeding. This procedure led to the establishment of a breeding colony in which 70 to 80% of the offspring proved to be unresponsive to short days, compared to only 20 to 30% nonresponders in the original laboratory population.

The physiological basis for nonresponsiveness in such animals was investigated and proved to be different than for the mice and voles discussed earlier. Nonresponsive Siberian hamsters failed to generate long-duration pineal melatonin signals in short days. These hamsters also displayed longer intrinsic circadian periods in DD and different phase angles of entrainment to short photoperiods compared to their photoperiod-responsive counterparts.

The short-day nonresponsive hamsters did display the winter phenotype when provided exogenous melatonin. Therefore, differences in circadian regulation of the pineal melatonin pattern, rather than in responsiveness of melatonin target tissues, were responsible for failure to respond to short photoperiod. Laboratory selection studies reveal the reservoir of genetic variation upon which natural selection may act.

In Siberian hamsters, both genetic variation and photoperiod history effects are important in determining whether animals will undergo reproductive inhibition in short days. Hamsters that have been exposed to very long days, such as LD 16:8 or 18:6, are far more likely to be unresponsive to subsequent short photoperiods than hamsters raised exclusively under an intermediate LD 14:10 day length. Projecting from this experimental result, one would predict that hamsters born early in the breeding season should be the ones most likely to remain reproductively active during winter because they have been exposed to the longest days of midsummer. In this relatively short-lived species, these would generally be the oldest animals in the population and thus the ones most likely to benefit by gambling on winter reproduction rather than saving resources for the unlikely event of their survival until the next year's breeding season. Whether similar effects of photoperiod history on short-day responsiveness are operative in other species is an interesting question.

SUMMARY

Many organisms show true (Type II) circannual cycles of physiological and behavioral changes when maintained in the laboratory under seasonally constant conditions. However, true circannual rhythms are not as universal as circadian rhythms. Type I species use seasonal timekeeping but fail to show persistent rhythms under constant conditions. Photoperiod is the most frequently employed cue for both Type I and Type II organisms. This chapter has emphasized photoperiodic cuing in annual biological rhythms because of the importance of photoperiodic cues and because the measurement of day length usually employs a circadian mechanism.

The circadian oscillators regulating insect photoperiodism may differ from those involved in the daily behavioral rhythms. In some insects, the circadian pacemakers involved in photoperiod time measurement may be damped rather than fully self-sustained. In these cases, carefully designed studies may be required to distinguish between circadian and hourglass mechanisms.

Photoperiodic time measurement in vertebrates is accomplished by a circadian clock. Both internal and external coincidence models have been proposed to explain this process. The pineal melatonin rhythm in mammals serves as an endocrine component of the photoperiodic mechanism. The rhythm of melatonin secretion is driven by signals from one or more circadian clocks in the suprachiasmatic nucleus. The duration of each nightly episode of melatonin secretion codes for day length in all photoperiodic mammals.

An accurate measurement of the current day length is an important part of photoperiodic regulation. In addition, photoperiodic history effects, such as the direction of change in day length, can also influence responses. Although laboratory photoperiodism studies have generally transferred organisms from one fixed photoperiod to another, organisms in nature are exposed to gradual, daily changes in day length. Consequently, some of the mechanisms available for fine-tuning the photoperiodic system may have been overlooked in laboratory investigations.

Nonphotic cues may help adjust the timing of reproduction to local conditions. Interaction of temperature, food availability, and social interactions may account for variations in the timing of breeding in natural populations from year to year. Successful breeding involves complex interactions between physical and social environmental factors. A great challenge for the future will be to understand the neuroendocrine mechanisms mediating the integration of these cues.

CONTRIBUTORS

Bruce Goldman, Eberhard Gwinner, Fred J. Karsch, David Saunders, Irving Zucker, and Gregory F. Ball wish to thank C. Robertson McClung and Herbert Underwood for their material contributions to Chapter 4.

STUDY QUESTIONS AND EXERCISES

1. How has an understanding of circadian clocks contributed to the understanding of the annual rhythms of many organisms? Organize a minisymposium on specific aspects of circadian time measurement in photoperiodic control of specific annual rhythms in a variety of animal species ranging from invertebrates to a spectrum of vertebrate species.

2. How have hypothetical models of the circadian system contributed to the formulation of models to explain photoperiodic time measurement?

3. A major difference between Type I and Type II species is the presence of a self-sustaining circannual rhythm only in Type II organisms. Among mammals, what are the similarities in the regulation of annual rhythms in Type I and Type II species?

4. Using examples of mammals, compare Type I and Type II species with respect to the following:

- Use of day length cues to establish synchrony between biological seasonal changes and local seasonal time
- Involvement of a circadian clock in photoperiodic time measurement
- Involvement of pineal gland and melatonin rhythm in photoperiodism
- Participation of an internal (endogenous) timing mechanism capable of "measuring" very long periods of time (several months) even when animals are held under seasonally constant conditions
- Occurrence of photosensitive and photorefractory phases during the annual cycle

5. Try defending the internal coincidence model for photoperiodic time measurement in mammals. Try to reconcile this model with the observations that the duration of nocturnal melatonin secretion conveys a day length signal in the form of a hormone rhythm.

6. How do photoperiod history effects contribute to the fine-tuning of seasonal biological rhythms?

7. What types of data would be required to provide convincing evidence that humans are photoperiodic?

8. What types of data would be necessary to support a hypothesis about circannual rhythmicity in humans? How difficult would it be to collect an adequate sample of data?

9. What types of overt end point responses are needed to test whether humans use photoperiod cues to regulate seasonal biological rhythms? Provide a rationale for your choice.

10. Compare the roles of the CO protein in plant photoperiodism to the role of melatonin in mammals. What is the relation of each of these molecules to the circadian systems and the photoperiodic responses of plants and mammals, respectively?

11. Writing technical papers on research is an integral part of a scientist's work. Review the study questions of Chapter 1, then check out a journal such as *The Journal of Biological Rhythms* to review the format. Write a brief paper in journal format, based on some of the data in this chapter. Include the following sections in your paper: Introduction, Material and Methods, Observations, Discussion, Summary, and References. Arrange a group presentation of your paper.

REFERENCES

1. Bae, H. H., R. A. Mangels, B. S. Cho, J. Dark, S. M. Yellon, et al. 1999. Ventromedial hypothalamic mediation of photoperiodic gonadal responses in male Syrian hamsters. J. Biol. Rhythms 14: 391–401.

2. Baker, F. R. 1938. The relation between latitude and breeding seasons in birds. Proc. Zool. Soc. Lond. A 108: 557–582.

3. Ball, G. F. 1993. The neural integration of environmental information in seasonally breeding birds. Amer. Zool. 33: 185–199.

4. Barrell, G. K., L. A. Thrun, M. E. Brown, C. Viguie, and F. J. Karsch. 2000. Importance of photoperiodic signal quality to entrainment of the circannual reproductive rhythm of the ewe. Biol. Reprod. 63: 769–774.

5. Bartness, T. J., and B. D. Goldman. 1989. Mammalian pineal melatonin: A clock for all seasons. Experientia 45: 939–945.

6. Berthold, P., and U. Querner. 1981. Genetic basis of migratory behavior in European warblers. Science 212: 77–79.

7. Billings, H. J., C. Viguie, F. J. Darsch, R. L. Goodman, J. M. Connors, et al. 2002. Temporal requirements of thyroid hormones for seasonal changes in LH secretion. Endocrinology 143: 2618–2625.

8. Bünning, E. 1960. Circadian rhythms and the time measurement in photoperiodism. Cold Spring Harb. Symp. Quant. Biol. 25: 249–256.

9. Daan, S., U. Albrecht, G. T. J. van der Horst, H. Illnerova, T. Rönneberg, et al. 2001. Assembling a clock for all seasons: Are there M and E oscillators in the genes? J. Biol. Rhythms 16: 105–116.

10. Danilevskii, A. S. 1965. Photoperiodism and Seasonal Development of Insects. Oliver and Boyd, Edinburgh, UK.

11. Dawson, A., and A. R. Goldsmith. 1983. Plasma prolactin and gonadotrophins during gonadal development and the onset of photorefractoriness in male and female starlings (*Sturnus vulgaris*) on artificial photoperiods. J. Endocrinol. 97: 253–260.

12. Dawson, A., V. M. King, G. E. Bentley, and G. F Ball. 2001. Photoperiodic control of seasonality in birds. J. Biol. Rhythms 16: 365–380.

13. Demas, G. E., and R. J. Nelson. 1998. Photoperiod, ambient temperature, and food availability interact to affect reproductive and immune function in adult male deer mice (*Peromyscus maniculatus*). J. Biol. Rhythms 13: 253–262.

14. Elliott, J. A. 1976. Circadian rhythms and photoperiodic time measurement in mammals. Fed. Proc. 35: 2339–2346.

15. Elliott, J. A., and B. D. Goldman. 1989. Reception of photoperiodic information by fetal Siberian hamsters: Role of the mother's pineal gland. J. Exp. Zool. 252: 237–244.

16. Freeman, D. A., and I. Zucker. 2000. Temperature-independence of circannual variations in circadian rhythms of golden-mantled ground squirrels. J. Biol. Rhythms 15: 336–343.

17. Freeman, D. A., and I. Zucker. 2001. Refractoriness to melatonin occurs independently at multiple brain sites in Siberian hamsters. Proc. Natl. Acad. Sci. USA 98: 6447–6452.

18. Garner, W. W., and H. A. Allard. 1920. Effect of the relative length of the day and night and other factors of the environment on growth and reproduction in plants. J. Agricult. 18: 553–606.

19. Goldman, B. D. 1999. The circadian timing system and reproduction in mammals. Steroids 64: 679–685.

20. Goldman, B. D. 2001. Mammalian photoperiodic systems: Formal properties and neuroendocrine mechanisms of photoperiodic time measurement. J. Biol. Rhythms 16: 283–301.

21. Goldman, S. L., K. Dhandapani, and B. D. Goldman. 2000. Genetic and environmental influences on short-day responsiveness in Siberian hamsters (*Phodopus sungorus*). J. Biol. Rhythms 15: 417–428.

22. Gorman, M. R., B. D. Goldman, and I. Zucker. 2001. Mammalian photoperiodism. In: Handbook of Behavioral Neurobiology Volume 12: Circadian Clocks, J. S. Takahashi, F. W. Turek, and R. Y. Moore (eds.), pp. 481–508. Plenum, New York.

23. Goss, R. J. 1969. Photoperiodic control of antler cycles in deer. I. Phase shift and frequency changes. J. Exp. Zool. 170: 311–324.

24. Gudermuth, D. F., W. R. Butler, and R. E. Johnston. 1992. Social influences on reproductive development and fertility in female Djungarian hamsters (*Phodopus campbelli*). Horm. Behav. 26: 308–329.

25. Gwinner, E. 1981. Circannual systems. In: Handbook of Behavioral Neurobiology Volume 4: Biological Rhythms, J. Aschoff (ed.), pp. 391–410. Plenum, New York.

26. Gwinner, E. 1986. Circannual Rhythms. Springer, Berlin.

27. Gwinner, E. 1991. Circannual rhythms in tropical and temperate-zone stonechats: A comparison of properties under constant conditions. Ökologie der Vögel. 13: 5–14.

28. Gwinner, E. 1996. Circannual clocks in avian reproduction and migration. Ibis 138: 47–63.

29. Gwinner, E., and V. Neusser. 1985. Die Jugendmauser europäischer und afrikanischer Schwarzkehlchen (*Saxicola torcruata rubicola* und *S.t. axillaris*) sowie von Fl-Hybriden. J. Ornithol. 126: 219–220.

30. Gwinner, E., and W. Wiltschko. 1980. Circannual changes in migratory orientation of the garden warbler, *Sylvia borin*. Behav. Ecol. Sociobiol. 7: 73–78.

31. Gwinner, E., J. Dittami, G. Ganshirt, M. Hall, and J. Wozniak. 1989. Endogenous and exogenous components in the control of the annual reproductive cycle of the European starling. In: Proceedings of the 18th International Ornithological Congress, pp. 501–515.

32. Hamner, W. M. 1971. On seeking an alternative to the endogenous reproductive rhythm hypothesis in birds. In: Biochronometry, M. Menaker (ed.), pp. 448–462. National Academy of Sciences, Washington, DC.

33. Hazlerigg, D. G., P. J. Morgan, and S. Messager. 2001. Decoding photoperiodic time and melatonin in mammals: What can we learn from the pars tuberalis? J. Biol. Rhythms 16: 326–335.

34. Hegstrom, D. D., and S. M. Breedlove. 1999. Social cues attenuate photoresponsiveness of the male reproductive system in Siberian hamsters *Phodopus sungorus*. J. Biol. Rhythms 14: 54–61.

35. Juss, T. S., S. L. Meddle, R. W. Servant, and V. M. King. 1993. Melatonin and photoperiodic time measurement in Japanese quail (*Conturnix coturnix japonica*). Proc. R. Soc. Lond. [Biol.] 254: 21–28.

36. Karsch, F. J., G. E. Dahl, N. P. Evans, J. M. Manning, K. P. Mayfield, et al. 1993. Seasonal changes in gonadotropin-releasing hormone secretion in the ewe: Alteration in response to the negative feedback action of estradiol. Biol. Reprod. 49: 1377–1385.

37. Karsch, F. J., J. E. Robinson, C. J. I. Woodfill, and M. B. Brown. 1989. Circannual cycles of luteinizing hormone and prolactin secretion in ewes during prolonged exposure to a fixed photoperiod: Evidence for an endogenous reproduction rhythm. Biol. Reprod. 41: 1034–1046.

38. Kenny, N. A. P., and D. S. Saunders. 1991. Adult locomotor rhythmicity as "hands" of the maternal photoperiodic clock regulating larval diapause in the blow fly, *Calliphora vicina*. J. Biol. Rhythms 6: 217–233.

39. Klein, D. C., R. Smoot, J. L.Weller, S. Higa, S. P. Markey, et al. 1983. Lesions of the paraventricular nucleus area of the hypothalamus disrupt the suprachiasmatic-spinal cord circuit in the melatonin rhythm generating system. Brain Res. Bull. 20: 647–652.

40. Lewis, D., D. A. Freeman, J. Dark, K. E. Wynne-Edwards, and I. Zucker. 2002. Photoperiodic control of oestrous cycles in Syrian hamsters: Mediation by the mediobasal hypothalamus. J. Neuroendocrinol. 14: 294–299.

41. Lincoln, G. A., and G. A. Richardson. 1998. Photo-neuroendocrine control of seasonal cycles in body weight, pelage growth and reproduction: Lessons from the HPD sheep model. Comp. Biochem. Physiol. C 119: 283–294.

42. Lumsden, P. J., and A. J. Millar. 1998. Biological Rhythms and Photoperiodism in Plants. Bios Scientific Publishers, Oxford, UK.

43. Lüning, K., and P. Kadel. 1993. Daylength range for circannual rhythmicity in *Pterygophora californica* (Alariaceae, Phaeophyta), and synchronization of seasonal growth by daylength cycles in several other brown algae. Phycologia 32: 379–387.

44. Majoy, S. B., and P. D. Heideman. 2000. Tau differences between short-day responsive and short-day nonresponsive white-footed mice (*Peromyscus leucopus*) do not affect reproductive photoresponsiveness. J. Biol. Rhythms 15: 501–513.

45. Maywood, E. S., R. C. Buttery, G. H. S. Vance, J. Herbert, and M. H. M. Hastings. 1990. Gonadal responses of the male Syrian hamster to programmed infusions of melatonin are sensitive to signal duration and frequency but not to signal phase nor to lesions of the suprachiasmatic nuclei. Biol. Reprod. 43: 174–182.

46. Nakamura, K., and M. Hodkova. 1998. Photoreception in entrainment of rhythms and photoperiodic regulation of diapause in a hemipteran, *Graphosoma lineatum*. J. Biol. Rhythms 13: 159–166.

47. Nicholls, T. J., B. K. Follett, A. R. Goldsmith, and H. Pearson. 1988a. Possible homologies between photorefractoriness in sheep and birds: The effect of thyroidectomy on the length of the ewe's breeding season. Reprod. Nutr. Dev. 28: 375–385.

48. Nicholls, T. J., A. R. Goldsmith, and A. Dawson. 1988b. Photorefractoriness in birds and comparison with mammals. Physiol. Rev. 68: 133–176.

49. Pengelley, E. T., S. J. Asmundson, B. Barnes, and R. C. Aloia. 1976. Relationship of light intensity and photoperiod to circannual rhythmicity in the hibernating ground squirrel, *Citellus lateralis*. Comp. Biochem. Physiol. 53A: 273–277.

50. Philip, A. R., J. M. Garcia-Fernandez, B. G. Soni, R. J. Lucas, and J. Bellingham. 2000. Vertebrate ancient (VA) opsin and extraretinal photoreception in the Atlantic salmon (*Salmo salar*). J. Exp. Biol. 203: 1925–1936.

51. Pittendrigh, C. S. 1981. Circadian organization and the photoperiodic phenomena. In: Biological Clocks in Seasonal Reproductive Cycles, B. K. Follett and D. E. Follett (eds.), pp. 1–35. John Wright, Bristol, UK.

52. Prendergast, B. J., M. R. Gorman, and I. Zucker. 2000. Establishment and persistence of photoperiodic memory in hamsters. Proc. Natl. Acad. Sci. USA 97: 5586–5591.

53. Prendergast, B. J., R. J. Nelson, and I. Zucker. 2002. Mammalian seasonal rhythms: Behavior and neuroendocrine substrates. In: Hormones, Brain and Behavior Volume 2, D. W. Pfaff, A. P. Arnold, A. M. Etgen, S. E. Fahrbacc, and R. T. Rubin (eds.), pp. 93–156. Elsevier, New York.

54. Provencio, I., G. Jiang, W. J. DeGrip, W. P. Hayes, and M. D. Rollag. 1998. Melanopsin: An opsin in melanophores, brain and eye. Proc. Natl. Acad. Sci. USA 95: 340–345.

55. Rowan, W. 1926. On photoperiodism, reproductive periodicity and the annual migration of birds and certain fishes. Proc. Boston Soc. Nat. Hist. 38: 147–189.

56. Ruby, N. F., J. Dark, H. C. Heller, and I. Zucker. 1998. Suprachiasmatic nucleus: Role in circannual body mass and hibernation rhythms of ground squirrels. Brain Res. 782: 63–72.

57. Saunders, D. S. 1990. The circadian basis of ovarian diapause regulation in *Drosophila melanogaster*: Is the period gene causally involved in photoperiodic time measurement? J. Biol. Rhythms 5: 315–331.

58. Saunders, D. S. 2002. Insect Clocks, 3rd Edition. Elsevier, Amsterdam.

59. Schwab, R. G. 1971. Circannual testicular periodicity in the European starling in the absence of photoperiodic changes. In: Biochronometry, M. Menaker (ed.), pp. 428–447. National Academy of Sciences, Washington, DC.

60. Schwartz, W. J., H. O. de la Iglesia, P. Zlomanczuk, and H. Illnerova. 2001. Encoding Le Quattro Stagioni within the mammalian brain: Photoperiodic orchestration through the suprachiasmatic nucleus. J. Biol. Rhythms 16: 302–311.

61. Silver, R., P. Witkovsky, P. Horvath, V. Alones, and C. J. Barnstable. 1988. Coexpression of opsin- and VIP-like immunoreactivity in CSF-contacting neurons of the avian brain. Cell Tissue Res. 253: 189–198.

62. Stetson, M. H., J. A. Elliott, and B. D. Goldman. 1986. Maternal transfer of photoperiodic information influences the photoperiodic response of prepubertal Djungarian hamsters (*Phodopus sungorus sungorus*). Biol. Reprod. 34: 664–669.

63. Suárez-López, P., K. Wheatley, F. Robson, H. Onouchi, and F. Valverde. 2001. CONSTANS mediates between the circadian clock and the control of flowering in *Arabidopsis*. Nature 410: 1116–1120.

64. Taiz, L., and E. Zeiger. 1998. Plant Physiology, 2nd Edition. Sinauer, Sunderland, MA.

65. Thomas, B., and D. Vince-Prue. 1997. Photoperiodism in Plants, 2nd Edition. Academic Press, London.

66. Underwood, H., and B. D. Goldman. 1987. Vertebrate circadian and photoperiodic systems: Role of the pineal gland and melatonin. J. Biol. Rhythms 2: 279–315.

67. Underwood, H., and L. L. Hyde. 1990. A circadian clock measures photoperiod time in the male lizard *Anolis carolinensis*. J. Comp. Physiol. A 167: 231–243.

68. Vaz Nunes, M., and J. Hardie. 1993. Circadian rhythmicity is involved in photoperiodic time measurement in the aphid, *Megaoura viciae*. Experientia 49: 711–713.

69. Vaz Nunes, M., and D. Saunders. 1999. Photoperiodic time measurement in insects: A review of clock models. J. Biol. Rhythms 14: 84–104.

70. Watson-Whitmyre, M., and M. H. Stetson. 1988. Reproductive refractoriness in hamsters: Environmental and endocrine etiologies. In: Processing of Environmental Information in Vertebrates, M. H. Stetson (ed.), pp. 219–249. Springer, New York.

71. Wingfield, J. C., and G. J. Kenagy. 1991. Natural regulation of reproductive cycles. In: Vertebrate Endocrinology: Fundamentals and Biomedical Implications Volume 4, P. Pang and M. Schreibman (eds.), pp. 181–241. Academic Press, New York.

72. Woodfill, C. J. I., M. Wayne, S. M. Moenter, and F. J. Karsch. 1994. Photoperiodic synchronization of a circannual reproductive rhythm in sheep: Identification of season specific cues. Biol. Reprod. 50: 965–967.

73. Yellon, S. M., and L. T. Tran. 2002. Photoperiod, reproduction, and immunity in select strains of inbred mice. J. Biol. Rhythms 17: 65–75.

74. Zivkovic, B. D., H. Underwood, C. T. Steele, and K. Edmunds. 1999. Formal properties of the circadian and photoperiodic system of Japanese quail: Phase response curves and effects of T-cycles. J. Biol. Rhythms 14: 378–390.

75. Zucker, I., and B. J. Prendergast. 1999. Circannual rhythms. In: Encyclopedia of Reproduction, E. Knobil and J. Neill (eds.), pp. 620–627. Academic Press, New York.

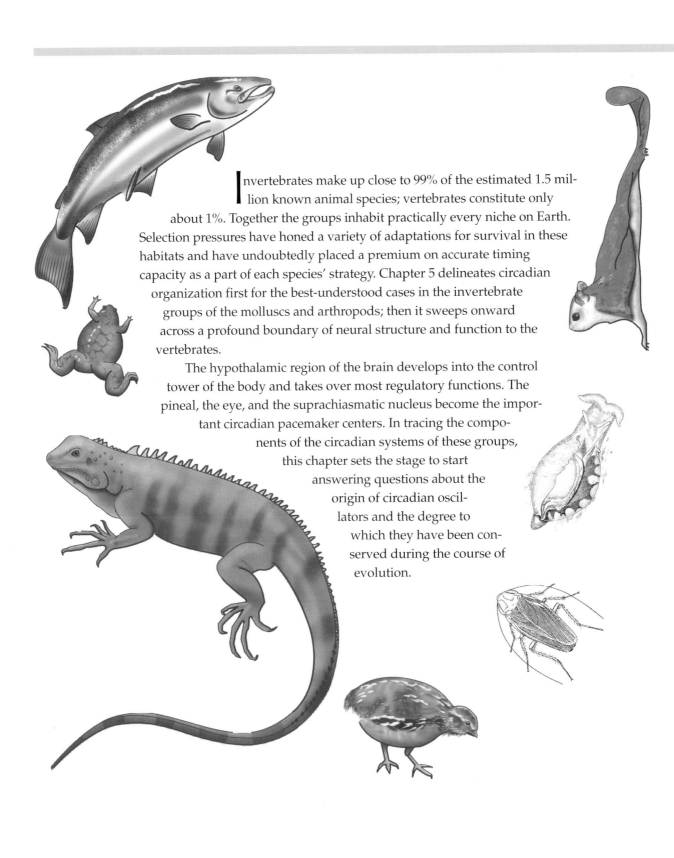

Invertebrates make up close to 99% of the estimated 1.5 million known animal species; vertebrates constitute only about 1%. Together the groups inhabit practically every niche on Earth. Selection pressures have honed a variety of adaptations for survival in these habitats and have undoubtedly placed a premium on accurate timing capacity as a part of each species' strategy. Chapter 5 delineates circadian organization first for the best-understood cases in the invertebrate groups of the molluscs and arthropods; then it sweeps onward across a profound boundary of neural structure and function to the vertebrates.

The hypothalamic region of the brain develops into the control tower of the body and takes over most regulatory functions. The pineal, the eye, and the suprachiasmatic nucleus become the important circadian pacemaker centers. In tracing the components of the circadian systems of these groups, this chapter sets the stage to start answering questions about the origin of circadian oscillators and the degree to which they have been conserved during the course of evolution.

Functional Organization of Circadian Systems in Multicellular Animals

*The illimitable, silent, never resting thing called
Time, rolling, rushing on, swift, silent, like an
all-embracing ocean tide.*

—Thomas Carlyle

Introduction: Diverse Structural Elements of the Circadian System Provide Multiple Patterns for Successful Living Clocks[22,28,29,31,43,48,57,65]

An immense number of organisms have been shaped both structurally and temporally by the winnowing hand of selection to fit the array of available habitats (see Plate 1 and Chapter 2). Of the estimated 20 to 30 million animals and half million higher plants, only about 1.5 million animals and 150,000 plants have been named and described. In almost every major group of these animals and plants, a parameter of circadian rhythmicity has been detected.

Many unicellular organisms, even though they lack specialized pacemaker organelles, exhibit three definitive features of circadian rhythms: free-running in constant conditions, entrainability, and temperature compensation. The ancient cyanobacteria, which may have been among the first organisms living on Earth, are now known to have precise circadian rhythms in some of their biochemical processes (Figure 5.1).

Unicells are valuable circadian models. Not only are unicell clocks interesting in their own right, but they cast considerable light on circadian rhythms of higher organisms. Single cells offer a relatively easy opportunity to look at rhythms. A population of unicells has the added advantage of being a more uniform substrate for biochemical analysis than an organ or tissues from a multicellular organism. In addition, favorable model species make possible a good comparative survey because they are distributed quite evenly in the major unicellular phylogenetic groups.

Among prokaryotes, *Synechococcus* is the star character of the cyanobacteria (see Figure 2.20). Of the eukaryotic protists, several species have been studied extensive-

(A)

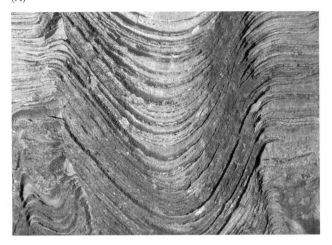

(B)

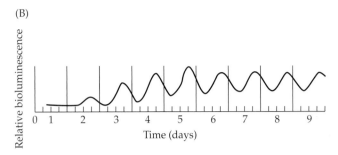

FIGURE 5.1 Circadian rhythms in bacteria. (A) Fossilized stromatolites containing cyanobacteria. (B) Circadian rhythm of a wild-type strain of the cyanobacterium *Synechococcus* that has been transformed with a luciferase reporter construct and maintained in continuous red light. (B from Johnson and Kondo, 2001.)

ly: *Euglena* in the euglenoids, *Paramecium* and *Tetrahymena* in the ciliates, *Gonyaulax* in the dinoflagellates, and *Acetabularia* in the chlorophytes (see Chapter 2). Unicells lie outside the perimeter of the organ-level performance that is the mission of this chapter because all life functions of unicells must by definition be carried out within the boundaries of a single cell. Unicells were introduced behaviorally in Chapter 2, and their molecular processes will be dealt with in Chapters 7 and 8. Similarly, plants lack multicellular circadian pacemakers, even though they have many circadian regulatory processes, as well as circadian photoreceptors scattered in their leaves. Like the unicells, they will not be treated in this chapter.

From the primitive unicells through the most advanced mammals, the complexity of body form and structure increases. The change in circadian component structure is particularly pronounced as the boundary between invertebrates and chordates is crossed. New structural elements, particularly neural ones, are often taken over by the circadian system as they evolve.

Although biological timing is widespread throughout the groups, documenting rhythmicity in some of the groups has been challenging. For many of the marine arthropod groups, complex life cycles with very small, fragile larval stages make a long-term study of rhythmicity very difficult. For example, success in rearing larval developmental stages of even the most common fiddler crabs has not been achieved until recently and can be accomplished only by means of demanding hand-rearing techniques. The result has been the narrowing of invertebrate model species to a small number of protists, molluscs, and arthropods. Even among vertebrates, many species are rare, taxing to keep alive in the laboratory, and difficult to observe in the field. Rhythmic functions of some species are hard to document in the laboratory in a noninvasive way. Consequently, most vertebrate information also comes from a handful of favorable, easy-to-study model species.

In light of the great diversity of natural history, as well as the structure and physiology of organisms, the many evolutionary solutions to the physical structure of living clocks in these groups are not surprising. Essential to any multicellular pacemaker system are three functional elements. First and foremost is an oscillator or pacemaker to give the basic drumbeat and set the period. If more than one oscillator contributes to the rhythms, a means of coupling individual oscillators may be part of the pacemaker. Second is a means of sensory input from the environment to correct the non-24 h basic periodicity to 24 h and adjust the phase to local time. Finally, the purely neural or humoral signals of the pacemaker neurons must transfer the rhythmic messages to effector organs of the overt rhythms. Learning the ingenious ways by which organisms have assembled the requisite components is one of the pleasures of studying chronobiology. Sometimes one element covers more than one task. In the sea hare, *Aplysia californica*, the eye is oscillator as well as photoreceptor for the circadian system, and it sends the output signal to the rest of the body.

The aim in this chapter is to select representative examples in a progression through the phylogenetic tree. Portraits will be painted for a few selected species of multicellular organisms whose mechanism of circadian rhythmicity is known. As much as available information permits, the anatomical components of these model circadian systems and their general function will be portrayed to illustrate the phylogenetic diversity of circadian systems.

The discussion in this chapter is limited to the model systems of multicellular organisms that have been most extensively studied. Such organisms have compartmented nervous systems with discrete pacemaker organs, as well as traceable input and output neural routes, and in many cases humoral output systems to the target organs regulated. The information should shed light on questions about when and why biological pacemakers first evolved and whether primordial clocks have been conserved throughout the evolution of successive groups.

The focus on the nervous system, however, may be somewhat misleading. The emphasis on neural pacemakers has arisen probably because behavioral rhythms are easily monitored and behavior is controlled largely by the nervous system. When efforts have been made to identify nonneural pacemakers, such as moth testes or insect cuticle, the results have been worthwhile.

Excellent Circadian Models Are Found in Two Invertebrate Groups: Molluscs and Arthropods[8,48]

Details of pacemaker and photoreceptor locations, as well as the presence of multiple oscillators and their coupling, are known for two molluscs[7,26,37,38,41,46,49]

THE EVOLUTIONARY RADIATION OF THE LARGE PHYLUM MOLLUSCA. With approximately 105,000 species, Mollusca is the second largest phylum, ranking only after Arthropoda. Great variability in form and function is found. The bivalve molluscs, such as oysters, clams, and scallops, have complex life cycles, with multiple minute larval forms whose circadian biology has not yet been studied. Adults are heavily armored with a pair of massive, calcium carbonate shells called valves to protect soft inner organs from predators. Mobility has been sacrificed to the almost impenetrable fortress of shell, and few circadian rhythms are known.

Similarly, in the gastropod snails, slugs, and sea hares, motility is reduced in most species, and only two species have attracted attention from chronobiologists. The most advanced group, the cephalopod molluscs such as octopi, nautilids, and squid are wondrously active. The octopus can sprint at high speed with its water jet, or dance delicately on tentacle tip while flashing iridescent moving color waves to attract a mate. The true brain of the octopus is the most complex known for any invertebrate, and the lateral eyes parallel the vertebrate eye in structure and function. Unfortunately, almost nothing is known about the octopus's circadian organs or function, probably because of its fragility in laboratory culture.

Well-developed circadian rhythmicity is known for only a handful of molluscan species, and only two of these species have been studied in detail. However, the great suitability of *Aplysia californica,* the sea hare, and *Bulla gouldiana,* the clouded bubble snail, has provided some of the most valuable insights available today about circadian organization and physiology from the whole organism level to the cell physiology level. Both *Aplysia* and *Bulla* are marine gastropod molluscs. Members of this group are relatively simple, both behaviorally and structurally. *Bulla* is carnivorous and night-active while *Aplysia* is herbivorous and day-active. Both organisms move chiefly by a creeping progression using the foot. Two simple eyes, each of which contains a circadian pacemaker, are embedded in the body wall.

THE COMPOUND ACTION POTENTIAL OF THE EYE AS THE MAJOR RHYTHM MEASURED. Although a weak rhythm of locomotor activity is found in both *Bulla* and *Aplysia,* the trait that makes both species invaluable for circadian study is the spontaneous compound action potential rhythm of the basal retinal cells of the simple eyes. In *Aplysia* and *Bulla* the eyes exhibit circadian rhythms of spontaneous nerve discharge activity that can easily be monitored with in vitro preparations along the optic nerve (Figure 5.2). Each eye in *Bulla* acts as a separate oscillator, but in the living animal both eye oscillators are coupled to each other. When the eyes are removed from the animal, they can be maintained in culture medium either separately or as a pair, and the nerve impulses can be recorded.

Since the discovery of precise circadian rhythms in isolated cultured eyes of *Aplysia* in 1969, and shortly thereafter in *Bulla,* study of the generation of circadian oscillation and coupling of the two eye pacemakers has made rapid strides. The rhythms are expressed in the frequency of compound action potentials (CAPs) in the optic nerve (see Figure 5.2A), which arise from a population of electrically coupled retinal cells in the base of the eye. At the peak of activity, CAPs are produced spontaneously at the rate of several per minute; at the trough, no CAPs are produced. When spike rate is recorded from the optic nerve and plotted over the course of the day, a very precisely defined peak is seen, with an abrupt rise at subjective dawn (see Figure 5.2B and C) and then continuing during an animal's subjective day.

A PACEMAKER IN THE RETINA OF THE EYE. The nervous systems of *Aplysia* and *Bulla* are organized similarly. The central nervous system is composed of a series of simple ganglia called, from anterior to posterior, the buccal, cerebral, pedal, pleural, and visceral ganglia (see the opening illustration of Chapter 11). Each eye communicates with the cerebral ganglia via an optic nerve. Anatomical and histological studies of the detailed structure of the eye have helped identify the components of the circadian system in the two molluscs.

The pacemaker has now been localized in specific cells of the eyes of the two species. The eye of *Bulla* is structurally simpler than that of *Aplysia,* making research with this species considerably easier (Figure 5.3). A small lens is surrounded by a distal or dorsal retinal layer containing about a thousand microvillous

(A) Compound action potentials (CAPS) from optic nerve of *Bulla* in DD

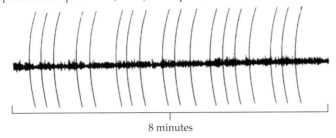

8 minutes

(B) Circadian rhythm of CAPS for *Bulla* in DD at 15°C

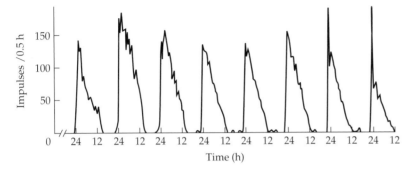

Time (h)

(C) Circadian rhythm of CAPS for *Aplysia* in DD at 15°C

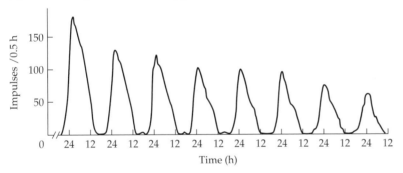

Time (h)

FIGURE 5.2 Spontaneous circadian rhythms of electrical activity measured along the optic nerve. (A) Electrical activity recorded from a *Bulla* optic nerve in DD for 8 minutes. (B, C) Circadian rhythms showing spontaneous CAP frequency from eyes in DD at 15°C for *Bulla* (B) and *Aplysia* (C). (From Blumenthal et al., 2001.)

toreceptor cells, the basal retinal neurons show a one-to-one correspondence in their firing rate with the CAP rhythm along the optic nerve, and they are electrically coupled to each other. All of these cells are similar in size and configuration, with their neural processes extending into the neuropil and their axons into the optic nerve. By careful surgical dissection, the number of basal retinal neurons can be reduced to about six cells in *Bulla* without loss of the compound action potential rhythm in the optic nerve. Because of their small size, 15 to 25 μm, it has not been possible to surgically reduce the cell number to a single neuron. In the reduced surgical preparation a rhythm persists that is progressively attenuated in amplitude as the number of basal retinal neurons is reduced.

The eyes can be removed and cultured in sterile seawater for up to 14 days. The rhythms of the intact eyes or reduced eye fragments persist in complete darkness, with peak impulse activity occurring in both species during an animal's subjective day. In darkness the rhythms free-run with a period close to but not exactly 24 h, thus indicating the circadian nature of the eye itself as an oscillator. Because the rhythm free-runs in constant conditions after explantation and entrains in culture in an LD schedule, a circadian pacemaker and the associated photoentrainment pathways must be located solely within the eyes.

Additional sophisticated techniques have recently been developed for maintaining single isolated basal retinal cells from *Bulla* in dispersed culture for extended periods of time. The results of recording from these preparations indicate that even a single *Bulla* basal retinal cell is capable of a persistent circadian rhythm of membrane conductance. Thus the pacemaker resides in single cells that in living animals must be electrically coupled for an integrated compound action potential rhythm.

Because of greater anatomical complexity of the eyecup of *Aplysia*, exact localization of the retinal circadian pacemaker in this species is more difficult. The lens is surrounded by a ring of microvillous distal photoreceptors, as in *Bulla*, but at least five microscopically distinct types of cells are present. The basal retinal zone of neurons in *Aplysia* has two distinct subtypes. Surgical reduction of the two zones demonstrates that the basal retinal zone is the pacemaker, and that a small number of neurons or possibly single neurons are competent as a pace-

photoreceptor cells. Of the two types of photoreceptor cells found, R and H, the R cells are more common. They do not spike spontaneously, and do not show a spontaneous rhythm in resting membrane potential. The less common H cells do discharge spontaneously, but their discharge pattern shows no relationship to the CAP rhythm. Total removal of the dorsal R and H photoreceptors in *Bulla* fails to change the CAP rhythm of a cultured eye.

Further searching reveals another group of neurons at the base of the retina, in the fibrous neuropil area of the eye, close to the exit of the optic nerve. A ring of approximately 130 cells called basal retinal neurons is found here (see Figure 5.3). In contrast to the dorsal R and H pho-

FIGURE 5.3 The eye of *Bulla*. Visible in this schematic drawing are the major structures of the eye: lens, distal retinal neurons, optic nerve, and basal retinal neurons. (From Jacklet and Colquhoun, 1983.)

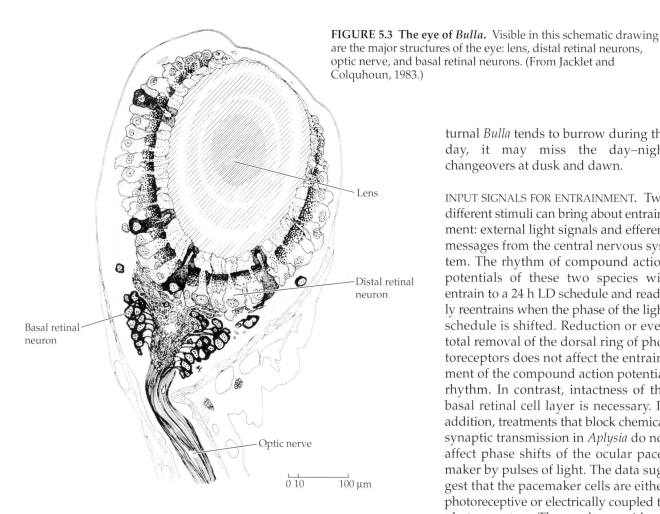

Lens

Distal retinal neuron

Basal retinal neuron

Optic nerve

0 10   100 µm

maker, although the work has not been carried out as completely as in *Bulla*.

MUTUAL COUPLING OF THE TWO EYES. Species differences have been observed in the degree of coupling of the two ocular pacemakers. In paired *Aplysia* eyes maintained in constant darkness, coupling between the eyes is weak or absent, and the two eyes soon desynchronize. Because *Bulla* eyes remain in phase in culture, the pacemakers appear to be strongly coupled. This species difference in the strength of coupling was confirmed in further studies in vitro. Following treatments to one eye that induced a phase difference of several hours between the ocular pacemakers, *Bulla* eyes tended to come back into phase at a rate of about 1 to 1.5 h per day, while in *Aplysia* phase separations were maintained after desynchronization of the two eyes. The uncoupling has little effect on the period, amplitude, or persistence of the rhythms in the eyes.

Possibly the primary function of the coupling pathway is to maintain synchrony between the pacemakers in the absence of environmental cues. Such internal synchrony might be more important for animals that are likely to miss environmental time cues. Because the noc-

turnal *Bulla* tends to burrow during the day, it may miss the day–night changeovers at dusk and dawn.

INPUT SIGNALS FOR ENTRAINMENT. Two different stimuli can bring about entrainment: external light signals and efferent messages from the central nervous system. The rhythm of compound action potentials of these two species will entrain to a 24 h LD schedule and readily reentrains when the phase of the light schedule is shifted. Reduction or even total removal of the dorsal ring of photoreceptors does not affect the entrainment of the compound action potential rhythm. In contrast, intactness of the basal retinal cell layer is necessary. In addition, treatments that block chemical synaptic transmission in *Aplysia* do not affect phase shifts of the ocular pacemaker by pulses of light. The data suggest that the pacemaker cells are either photoreceptive or electrically coupled to photoreceptors. The results provide an interesting case in which both pacemaker and photoreceptor reside in the same cell. The basal retinal cells act as oscillator and generator of the circadian CAP rhythm through spontaneous changes in membrane potential. They also play the role of photoreceptors for entraining the rhythm. The opsinlike immunoreactivity that has been detected in basal retinal somata and processes supports this interpretation.

The second part of the entrainment pathway involves neuromodulatory signals from the central nervous system. In *Aplysia* the message is serotonin, a biogenic amine. In *Bulla* a peptide, FMRFamide, is the effective agent. The substances can be traced histochemically from cell bodies in the cerebral ganglia along fibers entering the eyes. The specific mechanism by which phase shifting in photoentrainment takes place, either singly through photic signals or in concert with central nervous system signals, involves changes in membrane potential. Since such processes are by definition cellular, they belong to the domain of Chapter 6.

NEURAL AND HORMONAL OUTPUT SIGNALS. Another important question concerns the means by which circadian oscillators impose periodicity on the various physiological and behavioral processes they regulate. The existence of multiple alternative routes may complicate study of the pathways. Phase information, for example, might be transmitted through neural, hormonal, or a combination of messages. This chapter will emphasize structural routes, leaving most cellular details for Chapter 6.

The output of the ocular pacemaker to the central nervous system in both *Bulla* and *Aplysia* travels via the optic nerve, whose integrity is necessary for the normal circadian rhythms of behavior expressed by the molluscs. The most prominent rhythmic signal in the nerve is the rhythm in compound action potentials. In the *Bulla* optic nerve an additional rhythm involves smaller asynchronous impulses, which are 180 degrees out of phase with the rhythm in CAP. Recent results indicate that the rhythmic small-spike activity is produced by a subpopulation of distal photoreceptor cells whose activity is controlled by the basal retinal neurons via an inhibitory synaptic connection. How information arising from the two, out-of-phase cell populations is used in the central nervous system is completely unknown.

Structural elements and their function have also been localized in the insect group of the arthropods, but much more structural diversity has been found than in molluscs[15,16,17,36,46,48,50]

TYPES OF INSECT BEHAVIORAL AND PHYSIOLOGICAL CIRCADIAN RHYTHMS. Studies of circadian rhythmicity in insects reveal many physiological and behavioral examples, any one of which could potentially serve as an indirect marker for the state of the circadian clock. Certain outputs are more convenient and less invasive than others. The rhythms reported fall into about five basic groups. Daily locomotor activity rhythms are common in many insects and have been studied particularly in dipterans such as *Drosophila*, as well as in cockroaches (Figure 5.4), night-active moths, crickets, and beetles. As with most other terrestrial organisms, activity of arthropods is usually either nocturnal or diurnal, since animals are generally structurally best suited for

life in one of these two widely differing time frames of the daily cycle.

Some arthropods advertise during courtship with strongly diurnal or nocturnal rhythmicity. The calling of katydids on a summer night is unforgettable. Crickets also chorus nocturnally, using species-specific call coding for species recognition in courtship. The cricket *Teleogryllus commodus*, for example, calls at night during summer courtship, and this rhythm persists in isolated individuals in constant conditions in the laboratory as a very precise free-running rhythm (see Figure 2.8). Many female silk moth species release a scent at night to attract males from distances up to several kilometers (Plate 3, Photo D). The daytime chorus of 17-year locusts is also impressive.

Another behavioral category is gated population rhythms, such as pupal hatching rhythms of dipterans. Most insects undergo a complex life history. In the so-called incomplete metamorphic pattern, a sequence of egg, nymph, and adult stages occurs. In outward appearance the nymph is a wingless version of the small adult, which undergoes a series of molts and then expands its wings in the final molt. Orthopterans such as crickets, roaches, and grasshoppers are good examples. The alternate life history mode, complete metamorphosis, is displayed by a variety of other insects, including beetles, dipterans, and lepidopterans. For these species, the egg stage hatches into a larval grub or caterpillar, which feeds until mature and then enters a nonmotile stage, called a pupa. During the subsequent days the tissues of the larva are transformed into adult tissues. A dramatic transition called eclosion takes place when the adult breaks out of its pupal case, expands its appendages, and starts adult life (Figure 5.5). The eclosion rhythms of flies such as *Drosophila* have been the measurement parameter of choice for many fly studies (see Figure 2.11 in Chapter 2). Circadian rhythmicity acts as a gating process, allowing the emergence of new adults at the most appropriate time of day for maximal survival. Note that the eclosion rhythm is a population rhythm because it occurs only one time in the life of an individual. To study daily circadian rhythms of an individual fly, another function, such as locomotor activity, must be monitored.

Still another circadian behavior involves a continuously consulted biological pacemaker. In this case the clock is used to measure lapse of time to allow for compensation of the sun's movement of 15 degrees per hour across the sky when the sun is being used as a compass. Such time-compensated use of a clock in conjunction with a moving celestial compass for navigation is known for the honeybee. Long flights from the hive often entail several hours' absence from the hive, and accurate return requires a time-telling ability. The appearance of bees at flowers that open only at specific times of day is also well known (see Chapter 2). This

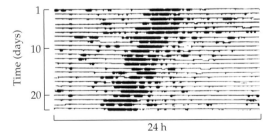

FIGURE 5.4 Circadian rhythms in insects. Locomotor activity represented in a single-plot actograph for a cockroach in DD. (From Barrett and Page, 1989.)

(A)   (B)   (C)   (D)   (E)

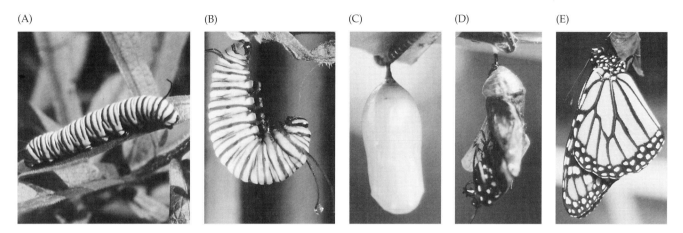

FIGURE 5.5 Life cycle of a monarch butterfly, shown in sequence: (A) larval caterpillar feeding on milkweed leaves, (B) caterpillar hanging up to pupate, (C) chrysalis or pupa, (D) adult emerging from chrysalis (eclosion), and (E) mature adult preparing to lay eggs. (Courtesy of P. DeCoursey.)

capacity, called Zeitgedächtnis, or "time memory," in bees also depends on a circadian timer.

Physiological responses are also known. One of these is the rhythmic change of retinal responsiveness to light. The response rhythm can be recorded as an electroretinogram. Many other physiological rhythms are not practical as clock hands because they would require very invasive documentation methods. For this reason, behavioral events such as locomotor activity and eclosion have been the most widely used circadian markers for insects.

LOCALIZATION OF CIRCADIAN PACEMAKERS IN INSECTS. The pacemakers regulating the circadian rhythmicity of insects have been studied in detail, and in several insects circadian pacemakers have been localized to discrete regions of the brain. The insect brain can be divided into two major areas. The optic lobes are paired bilateral structures that receive and process input from the compound eyes. Visual information is then transmitted into the central brain, which consists of the paired cerebral lobes.

The optic lobes generate circadian oscillations in many insect species (Figure 5.6). Removing both optic lobes of the cockroach *Leucophaea maderae* or disconnecting them from the rest of the brain resulted in loss of locomotor activity rhythms. Removing only one optic lobe had little effect on the rhythm. Similar results have been obtained in several other species of cockroaches and in crickets. The studies showed that at

least one intact optic lobe was necessary for the expression of locomotor activity rhythms and suggested that the optic lobes either contain the circadian pacemaker or are part of the output pathway of the pacemaker.

Two other pieces of evidence confirmed a pacemaker function for the optic lobes. The first arose from optic lobe transplants in cockroaches. When rhythms of the donor and host animals had different free-running peri-

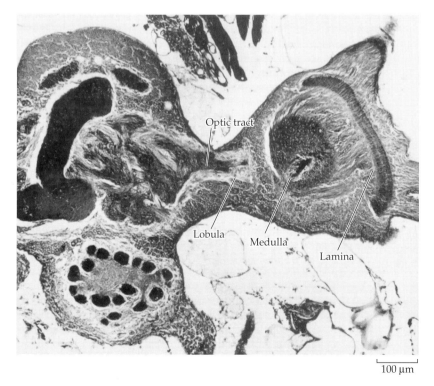

FIGURE 5.6 Photomicrograph of the brain (left), optic lobe, and compound eye (right) of a cockroach. Parts of the optic lobe include the lobula, the lamina, the medulla, and the optic tract. (From Binkley, 1990.)

ods, the preoperative period of the donor and the postoperative period of the host were strongly correlated. When optic lobes were transplanted from one animal to an animal whose own optic lobes had been removed, the rhythm of locomotor activity was restored. Highly convincing was the fact that the transplanted optic lobes not only restored rhythmicity but also imposed the period of the donor on the rhythm of the host. Finally, unequivocal evidence that the optic lobe contains a circadian pacemaker was provided by the demonstration that a lobe could continue to generate an oscillation even when isolated from the rest of the animal. If the optic lobes of either cockroaches or crickets were removed from the animals and maintained in organ culture, they expressed a circadian rhythm in spontaneous nerve impulse activity.

The region within the optic lobe that contains the circadian pacemaker for the locomotor activity rhythm in cockroaches has been more precisely identified through localized lesions. The optic lobes have three anatomically discrete regions. The lobula is the most proximal, the medulla is in the middle, and the lamina is near the compound eye (see Figures 5.6 and 5.7). The effects of small lesions suggest that structures responsible for generating the circadian signal are located in the proximal half of the optic lobe, most likely in a group of cells located ventrally between the medulla and the lobula.

The localization of circadian pacemakers that regulate locomotor activity raised the question of whether other rhythms are controlled by the same clock. The retinal sensitivity rhythm as measured by the electroretinogram is another circadian function found in many insects. The results of lesioning studies in crickets, beetles, and cockroaches indicate that the pacemaker involved is located in the optic lobes, and they suggest that the same pacemaker controls both the activity and the electroretinogram amplitude rhythms.

Not all insects use the optic lobe as circadian pacemaker. In contrast to cockroaches, crickets, and beetles, other insects, such as fruit flies and the developmental stages of some moths, are known in which the optic lobes are not necessary for rhythmicity and the pacemaker appears to reside in the cerebral lobes. In silk moths, both lesion and transplantation studies show that the circadian pacemaker controlling the eclosion rhythm is located in the cerebral lobes. In dipteran insects, such as flies and mosquitoes, locomotor activity rhythms continued after surgical lesions of the optic

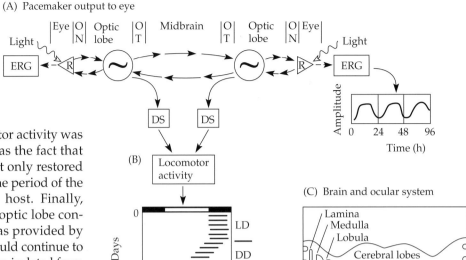

FIGURE 5.7 Cockroach brain and activity patterns. (A) Pacemaker output to eyes. (B) Pacemaker output to locomotor system. (C) Structure of brain and eye. DS, driven system; ERG, electroretinogram; ON, optic nerve; OT, optic tract; R, retina. (From Page, 1988.)

lobes but not after lesions of the cerebral lobes. Similar results have been obtained with genetic lesions in the fruit fly, *Drosophila melanogaster*. Circadian rhythms of locomotor activity in this species are present in a variety of neuroanatomical mutants with severely reduced or deformed optic lobes.

The anatomical details of the circadian system in *Drosophila* have been revealed by studies of the tissues and cells that express the *period* (*per*) gene. Extensive behavioral, cytochemical, and molecular genetic evidence supports a crucial role for the *per* gene in the circadian pacemaker controlling locomotor activity and eclosion rhythms in the fruit fly (see Chapters 7 and 8). The *per* gene is expressed widely in the head, thorax, and abdomen of wild-type flies, such that its complex expression pattern provides no definitive evidence regarding the location of circadian pacemakers controlling rhythmic behaviors (Figure 5.8A).Because this pattern has been altered by so many genetic and molecular manipulations, however, it has been possible to deduce the identity of the pacemaker cells in *Drosophila* by correlating *per* expression in specific cell types with the presence or absence of behavioral rhythmicity. The most compelling results are those involving several transgenic strains bearing a 7.2-kilobase (kb) fragment of *per* DNA that includes the entire coding sequence but lacks some upstream regulatory sequences. The expression pattern of this *per* transgene varies among different strains, pre-

(A)   (B)

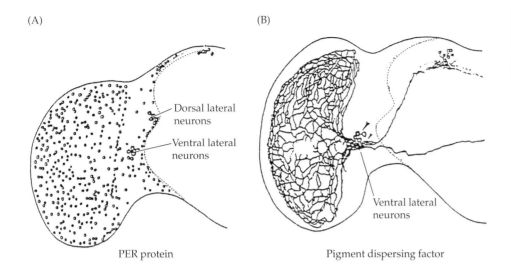

Dorsal lateral neurons

Ventral lateral neurons

Ventral lateral neurons

PER protein   Pigment dispersing factor

FIGURE 5.8 Expression of PER protein (A) and PDF (B) in the brain of *Drosophila*. For further explanation see the text. (From Helfrich-Forstser, 1995.)

sumably as a function of where in the genome the 7.2 kb *per* fragment is inserted.

Although most strains of 7.2 kb flies have no detectable expression of the transgene and are behaviorally arrhythmic, two 7.2 kb strains are rhythmic. One of these rhythmic strains expresses *per* only in a few neurons between the lateral protocerebrum and the medulla of the optic lobes, indicating that *per* expression in these lateral neurons is sufficient to generate locomotor activity rhythms. A circadian pacemaking function for these *per*-expressing lateral neurons is further supported by studies of neuroanatomical mutants. For example, the lateral neurons are present in small optic lobe mutants and in *sine occulis* double mutants, which retain behavioral rhythmicity. These lateral neurons seem to be absent in *disconnected* mutants, which are behaviorally arrhythmic.

Immunological studies have shown that cells in the pacemaker of fruit flies, cockroaches, and crickets can be labeled with an antibody to a crustacean peptide hormone called pigment dispersing hormone (PDF) (Figure 5.8B). Two distinct groups of lateral neurons express *per* in fruit fly brains: One is more dorsal and anterior with three to seven lateral neurons, and a second more ventral group of lateral cells is located near the anterior margin of the medulla. The ventral group but not the dorsal group is labeled with the PDF antibody (see also Chapter 8). In both cockroaches and crickets, PDF-immunoreactive neurons are located in the optic lobes in the same regions where lesion studies have indicated pacemaker activity. The numbers and projection patterns of the PDF-immunoreactive neurons are strikingly similar in *Drosophila*, cockroaches, and crickets, suggesting that they are structurally and functionally homologous. Although additional studies will be necessary, these results have led to the suggestion that PDF

immunoreactivity is a marker of circadian pacemaker cells in the insect brain.

INSECT CIRCADIAN PACEMAKERS IN TISSUES OUTSIDE THE NERVOUS SYSTEM. Efforts to find circadian oscillators in invertebrates have focused on the nervous system, but well-documented cases in insects substantiate the existence of circadian pacemakers outside the brain and optic lobe. In cockroaches, for example, the cuticle of the newly molted adult is secreted rhythmically in layers, one layer per day, and the rhythm persists in constant conditions with a circadian period that is temperature compensated. The rhythm continues in the animal after complete ablation of the optic lobes, and it can even be detected in pieces of leg tissue cultured in vitro. Thus the rhythm is driven by a circadian pacemaker system that is independent of the optic lobe pacemaker for activity. Possibly the epidermal cells that secrete the cuticular material are autonomously rhythmic.

The complex of testes and seminal ducts in gypsy moths also contains a circadian pacemaker. The release of sperm from the testes into the seminal ducts exhibits a circadian rhythm. The rhythm persists and can be phase-shifted by light in complexes of seminal ducts and testes that have been removed from the animal and maintained in organ culture. Thus this tissue contains both a circadian pacemaker and a photoreceptor for entrainment.

In another moth, *Samia cynthia*, preparation for pupation and metamorphosis involves a series of morphological and behavioral changes. One of these changes is an excretion of the contents of the gut, which occurs shortly after a cessation in feeding. Gut purge is regulated by a circadian clock located in the prothoracic gland, an endocrine organ that is responsible for secretion of the insect steroid hormone ecdysone. A small peak of ecdysone occurs in the hemolymph just prior to gut purge. A brain-centered clock is not necessary, since the circadian timing of the ecdysone peak remains in decapitated larvae. Other studies, involving transplantation of the prothoracic glands followed by their local illumination, indicated that the glands contain a photosensitive circadian clock that can regulate the timed release of ecdysone.

The widespread distribution of circadian pacemaker activity in insects has been shown in a remarkable fruit fly experiment in which the promoter for the *per* gene was coupled to a luciferase reporter, creating a fly that glows with a circadian rhythm (see also Chapter 11). A transgenic *Drosophila* can be constructed by fusion of the *per* gene with the luciferase gene *luc*. The *per*-driven bioluminescent oscillation can then be detected with photomultipliers and recorded as an actogram. Almost the entire fly glows. Similarly, green fluorescent protein can be driven by the promoter of the clock gene *per*. In vitro cultures of head, thorax, and abdomen tissues showed rhythmic circadian rhythms of luminescence. The phase of the rhythms of isolated pieces can also be reset by light. The results suggest that self-sustained circadian oscillators and their photoreceptors are widely distributed in the body of *Drosophila*.

DIVERSITY OF INSECT PHOTORECEPTORS FOR CIRCADIAN ENTRAINMENT. The photoreceptors responsible for entrainment of circadian rhythms in insects appear to be as varied as the pacemakers themselves. In at least two insect cases, the cockroach and the cricket, the compound eyes are the exclusive photoreceptive organs for entrainment. Severing the optic nerves between the eyes and the optic lobe eliminates the entrainment of locomotor activity rhythms to light cycles. Painting over the compound eyes has similar results.

In other species, extraretinal photoreception may be important. The photoreceptors for entrainment of the eclosion rhythm of the silk moth are located in the brain. In a definitive experiment, brains were removed and then either re-placed in the head region or transplanted to the abdomen. The pupae were subsequently placed in holes in a partition separating two chambers with light–dark cycles out of phase. The insect's eclosion was entrained by the light–dark cycle of the end of the pupae corresponding to the location of the brain. Additional evidence that extraretinal photoreceptors can entrain neural pacemakers has been obtained in a variety of butterflies, flies, and grasshoppers. Even though the compound eyes may not be necessary for entrainment, they may nevertheless participate, at least in some species. For example, blind or eyeless *Drosophila* mutants entrain well to light–dark cycles, indicating the existence of an extraretinal photoreceptor. However, the fact that wild-type flies are much more sensitive to entraining light pulses than are *eyeless* mutants, suggests that photoreceptors in the compound eye are important for entrainment in natural populations. Similarly, in the horseshoe crab the compound eyes contribute to entrainment along with photoreceptors in the tail.

NEURAL AND HORMONAL OUTPUT SIGNALS FOR COMMUNICATING TIMING INFORMATION. Another important issue concerns output by circadian oscillators for the various physiological and behavioral processes they control. Some progress has been made in understanding the coding and transmission of circadian information. Several plausible mechanisms have been proposed. Phase information within the individual could be represented by the level of a circulating hormone, by impulse frequency in specific neural circuits, by changes in general levels of neural excitability, or by a combination of these mechanisms.

Output rhythms have been documented for insect development. The secretion of a variety of insect hormones is under the control of the circadian system during development. Included are ecdysone, prothoracicotropic hormone, and eclosion hormone. These hormones are involved in the regulation of various developmental events, such as ecdysis. Most of the research has been carried out on various species of moths.

Experiments involving the transplantation of the silk moth brain, described earlier, provide the clearest demonstration of a hormonal link in the control of eclosion by the circadian system. Eclosion hormone is produced in neurosecretory cells located in the pars intercerebralis, near the dorsal midline of the brain. Circadian release of eclosion hormone triggers a motor program in the ventral nerve cord. The program then initiates a stereotyped behavioral sequence that ultimately results in the emergence of an adult moth from its pupal case.

Output signals have also been studied for adult behavior, but the role of humoral factors in the regulation of adult behavior patterns is less clear. The timing signal in cockroaches and crickets originates in the optic lobe and is transmitted to the brain via the optic tracts. Transmission to brain targets in the thorax requires that the connectives of the ventral nerve cord remain intact. Similarly in crickets, lesion studies demonstrate that axons of the optic tracts transmit the circadian signal from the optic lobe to the brain. On the other hand, evidence from brain transplant experiments in *Drosophila* show that locomotor activity rhythms can be imposed by a humoral signal from brains transplanted into the abdomens of arrhythmic host flies.

Other arthropods such as scorpions and horseshoe crabs have many circadian adaptations[3,15,16,17]

SCORPIONS. Opportunities abound with scorpions for examining the ecological and evolutionary relevance of daily endogenous timing, and for determining the neurological basis of the circadian pacemaker systems. These predatory venomous arachnids have evolved striking adaptations for desert life in their structure, behavior, metabolism, and daily rhythmicity. A very low metabolic rate, combined with a deadly stinger and a sit-and-wait strategy of hunting, help scorpions cope with food scarcity. Strict nocturnality and multiple sets of eyes are an effective temporal part of the total strategy.

The scorpion's stereotyped activity rhythm progresses daily from resting during daytime to awakening at twilight, followed by motionless doorkeeping until dark, then patrolling the territory for 2 to 3 h. The rhythm concludes with waiting motionless for 4 to 5 h in the vicinity of the scorpion's burrow to ambush any hapless prey (Figure 5.9A–C; see also Plate 2, Photo I).

The clear distinction between day and night states is controlled by a circadian clock system that can be monitored automatically under constant laboratory conditions. The strong autofluorescence of the scorpion cuticle under ultraviolet light facilitates field studies at night.

The central nervous system and sense organs are relatively simple in scorpions and can be studied by standard neurological methods. Ganglia are relatively small clusters of neuronal cell bodies and their neuropils. Included in the neuropils are synaptic regions of dendrites and nerve terminals. Neuropils are the main information-processing sites. In contrast to vertebrate neurons, those of invertebrates are mainly the source of neurochemicals, and are less relevant for the exchange of neural information. Scorpions have 12 eyes: two median eyes and five lateral eyes on each side. The median eyes serve

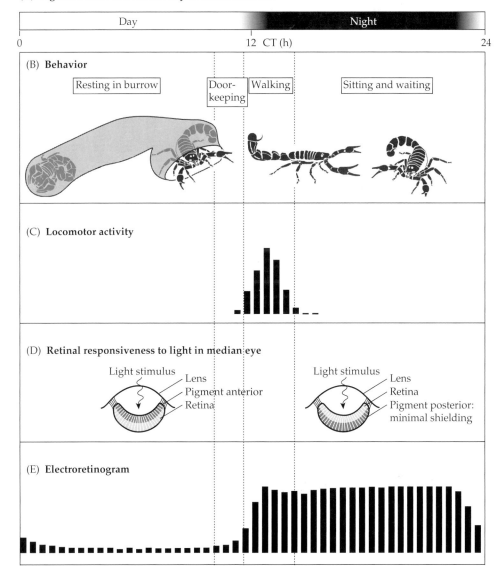

(A) **Light schedule and time of day in circadian hours**

(B) **Behavior**

(C) **Locomotor activity**

(D) **Retinal responsiveness to light in median eye**

(E) **Electroretinogram**

FIGURE 5.9 Circadian rhythmicity in a scorpion. (A) LD schedule and time scale. (B) Daily activity routine. (C) Locomotor activity during patrolling. (D) Retinal pigment position for median eyes during daytime (left) and nighttime (right). (E) Electroretinogram showing the daily rhythm of sensitivity to light. (Courtesy of G. Fleissner and G. Fleissner.)

in image formation; the lateral eyes are better suited for luminance detection and photon counting for circadian photoentrainment. Each individual optical structural unit, called an ommatidium, contains all the elements for light reception. In a compound eye the units are aligned for greatest light-gathering efficiency (see Figure 5.14). In both types of eyes, a retinal shielding pigment is withdrawn by night and migrates forward before dawn. The pigment movement results in a large-scale daily change in photosensitivity (Figure 5.9D). Pigment retraction at night facilitates vision in extremely low light, and the dispersal by day protects the fragile retinal cells from the intense sunlight of a desert habitat.

The daily rhythm of retinal pigment dispersal has been called the "sunglasses" phenomenon (Figure 5.9D and E), and it is measured in the laboratory by surface electrodes fixed to the lenses. A test light is programmed every half hour with a near-threshold intensity at less than 1 lux and with a duration of only 30 milliseconds. Because of its low intensity and short duration, the test light does not affect free-running rhythms or entrainment. The electrodes detect receptor potentials of the photoreceptor cells that have responded to the test light. The resultant electroretinogram records of responsiveness parallel the pronounced active and rest state. Both the median and lateral eyes have this sensitivity rhythm, although the amplitude is smaller in the lateral eyes.

Simultaneous recording of activity and electroretinogram rhythm can be continued for many months in the laboratory (Figure 5.10). The presence of a circadian pacemaker is reflected in the free-running locomotor activity rhythm and electroretinogram rhythm under constant dark conditions. In DD conditions the free-running period averages 23.8 h; in constant dim light, the period lengthens to about 26.1 h. Light–dark cycles are effective entraining agents.

The neuronal circuitry underlying circadian locomotor rhythmicity and pigment dispersal rhythm in scorpions has been described in detail. The sunglasses rhythm and behavioral rhythm are controlled by the same internal clock system. The oscillator cells have not yet been localized, but the sites appear to lie in the primitive brain, not in the eyes (Figure 5.11). The bilaterally symmetrical pacemakers are thought to lie next to the third, fourth, and fifth optic ganglia. Circadian signals are transmitted to the eyes by ten efferent neurons on each side. Their cell bodies lie at the lateral base of the brain, and their long axons connect the pacemakers to all retinas of both sides, and to the central body. Electroretinogram recording before and after optic nerve lesioning suggests that a series of low-frequency stimula-

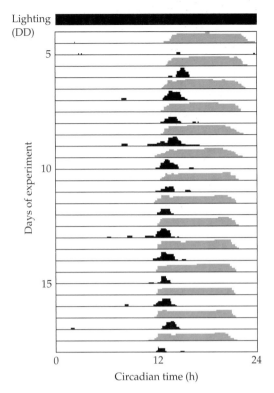

FIGURE 5.10 Simultaneous recording of patrolling activity (black) and electroretinogram rhythm (stippled) for one scorpion in total darkness. The free-running period is slightly shorter than 24 h. For further explanation, see the text. (From Fleissner and Fleissner, 2001b.)

tory impulses cause release of the neurotransmitter octopamine. Further pharmacological experiments demonstrate that octopamine induces the circadian night state.

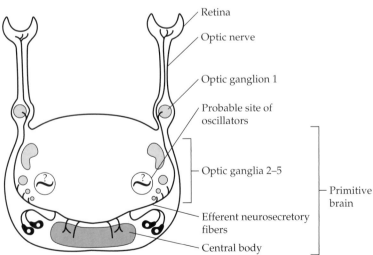

FIGURE 5.11 The scorpion brain and course of efferent neurosecretory fibers. The fibers shown in this schematic drawing transmit the circadian signal from the pacemakers to the retinas and to the central body. (From Fleissner and Fleissner, 2002.)

Scorpions use an intricate arrangement of visual light input. Imaging eyes and other photon-counting eyes supply information to a central pair of oscillators that in turn regulate a retinal sensitivity rhythm of the light receptor cells themselves. Such a system offers good opportunities for deciphering complex feedback loops in circadian pacemaker systems.

HORSESHOE CRABS. Related to scorpions is another ancient group of arthropods, the horseshoe crabs (Plate 4, Photo B). After spending most of their lives in shallow coastal waters, these mariners come ashore on sandy beaches in summer in mass aggregations for mating. Like their relatives the scorpions, horseshoe crabs also have highly interesting circadian ocular adaptations in their three pairs of eyes. Both the lateral eyes and the median ocellar eyes exhibit pronounced circadian rhythms of retinal sensitivity, as well as circadian patterns of spontaneous efferent action potentials.

Electroretinogram recording before and after optic nerve lesioning indicates that regulation comes from a circadian pacemaker located centrally in the brain (Figure 5.12). Another ocular circadian rhythm consists of daily shedding of the photosensitive membranes of the lateral eyes. Specialized retinal cell organelles capture and transduce photons to neural impulses. These organelles are composed of microvillous membrane arrays. The membranes are apparently unstable and are cast off and broken down in a brief dawn burst, then quickly reconstituted for the next day's activity.

Scorpions and horseshoe crabs have many advantages for analyzing ecological aspects of circadian adaptiveness and for analyzing the neurobiology of the circadian system. However, transplants and genetic studies of the rhythm generator will not be feasible until the specific oscillatory cells in pacemakers are localized.

Many neural changes accompanied the rise of the vertebrates[32,51]

In most bilaterally symmetrical invertebrates, centers of neural activity occur primarily as relatively small clusters of nerve cell bodies and neuropil tissue called ganglia. In primitive invertebrates, two ventral nerve cords run the length of the body, with cross-commissures between right and left sides, and with the ganglia ordered like beads along a chain. The entire nervous system is solid, without ventricles in either the cords or the ganglia. Molluscs have about seven pairs of ganglia along a central anterior–posterior axis. In relatively advanced gastropod molluscs such as *Bulla* and *Aplysia*, most of the ganglia have fused. The resultant ring-shaped primitive brain is located at the anterior head end (see the opening illustration of Chapter 11). Fusion of the ganglia in insects results in a neural mass above the esophagus and another below, termed, respectively, the supraesophageal and subesophageal ganglia. Sense organs are connected to the ganglionic brain by small nerve trunks, such as the optic nerve to the eyes. The true brain of the cephalopod molluscs is a complex structure with marked division of labor of the central brain into functional areas. Separate motor, visual, and learning centers are well developed. The cephalopod brain and eyes foretell the neural development that will come with the vertebrate chordates.

By comparison to invertebrates, the number of neurons in vertebrates has increased by several orders of magnitude. The number of synapses per neuron has also increased greatly. The complex organization of neurons into specific areas has become one of the most prominent vertebrate brain features. The basic plan of all vertebrate chordates is a central brain and spinal cord, running dorsally along the anterior–posterior axis. The brain and cord are hollow in all species and encased in a rigid cartilaginous or bony skull capsule. The flexible bony or cartilaginous vertebrae cover the spinal cord and permit the egress of the paired spinal nerves. The hollow spaces of brain and cord are enlarged into so-called ventricles and are filled with circulating cerebrospinal fluid. The system is most easily pictured in its primordial embryonic form as a simple tube. However, even as the neural folds of a fetus fuse into a tube, the anterior part of the tube enlarges and differentiates into five major brain regions (Figure 5.13). Such a stage is similar to the nervous system of the primitive protochordate *Amphioxus*.

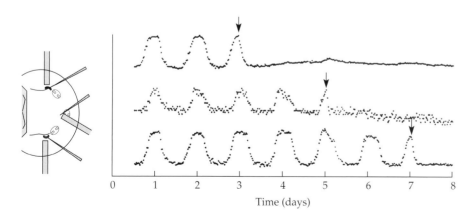

0   1   2   3   4   5   6   7   8

Time (days)

FIGURE 5.12 Circadian rhythms in the electroretinographic responses of the lateral eyes and median ocelli of the horseshoe crab. The responses are shown at right, with arrows indicating the time of sectioning of the optic nerve. The recording electrodes and fiber-optic light pipes for stimulation are shown at left. (From Barlow, 1983.)

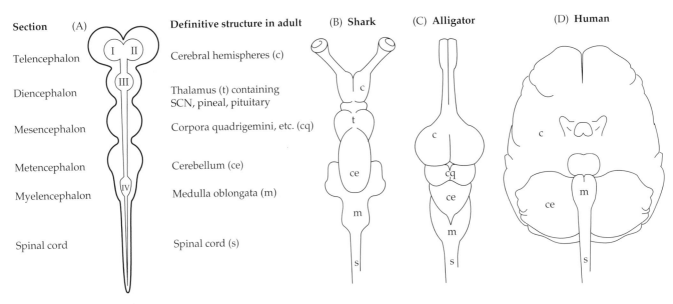

FIGURE 5.13 The basic vertebrate nervous system. (A) Primordial subdivisions of the brain in embryo. Ventral views of the brains of (B) an adult shark; (C) an adult reptile; and (D) an adult mammal, a human.

The brain increases greatly in complexity from lower to higher vertebrates. Certain regions enlarge differentially. Most prominent are the two bilateral lobes of the anteriormost telencephalon. These lobes develop into the cerebral hemispheres (see Figure 5.13) and take over the most advanced functions of association, memory learning, and higher processing. The second, or diencephalic, region becomes the housekeeping part of the brain. It deals with many of an organism's involuntary control functions, such as heartbeat, respiration, temperature control, and biological timing. From fish to mammals, the parts of the brain enlarge progressively.

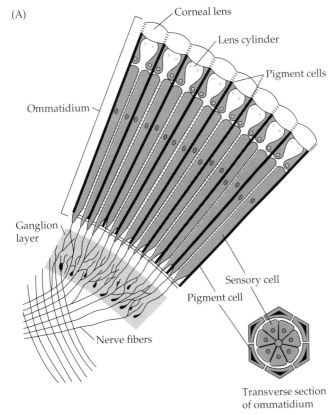

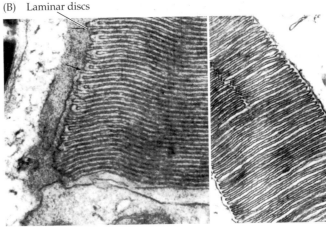

FIGURE 5.14 Complex visual organs. (A) Schematic of an arthropod compound eye. (B) Electron micrographs of the outer segment of a vertebrate retinal cone cell, showing laminar discs of the cone outer segment (left) and laminar folds of a vertebrate rod cell (right). (B from Leeson and Leeson, 1981.)

By the time the human stage has been reached, only two regions are still visible externally in dorsal view. The convoluted cerebral hemispheres at the anterior end overhang much of the brain. Posteriorly, the enlarged cerebellum representing the primordial metencephalon regulates locomotor activity.

A contrast is also seen between visual receptor organs of even the most advanced arthropod invertebrates and all vertebrates. Arthropods may have multiple simple eyes, called ocelli, as well as more complex compound eyes that contain many individual ommatidia (Figure 5.14A).

Several new neural photoreceptive organs have evolved in vertebrates for photoreception. Chief among these are the paired lateral eyes. Cephalopod molluscs and all eyed vertebrates uniformly possess a single pair of eyes with optically clear elements of the cornea and lens for aligning the image on the retinal sheet of rods and cones (see Chapter 11). In vertebrates, photons of light are transduced in photopigment molecules embedded in the lamellae of the rod or cone cells (Figure 5.14B).

Of great interest as the chief pacemakers for biological timekeeping are three key neural structures in or associated with the diencephalic region of the vertebrate brain. For some nonmammalian vertebrates, the retina of the eye is important as a pacemaker. In most amphibians, reptiles, and birds, the pineal body is also an important circadian pacemaker. The third component, not well developed until the mammalian level is reached, is the suprachiasmatic nucleus. In the pages that follow, the pacemaker machinery will be delineated for model species in each of the vertebrate classes: cyclostomes, fish, amphibians, reptiles, birds, and mammals. The survey will summarize trends in functional organization.

Nonmammalian Vertebrates Provide Insight into the Structural Diversity of Circadian Pacemaker Systems[60,62,63,64,65,67]

Many combinations of components reflect the variations in life history patterns of different species[60,62,65]

Most nonmammalian vertebrates have multiple circadian oscillators, and most have multiple photic input pathways as well. Menaker and his students have studied circadian regulation in many species in all vertebrate classes. He concludes,

> At the core of the vertebrate circadian system that controls and regulates the many overt rhythms are three structures together with their interconnections, which form a central circadian axis common to all vertebrates, even the most primitive. These are the retinas, the pineal and the suprachiasmatic nucleus of the hypothalamus. In one vertebrate species or another, each of these structures has been shown to contain circadian oscillators capable of sustaining

rhythmicity *in vitro* and/or conferring either phase or period when transplanted into lesioned hosts."[60]

The pacemaker unit involves a coupling of separate sites through neural and humoral pathways, to ensure a coherent regulation of the multiple overt rhythms of an individual. Environmental input for photic entrainment of the pacemaker system may come from the retina of the eyes, extraretinal photoreceptors of the brain, or pineal photoreceptors. Nonphotic entraining agents may also be important. Output messages may be humoral, such as melatonin, or neural. Some of these components have been remodeled extensively during the course of evolution from fish to mammals. The pineal gland and the SCN have changed progressively in dramatic fashion; the retina is conservative in comparison but has made at least one major physiological change. The pineal changes are best characterized by schematic diagrams of photic and endocrine functions of the pinealocyte major cells in vertebrate groups (Figure 5.15). The pineal of fish is a photosensitive endocrine gland that secretes melatonin.

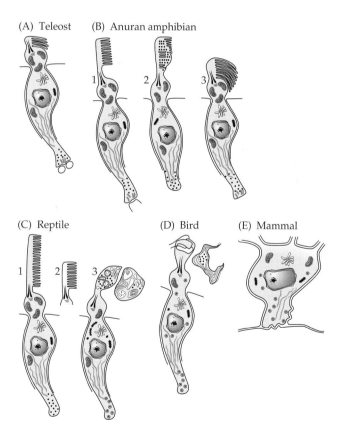

FIGURE 5.15 Representative vertebrate pinealocytes. Teleost (A), anuran amphibian (B), reptile (C), bird (D), and mammal (E), showing progression from photoreceptive pinealocytes to purely secretory conditions in mammals. Numbers 1 to 3 indicate variations from most photoreceptive to most secretory types. Cross section of the outer segment is shown for C and D adjacent to longitudinal sections. (From Menaker and Okshe, 1974.)

The relative contributions of the multiple components of these three aspects of a circadian system in an individual may vary on a seasonal basis, and the importance of each component may differ from one closely related species to another. As a result, the teasing apart of the contributions of circadian components has been very challenging even for a single species, and until recently progress has been slow. New immunocytochemical and other marking techniques hold out much promise for reinvigorating the search for components.

Information on circadian rhythmicity in cyclostomes and fish has increased dramatically in recent years because of new techniques for study[10,14,30,35,44,60]

CYCLOSTOMES. Hagfish and lampreys constitute the most primitive vertebrate group, the cyclostomes. The cyclostomes are fascinating from a circadian point of view because of their origin more than 450 million years ago. Most lamprey species are highly parasitic as adults, with extreme modifications for sucking the body fluids of bony fish. Related hagfish are scavengers. They often live practically buried in the flesh of large dead or dying fish in the ocean depths, and consume the flesh of the host until it is literally a bag of skin and bones. Eyes are extremely reduced, and the mouth is a formidable suckerlike disc armored with teeth for rasping holes in the host fish prior to consuming its flesh. Little is known about the larval stages. In contrast, the more advanced lampreys as adults are river or lake dwellers that swim around actively to locate and fasten onto healthy teleost fish. They then proceed to gouge great wounds in the side of the host and devour the host's flesh. After a time, they drop off and find a new host.

The ecological material just recounted will serve as the backdrop for the intriguing circadian story to follow. The cyclostome group illustrates the extreme malleability of circadian components that may occur in a very small taxonomic group of lower vertebrates. In the lamprey *Lampetra* sp., adults captured on their way upriver to spawn have precise free-running circadian rhythms of locomotor behavior under constant dark conditions in the laboratory. After pinealectomy, these lampreys become arrhythmic. The result clearly implicates the pineal organ as a primary circadian pacemaker. Further support for the pineal as pacemaker comes from the study of cultured lamprey pineal glands. Under in vitro conditions, the pineal glands of lampreys continue to secrete melatonin in circadian fashion. In another species of lamprey, *Petromyzon* sp., both the pineal gland and retina of adults continue in culture to secrete melatonin rhythmically. Presumably melatonin regulates the rhythmic locomotor activity. The pattern of multiple circadian pacemakers that secrete melatonin rhythmically will be seen in most other lower vertebrates.

Surprisingly, the more primitive hagfishes have evolved a different system. No pineal gland is present. Preliminary evidence suggests that the circadian pacemaker controlling locomotor activity is located in the ventral hypothalamus, at the site perhaps analogous to the mammalian SCN. Although surgical lesions of the optic tectum or the dorsal portions of the telencephalon–diencephalon of the hagfish did not affect the free-running rhythm, lesions in the ventral hypothalamus rendered them arrhythmic. Further studies are necessary to determine which hypothalamic nucleus in hagfish is the homolog of the SCN. Even more surprising is the use of the almost vestigial eyes as the sole circadian photoreceptor. The difference in the circadian axes of lampreys and hagfish has been attributed to their great difference in life history and photic exposure.

BONY FISHES. True fish, or teleosts, include a wide variety of jawed fish such as sunfish, trout, or salmon. The hunt for location sites, coupling, and mode of action of circadian system components was for many years especially slow in bony fish because of the dearth of clear-cut behavioral or physiological rhythms as markers for data collection. Locomotor rhythms are generally weakly expressed without clear measurement parameters in teleost fish, but one unusually clear-cut rhythm deserves mention. Gymnotid electric knife fish hide by day in crevices or under mud and sand in tributaries of the Amazon River and then emerge at night. Concomitant with their exclusively nocturnal lifestyle is the use of continuous trains of low-voltage discharges of species-specific waveform and repetition rate for electroperception and electrocommunication. The patterns may be remarkably stable in the group called "hummers," with levels of discharge that differ in frequency by day and by night. In the sandfish *Gymnorhamphichthys*, for example, the discharge is 10 to 15 hertz during the daytime but 60 to 100 hertz during nocturnal swimming. In the laboratory under constant conditions, the rhythm free-runs, and the fish readily reentrain to a new light schedule (Figure 5.16).

A new model has recently been developed in the zebra fish. Circadian rhythms of swimming activity and visual system function, as well as pineal melatonin synthesis, have been documented. The activity rhythms are well developed in only about 70% of adults, but they are more robust in larval fish (5 to 18 days of age). Mutants are being developed that should aid molecular analyses (see Chapter 7).

A few clues about possible circadian pacemakers are available for teleost fish. Technical difficulties have made the culture of retinas nearly impossible to achieve. Because interspecific variability in the capacity of reti-

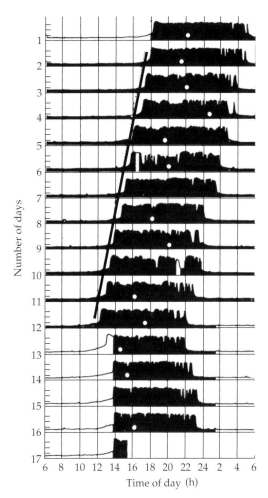

FIGURE 5.16 A circadian rhythmicity in a sandfish record-ed as changes in the electrical discharge frequency.
Lighting conditions include very dim light (0.5 lux) on days 1 to 12, followed by LD 12:12 on days 13 to 17. White circles indicate feeding. Oblique line indicates slope of the free-run-ning rhythm. For further explanation, see the text. (From Lissmann and Schwassmann, 1965.)

nas to generate self-sustained rhythmicity exists, extend-ing cyclostome observations to bony fish would be unwise. Direct evidence comes from pineal ablation and transplant studies, from studies of melatonin secretion in the cultured pineal, and from the entrainment of circadian rhythms by melatonin administration. Pinealectomy is followed by arrhythmicity in catfish, and by period change in three other fish species. The organ culture of pineal glands for 21 species of fish demonstrates that rhythms of N-acetyltransferase (NAT) activity or melatonin secretion can be entrained by 24 h light–dark cycles in all the species. The rhythms also persisted under constant conditions for 19 of the 21 species at least for a few cycles. In constant conditions, however, the rhythms of most fish species damp out quickly, casting some doubt on their self-sustained

nature. In addition to a pineal clock in bony fish, a homolog of the suprachiasmatic nucleus found ven-trally in the hypothalamus may play a circadian role in some species.

Input for photic entrainment of circadian pacemak-ers in fish comes from both extraretinal and retinal sources. The fact that entrainment persists after eye removal shows clearly that extraretinal photoreceptors play a role in some species. The pineal gland is an obvi-ous candidate. Originating as an outgrowth of the roof of the brain during development, in adult fish it remains attached to the brain by a stalk. A second part of the pineal complex, called the parapineal organ, is usually dorsal to the pineal, directly below the skull, occasionally beneath a skull aperture. This third eye is not known to have any circadian function for any ver-tebrate species.

The pineal of fish, like that of other lower verte-brates, contains large cells resembling rods or cones, the pinealocytes, and other more typical neurons. Two functions are attributed to the pinealocytes: photore-ception and melatonin secretion. In many fish the pineal photoreceptors resemble retinal photoreceptors with outer and inner segments plus a synaptic pedicle. Both light microscopy and electron microscopy confirm a photoreceptive structure for the pinealocytes. In addi-tion, immunocytochemical studies indicate the presence of photopigments. Electrophysiological studies of fish pinealocytes show that light inhibits their spike dis-charge activity.

Other deep-brain extraretinal photoreceptors must also exist because entrainment persists in some species after blinding and pinealectomy. The problem with locating these potentially multiple, minute detectors has been the difficulty in identifying them within the brain. Recent developments of new immunocytochemical methods in fish have greatly extended the possibilities for detection. Antibodies can be generated that react with the photoreceptive pigments of rod cells; these can then be injected into a fish to mark rod cells. Opsinlike staining has been found in the neurosecretory cells of the magnocellularis preopticus nucleus of fish. Future studies will be needed to determine whether these two sites function in circadian photoentrainment or pho-toperiodism.

The evidence for retinal participation by rods and cones in photoentrainment of fish is not very extensive. Blinded lake chubs entrain to an LD schedule, but only to a more limited spectrum, and to brighter intensity of light than fish with intact eyes do. As progress is made in discovering extraretinal photoreceptors, lesioning experiments could be carried out to delete the nonreti-nal photoreceptors and better judge the role of the eyes.

Output messages to coordinate multiple pacemak-ers could be either humoral or neural in nature in fish.

Similarly, either humoral or neural messages could regulate the many overt rhythms. The most widely studied substance has been melatonin, both in fish and in the other nonmammalian vertebrates. An important feature of pineal physiology is the existence of daily rhythms of enzymes and substrate concentrations, particularly those for the pathway leading from the initial indoleamine to the final melatonin product (see Figure 6.14). The activity of the rate-limiting enzyme in the synthesis of melatonin, NAT, is higher at night, and consequently melatonin levels are highest at night in both the pineal gland and the blood. The final enzyme in the synthesis pathway is hydroxyindole-O-methyltransferase, which occurs predominantly but not exclusively in pineal tissue.

In amphibians, the retina of Xenopus has been studied intensively as a circadian pacemaker[11,65]

FINDING A CIRCADIAN MARKER FOR AMPHIBIANS. Considerable effort to find convenient markers for an amphibian clock has turned up relatively few possibilities. The most promising and useful ones have been related to daily photoreceptor rhythms. One marked rhythm of the retina of amphibians is the retinal adaptation cycle that alters photoreceptive sensitivity to match changing environmental light levels. Environmental light intensities change about a billionfold during each day–night cycle. Apparently no single design for a light-sensing mechanism has evolved with a dynamic range large enough to detect contrast throughout the range of ambient light intensities encountered in the course of a single day.

One evolutionary solution has been a cycle of reliance on photopic vision using cone photoreceptors by day and scotopic vision using rod receptors at night. Daily shifts between cone-mediated and rod-mediated vision involve regulation at several levels of organization, including physiology and morphology of multiple retinal features. Examples include changes in retinal cell morphology called retinomotor movements. In some cases, the photosensitive rod and cone outer segments are repositioned.

Another strategy is movement of the dark melanin granules of the pigment epithelium to optimize either cone- or rod-mediated photoreception. The morphology and function of synapses between photoreceptors and second-order retinal neurons are also regulated by the circadian pacemaker to increase input from cones during the day and rods during the night. Still another circadian retinal function is the daily disc-shedding rhythm in which the photosensitive outer segments of the rod and cone photoreceptors are broken down and replaced by new segments. Discs are shed generally at the end of the period of active use in vision, and new discs form during the subsequent period of inactivity.

This brief survey indicates that in amphibians the alternation between daytime photopic vision and nighttime scotopic vision involves numerous biological events occurring at multiple levels of organization. Many of these cyclic changes occur just prior to dawn and dusk. Circadian regulation allows anticipatory preparation and may be a more effective way of providing temporal coordination of an organism and its photic environment than direct behavioral reactive responses to changing ambient light intensity.

XENOPUS MODEL. One of the most useful amphibian model species has been *Xenopus laevis*, the African clawed frog. The eyes of *Xenopus* each contain a circadian oscillator. A prime advantage of these eyes as a study system is the fact that they continue to function in culture after removal from the frog. The preparation involves removing cornea, lens, and vitreous humor from isolated eyes, leaving the neural retina, pigment epithelium, choroid layer, and scleral layer. The direct output of the eyecup oscillator was originally measured by the retina's production of retinal serotonin N-acetyltransferase (NAT) (see Chapter 8). The NAT assay utilizes homogenized groups of retinas for each time point, and it was possible for only 3 days. The phase of the circadian eyecup pacemaker can also be measured in terms of dopamine production. Recently developed is a third method that continuously measures melatonin release from individual eyecups cultured in a flow-through superfusion apparatus (Figure 5.17). The newer melatonin assay permits high-resolution measurement for time spans as long as 8 days.

The methodologies described here have been used to document the essential features of the circadian

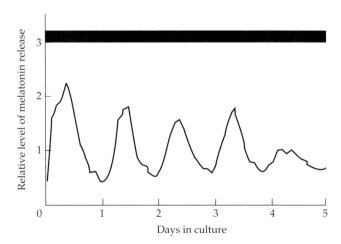

FIGURE 5.17 Circadian rhythm of melatonin release in the cultured retina of *Xenopus* in continuous darkness. (From Underwood, 2001.)

pacemaker system of the African clawed frog. *Xenopus* retina contains a circadian oscillator in each eye that free-runs in isolation under constant conditions (see Figure 5.17). All of the circadian pacemaker functions of the *Xenopus* eyecup are localized in the photoreceptor layer of the retina. Preparations consisting exclusively of the photoreceptor sheet can be made by mechanical removal of all other tissue. The photoreceptor sheet continues to produce melatonin rhythmically when cultured in constant darkness.

Pathways for entrainment are complex. The pacemaker cells in the photoreceptor layer can be entrained by cycles of light, as well as by administered dopamine. Apparently two parallel entrainment pathways bring input to the eye oscillator: one for light and one for dopamine. Dopamine may act as a signal for another entrainment stimulus, or it may modulate the effects of light.

Information available to date supports the hypothesis that all of the circadian functions of the eyecup are localized in the photoreceptor layer. The use of preparations that consist almost exclusively of rod or cone photoreceptors is particularly convincing. These data strongly suggest that the entrainment mechanism, the oscillator, and melatonin output are contained within individual photoreceptors.

The major circadian pacemaker of most reptiles is the pineal gland, but details of circadian function differ widely between species[18,19,21,24,27,59,60,62]

EARLY WORK ON REPTILE CIRCADIAN RHYTHMS. Circadian behavioral rhythms have been examined in reptiles with much interest almost from the beginning of the modern era of chronobiology. As part of the initial search for evidence of internal timing, Hoffmann reared two species of *Lacerta* lizards from eggs maintained in isolation in the laboratory under constant condition. After the eggs hatched, he monitored the locomotor activity of individual lizards in recording cages. The free-running periods in darkness for different individuals were very close to 24 h in all cases but never exactly 24 h. In LL the periods ranged from 22.25 to 23.5 h.

These findings indicate that biological timing is a genetic feature of organisms. A living physiological clock, and not merely a passive response to environmental changes or a learned response, is responsible. Many decades were to pass before the actual pacemaker was located or the mechanism understood. Because of the variations in pacemaker(s) and photoreceptors, the story is still unfolding today.

REPTILIAN DIVERSITY IN CIRCADIAN OSCILLATORS OF FIVE REPRESENTATIVE LIZARDS. Often large differences are seen in the role of components of the circadian system,

even for closely related species of lizards. Perhaps the easiest way to visualize this diversity is to outline the components for each of five extensively studied species, and then try to make some correlations and generalizations relative to the lifestyle of each.

Anolis carolinensis, the anole, is a small, highly active diurnal lizard in which pinealectomy abolishes locomotor activity completely. Culture of the pineal results in a clear, persistent rhythm of melatonin secretion. At the opposite end of the spectrum is *Dipsosaurus dorsalis*, the desert iguana. Neither pinealectomy nor removal of both eyes causes changes in the period of the free-running rhythm of locomotor activity, and cultured pineals only occasionally secrete melatonin rhythmically. On the other hand, lesions to the ventral hypothalamus that ablate 80% or more of the putative SCN site cause arrhythmicity. Although lesion studies are few in number, they strongly implicate the SCN as the site of an extrapineal circadian pacemaker in desert iguanas.

Two other lizards are intermediate in character. The ruin lizard, *Podarcis sicula*, is a small, agile, strongly diurnal lizard. Removal of its pineal gland or retinas, or of both pineal and retina produces significant changes in the free-running period of locomotion in DD but does not abolish rhythmicity. *Sceloporus occidentalis*, the fence lizard, resembles the ruin lizard. Either pinealectomy or enucleation causes changes in the free-running activity rhythm but does not abolish it.

Still another variation in components is seen in the green iguana, *Iguana iguana*. This species is a very large, tropical tree-dwelling lizard (Plate 2, Photo H). Not only does it have a locomotor activity rhythm in the laboratory under constant environmental conditions, but it is also unusual in having a small-amplitude rhythm of body temperature in the laboratory under constant conditions (Figure 5.18).

The green iguana possesses both a parietal eye and lateral eyes, as well as a pineal gland (Figure 5.19; see also Plate 3, Photo A). All three organs have been cultured, and all secrete melatonin rhythmically in culture, although only the pineal secretes melatonin directly into the bloodstream. Pinealectomy obliterates the rhythm of body temperature but not the locomotor rhythm. The parietal eye appears to have only a minor effect on the body temperature rhythm. Further evidence comes from melatonin injections. A single daily injection of melatonin given after the light–dark transition maintains the temperature rhythm, but comparable saline injections do not.

INPUT FOR ENTRAINMENT IN REPTILES. The input of environmental signals for entrainment is equally complex. Extraretinal structures are important for input into the circadian system because they are extremely sensitive to

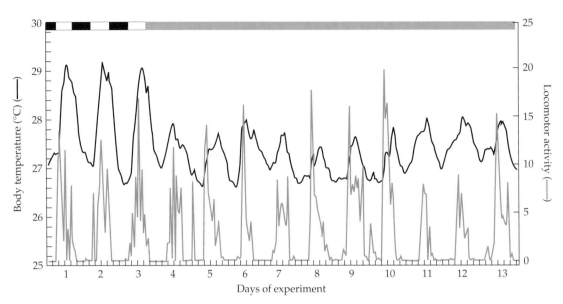

FIGURE 5.18 Circadian rhythms of body temperature (top curve) and locomotor activity (bottom curve) of *Iguana*. Light bar at top indicates LD schedule on day 13 and dim LL for remaining days, with white bars for bright light, black bars for darkness, and gray bar for dim light in LL. (From Tosini and Menaker, 1995a.)

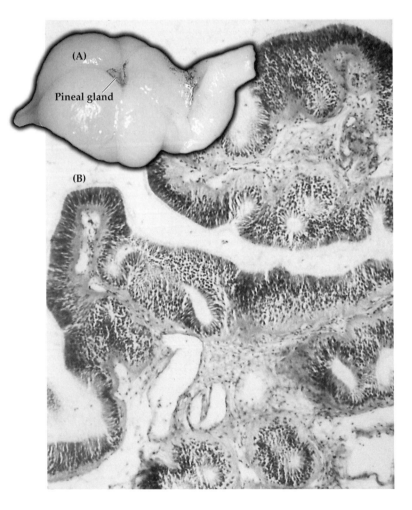

dim light. Immunohistological techniques have recently allowed localization of these receptors. The first vertebrate extraretinal photoreceptors were identified in the ventral hypothalamic area of birds, and shortly afterward, in 1996, in the brains of *Anolis carolinensis* and *Iguana iguana*. These so-called cerebrospinal fluid–contacting bipolar neurons (CSF-contacting neurons) were found exclusively in the paving cells of the lateral ventricle. They stained both for opsinlike activity and for vasoactive intestinal polypeptide (VIP) activity. Their unusual structure includes a dendritic process that extends toward the lateral ventricle of the brain. The bulbous end terminal protrudes into the ventricle; the axon extends ventrally and caudally. These cells lack the membranous stacks characteristic of retinal photoreceptors but are ciliated and contain numerous large vesicles. They share some of the structural features of pineal photoreceptors and of some invertebrate extraretinal photoreceptors. The deep-brain photoreceptors in *Anolis* are

FIGURE 5.19 Pineal gland of *Iguana iguana*. (A) Brain removed from the skull. (B) Low-power photomicrograph of the pineal gland. (Courtesy of M. Grace.)

good candidates for the extraretinal photoreceptors mediating entrainment, but the fact has not yet been unequivocally demonstrated. The relative contributions of retinal and extraretinal photoreceptors to photoentrainment have not yet been determined for any reptile.

A parietal organ is totally lacking in turtles and snakes. It is also absent in many species of lizards. Apparently the parietal organ plays little or no role in the circadian system of reptiles. The pineal gland, in contrast, is well developed and functional in many reptiles, not only as pacemaker but also as photoreceptor. Especially in lizards, the pineal photoreceptors strongly resemble retinal photoreceptors (see Figure 5.15). Visual stacks containing photopigment may be present in the outer segment, and they may come into contact with secondary neurons. In some groups, however, such as turtles and snakes, the photosensory cells of the pineal appear rudimentary or nonfunctional.

Although temperature is an effective entraining agent in many reptiles, the neural mechanism remains unknown. The chief entraining agent for melatonin and other indoleamine entities in homeotherms is the daily solar cycle, but in poikilotherms both light and temperature cycles are important. The duration of the light cycle or the amplitude of the temperature cycle can affect the amplitude, phase, or duration of the nocturnal melatonin pulse. Thus the pineal gland has been called a photothermal transducer that translates ambient light and temperature cycles into an internal signal, the melatonin rhythm.

One of the main mechanisms for coordinating the multiple oscillators with each other or the oscillators with target organs is melatonin. Although melatonin is produced by several sites in an organism, the pineal gland appears to be the main bloodstream source. Daily injections of melatonin can entrain the activity rhythm in *Sceloporus occidentalis*. Infusions that mimic physiological cycles of melatonin can be administered by programmable capsular devices called silastic implants. When delivered in this fashion, melatonin can cause period changes or arrhythmicity in several species such as *S. occidentalis, S. olivaceus,* and *A. carolinensis.* The data indicate that the daily rhythm of melatonin release from the pineal controls the phase and/or period of other elements of the system.

SYNTHESIS FOR REPTILES. In an extensive paper on the evolution of vertebrate circadian systems, Menaker makes a cogent argument combining elements of the ancient origin of the vertebrate circadian axis, the relative malleability of the axis to selective pressures in particular environmental niches, and rapid adaptation to particular photic niches. He wisely qualifies his statement by continuing that a correlation is not easy because of a variety of factors. These include the difficulty in

defining the photic niche of extinct animals, lack of information about the speed with which circadian organization can evolve under selective pressure, and the presence of other undefined pressures acting on an organism concurrently with photic pressures. Nevertheless, he concludes that the circadian axis of eye, pineal, and SCN pacemakers is as old as the vertebrates themselves, about 500 million years.

Menaker's hypothesis gains support from the commonality of elements among the nonmammalian classes. The most ancient hagfish and lampreys of the cyclostome group share basically the same elements of the circadian axis in terms of photoreceptive pineals and retinas with other nonmammalian vertebrate classes. Common to all, also, are pacemakers in eye, pineal, and basal hypothalamus, plus blood-borne melatonin as a transducer of environmental information. Circadian oscillators reside centrally, probably in the SCN area, and more peripherally in the pineal organ and eyes.

The pineal organ is linked to the rest of the system via melatonin; the eyes may be linked by melatonin or neurally. Light can entrain the pineal organ and eyes directly. It can entrain the central system via the extraretinal receptors in the brain, or it can entrain the central nervous system via the pineal organ and eyes. Rhythmicity in some elements of systems such as the SCN and pineal organ may damp in the absence of inputs from the rest of the system. This behavior may be due to a damping of the oscillators constituting the pacemakers or to an uncoupling of the multiple oscillators making up the pacemakers.

The principle of parsimony suggests that retention of the system from cyclostomes to birds is more logical than its de novo evolution in birds. Questions about the variability in multiple oscillators and environmental sensing structures for entrainment focus on the usefulness of having redundant circadian photoreceptors. Most non-ocular photoreceptors are dedicated for circadian light intensity detection, not image formation. Furthermore, the enigma of mammals remains because all species in this phylogenetically most advanced group are consistent in having a single primary pacemaker in the SCN and a single photodetection system in the retina.

Perhaps nonimaging photoreceptors are relatively easy to evolve. They may require only a light-absorbing molecule and a membrane channel. Such a feat has been accomplished many times in organisms ranging from higher plants and protists to vertebrates. The multiple photoreceptors may have resulted from natural selection because they do not need to form images, can be specially tuned, and can be located adjacent to the oscillators that require their input. The reason for the mammal condition will be discussed later.

Great diversity is seen in avian oscillators and input sensors[13,45,55,56,64,66,67,71,72,73]

VARIATION ON THEMES OF CIRCADIAN STRUCTURAL DIVERSITY. The circadian diversity seen in reptiles continues in birds. Avian circadian systems are composed of several interacting pacemaker sites. Species can vary significantly in the relative roles that the pineal, the SCN, and the eyes play as pacemakers. The importance of retinal and extraretinal photoreceptors in photoentrainment differs, and the range of neural and humoral (melatonin) output messages from the pacemakers of different species is also large. With such an assortment of systems, the simplest approach is to delineate a few dominant patterns found in birds. Four basic systems have been documented for components of the avian pacemaker, sensory input system, and output transducer. The starring roles in these four patterns are played by the house sparrow, the Japanese quail, the domestic chicken, and the domestic pigeon.

PASSERINE (PERCHING) BIRDS: THE HOUSE SPARROW. The first case is the pattern of all known passerine, or perching, birds. The study of house sparrows was one of the early successes in discovering the site of a circadian pacemaker. In 1968, Gaston and Menaker reported that deletion of the pineal organ of house sparrows resulted in loss of the daily rhythm of perch hopping and body temperature when the birds were placed in constant darkness (Figure 5.20). Such a result was not sufficient to demonstrate that the pineal was the pacemaker, for the pineal could merely have been part of the relay pathway. A second experiment clinched the case. When a pineal was transplanted to the anterior chamber of the eye, a rhythm was demonstrated that bore the phase of the donor bird. The pineal gland is located directly under the skull and meninges, sandwiched tightly between the cerebellum and cerebral hemispheres (Figure 5.21). Melatonin is the likely signal by which the pineal controls other oscillators in the system. The role of melatonin was also demonstrated by single daily injections at the light–dark transition that entrained the activity rhythm. Further experiments strongly supported the hypothesis of the pineal gland in sparrows as the primary pacemaker. These investigations included culturing of the pineal in vitro to detect a persistent circadian rhythm of melatonin secretion, or disrupting of the sympathetic neural connections of the pineal to show that rhythmicity persisted. However, the complete answer may be more complicated. Lesions of the suprachiasmatic area can disrupt rhyth-

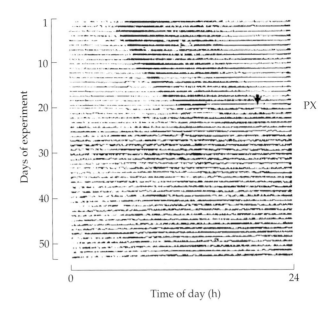

FIGURE 5.20 The function of the pineal gland in house sparrows. Loss of circadian rhythmicity is observed in a house sparrow after pinealectomy. PX marks the point at which the pineal gland was removed. For further explanation, see the text. (From Gaston and Menaker, 1968.)

micity, but it is not clear exactly which of two possible homologs of the mammalian SCN are involved. The lesions may have affected both sites.

Studies of the photoreceptors for photoentrainment in sparrows implicated extraretinal deep-brain receptors. A simple but ingenious method was used. Scientists showed that injection of India ink subcuta-

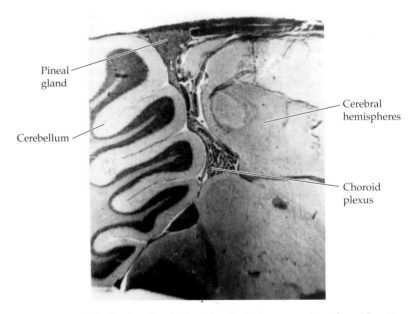

FIGURE 5.21 The brain of a chick. Histological preparation of a midsagittal section to show the pineal position directly under the skull and adjacent to the cerebral hemispheres and cerebellum. (From Binkley, 1990.)

neously on top of the skull prevented light from penetrating the skull and eliminated entrainment. Removal of the ink allowed the same bird to synchronize to a light schedule again. The specific cells have not yet been localized.

GALLIFORM BIRDS: THE JAPANESE QUAIL. A second model is the Japanese quail, a small galliform bird related to chickens and pheasants. Its daily locomotor rhythm is poorly developed, but other pronounced daily rhythms include rhythms of body temperature, egg laying, and melatonin secretion in the eye. The pacemaker system of the quail differs from that of passerine birds; pacemaker elements of the eye and a hypothalamic SCN component have been demonstrated. In contrast to the sparrow, the pineal organ does not appear to be a circadian pacemaker, and locomotor rhythms persist after removal of the pineal, at least in most individuals.

The retina of each eye of the quail has been shown to be the locus of a biological clock. Cultured eyes maintain a rhythm of melatonin secretion in isolation under constant conditions. Blinding by complete eye removal causes the activity and temperature rhythms to damp out and become arrhythmic over a period of 1 to 2 weeks. The results of blinding imply that central pacemakers cannot continue to cycle in the absence of daily input from the ocular pacemakers, or that the individual oscillator cells drift out of phase with each other in the absence of daily input from the ocular pacemakers. The SCN is a central oscillator coupled to the ocular pacemakers by hormonal outputs such as melatonin and by neural outputs. Photic input from the eye travels to the SCN by means of a direct monosynaptic retinohypothalamic tract.

Because of the predominant role of the SCN in mammalian circadian systems, considerable effort has been expended to determine whether a similar area exists in Japanese quail and other birds. Two nuclei of the avian hypothalamus have been identified as possible homologs, but attempts to identify which of these two nuclei are homologs of the mammalian SCN have not been conclusive. Key markers in mammalian SCN are neurosecretory cell groups containing neurochemicals such as neurophysin and/or vasopressin, along with others, such as vasoactive intestinal polypeptide (VIP), bombesin, and glutamic acid decarboxylase. Neither of the two avian nuclei contains all these neurochemical entities. Lesion studies have also been used to determine which areas in the avian hypothalamus are circadian pacemakers. Further research is needed to resolve the issues.

ANOTHER GALLIFORM BIRD VARIATION ON THE CIRCADIAN AXIS: THE DOMESTIC CHICKEN. A third widely studied avian species is the chicken. Although locomotor and temperature rhythms have been recorded, these are very poor markers for circadian studies because of their diffuse, imprecise nature. As a result, little is known about adult chicken circadian systems. Most studies have used dispersed-cell cultures of chick pineal cells (see Chapter 6). Culture methods have now been refined enough to allow 8-day recording under constant conditions for the continuous measurement of melatonin production by entire pineal glands or pieces. Even single pineal cells have been cultured. The role of the eye in adults is uncertain.

One of the most significant findings from in vitro culture is that single pineal cells in isolation show a circadian rhythm of melatonin release that can be entrained by LD cycles and persists for at least several cycles in DD. A single chick pineal cell, therefore, is the site of a circadian clock, a circadian photoreceptor, and the circadian melatonin-synthesizing output pathway. Accordingly, the pineal circadian clock of newborn chicks, at least during in vitro culture, is composed of individual cellular clocks that must interact to produce a single coherent output (see Chapter 6).

CIRCADIAN FEATURES IN THE DOMESTIC PIGEON. Still another combination of components is seen in the domestic pigeon, which provides the fourth scenario. The strong, precise locomotor rhythm of the adult serves as a good circadian marker for behavioral studies of its circadian system. Pinealectomy does not abolish the free-running locomotor rhythm in dim LL, but the rhythm becomes unclear and loses its precision. Normal cyclic locomotor activity continues in LD regimes after pinealectomy. Blinding by ocular enucleation does not destroy the activity or body temperature rhythms, but it makes them less precise. After pinealectomy plus enucleation, both the body temperature rhythm and the activity rhythm disappear in long-term dim LL maintenance.

Following pinealectomy and blinding, the amount of melatonin circulating in the blood drops to a very low level, suggesting that the output message driving the behavioral rhythms is melatonin from the pineal and retina. The role of melatonin was confirmed by infusions of physiological amounts and patterns of melatonin in birds whose pineals and eyes had been removed. The persistence of behavioral rhythms for a period of time after removal of the pineal and eye shows that other oscillators, possibly an SCN homolog, play a role in rhythmicity. However, the role of the SCN or other hypothalamic central oscillators has not yet been confirmed.

POSSIBLE PERIPHERAL AVIAN CLOCKS. The details of avian circadian organization that have been described here are incomplete insofar as they have portrayed only major, central oscillators of the nervous system. Little is known about peripheral oscillators in birds. Studies in other vertebrate classes have shown that multiple peripheral clocks exist in addition to the primary central ones. These peripheral clocks probably control local events within particular organs or tissues. Possibly they are synchronized and phased by neural or hormonal outputs from the central circadian system. Additional research is needed to decide whether the avian circadian system is hierarchical in arrangement, with central pacemakers controlling and coordinating the behavior of multiple peripheral oscillators.

INPUT PATHWAYS TO AVIAN PACEMAKERS. Environmental input for circadian system of birds is also multifaceted. The light–dark cycle is the most important source of information. In all vertebrates, an important neural photic route is the retinohypothalamic tract (RHT), which connects the retina to the SCN region. In mammals the RHT is the major or exclusive route by which photic entraining information reaches the SCN pacemaker. In birds, as well as other nonmammalian vertebrates, the RHT is assumed to play a similar role. Deep-brain extraretinal photoreceptors are important and have been found in every species of bird examined. The pineal glands of many birds are also photoreceptive. Sources of nonphotic input to the avian circadian system are not yet fully understood.

EFFERENT OUTPUT PATHWAYS IN BIRDS. Birds have many efferent output signals. High on the list of important transducers of pacemaker commands to the body is melatonin. A common thread of melatonin as a vital circadian output signal becomes evident for all nonmammalian invertebrates. Melatonin is secreted by the avian eye, by the pineal, and possibly by other sites. In birds such as the house sparrow, which become arrhythmic after pinealectomy, single daily injections of melatonin restore and entrain locomotor activity. Undoubtedly many other neural and humoral agents also play a role.

In Mammals the SCN Is the Main Oscillator[9,31,42,43,53,61,69,70]

A major mammalian circadian pacemaker has been localized to the suprachiasmatic nucleus of the ventral hypothalamus[25,33,52,54,58,68]

FUNCTIONS OF CIRCADIAN TIMEKEEPING IN MAMMALS. Circadian rhythmicity adapts mammals in a temporal sense to their particular environmental niche. The clock is an essential element of adaptation for daily or longer-term periodicity. It anticipates environmental change and allows an organism to be prepared for likely future demands. The circadian clock organizes physiology and behavior in such a way that the various processes necessary to life are scheduled to occur at the most appropriate time of the day or night, and in the most appropriate relationship to each other.

The list of daily circadian behavioral, physiological, and biochemical phenomena for mammals is very long. Metabolism, for example, switches between various nocturnal processes that contribute to growth and repair of the body, and diurnal processes that support energetic demand and psychological alertness. If humans try to execute physically or psychologically demanding complex tasks during the circadian night, performance is very poor and error prone. The sleep–wake cycle is a prominent circadian phenomenon in mammals (see Chapter 9). In parallel is the core body temperature, which starts to fall in the evening in anticipation of sleep and rises in the early morning in preparation for the active phase of a diurnal mammal. Rhythmic melatonin secretion is another important circadian function that occurs during nighttime, and the adrenal hormone cortisol is secreted primarily during the circadian day to help mobilize energy.

The second role of the clock is to regulate an organism's physiology and behavior over longer time scales. The estrous cycle of female hamsters is a complex example in which the duration of the cycle is not timed by a circadian clock but the ovulation event is gated by a circadian mechanism. Seasonal adaptations of temperate and polar species are other examples (see Chapters 2 and 4).

These few examples illustrate the intricacy of the circadian program. A series of interlocking, complementary cycles carry the body along an orderly progression and adapt it to the external world. Circadian outputs in mammals are often called circadian phenotypes or overt rhythms. Important questions concern the mechanisms of orchestration within the SCN and the channels of output communication used by the clock.

DATA SUPPORTING THE SCN AS THE PRIMARY PACEMAKER IN MAMMALS. Mammalian circadian systems differ from those of all other vertebrate classes. A single primary central oscillator, the SCN, is connected to a single photic sensor, the retina. Multiple modalities mediating output include efferent nerve tracts and humoral signals such as melatonin of the pineal or hormones of the pituitary gland. No longer is the pineal photoreceptive. No longer is the pineal an independent oscillator; instead its rhythmicity is driven by the SCN. The eye has become a local pacemaker.

Evidence that the SCN is the chief circadian pacemaker comes from many reinforcing sources. Early evidence resulted from experiments that isolated the SCN neurally from other parts of the brain in vivo preparations by knife cuts of all of its afferent and efferent

(A)

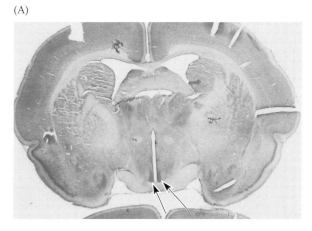

(B)

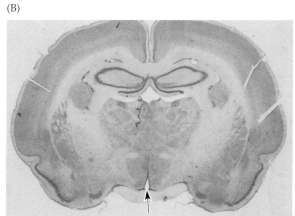

(C)

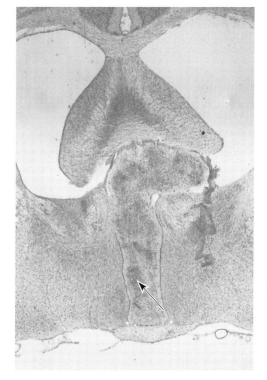

FIGURE 5.22 Transplantation of the SCN pacemaker in golden hamsters. (A) Cross section of an intact hamster brain at the level of the SCN (arrows). (B) Comparable section for a lesioned animal showing lesion cavity (arrow). (C) Comparable section for an SCN-transplant animal showing dark-staining SCN in graft (arrow). (Courtesy of P. DeCoursey.)

nerves. The resultant hypothalamic island was correlated with loss of overt circadian rhythms. Similarly, ablation of the SCN by electrolytic or chemical lesions (Figure 5.22; see also Plate 3, Photo B) invariably results in the permanent loss of most overt circadian rhythms.

Considered alone, this criterion is not sufficient proof for several reasons. Partial lesions with as little as 10% of SCN tissue remaining are adequate to sustain rhythmicity. Careful histological assessment following a lesioning experiment is important. Because the SCN is not encapsulated by distinct anatomical boundaries, immunocytochemical staining for peptides limited in the hypothalamus to the SCN is the method of choice (see Chapter 6).

Another uncontested piece of evidence comes from SCN transplant studies. Techniques are now available for transplantation of fetal SCN into the third ventricle of a lesioned (arrhythmic) adult host animal (Figure 5.23). Grafted SCN tissue restores circadian activity rhythms with a period that is specific to the genotype of the donor graft. Care must be taken to assess the completeness of the lesion at the termination of the experiment because even a few cells of host SCN might sustain rhythmicity in the host.

In light of this problem, experiments using SCN tissue from donor *tau* mutant hamsters have been particularly convincing. In golden hamsters, *tau* is a semidominant circadian period mutation that occurred spontaneously in a laboratory colony in 1982 (see Chapter 7). The period of free-running heterozygotes in DD is 22 h, that of homozygotes is 20 h, and the wild-type period is 24.1 h in DD. The great advantage of *tau* mutants for transplantation over normal wild-type transplants is that the transplant from a 22 h or 20 h donor to a wild-type host results in reestablishment of a 22 h or 20 h rhythm, respectively, in the host. The result eliminates any possibility of ambiguity from a residual host rhythm (see Figure 5.23).

Another piece of evidence for the pacemaker role of the SCN of mammals comes from explant culture and dispersed-cell culture. Details will be considered systematically in Chapter 6. Culture of the SCN explants in vitro has demonstrated the persistence of free-running rhythms for extended periods of time. The favored recording method is extracellular recording of the average spike activity of SCN maintained in culture. In addition, free-running cir-

(A) Transplant experiments with homozygous *tau* mutants

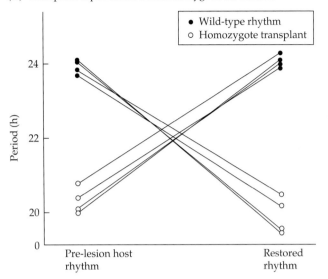

(B) Transplant with heterozygous *tau* mutants

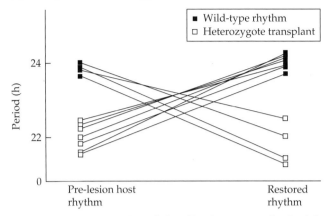

FIGURE 5.23 Restoration of circadian locomotor rhythmicity in SCN-lesioned *tau* mutant hamsters and wild-type controls with SCN transplants. (A) Results from homozygote experiments. (B) Results from heterozygote experiments. For further explanation see text. (From Ralph et al., 1990.)

cadian rhythms have been documented for single SCN cells in a dispersed-cell culture (see Chapter 6). Finally, the SCN pacemaker has been linked anatomically to a photoreceptor by tract tracing. The fact that severing the RHT connection between eye and SCN results in failure of entrainment shows the functional link.

STRUCTURAL ORGANIZATION OF THE MAMMALIAN SCN. Structurally the SCN pacemaker consists of a pair of ovoid bodies containing about 16,000 neurons and other support cells. The SCN lies directly above the optic chiasm. The neurons are among the smallest of the entire brain, and they lie closely packed. The SCN consists structurally as well as functionally of a shell and a core (Figure 5.24). The shell lies for the most part dorsal to the core, but in the middle region of the SCN it occupies an almost vertical or medial position. The core, in contrast, is ventral or ventrolateral in its position. Shell neurons are slightly smaller and crowded more closely than core neurons. The core neurons are more densely filled with organelles and endoplasmic reticulum, and the nucleoli are less marginated than in shell neurons. Although structural differences between core and shell neurons are relatively slight, the cytochemical differences are very pronounced, giving rise to a highly characteristic chemoarchitecture of the SCN core and shell. For further cellular details of the SCN, see Chapter 6.

EXTRA-SCN OSCILLATORS. Peripheral clocks in mammals and their interaction with the SCN pacemaker are an important topic in current circadian research. Although extra-SCN oscillators have been suspected in mammals for some time, a direct demonstration was first provid-

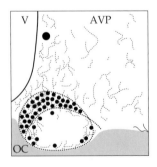
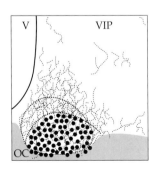
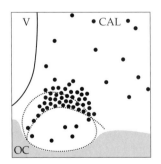
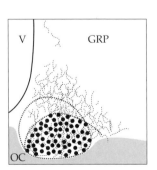

FIGURE 5.24 Schematic representation of peptide immunoreactivity in coronal sections of the middle of rat SCN. The neuropeptides are concentrated either in the shell (AVP and CAL) or in the core (VIP and GRP). V, third ventricle; OC, optic chiasm; AVP, vasopressin; VIP, vasoactive intestinal polypeptide; CAL, calretinin; GRP, gastrin-releasing peptide. (From Moore and Leak, 2001.)

ed by measurement of a circadian rhythm of melatonin synthesis in hamster retinas in vitro. Since then, a variety of cultured mammalian tissues has been shown to exhibit damped oscillations. The sustained cycling of these peripheral oscillators probably depends on the SCN (see Chapter 6).

Input pathways to the SCN include both neural projections and humoral feedback signals[1,2,5,12,39]

PHOTIC PATHWAYS FOR INPUT OF ENVIRONMENTAL INFORMATION. Although the circadian clock in the mammalian SCN can function without entraining cues, it is normally modulated by various environmental, physiological, and behavioral conditions. Particularly important is the timing and intensity of light. The effects of these modulating factors on the circadian clock are revealed through changes in the phase, amplitude, and free-running period of the expressed rhythms. Intense research has focused on the input factors affecting the SCN clock, the pathways through which they reach the clock, and the mechanisms underlying their effects (Figure 5.25).

The best-characterized SCN input is a direct neural projection from the retina of the eyes. Because light is the most salient feature of the environment affecting circadian rhythms, a direct neuronal projection to the SCN from the eyes is not surprising. The fact that animals deprived of their eyes free-run shows that the eyes are necessary for photoreception in the SCN. The details of circadian retinal physiology will be deferred to Chapter 6. For now, suffice it to say that the retinal photoreceptor transmission pathways leave the eye by way of the optic nerve and travel to the SCN.

A variety of anatomical and functional techniques have been used to characterize this pathway, which is called the retinohypothalamic tract (RHT). One very effective method involves injecting an enzyme called horseradish peroxidase into the eye. The enzyme is taken up by retinal ganglion cells and transported down the axons to their nerve terminals. After the tissue is sectioned and processed appropriately, the axonal projections are visible as they travel through the tissue and identify the place where they terminate. The RHT terminates primarily in the ventrolateral portion of the SCN, where cells respond immunoreactively for vasoactive intestinal polypeptide and gastrin-releasing peptide.

A newer technique using herpesvirus for labeling pathways demonstrates that the RHT originates from a distinct subpopulation of retinal ganglion cells that are distributed relatively uniformly across the retina and constitute a subset called W cells. The RHT appears to use the excitatory amino acid glutamate and pituitary adenylate cyclase–activating peptide as its neurotransmitters. In addition to this anatomical and physiological evidence, a variety of functional studies indicate that light phase-shifts the SCN at least in part through stimulating glutamatergic pathways.

The SCN also receives indirect retinal input from the intergeniculate leaflet of the thalamus. This structure is a thin layer of cells within the lateral geniculate body that receives the bulk of the retinal input used for visual imaging. Thus the projection to the SCN probably carries retinal information that has been modified in the geniculate body. Although the geniculate tract to the SCN carries photic information, its involvement in light-induced phase shifts for photic entrainment remains unsolved. Some data suggest that the tract input increases the sensitivity of the clock to light input received by the main tract, the RHT.

NONPHOTIC PROJECTIONS TO THE SCN FOR INPUT OF ENVIRONMENTAL INFORMATION. The SCN pacemaker of mammals is also sensitive to nonphotic signals. A major neural input to the SCN comes from the raphe nuclei located in the midbrain. The function of this projection is, again, unclear. Although the raphe nuclei receive a small amount of retinal input, primarily they relay information about the general arousal state of the animal. Raphe input to the SCN is not necessary for the expression of circadian rhythms, yet it modulates SCN clock activity because raphe lesions tend to decrease the amplitude and regularity of wheel-running rhythms and shorten the period of these rhythms in constant darkness.

The raphe projection to the SCN terminates along with the photic inputs to the SCN in the ventrolateral SCN. The raphe also projects to the intergeniculate nucleus, thus allowing the photic information carried by the geniculate target to be modified by the raphe. The

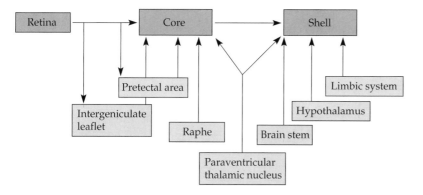

FIGURE 5.25 Neural inputs to the mammalian SCN pacemaker. The diagram emphasizes the diversity of afferent projections to the SCN and the specificity of the target neurons of either the shell or core areas. (From Moore and Leak, 2001.)

neurotransmitter associated with the raphe projection is serotonin, and serotonin modulates SCN clock activity. Recent data have shown that increased locomotor activity causes the circadian clock to be phase-shifted (see also Figure 8.7). The results indicate that the pathway through which behavioral activity phase-shifts the SCN circadian clock involves both the raphe and intergeniculate projections to the SCN.

HUMORAL FEEDBACK LOOPS TO THE SCN.   Another form of nonphotic modulation of the SCN clock phase involves melatonin, which is produced by the pineal gland during the nocturnal phase of a mammal's circadian cycle. The pineal gland of mammals is no longer photoreceptive (see Figure 5.15E) and has become instead a neuroendocrine organ. The lamellae characteristic of rod and cone cells have been lost, and instead the mammalian pinealocytes appear secretory in structure. Gone, too, are the cilia characteristic of bird pinealocytes. The SCN contains one of the highest concentrations of melatonin receptors in a mammal's body, suggesting that it may be an important site of melatonin activity. Similarly, daily injections of melatonin can entrain behavioral rhythms in vivo, and melatonin administration can shift the phase of SCN brain tissue slices cultured in vitro.

The ability of melatonin to entrain the SCN circadian clock may be especially important during fetal development of the SCN. In rodents the SCN begins functioning several days prior to birth, before SCN neurons have developed functional synapses. During this stage of development the mother somehow communicates circadian phase to the fetuses and synchronizes their clocks to the environment of the outside world. The means of communication may be melatonin.

Each of these nonphotic phase signals forms part of a feedback loop that involves the SCN circadian clock. Behavioral activity, melatonin secretion, production of a substance known as neuropeptide Y, and serotonin release are all diurnal rhythms. In other words, they are overt hands of the clock, and yet each has clearly been shown to modulate SCN clock phase. Therefore, these daily rhythms themselves can act as input signals. Because the SCN clock can function properly without them, they must form feedback loops external to the clock itself. A reasonable supposition is that the SCN clock regularly uses its own output signals to increase its stability or accuracy. As more is learned about the complex mammalian circadian system, other examples of circadian feedback loops will undoubtedly be found.

SCN output pathways for neuronal and humoral signals are numerous and depend on both target organ and function[42,43]

GENERAL ASPECTS OF SCN OUTPUT PATHWAYS.   SCN outputs and their mode of controlling overt rhythms are better known for mammals than for any other pacemaker system, for both daily and longer-term rhythms. The mammalian SCN pacemaker system drives circadian rhythms through a combination of neural and endocrine messages. The SCN has direct access to a wide range of neuroendocrine and behavioral control systems via many neural and endocrine linkages, especially in the hypothalamus, brain stem, and pituitary. Most connections are local in the hypothalamus, but secondary neural relays and humoral agents reach out to all parts of the body.

METHODS FOR DETERMINING SCN OUTPUT CONNECTIONS.   Two important means of following the neural connections of the SCN to the rest of the brain have been the use of immunocytochemical staining procedures and anatomical tract-tracing techniques (see Chapter 6). Immunocytochemistry identifies SCN cells on the basis of their neurochemical content. All SCN neurons are GABA-ergic, which means that they use gamma-aminobutyric acid as a neurotransmitter. GABA is likely to inhibit activity in target areas, although it is not yet clear how this inhibition affects the final output of any SCN target. More selective information can be obtained by detecting peptidergic SCN cells, those that use vasopressin, gastrin-releasing peptide, and vasoactive intestinal polypeptide as transmitters. These peptidergic cells send axons into the adjacent hypothalamus and to more distant targets.

Marker stains have followed efferent neurons of the SCN to specific target areas. By applying various tracers, it has been possible to identify the types of neurochemical input received by particular target areas of the SCN. Then immunocytochemical methods can be used to determine the neurochemical phenotype of cells in receipt of SCN efferents. Both immunocytochemical and tracer techniques have weaknesses, but by using them in combination, a picture has emerged of the pathways that convey neural signals from the SCN.

The SCN graft technique can provide significant information about the nature of the circadian neural signal arising from the SCN. Temporal chimeras were constructed using an SCN transplant from a fetal *tau* mutant hamster with a period of 20 h. Circadian activity was examined in wild-type hamsters that had received a partial lesion of the SCN plus a *tau* mutant graft. The resulting activity profiles expressed periods dependent on both the host and the grafted SCN, such that activity bouts with a 20 h and a 24 h component could be identified. If the usual output of the clock is predicted to be a periodic inhibition of activity, then activity in the chimera would only occur when the active phase of both elements coincided; at all other times, one of the two signals would suppress locomotion. Conversely, if the SCN signal is predicted to be a periodic stimulation of locomotor centers, then activity bouts would occur at both active phases without interaction. The experimental results are more complex than either of these two predictions and are consistent with a hypothesis that both SCNs generate a signal with alternating stimulatory and inhibitory components.

The SCN graft preparation is also excellent for addressing the controversial question of neural versus humoral outputs. Knife cuts have been used to interrupt output pathways of the SCN and block the expression of particular rhythms. The results could possibly be considered compelling evidence of neural control. SCN grafts, however, only restore locomotor activity cycles and sleep–wake cycles. None of the endocrine or autonomic rhythms such as daily melatonin or corticosteroid secretion are present in SCN-lesioned animals bearing successful grafts. Early graft studies also showed that the transplanted SCN had relatively sparse outgrowth into the host tissue and that grafts placed in atypical positions were still able to rescue behavioral rhythms. The finding would not be expected if proximity to target and the consequent ease of SCN rewiring were decisive factors.

Another important feature of the graft protocol is that it allows assessment of humoral mechanisms. Since neonatal or fetal SCN held in vitro is able to secrete certain chemical messengers in a circadian fashion, the case for humoral output appears strong. SCN tissue can be placed in small capsules that allow diffusion of chemical messengers from the SCN cells but do not permit neural outgrowth. These capsules can then be transplanted into the third ventricle of SCN-lesioned hosts. Even encapsulated SCN transplants are effective in restoring rhythmicity in SCN-lesioned animals. Perhaps this reflects a selective form of output coupling in which some functions require humoral regulation but others are dependent on hard-wired neural connections.

IMPORTANT SCN NEURAL PROJECTIONS. To understand how the circadian signal is able to influence an extensive array of rhythms, it is necessary to briefly review the pathways by which the SCN communicates with the rest of the brain. Neural connections of the SCN will be emphasized since these are better known than the humoral ones. As might be anticipated for a system with pervasive influence, the connections of the SCN are numerous and varied. Many projections are intrinsic within the SCN. The initial SCN messages are carried in concentric rings of relays to distant targets throughout a mammal's body. The first extrinsic relays are close to the SCN, situated primarily in the adjacent hypothalamus. Especially prominent as a relay station is the subparaventricular area immediately dorsal to the SCN and the caudal continuation of this zone, called the retrochiasmatic area. In fact, the densest external fiber connections from the SCN are found in these two areas. Other primary targets close at hand in the hypothalamus include the paraventricular nucleus and the intergeniculate leaflet. More distant targets include the preoptic areas of the hypothalamus and the tuberal posterior hypothalamus. A few direct, long-distance neural links may be important in the timing of endocrine events, but direct enervation of distant targets by the SCN is sparse. Only a few SCN fibers project to the midline thalamus or the basal forebrain. Instead, the neurons in the primary relay areas project widely throughout the brain and spinal cord and are able to facilitate circadian regulation in two ways. First, they overlap the distribution of the sparse direct enervation from the SCN and amplify the signal. Second, because their connections are extensive, they provide an important indirect route for circadian information.

Another important feature of the SCN efferent connections is the strong reciprocal enervation between the SCN and its local relays. The arrangement entails feedback regulation that may contribute to the stability of the circadian signal or facilitate the integrated control of diverse outputs. The SCN and its relays may therefore be looked on as a circadian neural network, with the SCN as the principal regulator. The SCN provides a signal that is filtered via the relays and converted prior to transmission to forms most appropriate for the specific targets.

This independent distribution of a circadian signal along various neural pathways has been demonstrated experimentally by the use of selective lesions to SCN efferents and targets. For example, lesions of the paraventricular nucleus of the dorsal hypothalamus block the nocturnal secretion of melatonin, while lesions placed rostral to the SCN block the ovulatory LH surge but have no effect on melatonin rhythms. Circadian activity rhythms survive both types of lesion but are ablated by retrochiasmatic lesions and/or knife cuts around the SCN. Thus, multiple representations of the circadian signal exist, which raises the question of how the signals vary, and whether all of these representations of the signal are identical. A few selected examples will illustrate the far-reaching contacts of the SCN pacemaker and the individuality of regulation for particular functions.

EXAMPLES OF SCN REGULATION. The entire account of circadian regulation in mammals is very lengthy and can only be sketched briefly in this book. The following examples will provide a tantalizing glimpse of the complexity of circadian regulation in mammals and a view of some of the multiple output signals by which its tasks are accomplished.

The sleep–wake cycle relies on neural communication between the SCN and sleep centers in the reticular formation of the midbrain, in the pons, and in the medulla. These centers in turn have widespread neural connections through which they can influence many parts of the nervous system. Within the spinal cord, these neural connections regulate muscle tone and reflexes and the transmission of sensory information to the brain. Rostral projections from the reticular centers to the forebrain control consciousness and the patterning of sleep. The rhythm of core body temperature, which reflects circadian control over the balance between heat generation and heat loss, is also dependent on neural connections. The body temperature rhythm is also tightly linked to the sleep–wake rhythm. Control is brought about in part by changes in basal metabolic rate, which indicates that the clock is able to communicate with those components of the autonomic nervous system that are responsible for regulation of metabolism.

Another neural link that involves SCN regulation of both the peripheral and the central nervous systems is the control of circadian release of melatonin. The pineal gland varies in function dramatically between species. In nonmammalian vertebrates it plays a central circadian role as photoreceptor, pacemaker, and generator of the melatonin output signal. In mammals the gland is not photoreceptive, but rather is driven by the SCN. For mammals, the pineal gland's nocturnal output of melatonin is a critical signal that is involved in functions such as amplification of the daily body temperature rhythm, facilitation of sleep, and control of seasonal photoperiodic responses, as well as entrainment of the developing clock of the fetus. Rigid temporal control in mammals is therefore a key requirement for the secretion of melatonin, and it is accomplished in several ways. A dedicated neural projection runs from the SCN to the pineal gland along autonomic nerve fibers (see Figure 4.17). In addition, melatonin is secreted into the bloodstream immediately after synthesis, without storage. The biochemical regulatory events for the synthesis and secretion of the hormone can be activated very rapidly. A magnification step is also involved that provides a high-amplitude rhythm. Light during circadian night immediately truncates the synthesis of melatonin. Finally, melatonin is cleared very rapidly from the circulation, with the result that the rhythm has a very sharp onset and offset, one that is consistent for an individual mammal from one night to the next. In these ways the melatonin profile provides an unambiguous reference for time of day and for season.

A further important example of neuroendocrine regulation concerns the circadian link between the SCN and the anterior pituitary gland. The SCN controls the production of corticotropin-releasing hormone from the hypothalamus by neural projections. Daily release of this hormone regulates rhythmic release of adrenocorticotropic hormone (ACTH) by the anterior pituitary gland. In turn, ACTH modulates circadian release of glucocorticosteroids and minerocorticosteroids from the adrenal cortex. These steroid hormones are vital for many mammalian physiological functions.

Neuroendocrine links are also important in the estrous cycle of female rodents, the timing of which is accomplished by interactions between the neural signals of the circadian clock and the endocrine signals of the ovary. Since a female mammal produces viable eggs cyclically, she is fertile for a restricted period of time. In order to maximize reproductive success, mating must occur during the fertile window. In nocturnal rodents such as hamsters and rats, mating can only occur at night, during the active phase of the circadian cycle, at times when encounters between male and female are most likely. Consequently, nocturnal rodents exhibit a very precise linkage between the circadian clock and the neuroendocrine mechanisms that control ovulation and copulation. Typically a female hamster ovulates once every 4 days at about midday. Ovulation is triggered by a rise in the circulating levels of estrogen secreted by the ovary, and it occurs in association with enhanced sexual receptivity during the following night. Receptivity, in turn, depends on a rise in progesterone secretion by the ruptured follicle. The complete 4-day behavioral and neuroendocrine cycle is thus generated by an interplay between circadian and ovarian timers and is tightly phase-locked to the rhythm of light and dark by entrainment of the SCN. The neural pathways that facilitate the interaction between circadian and ovarian cues are only partly known.

A streamlined mammalian circadian axis has evolved[39,42,60]

Several factors have favored a hierarchical circadian system in mammals. A single primary central SCN pacemaker and a single input for light information carry out the task. Specialized photon counter cells of the retina are connected by a direct connection from the retinohypothalamic tract to the SCN. Such a pattern appears consistent in all mammals and contrasts with the multioscillator, multiphotoreceptor pattern seen in all other vertebrate groups. The departure of mammals from the standard vertebrate pattern may be related to their exclusively nocturnal activity in their early evolutionary history, when the dominant reptilian predators were mostly day-active. It was only after the decline of reptiles that some mammals secondarily adapted to daylight activity.

Many factors are involved in such a widespread nocturnal life-history pattern. Initially it must have provided an advantage for day-active nonmammalian vertebrates to possess multiple possible dedicated photoreceptors, many of which were nonimaging light detectors. Multiple circadian photoreceptors may have been selected for specifically because they did not form images but instead were specially tuned to respond to different parameters of the photic environment. Consequently they could be located adjacent to their oscillators. Nocturnal mammals, in contrast, most often experienced only the dim light of dusk and dawn. Highly sensitive photoreceptors were necessary for effective function. The presence of multiple photoreceptors at deep locations such as pineal and/or interior brain sites might give conflicting signals about the time of day. Such a liability might be eliminated rather quickly by selection. Circadian photoreceptors have been present in the vertebrate retina for half a billion years, but exclusive dependence on them may be a mammalian innovation. Consequent changes in the photodetection system may have precipitated other changes in the pacemaker system. The end result is an exclusive retinohypothalamic connection to the SCN and the dependence of mammals on the SCN as primary circadian pacemaker. One striking photoreceptor change in mammals has been the use of retinal ganglion cells for photic input rather than rod and cone photoreceptor input (see Chapter 6).

SUMMARY

The majority of information about circadian organization in animals comes from relatively few model organisms. Circadian systems of all multicellular animals consist of one or more pacemakers for generation of daily rhythms, an environmental input for sensory information needed for entrainment, and output messages to regulate daily physiological and behavioral rhythms. Many variations are seen in clock components, as well as input and output features. Searches for pacemakers have focused on nervous systems, but several tissues outside the nervous system produce robust circadian rhythms. Research in progress suggests that rhythmicity may be a property of many types of cells in a great variety of organisms. More information is needed on the relationship between these peripheral clocks and central nervous system pacemakers.

Circadian research on invertebrates has emphasized the nervous systems of two species of gastropod molluscs and of several species of arthropods. In the marine molluscs *Bulla gouldiana* and *Aplysia californica*, precise, persistent circadian rhythms of compound action potentials free-run in cultures of eyes under constant conditions in the laboratory. Large basal retinal neurons of the eye serve as pacemakers and photoreceptors for entrainment. Future research will probably expand the number of molluscs known to have well-developed circadian systems. The best known arthropod circadian systems occur in insects, scorpions, and horseshoe crabs. For

insects the primary pacemakers are found either in optic lobe sites of crickets, cockroaches, and beetles or in cerebral lobe sites of moths and flies. Retinal and extraretinal photoreceptors provide photic input for entrainment to local environmental conditions. The scorpion clock lies in the primitive brain. Five sets of lateral eyes serve in luminance detection for circadian entrainment. Other secondary pacemakers of arthropods, such as the circadian clock in the testes of gypsy moths, lie outside the central nervous system, and they appear to have their own photoreceptive cells.

Primary circadian pacemakers in nonmammalian vertebrates often involve the pineal gland, the retina, and the SCN. The pineal gland is photoreceptive in lower vertebrate classes, and may serve, along with the retina, for photoentrainment. Several other peripheral photoreceptors also occur in these groups, and nonphotic entrainment is important in some species. In mammals the primary pacemaker is the SCN, and photic input for entrainment is provided exclusively by the retina via the retinohypothalamic tract. Output signals for vertebrates include both neural and humoral signals. In addition to primary central neural pacemakers, a variety of local peripheral pacemakers outside the central nervous system have been demonstrated for vertebrates.

CONTRIBUTORS

Patricia J. DeCoursey wishes to thank Gene Block, Gerta Fleissner, Günther Fleissner, Michael Menaker, Robert Y. Moore, Terry Page, Kathleen Siwicki, and Herbert Underwood for their material contributions to Chapter 5.

STUDY QUESTIONS AND EXERCISES

1. A wide variety of circadian systems is found in fish. One of the most primitive fish, the hagfish, uses a ventral hypothalamic oscillator, which possibly is an SCN homolog. Its close relative in the cyclostome group (the lamprey) and probably most teleost fish use primarily a pineal oscillator.

    a. How does the hagfish compare in its circadian system components with mammals? Discuss in terms of oscillator, input components, and output components.

    b. Why do you think the hagfish and the lamprey differ?

2. A clue to the preceding questions comes from a consideration of ecosystems. For a review of ecological aspects of ecosystems, try the following exercise. Read again the material on ecosystems in Chapter 2. Recall that ecosystems can usually be defined by a unique combination of four or five basic features: temperature in daily and annual range, moisture availability, altitude above or below sea level, and substrate.

Define the following terms. Be sure to mention any temperature, moisture, and substrate requirements.

Tropical rain forest
Prairie
Seacoast
Marsh
River

3. Think of examples in which the ecosystem features of an organism's habitat have been selective agents for the type of timing system. (*Hint*: Do deep cave species or deep ocean forms have circadian rhythms?) Explain.

4. Can you think of organisms living in moving water such as rivers, coastal ocean waters, or waterfalls that have well-developed, precise pacemakers. If not, why not?

5. Another explanation for the terrestrial-mammal type of circadian system is based on the so-called bottleneck hypothesis. Several major extinctions of plants and animals have occurred in the course of evolution of life on Earth (see Chapter 11). How would these affect the evolution of circadian pacemakers in organisms?

6. Summarize the main components of the circadian systems of four representative birds by filling in the table below. Make a similar table for reptiles and complete the chart.

Response of the Circadian Marker

Species	Deletion	Body temperature	Locomotor activity	Blood melatonin
Sparrow	Pineal			
	Eye			
	SCN			
	Multiple			
Japanese quail	Pineal			
	Eye			
	SCN			
	Multiple			
Chicken	Pineal			
	Eye			
	SCN			
	Multiple			
Pigeon	Pineal			
	Eye			
	SCN			
	Multiple			

REFERENCES

1. Arendt, J. 1995. Melatonin and the mammalian pineal gland. Chapman Hall, New York.

2. Barinaga, M. 2002. How the brain's clock gets daily enlightenment. Science 295: 955–957.

3. Barlow, R. B., Jr. 1983. Circadian rhythms in the *Limulus* visual system. J. Neurosci. 3: 856–870.

4. Barrett, R. K., and T. L. Page. 1989. Effects of light on circadian pacemaker development: I. The freerunning period. J. Comp. Physiol. A 165: 41–49.

5. Berson, D. M., F .A. Dunn, and M. Takao. 2002. Phototransduction by retinal ganglion cells that set the circadian clock. Science 295: 1070–1073.

6. Binkley, S. 1990. The Clockwork Sparrow: Time, Clocks, and Calendars in Biological Organisms. Prentice-Hall, Englewood Cliffs, NJ.

7. Block, G. D., and D. G. McMahon. 1984. Cellular analysis of the *Bulla* ocular circadian pacemaker system. III. Localization of the circadian pacemaker. J. Comp. Physiol. A 155: 387–395.

8. Blumenthal, E. M., G. D. Block, and A. Eskin. 2001. Cellular and molecular analysis of molluscan circadian pacemakers. In: Handbook of Behavioral Neurobiology Volume 12: Circadian Clocks, J. S. Takahashi, F. W. Turek, and R. Y. Moore (eds.), pp. 371–394. Kluwer Academic/Plenum, New York.

9. Buijs, R. M., and A. Kalsbeek. 2001. Hypothalamic integration of central and peripheral clocks. Nature Rev. Neurosci. 2: 521–526.

10. Cahill, G. M. 2002. Clock mechanisms in zebrafish. Cell Tissue Res. 309: 27–34.

11. Cahill, G. M., and J. C. Besharse. 1991. Re-setting the circadian clock in cultured *Xenopus* eyecups: Regulation of retinal melatonin rhythms by light and D_2 dopamine receptors. J. Neurosci. 11: 2959–2971.

12. Daan, S. 2000. Colin Pittendrigh, Jürgen Aschof, and the natural entrainment of circadian systems. J. Biol. Rhythms 15: 195–217.

13. Ebihara, S., I. Oshima, H. Yamada, M. Goto, and K. Sato. 1987. Circadian organization in the pigeon. In: Comparative Aspects of Circadian Clocks, T. Hiroshigi and K-I. Honma (eds.), pp. 84–103. Hokkaido University Press, Sapporo, Japan.

14. Falcon, J., and J. P. Collin. 1989. Photoreceptors in the pineal of lower vertebrates: Functional aspects. Experientia 45: 909–912.

15. Fleissner, G., and G. Fleissner. 2001a. Neuronal organization of circadian systems. In: Scorpion Biology and Research, P. Brownell and G. A. Polis (eds.), pp. 107–137. Oxford University Press, New York.

16. Fleissner, G., and G. Fleissner. 2001b. The scorpion's clock: Feedback mechanisms in circadian systems. In: Scorpion Biology and Research, P. Brownell and G. A. Polis (eds.), pp. 138–158. Oxford University Press, New York.

17. Fleissner, G., and G. Fleissner. 2002. Perception of natural zeitgeber signals. In: Biological Rhythms, V. Kumar (ed.), pp. 83–93. Narosa, New Delhi, India.

18. Foà, A. 1991. The role of the pineal and the retinae in the expression of circadian locomoter rhythmicity in the ruin lizard, *Podarcis sicula*. J. Comp. Physiol. 169: 201–207.

19. Foà, A., M. Flamini, A. Innocenti, L. Minutini, and G. Monteforti. 1993. The role of extraretinal photoreception in the circadian system of the ruin lizard *Podarcis sicula*. Comp. Biochem. Physiol. A 105: 223–230.

20. Gaston, S., and M. Menaker. 1968. Pineal function: The biological clock in the sparrow. Science 160: 1125–1127.

21. Grace, M., V. Alones, M. Menaker, and R. G. Foster. 1996. Light perception in the vertebrate brain: An ultrastructural analysis of opsin- and vasoactive intestinal polypeptide immunoreactive neurons in inguanid lizards. J. Comp. Neurol. 367: 575–594.

22. Hastings, J. W., B. Rusak, and Z. Boulos. 1991. Circadian rhythms: The physiology of biological timing. In: Neural and Integrative Animal Physiology, C. L. Prosser (ed.), pp. 435–546. Wiley, New York.

23. Helfrich-Forstser, C. 1995. The period clock gene is expressed in central nervous system neurons which also produce a neuropeptide that reveals the projections of circadian pacemaker cells with the brain of *Drosophila melanogaster*. Proc. Natl. Acad. Sci. USA 92: 612–616.

24. Hoffmann, K. 1957. Angeborene Tagesperiodik bei Eidechsen. Naturwissenschaften 44: 359–360.

25. Inouye, S. T., and H. Kawamura, H. 1979. Persistence of circadian rhythmicity in a mammalian hypothalamic "island" containing the suprachiasmatic nucleus. Proc. Natl. Acad. Sci. USA 76: 5962–5966.

26. Jacklet, J. W., and W. Colquhoun. 1983. Ultrastructure of photoreceptors and circadian pacemaker neurons in the eye of a gastropod, *Bulla*. J. Neurocytol. 12: 373–396.

27. Janik, D. S., G. E. Pickard, and M. Menaker. 1990. Circadian locomoter rhythms in the desert iguana II: Effects of electrolytic lesions to the hypothalamus. J. Comp. Physiol. 166: 811–816.

28. Johnson, C. H. 2001. Endogenous timekeepers in photosynthetic organisms. Annu. Rev. Physiol. 63: 695–728.

29. Johnson, C. H., and T. Kondo. 2001. Circadian rhythms in unicellular organisms. In: Handbook of Behavioral Neurobiology Volume 12: Circadian Clocks, J. S. Takahashi, F. W. Turek, and R. Y. Moore (eds.), pp. 61–74. Kluwer Academic/Plenum, New York.

30. Kabasawa, H., S. Ooka-Souda, and F. Takashima. 1993a. Entrainability of locomoter activity rhythm to the reversed light-dark cycle in the hagfish. Nippon Suisan Gakkaishi 59: 1147–1150.

31. Klein, D. C., R. Y. Moore, and S. M. Reppert. 1991. Suprachiasmatic Nucleus: The Mind's Clock. Oxford University Press, New York.

32. Korf, H. W., C. Schomerus, and J. H. Stehle. 1998. The Pineal Organ, Its Hormone Melatonin and the Photoneuroendocrine System. Springer, New York.

33. Lehman, M. N., R. Silver, W. R. Gladstone, R. M. Kahn, M. Gibson, and E. L. Bittman. 1987. Circadian rhythmicity restored by neural transplant. Immunocytochemical characterization of the graft and its integration with the host brain. J. Neurosci. 7: 1626–1638.

34. Leeson, T. S., and C. R. Leeson. 1981. Histology, 4th Edition. Saunders, Philadelphia.

35. Lissmann, H. W., and H. O. Schwassmann. 1965. Activity rhythm of an electric fish, *Gymnorhamphichthys hypostomus* Ellis. Z. vergleichende Physiol. 51: 153–171.

36. Loher, W. 1972. Circadian control of stridulation in the cricket *Teleogryllus commodus* Walker. J. Comp. Physiol. 79: 173–190.

37. McMahon, D. G., and G. D. Block. 1987. The *Bulla* ocular circadian pacemaker. I. Pacemaker neuron membrane potential controls phase through a calcium-dependent mechanism. J. Comp. Physiol. A 161: 335–346.

38. McMahon, D. G., S. F. Wallace, and G. D. Block. 1984. Cellular analysis of the *Bulla* ocular circadian pacemaker system. II. Neurophysiological basis of circadian rhythmicity. J. Comp. Physiol. A 155: 379–385.

39. Meijer, J. H. 2001. Photic entrainment of mammals. In: Handbook of Behavioral Neurobiology Volume 12: Circadian Clocks, J. S. Takahashi, F. W. Turek, and R. Y. Moore (eds.), pp. 183–210. Kluwer Academic/Plenum, New York.

40. Menaker, M., and A. Oksche. 1974. The avian pineal organ. In: Avian Biology, D. S. Farner, J. R. King, and K. C. Parkes (eds.), pp. 79–118. Academic Press, New York.

41. Michel, S., M. E. Geusz, J. J. Jaritsky, and G. D. Block. 1993. Circadian rhythm in membrane conductance expressed in isolated neurons. Science 259: 239–241.

42. Moore, R. Y. 1999. Circadian timing. In: Fundamental Neuroscience, M. J. Zigmond, F. E. Bloom, S. C. Landis, J. L. Roberts, and L. R. Squire (eds.), pp. 1189–1206. Academic Press, New York.

43. Moore, R. Y., and R. K. Leak. 2001. Suprachiasmatic nucleus. In: Handbook of Behavioral Neurobiology Volume 12: Circadian Clocks, J. S. Takahashi, F. W. Turek, and R. Y. Moore (eds.), pp. 141–171. Kluwer Academic/Plenum, New York.

44. Morita, Y., M. Samejima, and U. Katsuhisa. 1987. The role of direct photosensory pineal organ in the LD and circadian rhythm. In: Comparative Aspects of Circadian Clocks (Proceedings of the 2nd Sapporo Symposium on Biological Rhythms), T. Hiroshige and K-I. Honma (eds.), pp. 73–81. Hokkaido University Press, Sapporo, Japan.

45. Norgren, R. B., Jr., and R. Silver. 1989. Retinohypothalamic projections and the suprachiasmatic nucleus in birds. Brain Behav. Evol. 34: 73–83.

46. Page, T. L. 1982. Transplantation of the cockroach circadian pacemaker. Science 216: 73–75.

47. Page, T. L. 1988. Circadian organization and the representation of circadian information in the nervous systems of invertebrates. Adv. Biosci. 73: 67–79.

48. Page, T. L. 2001. Circadian systems of invertebrates. In: Handbook of Behavioral Neurobiology Volume 12: Circadian Clocks, J. S. Takahashi, F. W. Turek, and R. Y. Moore (eds.), pp. 79–103. Kluwer Academic/Plenum, New York.

49. Page, T. L., and K. G. Nolovic. 1992. Properties of mutual coupling between the two circadian pacemakers in the eyes of the mollusc *Bulla gouldiana*. J. Biol. Rhythms 7: 213–226.

50. Plautz, J. D., M. Kaneko, J. C. Hall, and S. A. Kay. 1997. Independent photoreceptive circadian clocks throughout *Drosophila*. Science 278: 1632–1635.

51. Purves, W. K., D. Sadava, G. H. Orians, and H. C. Heller. 2001. Life: The Science of Biology, 6th Edition. Sinauer, Sunderland, MA.

52. Ralph, M. R., R. G. Foster, F. C. Davis, and M. Menaker. 1990. Transplanted suprachiasmatic nucleus determines circadian period. Science 247: 975–982.

53. Refinetti, R., and M. Menaker. 1992. The circadian rhythm of body temperature. Physiol. Behav. 51: 613–637.

54. Stokkan, K-A., S. Yamazaki, H. Tei, Y. Sakaki, and M. Menaker. 2001. Entrainment of the circadian clock in the liver by feeding. Science 291: 490–493.

55. Takahashi, J. S. 1987. Cellular basis of circadian rhythms in the avian pineal. In: Comparative Aspects of Circadian Clocks, T. Hiroshige and K-I. Honma (eds.), pp. 3–15. Hokkaido University Press, Sapporo, Japan.

56. Takahashi, J. S., N. Murakami, S. S. Nikaido, B. L. Pratt, and L. M. Robertson. 1989. The avian pineal, a vertebrate model system of the circadian oscillator: Cellular regulation of circadian rhythms by light, second messengers, and macromolecular synthesis. Recent Prog. Horm. Res. 45: 279–352.

57. Takahashi, J. S., F. W. Turek, and R. Y. Moore (Eds.) 2001. Handbook of Behavioral Neurobiology Volume 12: Circadian Clocks. Kluwer Academic/Plenum, New York.

58. Tosini, G., and C. Fukara. 2002. The mammalian retina as a clock. Cell Tissue Res. 309: 119–126.

59. Tosini, G., and M. Menaker. 1995a. Circadian rhythm of body temperature in an ectotherm (*Iguana iguana*). J. Biol. Rhythms 10: 248–255.

60. Tosini, G., and M. Menaker. 1995b. The evolution of vertebrate circadian systems. In: Circadian Organization and Oscillatory Coupling, K-I. Honma and S. Honma (eds.), pp. 39–52. Hokkaido University Press, Sapporo, Japan.

61. Tosini, G., and M. Menaker. 1996. Circadian rhythms in cultured mammalian retina. Science 272: 419–421.

62. Tosini, G., and M. Menaker. 1998. Multioscillatory circadian organization in a vertebrate, *Iguana iguana*. J. Neurosci. 18: 1105–1114.

63. Underwood, H. 1985. Pineal melatonin rhythms in the lizard *Anolis carolinensis*: Effects of light and temperature cycles. J. Comp. Physiol. 157: 57–65.

64. Underwood, H. 1990. The pineal and melatonin: Regulators of circadian function in lower vertebrates. Experientia 46: 120–128.

65. Underwood, H. 2001. Circadian organization in nonmammalian vertebrates. In: Handbook of Behavioral Neurobiology Volume 12: Circadian Clocks, J. S. Takahashi, F. W. Turek, and R.Y. Moore (eds.), pp. 111–135. Kluwer Academic/Plenum, New York.

66. Underwood, H., R. K. Barrett, and T. Siopes, T. 1990. The quail's eye: A biological clock. J. Biol. Rhythms 5: 257–265.

67. Underwood, H., C. T. Steele, and B. Zivkovic. 2001. Circadian organization and the role of the pineal in birds. Microsc. Res. Tech. 53: 48–62.

68. Welsh, D. K., D. E. Logothetis, M. Meister, and S. M. Reppert. 1995. Individual neurons dissociated from rat suprachiasmatic nucleus express independently phased circadian firing rhythms. Neuron 14: 697–706.

69. Yamazaki, S., V. Alones, and M. Menaker. 2002. Interaction of the retina with suprachiasmatic pacemakers in the control of circadian behavior. J. Biol. Rhythms 17: 315–329.

70. Yamazaki, S., R. Numano, M. Abe, A. Hida, R. I. Takahashi, et al. 2000. Resetting central and peripheral circadian oscillators in transgenic rats. Science 288: 682–685.

71. Yoshimura, T., S. Yasuo, Y. Suzuki, E. Makino, Y. Yokota, et al. 2001. Identification of the suprachiasmatic nucleus in birds. Am. J. Physiol.: Regul. Integr. Comp. Physiol. 280: R1185–R1189.

72. Zatz, M. 1996. Melatonin rhythms: Trekking toward the heart of darkness in the chick pineal. Sem. Cell Dev. Biol. 7: 811–820.

73. Zivkovic, B. D., H. Underwood, C. T. Steele, and K. Edmonds. 1999. Formal properties of the circadian and photoperiodic systems of Japanese quail: Phase response curve and effects of t-cycles. J. Biol. Rhythms 14: 378–390.

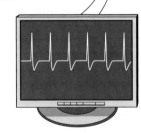

As the view of pacemakers is magnified from organ level to the structure of pacemaker cells and their chemical processes, a new modern understanding of biological timekeeping emerges.

Important questions concern generation of the self-sustained oscillatory processes in single individual cells, coupling of the individual cellular oscillations in the pacemaker, input and processing of sensory information for entrainment, and generation of output commands. Highly sophisticated equipment and techniques are required: elaborate light and electron microscopes (see Figure 6.1), specialized recording electrodes with computerized oscilloscopes for rapid data collection and analysis.

Earlier histological methods for studying cells have developed into precise immunocytological approaches for visualizing neuropeptides and other biochemical entities within cells. Protocols have been developed for explanting formerly intractable nerve cells, either as slices or individual dispersed cells, into long-term culture. These are but a few of the versatile tools available now for chronobiology research. Some of these methods are highly specialized and technical. At the same time, they are essential for understanding progress in biological timekeeping. For these reasons a tutorial is provided in Chapter 6 to help provide background on specialized cell physiology techniques.

Cell Physiology of Circadian Pacemaker Systems in Metazoan Animals

Dost thou love life, then do not squander time, for that's the stuff life is made of.

—Benjamin Franklin

Introduction: The Important Cellular Questions Require an Appropriate Organism and Preparation[6,8,11,46,48,59,69]

Studying cellular processes of circadian pacemaker cells has been the most challenging level of research in chronobiology. Highly specialized tools have been required, and many of the results are still preliminary. Conceptually one of the most difficult aspects for readers will be to distinguish generation of the rhythm from both input signals and cellular output, all of which go on simultaneously in single independent clock cells. In most cases the greatest success has been achieved by culturing of single clock cells or dispersed cells.

An excellent review by Takahashi et al.[29] points out the idealized basic three criteria of a model system for analyzing circadian rhythms at the cellular level. First, the assumed pacemaker should tolerate isolation in culture and exhibit persistence of circadian overt output rhythms without damping for extended periods of time in order to demonstrate the convincing presence of a circadian oscillator. Second, the circadian pacemaker must be entrainable in culture by 24 h entraining agents, such as a light–dark cycle. Third, the cellular model system under consideration should be suitable for perturbation analysis by pulse or step inputs, which allow dissection of the sequence of oscillation generation, its entrainment, and its output. Other criteria can be added to this list. A reliable, continuous, and affordable assay of rhythmicity at the behavioral and especially at the cellular level is vital. The pacemaker ideally should consist of abundant, reasonably homogeneous material and should be suitable for biochemical and genetic analysis.

In the sections that follow, three well-established cellular models will first be briefly introduced. A short tutorial on the methodologies used in studies of circadian cell physiology will precede the discussion of the models. Information that has accumulated about cellular processes will then be organized for each model around its oscillators at the tissue and single-cell levels. Finally, the commonalities and differences of the models will be compared.

Cellular questions concern generation of circadian oscillations, coupling, input information, and output commands. With accelerating pace each successive year, steps are being taken to locate and identify the cellular component parts of circadian pacemaker systems and to understand their physiology. Progress has been made at the cellular level in learning how rhythms are generated within pacemaker cells, how individual pacemaker cells are coupled to each other, how pacemaker cells process information from the sensory messages for entrainment, and how they generate commands and messages to drive the overt rhythms. No rigid chapter boundaries exist in a living organism between the organ, cellular, and molecular levels of operation, but out of necessity the authors needed to make some arbitrary decisions about where to place the material for Chapters 5 through 8.

Even the first step in circadian cell physiology, locating a pacemaker and ensuring that it is necessary and sufficient for expression of rhythmicity in an organism, is not simple. The primary strategy for identifying circadian pacemakers in whole animals requires initially documenting arrhythmia induced by lesioning and subsequently transplanting a replacement pacemaker to restore rhythmicity (see Chapter 5). Cellular questions become more penetrating, and the requirements are more stringent. Is more than one central pacemaker involved? Can any of the multiple oscillators be singled out for study to reduce the complexity? Can individual pacemaker cells act as oscillators, and how are they coupled?

Measuring Cellular Rhythmicity Requires Sophisticated Assays[30,71]

The list of favorable models for circadian cell physiology has been narrowed to three[56,59,69]

The eye of the clouded bubble snail *Bulla*, the chick pineal in culture, and the rodent SCN all continue to oscillate robustly for many cycles in vitro, and each has a special set of experimental advantages (Table 6.1). Reliable indicators such as these are needed as markers for studying circadian cell physiology. Ideally, a technique should demonstrate and allow recording of the period, amplitude, and phase (see Chapter 1) of the pacemaker's oscillation over extended increments of time without induction of artifacts. The measurement should be made as close as possible to the pacemaker because indirect observation of the hands of the clock can introduce many sources of error.

Methods should overlap and be artifact free[31]

Multiple outputs of the clock should be measured simultaneously because some manipulations may eliminate or distort one output rhythm of pacemaker system cells but not others. Although this specification is highly desirable, in practice it is also very difficult and often not feasible. The reader should also constantly keep in mind that the three essential functions of the circadian system of individual pacemaker cells consist of generating oscillation, processing environmental entraining information, and formulating output commands for the behavioral effector system.

Such a directive becomes particularly challenging for pacemaker systems like *Bulla*, in which all the elements can be found in a single basal retinal nerve cell. Confusion can result because the oscillator's properties are

---

TABLE 6.1   Features of three established model systems

Features of the three systems	Model species		
	Bulla	Chick	SCN
Longevity of the preparation in organotypic (slice) culture	10–14 days	8 days	Months
Large size neurons for easy recording	•		
Ease in "reducing" the preparation	•	•	
Successful dissociated cell lines	•	•	•
Availability of immortalized cell lines			•
Dual parts allowing simultaneous control and experimental set-up in same animal	•		
Heterogenicity of neuronal pacemaker cell types			•
Oscillator, photoreceptor and efferent in same cell	•	•	
Separate photoreceptor			•
Location of rhythm generating neurons known	•	•	

most easily measured indirectly by means of discharges along the optic nerve, and in reality the discharges are the output signal from the pacemaker. These discharges are known to be the output signal because optic nerve impulses, for example, can be blocked reversibly for days through the use of pharmacological toxin methods. When the toxin is washed off and the discharges return, the phase is the same as if the oscillator had continued to run without the treatment. Similar conceptual problems are encountered in single-cell chick pineal culture; all elements are found in a single pinealocyte, and rhythmic properties of the pacemaker itself are measured in terms of the melatonin output signals.

Some of the greatest challenges in learning circadian cell physiology come from the diversity of interdisciplinary techniques used. Some tools used in cell physiology are shown in Figure 6.1. Readers familiar with the techniques, including their unique values and their pitfalls and deficiencies, can jump to the case studies of *Bulla*, chick pineal, and mammalian SCN.

Two general approaches include in vivo and in vitro cell culture[16,17,38,40]

A distinction will be made from the very start between in vivo and in vitro approaches. In vivo research examines events in the whole living animal and has many advantages (Figure 6.2). The likelihood of artifact is reduced because all cells remain in place, bathed by the normal physiological solutions of the animal's body fluids. If the recording devices are not too invasive, the length of time that the measurement can be sustained may be much greater in the living animal than in dish culture. However, because the structure under study remains attached by all its physiological ties, in vivo preparations are usually much more complex than in vitro preparations, and feedback from other parts of the nervous system may modify circadian cell function. In vitro methods are usually more applicable to cell physiology studies, but ideally both modes should be studied and differences between the two carefully noted.

In general, circadian studies are easier to interpret when data are continuously recorded from individual cells, tissues, or animals, each monitored over many circadian cycles. More confusing results can arise when a population is pooled and sampled at only a few phases

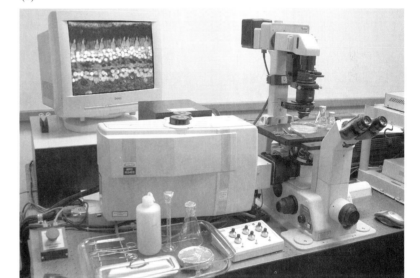

(A)

(B)

FIGURE 6.1 Specialized microscopes used to study cell structure and physiology of circadian pacemakers. (A) Scanning electron microscope. A neuron is shown on the monitors. (B) Laser scanning confocal imaging system, with rods and cones shown on the monitor. (Courtesy of P. DeCoursey.)

of a single cycle, especially if the rhythms among the individuals are not synchronous. For example, apparent arrhythmicity of a population may be attributed either to a loss of rhythmicity of each individual or to a desynchronization of rhythmicity between individuals because each of the individuals expresses a rhythm that differs in phase or period.

In vitro culture maintains tissues or cells in dish culture in a meticulously prepared sterile artificial medium with appropriate pH and oxygen levels rigorously maintained (Figure 6.3). Although culture methods have improved greatly in the past decade, they are still chal-

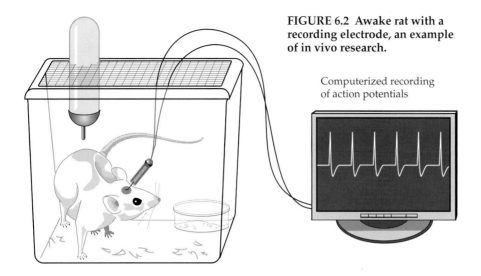

FIGURE 6.2 Awake rat with a recording electrode, an example of in vivo research.

Computerized recording of action potentials

Computerized recording of action potentials

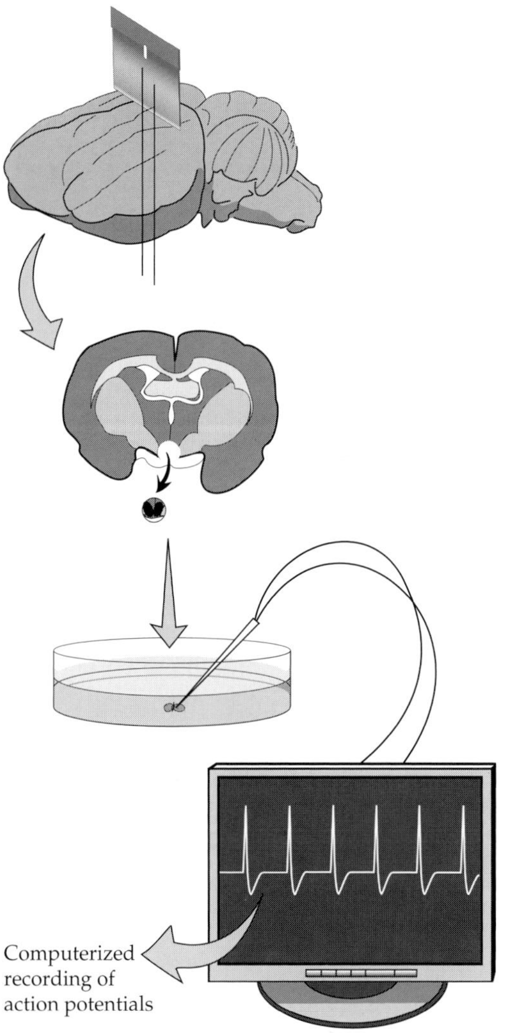

FIGURE 6.3 Explant culture of an SCN slice, an example of in vitro studies.

lenging, especially for neuronal culture, and the degree of long-term success is highly variable from one research laboratory to another. Explant culture involves a slice containing the desired cells. The slice technique is one of the most widely used culture methods in the electrophysiological study of the mammalian SCN.

In reduced preparations, the organ, such as the pineal gland, is cut successively into smaller and smaller pieces, usually in an attempt to isolate a single pacemaker cell. Pacemaker reduction has been used extensively in the chick pineal and in *Bulla* preparations, but not in the cellularly heterogeneous SCN.

The culture of dispersed cells or single pacemaker cells is demanding. A mechanical or enzymatic method is generally used to isolate single cells before they are transferred to a culture chamber designed to accommodate the type of measurement desired. For melatonin assays of single chick pineal cells, for example, the dispersed cells are placed in a flow-through chamber, not only to supply a constant level of nutrients and oxygen but also to obtain the effluent for melatonin determinations. The substrate chosen is often critical to the success of single-cell culture. Sometimes a layer of fibroblast cells is laid down as a type of basement membrane. In chick pineal culture, small beads for pineal cells to climb upon were critical in the isolation and transfer of single cells to a culture chamber.

Culture dishes are currently available with embedded microarrays of electrodes for recording from SCN slices and dispersed-cell cultures. After a sample of tissue or dispersed pacemaker cells is placed in the chamber, the cells settle on the substrate and make contact with an electrode. In this way the timing of neuronal firing patterns for multiple individual cells can be recorded simultaneously (Figure 6.4).

Many highly specific methods are available[16,17,18,31,37,54,58,71]

OVERVIEW. Some of the more common cell physiology methods used in circadian research today will be described here to give a feel for the multiple approaches that the cell physiologist must use to demonstrate the workings of pacemaker cells. For convenience, the specific methods can be grouped into several broad categories, as outlined in Table 6.2 Many approaches use a battery of methods to test a hypothesis. Later in this chap-

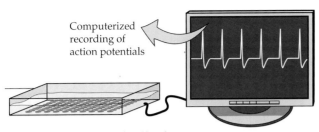

Computerized recording of action potentials

FIGURE 6.4 Dispersed-cell culture.

ter the recent localization of circadian photoreceptors in mammalian retinal ganglion cells will be described. This discovery illustrates the effectiveness of using many high-tech methods to study circadian cell physiology.

STAINING. A versatile and widely employed set of techniques for visualizing the physiology of pacemaker cell processes involves preserving and staining the cells. Stains can unveil minute details of the architecture of a cell, thereby revealing structural specialization for specific physiological functions. In some cases a so-called vital stain can be used to detect the presence of active substances in living cells. The methods are referred to as histological if simple dye stains are used, and as immunocytological if antibodies are generated against protein targets in the cell.

Many stains are available. Some are quite general; others may highlight specific substances. Nissl stain, for example, is widely used in SCN transplant studies because the cresyl violet stain is absorbed selectively by the rough endoplasmic reticulum in SCN neurons. The techniques may be carried out at specific phases of the pacemaker rhythm, thus allowing a series of timed snapshots throughout the cycle of a rhythmic cellular process. Multiple groups of animals are sacrificed at intervals throughout a circadian day (for example, every 3 h), to track the concentration of metabolites in a pacemaker cell.

More physiological detail is visualized through antibody localization/immunostaining procedures on tissue sections. Antibodies for almost any neuropeptide antigen, as well as other components in cells, can be generated and used as markers. With immunocytochemistry, the choice of a particular antibody, as well as the timing, temperature, and method of applying the reagents is critical. Visualizing the SCN through localization of an antibody requires that the targets are fixed within the brain tissue by a chemical such as paraformaldehyde. An example of a target might be neuropeptides located in and around the SCN. Thin serial sectioning is then carried out with a special cutting instrument such as a vibratome, a microtome, or a cryostat. Sections are processed through a series of antibody attachment steps. The primary antibody attaches to the desired neuropeptide antigen, and excess unbound primary antibody is washed off. A secondary antibody attaches to the surface of the primary antibody. One example is the use of a primary antibody from rabbit, and a secondary anti-rabbit globulin. The secondary antibody may include a fluorescent marker, or it can be reacted chemically to yield a colored precipitate. In this way, many neuropeptides have been identified and localized in the SCN (Figure 6.5).

Two or more antigens may be co-localized to a single cell by the use of primary antibodies made in different animals such as rabbit and mouse. The appropriate secondary antibodies are conjugated to markers that fluoresce with different colors when excited by light of the appropriate wavelengths. For example, anti-rabbit could fluoresce green and anti-mouse could fluoresce red. Sections can be viewed in a confocal microscope, which makes virtual, optical sections less than 1 μm thick through the tissue. The levels of some of these substances oscillate over 24 h, through either rhythmic synthesis or degradation. Consequently, immunocytochemical tech-

TABLE 6.2	Categories of cellular methods
Category	Description
Staining	Staining cell organelles, hormones, receptors, gene products, or processes; used in standard histology, immunocytochemistry, in situ hybridization
Electrophysiology	Electrophysiological studies for recording spikes, for lesioning, and for tracking ion fluxes
Neuropharmacology	Neuropharmacological studies with slices or single cells
Neuronal tracers	Following neuronal projections (pathways) by means of anterograde and retrograde tracers
Metabolic rate	Measuring the metabolic rate of pacemaker cells with 2-DG
Transgenic bioluminescent markers	Using transgenically engineered pacemaker cell technology for bioluminescent indicators of rhythmicity
Action spectra	Constructing action spectra for identifying photopigments in vivo

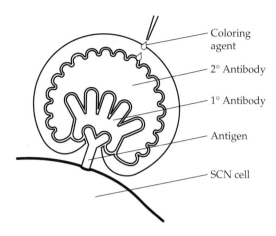

- Coloring agent
- 2° Antibody
- 1° Antibody
- Antigen
- SCN cell

FIGURE 6.5 Immunocytochemical staining.

niques have provided markers for the presence of rhythmic SCN cells in transplant studies, as well as assays of clock-controlled biochemical processes.

Another powerful staining technique for localizing and assaying SCN gene products is in situ hybridization, which also relies on the processing of serial sections. The method visualizes messenger RNA (mRNA), the intermediate step between DNA transcription and protein translation. Probes are made that bind to specific cellular mRNAs via complementary base pairing. Then the resulting hybrids are usually imaged autoradiographically through the use of radioactively labeled probes. The in situ method has been particularly useful for tracking spatial and temporal patterns of SCN gene expression (see Chapter 8).

Many other specialized methods of visualizing cell physiology processes are available. The so-called tracer techniques mark the polarized path of travel along a neuron and allow identification of neuronal connections called neural projections. Both anterograde and retrograde tracers have proven very useful in SCN studies.

ELECTROPHYSIOLOGY. Because most pacemaker cells are neurons, they are highly specialized for maintenance of a resting potential across their membranes, and for conducting electrical discharges or spikes down the length of their specialized axonal processes. Electrophysiological study of pacemaker neurons, particularly with in vitro methods, is one of the most prevalent cell physiology techniques used in studying circadian pacemakers. A variety of electrode types are available, suitable for detecting different parameters of the membrane potential and its discharge. Some electrodes record extracellularly from populations of neurons, others from single neurons. One specialized type is a patch clamp electrode, which uses a suction electrode to gently pull a piece of membrane from the cell for studying transmembrane properties.

Understanding how nerve cells conduct electrical messages is essential for interpreting electrophysiolog-

ical data. All nervous systems are composed of conducting cells called neurons, plus glial cells, and several types of connective tissue. Although neurons come in a variety of sizes and shapes, all have specialized structures for communicating with each other and for ensuring a one-way flow of messages. Even among the conducting type of neurons, great variation is found (Figure 6.6). Some types have extensive arborizations for contacting other neurons (see Figure 6.6C). Others, such as the SCN neurons, are very small and closely packed, with short processes. Some are specialized for both neu-

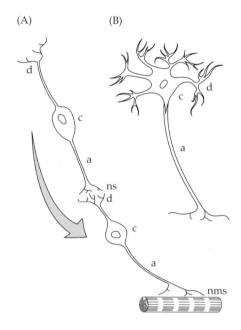

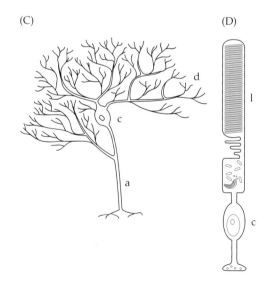

FIGURE 6.6 Neuron structure and function. (A) Two simple neurons with a neural synapse (ns) and neuromuscular synapse (nms). (B) Parts of a neuron. (C) A complex neuron. (D) Rod cell. a, axon; c, cell body; d, dendrites; l, lamina of photoreceptor. The arrow indicates the unidirectional flow of a nerve impulse in the neuron.

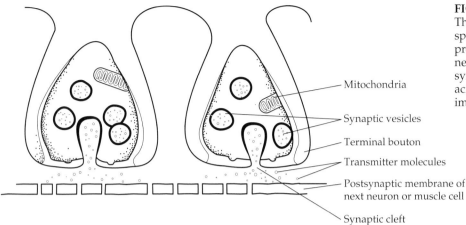

FIGURE 6.7 Crossing the synapse. The passage of a nerve discharge spike into the terminal bouton of a presynaptic neuron causes release of neurotransmitter substances into the synaptic cleft. The transmitters diffuse across the synapse and initiate a nerve impulse in the postsynaptic neuron.

Mitochondria

Synaptic vesicles

Terminal bouton

Transmitter molecules

Postsynaptic membrane of
next neuron or muscle cell

Synaptic cleft

ral and endocrine functions. The sensory neurons are perhaps the most atypical. Photoreceptive rods and cones lack axons and have highly specialized units of lamellar discs and folds to align the photopigments maximally for light absorption (see Figure 6.6D).

The discharge of a neuron can normally propagate in only one direction because of its structure. In this regard an action potential is said to be polarized with unidirectional flow. At the far end of a neuron is a synaptic cleft or space that the electrical message must cross by means of a synapse. A chemical neurotransmitter is released by the presynaptic neuron when a discharge reaches the terminal projections of the axon called boutons. After diffusing across the synaptic space, the chemical neurotransmitter activates receptors on the postsynaptic neuron and thereby initiates a discharge in the postsynaptic neuron (Figure 6.7). In this electrochemical fashion, a message travels ionically along a neuron and chemically across a synapse. Backward propagation with resulting canceling of membrane potentials is prohibited. In addition to the synapse, other mechanisms also prevent backward propagation.

A nerve message is propagated down the length of the axon at high speed by a complex process that can be compared very simplistically to water flowing down a garden soaker hose when the water flow is started (Figure 6.8). The ionic gates are superficially analogous to the tiny holes in the tube of the soaker hose. When at rest, a neuron maintains an electrical gradient across its membrane. The resting membrane potential of excitable

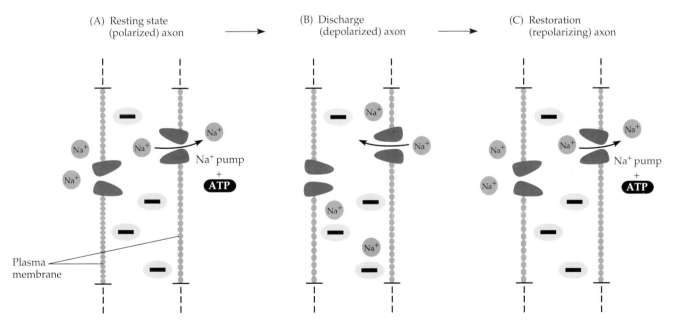

FIGURE 6.8 Conduction of a nerve impulse in three sequential states of an axon.
Segments of an axon are shown in the three sequential stages of discharge:
(A) Resting. (B) Discharge. (C) Restoration. For further explanation see text.

cells is negative, about –70 millivolts, and it is influenced mainly by the resting conductances for potassium and sodium ions. Intracellular potassium is high and sodium is low compared to their concentrations in extracellular fluid. The cytoplasm also contains organic anions, especially proteins.

These concentration gradients are required for excitation and are maintained by ion pumps or exchange mechanisms. Initial depolarization of the membrane opens voltage-gated sodium channels and produces an inward sodium current, causing further depolarization and opening more channels. An action potential is triggered in all-or-none fashion because the inward sodium current overcomes the stabilizing outward potassium current and drives the membrane potential toward the equilibrium potential for sodium, which is about +60 millivolts.

The action potential is terminated by the inactivation of sodium channels to a refractory state. At the same time, voltage-dependent potassium channels open. Potassium channels activate more slowly than sodium channels and repolarize the membrane toward potassium's negative equilibrium potential. In brief, the currency of nerve cells is primarily coding of messages by the frequency and amplitude of their electrochemical messages in unidirectional waves along neurons and across synapses.

LESIONING. One specialized aspect of pacemaker electrophysiology involves electrolytic lesioning to delete tissue. A sturdy electrode is used to apply known amounts of current in a highly precise fashion for destroying specifically pinpointed areas in the brain. Considerable experience is required to reach the desired point consistently and destroy a small target location with near 100% accuracy and with minimal damage to surrounding tissue. The problem is particularly acute for SCN lesioning because these nuclei lie in shallow depressions on the dorsal surface of the optic chiasm. The consummate skill required to lesion all of the SCN tissue without damaging the optic nerve is a major problem for SCN transplant surgery.

In initial steps of vertebrate brain surgery, skull landmarks are used to locate the exact site. A bone plug is then removed by drilling to gain access for the lesioning electrode into the brain. One important landmark is bregma, the midline intersection of the sutures of frontal and parietal bone plates on the dorsal surface of the skull. Bregma works well for animals that retain the suture lines as adults, but in many common wild species of rodents, the sutures fuse and eventually disappear; even if present, suture lines may be quite vari-

able and lead to error. For animals lacking suture lines, another reliable landmark strategy must be devised.

All mammalian pacemaker lesion surgery is blind surgery, without a view of the target. The electrode must be lowered through brain tissue to reach the SCN target. A highly exact method for lowering the electrode accurately to the target site is essential. A precise three-dimensional measuring device called a stereotaxic instrument is used to position the lesioning electrode in the lateral and anterior–posterior planes and then lower it to the correct coordinates of the dorsal–ventral plane (Figure 6.9).

Detailed brain road maps called brain atlases are commercially available for species such as laboratory rat and mouse. In these cases, brains have been serially sectioned in a cross-sectional plane, and the structures of the brain have been mapped out in detail on separate anterior–posterior maps at about 25 μm intervals along the length of the brain. For species lacking such a commercial atlas, the maps must be laboriously constructed by the researcher by means of standard histological stains. Lesioning and stain techniques have been especially useful in studies of the SCN pacemaker.

NEUROPHARMACOLOGY. In pharmacological approaches for understanding neurons, a specific chemical is applied to cells to either inhibit or enhance a cellular bio-

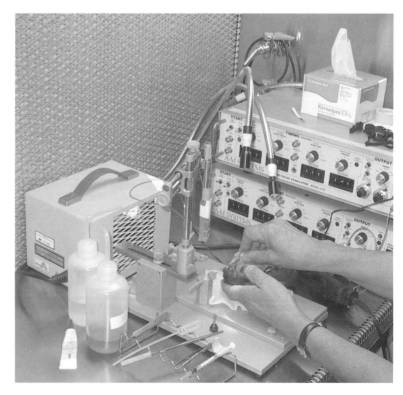

FIGURE 6.9 Electrolytic lesioning of the SCN. A squirrel is shown being mounted in the stereotaxic instrument (right foreground); the electronic stimulator apparatus is in the right background. (Courtesy of P. DeCoursey.)

chemical process in the pacemaker. The approach is most often used on slices or cultures of neurons to find the chemical entities and sequences by which a process operates. This chapter will consider examples concerning pacemaker mechanism, entrainment processes, and output signals of all the models. Much of the work is more technical than the scope of this text will permit.

METABOLIC MARKERS. Because working tissues consume energy, the degree of activity in a cell is closely coupled to its energy metabolism. Experiments in the SCN utilize an autoradiographic method for the in vivo determination of the rates of glucose utilization in brain cells. Radioactively labeled aliquots of a competitive glucose analog, 2-deoxy-D-[1-^{14}C]glucose (2-DG), are injected into the animal. The SCN is harvested shortly afterward, and slices of the tissue are then placed on X-ray film in the dark to obtain an image of the localization and concentration of the radioactive 2-DG in the SCN. The autoradiographic method has been a useful general assay of SCN rhythmicity. It has also been used under special circumstances, such as examining rhythmicity in the fetus while it is still in the uterus.

Glucose utilization is high during the subjective day and low during the subjective night in both nocturnal and diurnal animals (Figure 6.10). Despite this correla-

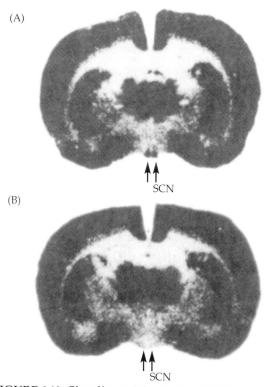

FIGURE 6.10 Circadian 2-DG activity of SCN pacemaker cells. (A) Daytime. (B) Nighttime. The arrows point to the position of the SCN in the coronal section of a hamster brain. (From Schwartz, 1991.)

tion, action potential generation probably does not account for the rhythm of energy demand. Electrical spike activity is estimated to account for less than 3% of brain energy consumption, while active ion transport is believed to be the single most energetically expensive neural process. The activity of the sodium/potassium ATPase pump is responsible for about 40% of the brain's total respiration. A strong disadvantage is that the technique does not ordinarily distinguish individual cells.

TRANSGENICALLY ENGINEERED PACEMAKER CELLS. A powerful molecular technique uses a reporter gene introduced into an organism's genome to follow circadian cellular processes. Rhythmic genes that express output signals in the SCN, for example, have been linked to a bioluminescent luciferase system or to a green fluorescent protein in order to track certain circadian processes. Recording of the green glow of an activated luciferase marker requires image intensifiers and sophisticated cameras.

A strong advantage of reporter genes is the ability to follow gene expression longitudinally without having to sacrifice the animal tissue as must be done with in situ hybridization. However, some limitations also exist. Bioluminescence from luciferase is dim and requires continuous addition of luciferin substrate to the region under study. Although light emission from the green fluorescent protein reporter can be documented photographically in single cells, excitation of the activated reporter requires repeated illumination with potentially damaging ultraviolet radiation.

ACTION SPECTRA FOR IDENTIFYING PHOTOPIGMENTS IN PHOTORECEPTOR PHYSIOLOGY. Every phototransduction process, such as circadian phase shifting or photoinduction of flowering in plants, requires a photopigment localized in specialized cells. A complex organ like the vertebrate retina contains many types of cells. Until very recently, direct methods for localization of individual photoreceptors was not possible.

A commonly used strategy for characterizing and identifying a distinct photopigment associated with a specific function is to measure its action spectrum. The method is based on the fact that different photopigments absorb light best at different colors or wavelengths of light. For constructing an action spectrum for circadian photoreceptors, the phase shifting of locomotor activity at a specific phase point by equal intensities of different wavelengths of light has been employed.

In one example, the effect of a 15-minute pulse of dim monochromatic light at circadian time (CT) 18 on wheel running of a free-running golden hamster in constant darkness was the method chosen (Figure 6.11A). CT 18 is the phase point that occurs 6 h after onset of activity on the hamster phase response curve, and it is

(A)

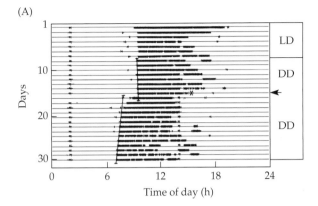

(B)

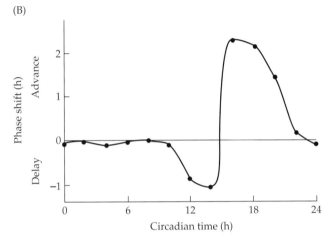

(C)

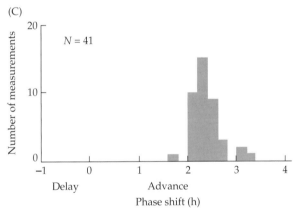

FIGURE 6.11 Action spectrum method for identifying the photoreceptor for phase shifting of hamster circadian locomotor activity. (A) Light pulse method using a 15-minute pulse of monochromatic light at CT 18. Black guidelines indicate the slope before and after the light pulse; the arrow shows the time of the pulse on day 16. (B) Phase response curve of hamsters based on 15-minute white light pulse administered to a hamster free-running in DD. Data points represent means. (C) Histogram of the amount of phase advance from a light pulse at CT18 in 41 test cases. (From Takahashi et al., 1984.)

points at CT 18 is an indication of the high precision of the response (Figure 6.11C).

The summary curve of phase-shift response versus intensity of light is a type of dose response curve called a fluence response curve because it plots the phase-shift amplitude of response as a function of light intensity for each particular wavelength assayed. When the half-maximum phase-shift amplitudes for seven wavelengths were plotted, an action spectrum was produced (for example, see Figure 6.27). A requisite part of an equal intensity action spectrum is the systematic construction of full dose response curves for selected wavelengths across the visible range of light. Without demonstration of parallel curves at the representative wavelengths, ambiguity could exist about the occurrence of multiple overlapping photopigments. The resultant action spectrum curve is then matched with standard known curves for pure solutions of photopigment. Although simple in theory, the experiment is extremely labor intensive, requires many animals and much time, and may have problems if multiple photopigments are involved. In addition, the specific cell or site of the photoreceptor is not known for the complex tissue of the mammalian retina. The fluence curves and action spectrum for circadian phase shifting in the golden hamster will be portrayed later in the chapter, in the section on identification of retinal ganglionic photoreceptors in mammals.

The Mollusc *Bulla* Has Been a Highly Useful Invertebrate Model for Studying Cellular Circadian Physiology[8,9,10,11,48]

Bulla eyes have many favorable properties for circadian physiology studies[7,8,11,28]

Although *Bulla* and *Aplysia* are quite similar in their pacemaker system properties and behavior, they show small differences, probably related to ecological differences. *Aplysia*, which has a more complex eye structure, has more unknowns than *Bulla*. For the sake of clarity for beginning chronobiologists, the story will be limited to *Bulla*. As pointed out in Chapter 5, the primary pacemaker system is found in the specialized group of about 130 large basal retinal neurons at the base of each eye. These cells are very similar, ranging in diameter from 15 to 24 μm. Their axons travel by way of the optic nerve. Such a location, involving paired eyes, is a unique advantage for study. In a pair of cultured eyes, no feedback from the remainder of the nervous system is present. The study method of choice for cell physiology is the recording of spontaneous nerve impulses from the optic nerve.

Another positive quality of the *Bulla* in vitro system is its durability. The 8-day survival time of the isolated eye and persistence of its electrical discharge rhythm in

the point of maximum phase advance (Figure 6.11B). It is thus the most sensitive and accurate point for measurement. The summary of results for 41 replicate test seawater culture is very favorable for studying repetitive 24 h cellular processes. In addition, the waveforms of the two eyes of a particular individual are almost identical. This property allows one eye to serve as a control for the experimental eye when the effects of ionic manipulation or pharmacological treatment are being tested.

The pacemakers have been localized in Bulla at the tissue and single-cell level[8,38,43]

Continuous intracellular recordings from single basal retinal neurons for over 74 h demonstrate clear circadian rhythms in both membrane potential and action potential frequency under constant conditions. The circadian rhythm of optic nerve impulse activity is due to the spontaneous firing of the electrically interconnected basal retinal neurons. These impulses are called compound action potentials because they consist of multiple individual axonal action potentials. Firing rate increases at subjective dawn (CT 0), remains high during subjective day, and drops off by subjective dusk (CT 12). The discharge rates of the optic nerves and of individual basal retinal neurons correspond one to one. Long-term recordings of action potentials in DD show a circadian rhythm close to 24 h (Figure 6.12).

The cyclic discharge is driven by a rhythm of membrane potential. Intracellular recordings obtained from basal retinal neurons in a semi-intact retina have revealed a circadian rhythm of membrane potential in which the membrane is relatively depolarized during subjective day and hyperpolarized during subjective night. The membrane potential rhythm appears to be driven primarily by a rhythm in membrane conductance, with conductance relatively high during the subjective night and reduced by approximately 50% during the subjective day. The change in conductance appears to be due primarily to a potassium conductance rhythm.

The history of progress in the study of basal retinal cell physiology has unfolded with many surprises. Although early predictions projected that the circadian period would be an emergent property of a network of higher-frequency oscillators, the facts are very different. Each individual basal retinal cell can act as a 24 h pacemaker. Surgical reduction of the cell mass to six cells shows that the circadian output persists. The amplitude is proportional to the number of cells. More recently, single basal retinal cells have been cultured individually and shown to oscillate with a period close to 24 h in constant conditions (see Figure 6.12). The means by which autonomous oscillations of 24 h periodicity are generated in the *Bulla* basal retinal neurons is still unclear, but undoubtedly it involves translation and transcription in the synthesis of oscillator entities. Such research lies in the molecular realm and is still in its infancy in *Bulla*.

Bulla ocular pacemakers show two distinct levels of coupling[9,10,29,49,52,53]

Individual cells of each eye oscillator are coupled to each other. Many of the details of cellular coupling within the basal retinal cells of an eye are still unknown, but electrical coupling is an important element. Electrical coupling brings about mutual entrainment among the population of separate circadian oscillators in the eye to maintain synchrony among the population. Thus the electrical coupling between basal retinal neurons that leads to generation of the compound action potential ensures that the pacemakers of the many basal retinal neurons stay synchronized.

Besides the coupling of cells within each eye, the phase of the two eyes is also coupled. The coupling is most likely mediated by depolarization because compound action potentials in one eye can trigger compound action potentials in the other eye. The mechanism of coupling is probably also via electrical synapses, but chemical communication has not been entirely ruled out.

Light input for entrainment also involves the basal retinal neurons[11,48]

The basal retinal neurons of the eye are also the chief photoreceptors for entrainment of the ocular rhythm. However, diurnal activity in eyeless animals suggests that extraretinal photoreceptors play a small role in the temporal regulation of activity. Some ganglia have been shown to be photoreceptive, but little is known about their role in circadian entrainment.

Progress has been made on the cellular mechanisms in the basal retinal cells. Light shining on the eye has two effects. It acutely excites the retina and results in an increase in the firing rate of the compound action potential rhythm throughout the circadian day. In addition, per-

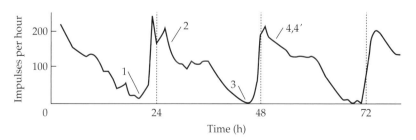

FIGURE 6.12 Rhythm of compound action potential frequency from an isolated *Bulla* eye. Solid line shows average from multicellular recording and numbers indicate recordings from 4 isolated single cells. Recordings were made on two occasions, 4 and 4′, for the fourth cell. (From Michel et al., 1993.)

manent phase shifts of the circadian spontaneous firing rate of the compound action potential rhythm are produced. The magnitude and direction of phase shifts depend on the circadian time of day. The phase response curve (see Chapter 3) resembles that of many other organisms, with delays in late subjective day and early evening, advances in late subjective night, and almost no effect during the remaining subjective daytime.

Light depolarizes the basal retinal cells. An interesting way to demonstrate this is to bathe an eye in higher-than-normal potassium concentrations. High potassium concentration depolarizes all eye cells. The potassium treatment causes phase shifts comparable to pulses of light. Conversely, lowering potassium hyperpolarizes the cells, thereby preventing phase shifts of the rhythm by light. A further demonstration of the depolarizing effect of light on membranes of the pacemaker neurons entailed the injection of depolarizing current into a single basal retinal cell. Because all 130 basal retinal cells are electrically coupled, the entire population was depolarized and phase-shifted. The sum total of all these demonstrations is that depolarization of the membranes of pacemaker neurons is both necessary and sufficient to phase-shift the *Bulla* circadian eye clock.

Phase shifting by membrane depolarization is related to calcium movement into the pacemaker cells. The calcium ions must apparently traverse the cell membrane into the intracellular compartment through voltage-gated channels that are opened by membrane depolarization. The next question concerns the function of the Ca^{2+} ions. Possibly protein phosphorylation acts via a Ca^{2+}-dependent kinase, but little is known yet about which specific proteins may be important. An interesting parallel is seen in mammals. Transmitter substances such as serotonin have input to the clock and can modify phase shifting by light. In neither case is the functional role of the input pathway known.

Circadian output in Bulla *is mediated by electrical discharges of the basal retinal neurons*[11]

Membrane potential and intracellular calcium concentration are key parameters that influence the phase of the molluscan circadian clock cells. The basal retinal cells also generate the output signal that regulates the physiological and behavioral processes of *Bulla*, and these output signals are also electrical. The output cer-

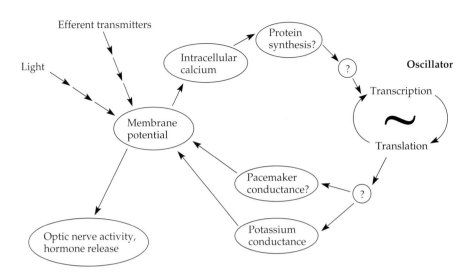

FIGURE 6.13 Model of the feedback from output to input in the molluscan ocular pacemaker. (From Blumenthal et al., 2001.)

tainly affects intracellular calcium concentration via the voltage-gated calcium channels that open during action potentials. It follows that the output of the clock could feed back through voltage-dependent calcium channels and affect the clock. The functional importance of this potential feedback loop is unknown (Figure 6.13).

In summary, the value of *Bulla* eye pacemakers as a model system for studying circadian cellular function relates largely to the simplicity of the pacemaker. Each ocular pacemaker system consists of about 130 large, individual basal retinal cells. Each single basal retinal neuron can act as an independent pacemaker system and carries out all three aspects of a circadian system: generation of oscillation, input for entrainment, and output to direct the circadian overt rhythms. Each basal retinal cell can act as an entire pacemaker system, but all 130 are coupled electrically probably by their neurite processes for a unified output that has a period close to 24 h.

The Chick Pineal Circadian System Has Many Excellent Attributes for a Vertebrate Model[22,45,46,59,60,61,62,69]

The cultured chick pineal is more favorable for circadian studies than are adult chickens[59]

The adult chicken has several semiautonomous pacemakers, including a retinal clock, a pineal clock, and possibly an SCN pacemaker. A location just below the skull makes the pineal gland easily accessible. As a result, it can be removed surgically with minimal disturbance to other brain functions. The poorly developed behavioral rhythms render interpretation of data difficult, but pineal removal in adult chickens does not eliminate locomotor or feeding activity. Although the biosynthesis of melatonin is strongly rhythmic in adult living chickens, evidence of the rhythm must be obtained by sequential sampling over days. As a result, the procedure is very invasive, assays are expensive, and results are often erratic. Removal of both the retina and the pineal results in arrhythmicity.

On the other hand, pineal glands from newly hatched chicks can be kept alive relatively easily.

Because the gland is fairly homogeneous in cell structure, small pieces can be used for reduced preparation cultures. Melatonin rhythms are expressed in chick pineal pieces, in dispersed pineal cell culture, and even in single pinealocyte cells in culture.

Historically the development of satisfactory measurement methods for chick pineal rhythms has been difficult. The initial labor-intensive method of hand harvesting cells from the cultures for melatonin analy-

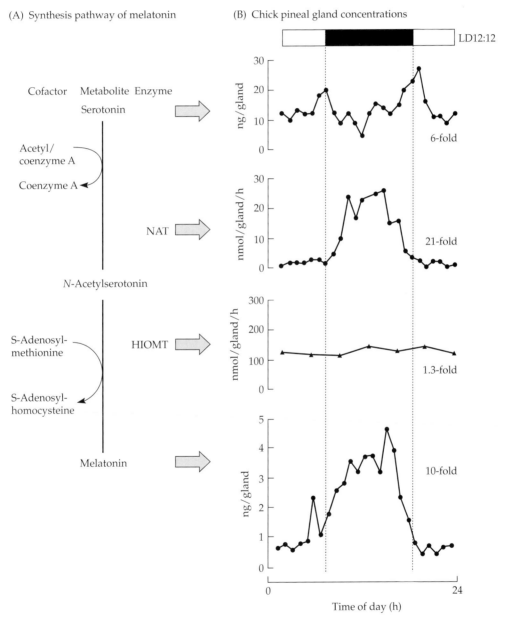

FIGURE 6.14 The pathway of melatonin synthesis in avian pineal glands in vitro. (A) Synthesis proceeds from serotonin to N-acetylserotonin under the action of the enzyme N-acetyltransferase (NAT) to melatonin under the action of the enzyme hydroxyindole-O-methyltransferase (HIOMT). Substrates are indicated by the vertical bar, with enzymes at right and cofactors at left. (B) Arrows point to 24 h actographs of four components of the pathway that have been evaluated as assay possibilities for circadian rhythms of melatonin production under the LD 12:12 light schedule shown by the light bar at top. Numbers at right indicate day–night amplitude change. (From Binkley, 1990.)

ses was improved by automation. A flow-through system enabled timed collection of effluent samples from superfused pineal cultures. A second set of problems involved quantitative detection of miniscule amounts of melatonin or one of the products in its synthesis pathway (Figure 6.14). Melatonin assays are expensive and sensitive to artifacts. Several laboratories attempted to measure N-acetyltransferase (NAT). The false assumption was that the first step in synthesis would indicate the final melatonin concentration. The current consensus is that melatonin is the most accurate measure of circadian rhythmicity in chick pineal culture. These rhythms generally damp out after 4 to 8 days under constant conditions, probably due to desynchronization among a group of pacemaker cells. In summary, chick pineal cells are easy to keep alive in culture, but rhythmicity under constant conditions lasts only a few days, and the measurement of melatonin is not an ideal marker.

The pinealocytes of chick pineal gland in vitro are circadian oscillators[2,45,46,69]

OVERVIEW.   The cellular structure of the intact chick pineal gland gives insight into its circadian cell function. The pineal gland is encapsulated in a connective-tissue encasement that separates it from the adjacent cerebral cortex and cerebellum (see Figure 5.22 in Chapter 5). The tissue is richly vascularized and lies above a large neural sinus. Although the gland is relatively homogeneous, it has several different cell types. The pinealocytes, which are large neuroendocrine cells with highly interesting properties, are the cells of primary interest to chronobiologists. Lamellae resembling the plates and folds of the mammalian rod and cone cells are seen in the cytoplasm, and these have been shown to contain a photopigment and to be photoreceptive in the visible range of light (Figure 6.15). Avian pinealocytes are derived from neural primordium and are photoreceptive, yet they are not true neurons, for they do not generate or conduct action potentials.

(A)

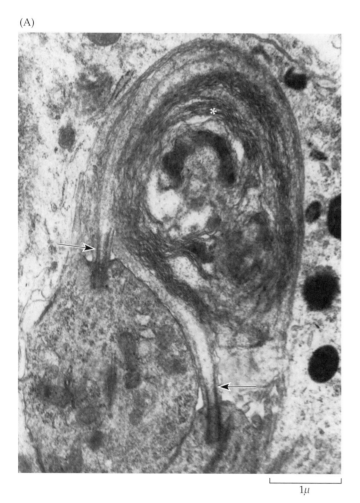

(B)

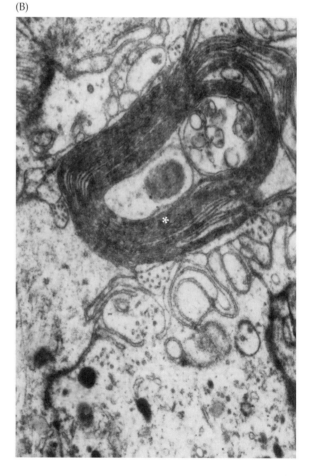

FIGURE 6.15 **Concentric lamellar body formed by the cilia and layered membranes in avian pinealocytes.** (A) House sparrow. Asterisk indicates lamellar folds; each arrow points to the base of a cilium. (B) Domestic duck. Asterisk indicates lamellar folds. (From Menaker and Oksche, 1974.)

(A)

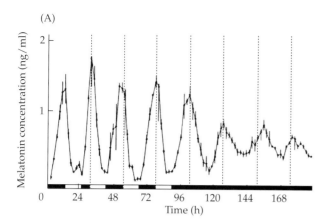

(B)

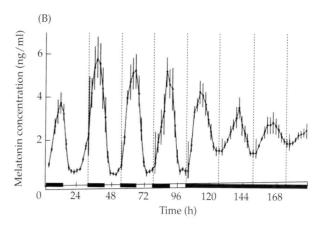

FIGURE 6.16 Circadian pacemaker function in terms of the melatonin secretion rhythm of chick pineal cells. (A) Melatonin secretion of superfused chick pineal gland in vitro for 4 days in LD 12:12 with a 6 h advance in light schedule, followed by 4 days in DD. (B) Comparable data with a 6 h delay in the light schedule. (From Takahashi et al., 1989.)

PACEMAKER FUNCTIONS. Current methods allow continuous recording of melatonin from dispersed-cell culture for approximately 1 week. Both reduced preparations and dispersed-cell cultures are precisely rhythmic in terms of spontaneous production of melatonin. Free-running rhythms of melatonin synthesis persist in dispersed-cell cultures of chick pineal under constant dark conditions (Figure 6.16).

Generation of circadian rhythmicity is a cellular process that does not require normal tissue organization because the capacity is found in individual pineal clock cells. The rhythm free-runs with a period close to 24 h, then damps out. When records of several cells are superimposed, small individual differences in period are seen. If a culture is returned to LD 12:12 after damping out, a clear oscillation is again apparent. The results suggest that damping of the rhythm in DD is caused by desynchronization of a population of cells, not by the lack of an oscillator.

Light input for in vitro chick pineal cell cultures involves at least two input pathways[46,68,69]

Light input may be carried by several routes in adult chickens. Retina, pineal gland, and extraretinal photoreceptors are all possibilities. In isolated cell culture, however, the sole possibility is the pinealocyte. The lamellar structures serve as the framework for efficient packing and alignment of the photopigment molecules involved in phototransduction (see Figure 6.15).

In chick pineal culture, light has two separate effects on the melatonin rhythm: acute and phase-shifting. In the acute reaction, melatonin synthesis is inhibited and levels in the bathing medium rapidly fall off. During subsequent circadian cycles, melatonin synthesis and release again rise sharply during the dark or subjective night period. In the second, entraining effect, the phase of the melatonin cycle is permanently shifted.

Under LD conditions, the melatonin synthesis rhythm persists in dispersed-cell cultures at least 6 days. When the schedules are advanced or delayed 6 h, the phase of melatonin release is shifted accordingly. Entrainment depends primarily on light-induced phase shifts of the circadian rhythm, with the amount of shift related to the subjective timing of the administered light pulse (see Chapter 3). Plotting the magnitude of these shifts relative to subjective time shows circadian responsiveness as a phase response curve. Most light phase response curves are similar for all vertebrates and even for many invertebrates for a variety of overt rhythms. The phase response curve for phase shifting of melatonin concentration by light has phase delays in early subjective night, phase advances in late night to early morning, and few or no phase responses in early day. In addition to the phase-shifting effect of light, the amplitude of the rhythm decreases following a pulse exposure to light entrainment.

The issue of photopigments for the two routes is not completely resolved. The pigment for entrainment is thought to be unique to birds and is called pinopsin, or P-opsin. The properties of P-opsin are chemically intermediate between cone pigments and rhodopsin of rods. Much about the circadian role, if any, of P-opsin in the melatonin rhythm remains to be discovered. A rhodopsin-like photopigment is involved in the acute response.

Pharmacological data indicate that these two processes use different pathways. Pertussis toxin blocks the acute inhibitory effects by light but not phase shifting by light. The selective blockage indicates that acute regulation operates through cAMP (cyclic adenosine monophosphate), but the phase-shifting route uses the clock itself for circadian regulation (see Figure 6.17).

The circadian output rhythm in the chick pineal culture involves melatonin synthesis and secretion[4,23,32,46,47,67,68,69,70]

The single measurable output rhythm from the chick pineal culture is melatonin secretion. When light input is used to track circadian processes to the melatonin output, both acute and phase-shifting pathways eventually converge on the conversion route from serotonin to melatonin (see Figure 6.14). In phase-shifting effects, the circadian clock acts primarily in regulating mRNA levels. In contrast, for the acute inhibition effects, cAMP acts both on mRNA levels and other distal steps. As the data lead from the cellular level to the molecular basis of the oscillation, it becomes clear that protein and RNA synthesis are important in melatonin synthesis. Inhibitors of protein and RNA production also block melatonin production.

Extracellular calcium is required for melatonin production in the chick pineal. Calcium ions are involved in both acute and phase-shifting processes, but the details are not yet entirely clear. A novel cation channel found in chick pineal cells, I_{LOT}, is permeable to calcium ions. The channel is open for an unusually long time, most often in the dark of an LD cycle or in subjective night in DD. It may be involved in the regulatory output for the melatonin synthesis rhythm. Much of the information on light effects in chick pineal cultures comes from complex pharmacological experiments that are beyond the scope of this book. For summary purposes, the proposed schemes for light action on the melatonin rhythm can best be represented schematically (Figure 6.17).

The data indicate that the pinealocytes of the chick pineal gland, like the *Bulla* eye, contains all pacemaker components. These separate elements include the central oscillator, the input pathways for environmental entrainment, and the output pathways to regulate overt circadian rhythms at the cell, tissue, and whole gland level.

The Mammalian SCN Circadian Pacemaker System Has Been Studied More Extensively Than Any Other Pacemaker System[12,15,21,25,31,35,36,39,41,51,56,63,64]

The circadian system in mammals has many advantages for study[18,19,26,34,44,50]

OVERVIEW. The suprachiasmatic nucleus (SCN) consists of two small, densely packed clusters of nerve cells located just above the optic chiasm in the anterior ventral hypothalamus of the mammalian brain. Only about 10,000 neuronal cells are found in each cluster, along with glial cells and fibroblasts. The SCN lies at the base of the third ventricle, dorsal to the optic chiasm. The fusion of the two optic nerves forms the optic chiasm. Axons project from the retinas of the eyes to the SCN and to the lateral geniculate nucleus, which is located posteriorly in the brain. Further details of structural and functional circadian organization are covered in Chapter 5 (see also Plate 3, Photo B).

The term *cell physiology* is used in a broad sense in this chapter, encompassing traditional biochemistry and electrophysiology of cells, wherever these aspects are known. In addition, the term will cover the function of individual specialized cells such as photoreceptive ganglion cells. Because much of the cell

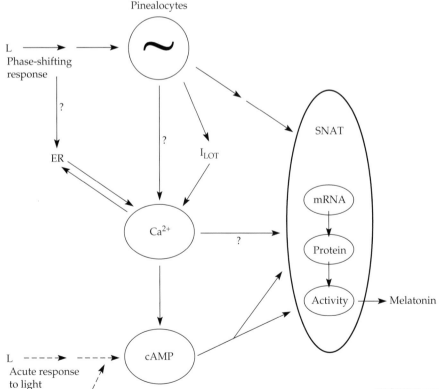

FIGURE 6.17 Pathways for the acute and entraining effects of light on chick pineal cells in culture. ER, endoplasmic reticulum; NE, norepinephrine; SNAT, serotonin *N*-acetyltransferase. (From Zatz, 1996.)

biochemistry depends on detailed neuropharmacology research beyond the scope of this text, schematic diagrams will replace technical details in many cases.

DIFFERENCES BETWEEN AVIAN AND MAMMALIAN CIRCADIAN SYSTEMS. A comparison of the mammalian and the prior avian system will help review and highlight major differences (Figure 6.18). In the chicken, two major, semi-autonomous circadian oscillators are found in retina and the pineal gland, respectively, and a third probable oscillator is found in the SCN. The homology of the SCN gland in birds, however, is not entirely clear. Photoreception for circadian entrainment in birds takes place in the retina but also in diffusely scattered, extraretinal photoreceptors. The cerebrospinal fluid–contacting neurons of the lateral ventricle walls of the green iguana contain opsin photopigments and are putative extraretinal photoreceptors; very similar cells have been found in analogous sites in birds.

In contrast, the SCN is the primary oscillator for all mammals. The pineal and other oscillators are reduced to peripheral oscillators or at least to very local oscillators, such as the retinal pacemaker of the hamster. Photoreception takes place only in the retina of the eyes.

EVALUATION OF THE SCN SYSTEM AS A MODEL. The attributes of an ideal model system were enumerated at the beginning of the chapter. The necessary and sufficient criteria for a pacemaker have been accomplished with great clarity in mammals through the use of natural and induced circadian mutants in the SCN-lesion/transplant paradigm (see Chapter 5). Early SCN transplant studies used fetal tissue from wild-type hamsters. These experiments were not totally definitive because the periods of host and transplant were the same. The more recent *tau* mutant strain of hamsters has offered a solution. The restoration of rhythmicity in SCN-lesioned wild-type hosts by *tau* mutants has demonstrated unequivocally that both phase and period of the donor graft determine circadian parameters in the arrhythmic host.

One asset of the mammalian SCN system for cell physiology studies is that the oscillator, the retinal environmental sensor, and neural output signals are all separate entities. The discrete nature of the components makes some parts of the analysis easier. The remaining assets are all specific to studying cell physiology of the model. The model must be able to tolerate in vitro culture, with continued expression of one or more circadian oscillations as marker output rhythms for measurement purposes. The SCN scores highly here, even though other types of neuronal tissue are notorious for their difficulty in culturing. Both in vivo and in vitro studies have been possible, but much greater emphasis has been placed on in vitro approaches.

The precise rhythm of neuronal discharge rate serves as a good measurement end point. The ideal in vitro preparation should also have the capability for entrainment. Here the SCN slice or cell culture falls short because SCN neurons are not photosensitive. The system should also respond to perturbations such as pulses of chemicals in order for phase response curves to be constructed. The SCN system is excellent in this respect. The system should be reliable, continuous, and economical. Electrophysiology is expensive in terms of equipment and requires considerable skill. Nevertheless, it is manageable with experience and a modest research grant. For some cellular analyses, the system should be composed of a considerable quantity of homogeneous tissue. From this angle, the SCN system poses major hurdles. The heterogeneity of cell types has

(A) Bird pacemaker system
(composite of several species)

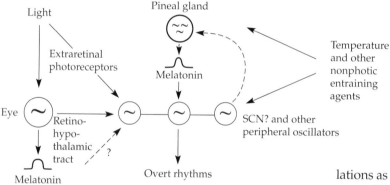

(B) Mammal pacemaker system

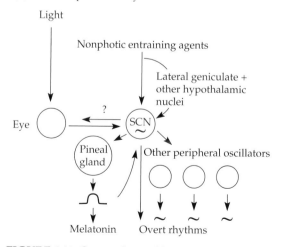

FIGURE 6.18 **Comparison of bird and mammal pacemaker systems.** (A) Avian system. (B) Mammalian system. Possible feedback loops are simplified or eliminated for clarity. (A after Underwood, 2001; B courtesy of P. DeCoursey.)

hindered all attempts to identify the cellular site of rhythm generation.

The SCN in slice culture is a self-sustained oscillator[19,20]

The capacity of the SCN to generate free-running signals that regulate numerous rhythmic functions in mammals is widely accepted. The neuronal firing rate of populations of SCN neurons can be recorded in several ways. For slices, large electrodes called multiunit electrodes give an average picture of the several types of neuron waveforms and frequencies. With single-unit recording, the operator randomly introduces the electrode into the SCN slice to sample the firing rates of individual neurons and plot them on an ensemble graph. In vitro recordings agree well with results from in vivo recordings. The resultant plot for slices over daily time spans produces a circadian graph with the highest firing rate during subjective day and the lowest at night (Figure 6.19).

Single SCN cells in dispersed-cell culture are self-sustained oscillators[26,27,66]

One of the most innovative ways to study circadian physiology of the SCN cells is to use a dispersed-cell culture. SCN cells are first deaggregated and plated out on a multielectrode chamber (see Figure 6.4), then allowed to stabilize. After a few days the cells settle in

TABLE 6.3	Immunolabeling of neurons in SCN cultures
Antigen	Percentage of immunopositive neurons
Gamma-aminobutyric acid (GABA)	69.4 ± 1.9
Vasopressin neurophysin	13.5 ± 0.7
Vasoactive intestinal peptide (VIP)	8.7 ± 0.5
Somatostatin	1.0 ± 0.1

one spot on top of an electrode and reestablish synaptic connections with adjacent neurons. The firing rate of each neuron in contact with an electrode can be recorded automatically and continuously for many days. Its firing rate can be measured exactly and compared to other neurons in the culture at known locations. After 2 weeks, the cells can be fixed and stained individually with immunocytochemical methods, and the cell type can be determined by its neuropeptide content (see Figure 6.5).

Immunolabeling shows that almost all SCN cells are GABA-ergic. They label also for the neuropeptides vasopressin, vasoactive intestinal peptide (VIP), and somatostatin. The latter three are subpopulations of the GABA neurons (Table 6.3). The percentages shown in Table 6.3 are about half of the percentages found in vivo. Identification of the actual clock cells is a question of high priority. About 50% of cultured cells were active neurons. Therefore, the capacity for oscillation cannot be the exclusive domain of vasopressin, VIP, or somatostatin cells because they constituted in total only 23% of the cells in cultures.

Rhythmically firing SCN neurons are called clock cells. In dispersed-cell culture in multielectrode chambers, the simultaneous recording of many neurons indicates that each separate clock cell has a characteristic circadian period, close to but not exactly 24 h in period. Individual SCN clock neurons in the same culture thus have the capacity for acting as independent pacemakers with different circadian periods and with different phases relative to each other (Figure 6.20).

In one case of four neurons fairly dispersed on the plate, the phases differed widely (Figure 6.20A,B). In another case, two neurons lying very close together had oppositely phased rhythms (Figure 6.20C). Unlike the results from slice or in vivo recording, the firing rhythms of clock neurons in the same culture were not synchronized in spite of abundant working synapses. Even though synapses between cells had formed in vitro, they were apparently neither necessary for the oscillation of single cells nor sufficient for synchronizing them. Possibly another rhythmic factor, such as diffusible

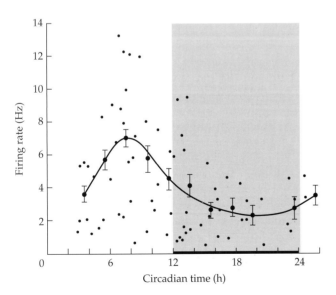

FIGURE 6.19 Circadian rhythm of spontaneous neuronal discharge rate in slice culture of a rat SCN. Small circles represent the mean rate for single neurons; large circles, the mean rate for all neurons measured in each 2 h time period. The gray area indicates circadian night, the white area circadian daytime. (From Gillette, 1991.)

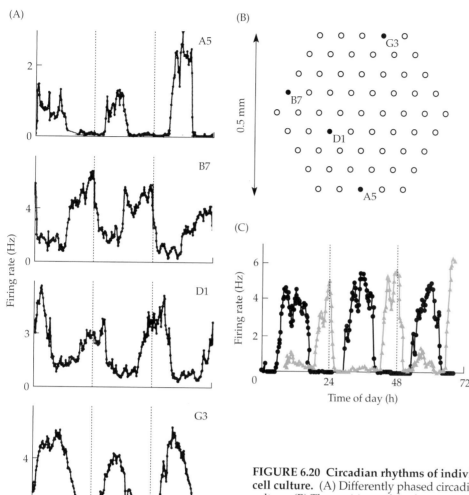

FIGURE 6.20 Circadian rhythms of individual SCN neurons in dispersed-cell culture. (A) Differently phased circadian rhythms of SCN neurons in one culture. (B) The positions of the four neurons, labeled A5, B7, D1, and G3, relative to each other on the electrode grid are shown. (C) Oppositely phased circadian rhythms of two adjacent SCN neurons in one dispersed-cell culture. (From Welsh et al., 1995.)

melatonin rhythms for in vivo recording, synchronizes all cells of a population. The firing pattern of an intact SCN could be an emergent network pattern of the entire population of cells, or a coupling of individual single-celled oscillators. The multielectrode recording chambers for dispersed-cell culture suggest that the latter hypothesis is correct.

One highly interesting result of the dispersed-cell culture is that the firing rate of clock neurons is not an integral part of the clock mechanism. When firing is prohibited by addition of a blocking toxin such as tetrodotoxin, clock function continues. Evidence comes from washing out the action potential blockade toxin to reinitiate firing. The phase of the free-running rhythm continues unshifted (Figure 6.21A and B).

The firing rhythm of a single individual neuron in dispersed-cell culture was recorded continuously for 3 days (Figure 6.22). The characteristic spike is easily distinguishable above the background noise level (Figure

6.22A). When the total number of spikes are tallied and plotted for 4 h intervals, a precise circadian rhythm is evident (Figure 6.22B). The start of spontaneous firing is very sharply delineated (see Figure 6.22B, point c, and 6.22C, trace c). Firing rate is very uniform during the active period of the neuron (see Figure 6.22C, traces c and d), then it drops off abruptly to the resting minimal firing rate (see Figure 6.22C, traces a, b, e, and f). The waveform of the single neuron is indistinguishable from the waveform of a population of SCN clock cells (see Figure 6.19).

Single SCN cells appear competent to restore rhythmicity in SCN hosts[57]

Several types of experiments have shown that structural integrity of an SCN transplant is not necessary for restoration of circadian rhythmicity in arrhythmic hosts. Even dispersed single SCN cells are competent. In one

(A)

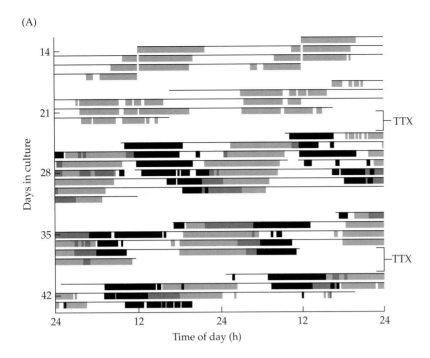

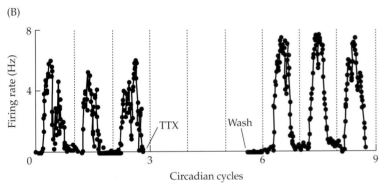

(B)

FIGURE 6.21 The effect of blocking action potentials by tetrodotoxin (TTX) on phase and period of clock neurons in dispersed-cell culture. (A) Doubleplot of two neurons 350 μm apart in the same culture with bars showing times when firing rate was above mean for each row. The free-running period for Cell 1 (gray) was 23.25 h, and for Cell 2 (black) was 25.5 h. Dark gray indicates the overlap in firing of Cell 1 and Cell 2. Absence of a thin baseline indicates a gap in data collection. During the second and fourth gap, all action potentials were reversibly blocked by bath application of 300 nmol TTX. (B) Circadian firing rate of Cell 1 before and after TTX, showing that the recovered rhythm is in phase with the pretreatment rhythm. "TTX" indicates the time of addition of the blocking agent; "Wash" indicates the time of its removal by flushing with fresh culture medium. The horizontal time axis is marked in multiples of the cell's circadian period length of 23.25 h. (From Welsh et al., 1995.)

also not necessary. Locomotor rhythmicity has been reestablished by SCN tissue implanted within semipermeable polymeric capsules. Neural outgrowth is prevented by the capsules, but substances with a mass of less than 500 kDa can diffuse across them. The effectiveness of these implanted capsules suggests that a humoral factor is responsible.

The intercellular coupling mechanism in the SCN may involve GABA[13,65]

The precision of the free-running period characteristically expressed by whole animals appears to be determined by the average period of the population of coupled SCN single-cell oscillators. Such a means of control contrasts with many other rhythmic tissues. In the heart, for example, the fastest cells in the sinoatrial pacemaker node set the heartbeat frequency for the entire organ.

One study of golden hamsters illustrated the point by comparing the period and precision of individual cell periods in dispersed-cell cultures to the period of locomotor wheel running of intact animals. Much greater variability was seen for individual SCN clock cells than for intact animals. The free-running period of the wheel-running rhythm of intact animals was compared to the average period of the firing-rate rhythm of dissociated SCN cells from wild-type, heterozygous, and homozygous *tau* mutants. Wild-type periods for firing rate averaged 23.43 ± 1.34 h, heterozygous periods averaged 20.97 ± 2.14 h, and homozygous periods averaged 19.28 ± 1.69 h. Periods for wheel running were 24.15 ± 0.06 h for the wild type, 22.39 ± 0.28 h for heterozygotes, and 20.75 ± 0.39 h for homozygotes.

type of experiment, dispersed-cell suspensions of fetal hypothalamus containing the SCN were injected stereotaxically into the third ventricle of the brain of arrhythmic host hamsters. After several weeks rhythmicity gradually reappeared.

A similar result was obtained by a technique called cell immortalization. The process involves infection of a culture of SCN cells with a retroviral vector encoding an adenovirus gene. These so-called SCN 2.2 cells are very long-lived and are capable of cell division. After these cells were transplanted into the brains of arrhythmic SCN-lesioned rats, circadian rhythmicity was restored in 50% of the animals. As a control for the experiment, other rats were injected with cells called NIH 3T3 fibroblasts. These cells constitute a cell line in which circadian oscillation can be induced by serum shock techniques but the rhythm damps out after a few cycles. Rhythmicity was not restored in any of the control rats injected with NIH 3T3 fibroblasts.

Direct neuronal contact of SCN cells that have been injected into the third ventricle of arrhythmic animals is

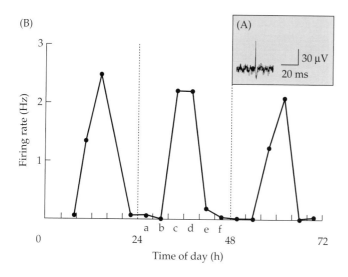

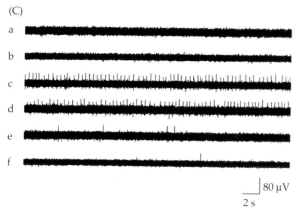

FIGURE 6.22 Firing rate of a single SCN neuron in dispersed-cell culture. (A) Raw data for a single spike showing characteristic waveform above background noise. (B) Three-day record of spontaneous spikes per 4 h interval to indicate sharp onset of firing. (C) Representative traces for points a–f, shown in (B). (From Welsh et al., 1995.)

The neurotransmitter GABA may couple the individual SCN clock cells. When dissociated SCN cells in culture are exposed to a daily 3 h GABA pulse at the same time of day for 5 days, the phases of their firing-rate rhythms become synchronized. This effect has been ascribed to the activation of GABA receptors, which conduct mainly chloride ions. Recent data suggest that the chloride rhythm is due to circadian modulation of the efficiency of a chloride transport system. Such a switching mechanism could serve a feedback function to both amplify SCN activity during the day and suppress it at night.

SCN cells may also be synchronized via calcium-independent, nonsynaptic bursts of action potentials in solutions with low calcium levels. Furthermore, a circadian rhythm of SCN glucose utilization is present in the fetus at an age before synapses form in its developing SCN. These findings have led to the idea that SCN cells might be joined by direct electrical connections formed

by gap junctions. Ions and other small molecules are capable of passing across gap junction channels between coupled cells, thus providing electrical and metabolic connections.

In SCN slices, such direct connections have been demonstrated by injection of individual cells with the dye biocytin and then documentation of the presence of intercellular dye coupling. The degree of coupling is affected by GABA mechanisms and also depends on circadian phase. Extensive coupling occurs during the subjective day when firing rate is high, and minimal coupling during the subjective night when firing rate is low. SCN cells called astroglial cells are also interconnected via gap junctions and can support intercellular waves of calcium release. Much still remains to be learned about coupling of the individual clock cells within the SCN.

Mammalian circadian photoreceptors lie in photoreceptive retinal ganglion cells[1,3,5,24,58]

HISTORY OF THE MAMMALIAN CIRCADIAN PHOTORECEPTOR SEARCH. Early measurements of circadian photoentrainment in mammals indicated that the eye was necessary for entrainment in all mammals studied. Blinding by enucleation of the eyes resulted in free-running rhythms of animals on normal LD schedules. Then an astute guess about optic projections led to discovery of the SCN. A tracer injected into the eyecup marked the main neuronal pathway from the retina, back along the optic nerve and primary optic tract to normal imaging centers of the brain. Some fibers, however, ended at the SCN. The SCN was immediately lesioned, and the answer was spelled out in a few days. The SCN-lesioned animals were arrhythmic. However, as soon as the mammalian circadian pacemaker was localized, other circadian biologists began a search for the photoreceptors involved in photoentrainment of the free-running clock. As easy as the answer seemed, scientists were to struggle for 30 years to find the answer. An array of almost every high-tech cellular technique described earlier was needed for the job.

FUNCTIONS OF THE VERTEBRATE EYE. An overview of eye function relative to retinal structure and physiology will help explain why the initial trails to seek circadian photoreceptors led to dead ends. The outer parts of the vertebrate eye are optically transparent structural elements to allow the fragile, nonreplaceable photoreceptive neurons to do their work safely. The retina does the actual physiological phototransduction of light to neuronal signals. One group of functions of the eye revolves around image formation. The seeing part of the eye's workday is proverbial and expected. Another less known, involuntary function is oculomotor: positioning of the eyeball in the direction of movement. Still another function is light-activated pupillary constriction, to reduce the amount of potentially damaging light that falls on the retinal cells. Finally, the retina functions as a photon counter for the circadian system to relay information about intensity of light and time of day.

The imaging function requires high-resolution definition of details and contrast, for both daytime and nighttime vision. The remaining three functions depend on luminance or irradiance information about the mean level of illumination averaged over relatively long periods of time. Imaging processes deal with information over milliseconds; luminance-related events are con-

cerned with light over minutes or hours. If every flash of brilliant lightning in a crackling nighttime thunderstorm, for example, phase-shifted the circadian pacemaker of a flying squirrel foraging in the forest, the squirrel would have serious circadian problems. Similarly, if moonlight reset the squirrel's light-sensitive imaging rods all night long, its photic input system might be maladaptive for circadian functions.

In the course of purely physical tasks carried out by an eye, photons of light fall in highly focused fashion on the surface of the retina and travel passively through the

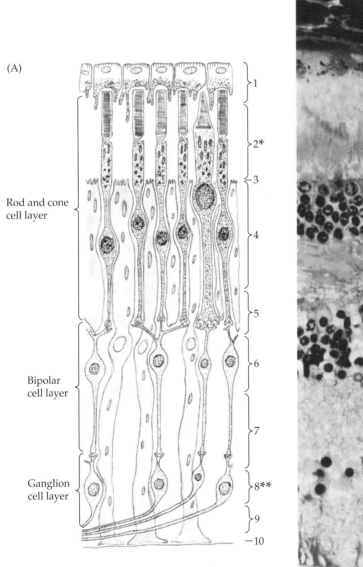

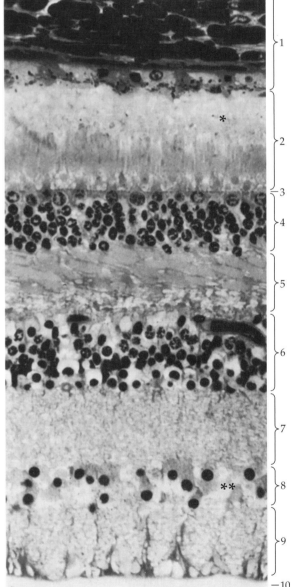

FIGURE 6.23 Cellular structure of the mammalian retina. Schematic drawing (A) and photomicrograph (B) of retina. Numbers indicate the visible layers in the histological preparation: 1, pigment epithelium layer; 2, outer segments of rods and cones; 3, outer limiting membrane; 4, outer nuclear layer; 5, outer plexiform layer; 6, inner nuclear layer; 7, inner plexiform layer; 8, layer of retinal ganglion cells; 9, layer of optic nerve fibers; 10, inner limiting membrane. * marks the level of rods and cones where phototransduction for visual imaging takes place; ** marks the phototransduction zone of ganglion cells for circadian phototransduction. (From Leeson and Leeson, 1986.)

three layers of neuronal cells of the neural retina. The innermost sheet of the visual retina consists of traditional rod and cone photoreceptors (Figure 6.23). Books have been written about rods and cones and their role in visual imaging. Nobel Prizes have been awarded to several scientists for work on visual fields in imaging by rod–cone complexes and their connectives. The extraordinary specialization of rod and cone cells for transducing light energy into the electrochemical action potentials of the optic nerve had fascinated scientists for decades. The patterns by which photopigments such as rhodopsin are systematically embedded in unique parallel lamellae, the better to capture every photon, were well known (see Figure 5.14). Nowhere else in the mammalian body are cells so admirably adapted for phototransduction. The photoreceptors synapse in the retina with short bipolar cells. These in turn synapse with the large retinal ganglion cells (see Figure 6.23). The long axons of the retinal ganglion cells extend in the optic nerve to optic processing centers in the midbrain or to the SCN.

FUNCTIONS OF THE KNOWN PHOTORECEPTORS. Until recently, only rods and cones seemed capable of phototransduction in mammals. However, doubt was cast on that theory by an article published in 1984. It contained an action spectrum (see Figures 6.11 and 6.24) for the photopigment involved in circadian phase resetting and entrainment. The precise parameter of the amount of phase shifting at CT 16 by a 15 minute pulse of monochromatic light was chosen because that is the part of the phase response curve in the hamster where large advance phase shifts occur (see Figure 6.11). The method quantified the spectral sensitivity of the photoreceptor process involved (Figure 6.24), and the method also allowed determination of the threshold for phase shifting at CT 16 by green light, which was the most effective wavelength of the action spectrum.

The results showed that the photoreceptors had several novel properties. The scientists wrote:

> Although the spectral sensitivity is consistent with a rhodopsin-based photopigment, two features of the photoreceptive system that mediates entrainment are unusual. The threshold of the response is high, especially for a predominantly rod retina like that of the hamster, and the reciprocal relationship between intensity and duration holds for an extremely long time (up to 45 minutes). These results suggest that the photoreceptive system mediating photoentrainment is markedly different from that involved in visual image formation and that even when clock photoreception is performed by the image-forming eyes, the rules that govern it and perhaps even the retinal cells that mediate it may be very different from those involved in image formation.[58]

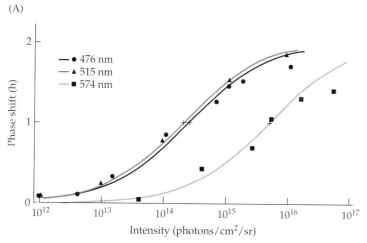

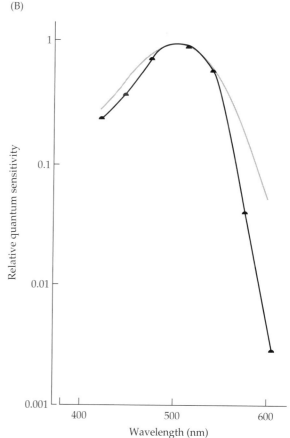

FIGURE 6.24 **Determination of the photoreceptor pigment responsible for phase shifting of circadian locomotor activity by means of an action spectrum.** (A) Complete intensity response (fluence) curves based on 15 minute pulses of monochromatic light at 3 wavelengths for different intensities from threshold to saturation phase shifting, with + indicating the half-maximum phase shift of 1 h. (B) Action spectrum curve for phase-shifting shown as a best-fitted curve through the half-maximum responses for 7 wavelengths (black half-moon symbols and curve) and the curve for extracted hamster retinal photopigment (gray curve). (From Takahashi et al., 1984.)

NEW CONFIRMATIONS OF NOVEL CIRCADIAN PHOTORE-CEPTORS IN RETINAL GANGLION CELLS.   Regrettably, the action spectrum methods that yielded the cogent prediction just described had been pushed to their limit, and further discovery had to await a new technique. Soon it appeared in the form of a genetic mutation called *rodless* in mice. In homozygous recessive individuals, the rod cells deteriorated within a few weeks of birth, followed by most of the cones. Still, the mice entrained to dim LD schedules. Skeptics dismissed the results by saying that a few undetected photoreceptors remained. However, in 1999 a laboratory in England showed uncontestably that circadian rhythms were photoentrained in mice that completely lacked rods and cones.

Several new potential photopigments, cryptochrome and melanopsin, were identified in mammalian retinal cells. Double staining of the retinal ganglion cells projecting to the SCN and the melanopsin-containing cells of the retina. The two coincided, and a ganglion cell, not a rod or cone, was confirmed as the site of a candidate pigment for photoreception. In 2002, scientists used electrophysiology and vital dye marker techniques to show that the retinal ganglion cells extending via the retino-hypothalamic tract to the SCN responded with induced spike discharges when illuminated with light. The action potentials occurred even when drugs were used to block all neuron-to-neuron connections in the retina. Furthermore, scientists were able to isolate single retinal ganglion cells projecting to the SCN in culture and show that they were light sensitive (Figure 6.25; see also Plate 3, Photo F)). All that is needed now is to identify

the photopigment in light-sensitive retinal ganglion cells.

SIGNIFICANCE OF THE NEW-FOUND CLASS OF GANGLIONIC PHOTORECEPTORS IN MAMMALS.   Part of the excitement of the new retinal ganglionic photoreceptors is that they don't just project to the SCN, but have implications for widespread functions beyond the circadian clock. The evidence is mounting that these ganglionic cells spread their dendritic processes widely over the retina, perhaps literally to average the number of photons arriving over relatively long periods of time and to supply that information to multiple areas of the brain. Thus they may be ideal both for luminance-measuring functions for the circadian system and for other functions, such as light-induced pupillary constriction. These photoreceptive ganglionic neurons may even be related to the well-known inhibition of melatonin by light, in which a luminance measurement system may be more suitable than an imaging system.

Recent research has pointed out some important functional aspects of circadian photodetection in mammals[39,40,41]

Afferent incoming photic information affects the activity of SCN neurons in many ways. One way is to cause the release of a neurotransmitter at a nerve ending. The release results in the opening or closing of ionic channels and consequently in a change of the membrane potential of a neuron. Changes in cell membrane potential are reflected by changes in electrical activity of the neuron, including changes in the rate of neuronal spike discharge.

In studying phototransduction processes in the SCN, an intracellular measure is theoretically the best method. However, serious problems exist. To record intracellular changes in membrane potential, one recording electrode must contact the cell's internal environment, and a reference electrode must be placed outside of the neuron. This recording method is quite damaging for SCN neurons because they are among the smallest of the central nervous system. For that reason, most recordings of SCN neurons have been performed by monitoring of the extracellular change in the cell's electrical activity. The neuronal discharge can be picked up outside the SCN cell by an electrode in a noninvasive way, and the recording can be performed for extended periods of time.

Extracellular recordings of SCN neurons in nocturnal rodents have revealed that about one-third of the neurons are light responsive. These neurons change electrical activity when a light pulse is presented or when light intensity is gradually changed. The location of these light-responsive neurons fits well with the known afferent projection area of the retinohypothalamic tract (RHT) from the retina to the ventral and lateral SCN of rats and hamsters. Most light-responsive SCN

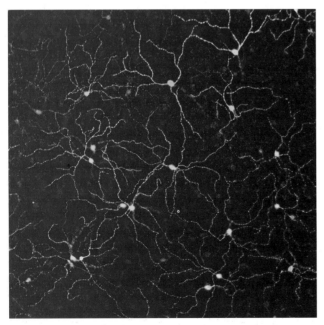

FIGURE 6.25 Photoreceptive retinal ganglion cells. (Courtesy of I. Provencio and A. Castrucci.)

neurons increase discharge rate in response to light and are called light-activated cells. A minority of SCN neurons decrease discharge rate in response to light and are called light-suppressed cells.

The light response profile of SCN neurons is characteristic and different from the response pattern for image-forming neurons. Instead of exhibiting a short transient response to light, SCN neurons change discharge rate for the full duration of retinal illumination. The response to light is called tonic, or sustained. Functionally this response type makes sense with respect to the function of the SCN neurons. They must inform the pacemaker of the photic condition in the environment for as long as the light conditions prevail.

For a long time it was difficult to explain the sustained response characteristics of SCN neurons. How could they continue responding when the light responsiveness of both rods and cones is transient? How could transiently responding photoreceptors cause a sustained response to light that could last for many hours? With the recent notion that other types of visual pigments may mediate photic input to the SCN, a new explanation has become available. In fact, the response characteristics of melanopsin-containing ganglion cells are also sustained. The properties of melanopsin and cryptochrome fit very well with the response characteristics of SCN neurons (as mentioned already in the discussion of photoreceptive retinal ganglion cells).

The range of light intensities in the environment is large, extending from almost zero lux on a moonless night to half a million lux or more on a sunny day. SCN neurons are able to code for day length only within a restricted range of light intensities. For example, light intensities below 0.1 lux do not result in changes in discharge rate in the SCN of the hamster. For the albino rat, the threshold intensity is a little lower.

These threshold intensities are high compared to the threshold for vision. They correspond well instead with threshold intensities for inducing a phase shift. In a nearly linear way, an increase in light intensity results in an increase in neuronal discharge rate. The light response becomes saturated at about 1,000 lux. Whereas threshold intensities correspond with the light intensities at early dawn or late dusk, saturating intensities match light intensities found during a cloudy, dark day. The circadian system cannot discriminate between a cloudy and a sunny day, but it seems perfectly equipped to judge the difference between day and night.

The response of diurnal animals to light is very similar to that of nocturnal animals. They show phase delays during the beginning of the night, and they respond with phase advances during the second part of the night. Their light-responsive SCN neurons show response characteristics that resemble those of nocturnal animals to some extent. First, and most important, their respon-

siveness to light is sustained. The proportion of light-activated cells in diurnal squirrels and degus, however, is much lower than in nocturnal species, and it is similar to the proportion of light-suppressed neurons. The net effect of light on the neuronal discharge of the SCN is therefore much smaller. Yet diurnal animals entrain as well as nocturnal animals, and it would be intriguing to know why these different response types, light-activated and light-suppressed, exist at all. Are they both involved in phase shifting by light, and are they both involved in transmitting information to the pineal gland to suppress melatonin? The answer is unknown.

Entrainment of animals relies strongly on the fact that phase delays occur during early night, and phase advances during late night. The question of whether this qualitative difference in behavioral response is reflected by changes in neuronal response was investigated by implantation in rats of a stationary electrode that was connected to a rotating contact. With such an implant, the rats could move freely in their cage while electrophysiological recordings of discharge rate were made from the SCN (see Figure 6.2). This technique allowed scanning of the light responsiveness of the SCN in a fixed place within the SCN throughout the circadian cycle. Simultaneous behavioral recordings of circadian activity from the animal documented the circadian phase. The results showed that SCN neurons have maximum sensitivity to light during the middle of the night and minimum sensitivity during the middle of the day. In other words, light input is gated and results in maximal response during those hours in which the pacemaker responds with a phase shift to light. No qualitative differences were observed, however, between early night and late night, despite the occurrence of delays and advances. The mechanism for determining the direction of the phase-shifting response must lie farther downstream in the light transduction process.

A current conception of light-induced signal transduction in the SCN involves a cascade of events. The first step is the release of glutamate from RHT terminals. The postsynaptic membrane then depolarizes and a calcium influx occurs. Activation of intracellular serine–threonine protein kinases and nitric oxide synthase is the next step. The consequent activation of transcription factors and the regulation of target gene expression can then occur (see Chapters 7 and 8). More research must be done, however. For some of these steps, relatively few data are available on their relationship to one another or to light-induced phase shifting.

The emerging complexity of cellular biochemistry, including networks of signaling and regulatory molecules with extensive cross talk between metabolic cascades, complicates this work. Moreover, glutamatergic neurotransmission may be significantly modified by colocalized neuroactive peptides such as substance P and

PACAP (pituitary adenylate cyclase–activating peptide). Finally, SCN cells constitute a heterogeneous population, and it remains uncertain which of the intervening steps that have been described in this chapter are intracellular and which are intercellular.

Progress is being made on cell physiology aspects of the output signals of mammalian SCN pacemaker systems[14,34]

Immunocytochemical localization of the neuropeptides vasopressin, vasoactive intestinal peptide (VIP), neuropeptide Y, and serotonin in rodent hypothalamus tissue demonstrates different spatial distribution patterns of these neuropeptides within the SCN. The distinct organization of these neuropeptides within the SCN hints at a functional organization of the cells. Secretory rhythms of neuroactive peptides synthesized in the SCN have also been measured in vitro, and a circadian rhythm of arginine vasopressin (AVP) levels has been studied extensively. Levels of AVP are highest during the subjective day of both nocturnal and diurnal animals. The role of these neuropeptides in the function of the SCN as a biological pacemaker, however, is still not fully clear. Identifying such a peptide rhythm is an important step in analyzing rhythmicity because it provides a window on clock-controlled gene expression (see Chapter 8). If the biochemistry of a circadian-regulated process is known, then the molecules involved in its control should have a clear relationship to the oscillatory mechanism of the pacemaker itself.

Although single cells can generate circadian rhythms, other circadian properties, such as entrainment and history dependence, might arise at a higher, tissue level of organization. SCN neurons are heterogeneous and appear to be compartmentalized into two topographically distinct subdivisions: a dorsomedial shell and a ventrolateral core (see Chapter 5). The spatial arrangement by neurochemical content suggests a functional organization. The neuropeptides AVP and somatostatin are synthesized within some of the dorsal neurons, and the levels of mRNAs and peptides for these entities exhibit circadian rhythmicity, with higher levels during the subjective day.

Some of the ventral neurons synthesize VIP or GRP (gastrin-releasing peptide), and their mRNA and peptide levels show oppositely phased rhythms during the light–dark cycle, with high levels of GRP during the light and high levels of VIP during the dark. VIP and GRP are co-localized in some of these neurons. At least in rats and hamsters, in which SCN subdivisions are clearly defined, the photic and circadian regulation of clock genes appears to occur in separate ventral and dorsal cell populations, respectively, suggesting that their function in circadian timekeeping is likely to be cell specific. In hamsters, a cluster of cells expressing cal-bindin seems to be crucial for the expression of locomotor rhythmicity.

The rodent circadian clock has been modeled as a complex pacemaker consisting of two mutually coupled oscillators. Variability in the phase relationship between two oscillators could account for several circadian properties. A morning oscillator could be accelerated by light and synchronized to dawn, and an evening oscillator decelerated by light and synchronized to dusk. Dual oscillators might explain the appearance of bimodal circadian activity patterns, the ability to measure seasonal changes of day length (photoperiod), and the dependence of circadian period on an animal's previous exposure to light–dark cycles (aftereffects). Among other possibilities, a dual circadian organization might originate from an intercellular network interaction within SCN tissue. Such a mechanism accounts for the phenomenon known as splitting, in which an animal's single daily burst of locomotor activity dissociates into two components that each free-run with different periods until they become coupled 180 degrees apart. Splitting behavior appears to be a consequence of a paired SCN that has become reorganized into two oppositely phased, left- and right-sided circadian pacemakers.

One way to test the idea of right- and left-sided SCN oscillators uses the plasma clot culture method. When rat SCN slices are embedded in plasma clots on coverslips and incubated for weeks in rotating roller tubes, they lose about 70% of their neurons and flatten to a few cell layers thick. Such organotypic cultures exhibit circadian rhythms of the release of both AVP and VIP. If the cultures are treated with antimitotics for 24 h on day 2, the two neuropeptide rhythms appear to free-run separately with different circadian periods, suggesting that they represent a chemical manifestation of two independent oscillators. However, the effect may represent merely an artifact of culturing. The SCN neurons in these cultures may have developed connections different from those present in vivo. Perhaps multiple cellular oscillators within the SCN can align themselves in distinctly different configurations: left–right in splitting versus dorsal–ventral in normal rhythmicity. Whether this kind of plasticity might be physiologically significant for encoding a variety of temporal programs or is merely a consequence of cells with multiple couplings within a tissue will be a challenging question for future research.

Synthesis: Comparison of the Circadian Properties of the Three Model Systems Highlights Functional Similarities and Differences[6,8,11,59,66,69]

Individual invertebrate and vertebrate pacemaker neurons and pinealocytes contain autonomous circadian oscillators. The three established model systems illustrate that single cells can generate circadian rhythms. Individual basal retinal neurons in *Bulla* have been isolated as reduced preparations or single cells in culture. Microelectrode measurements of membrane conductance in LL conditions show high conductance in late subjective night and decreases near projected dawn, at the phase where action potential frequency increases. In the chick pineal gland in culture, progressive tissue reduction has no effect on melatonin rhythmicity. Reduced pieces and even single cells maintain a self-sustained circadian periodicity. SCN slices and single cells in dispersed-cell cultures show a self-sustained rhythm, with individually varying free-run values for single clock cells. Thus, multicellular pacemakers in higher animals are based on single-celled circadian oscillators, as in simpler unicellular organisms. Circadian rhythmicity may be a general property of most cells.

Single-cell oscillators in multicellular pacemakers are mutually coupled. The presence of coherent pacemaker rhythms of electrical activity, glucose utilization, and neurochemical output implies that the ensemble of individual cellular oscillators is synchronized in some way, either by mutual coupling or by inputs imposed from elsewhere in the organism. The basal retinal neurons in *Bulla* are electrically coupled to one another. In the SCN, the period of the free-running rhythm expressed by whole animals appears to be determined by the average period of the population of coupled cellular oscillators. The picture is not yet clear for chick pineal pacemakers. Even though individual clock cell periods are not very accurate and differ widely from the mean value of the intact organism, the rhythm's precision is apparently collectively enhanced by coupling of the cellular population.

Photic input for entrainment travels by multiple routes in nonmammalian vertebrates but exclusively by retinal photoreceptive ganglion cells in mammals. In *Bulla* the chief input route is retinal, although extraoptic routes have not been ruled out completely. In most birds the retina, the pinealocytes, and extraretinal photoreceptors all have photic input to the circadian system. These neurons contain an opsinlike pigment and are considered putative extraretinal photoreceptors, but their role in circadian photoentrainment has not been established. In mammals the exclusive photic projection route for environmental resetting of the SCN pacemaker is retinal via the retinohypothalamic tract. Specialized ganglionic neurons, rather than rod or cone cells, act as luminance detectors for photon counting in entrainment.

In the two neuronal cases of *Bulla* and mammalian SCN, photic shifts are mediated by membrane depolarization. In the mollusc eye, light during the sensitive phase of the circadian cycle depolarizes the basal retinal neurons. Membrane depolarization, either by direct current injection or by change in the extracellular potassium concentration, generates phase shifts of the discharge

rhythm similar to those produced by light pulses, and blocking light-induced membrane depolarization also blocks the phase-shifting effects of light. Depolarization stimulates a voltage-activated calcium conductance, and the removal of extracellular calcium or the application of calcium channel blockers prevents light- or depolarization-induced phase shifts. In the rodent SCN, a depolarization-induced neuronal excitation has also been inferred. The excitatory amino acid glutamate appears to be the primary retinohypothalamic tract (RHT) neurotransmitter responsible for mediating the circadian actions of light. Glutamate is localized to RHT terminals innervating SCN neurons, and it is released by optic nerve stimulation of SCN slices in vitro. Glutamate application excites SCN neurons in vitro, with resultant phase shifts that mimick a photic-type phase response curve. Receptor blockers inhibit both the phase-shifting actions of light in vivo and the electrophysiological effects of optic nerve stimulation of slices in vitro.

Pinealocytes are nonneuronal, in contrast to *Bulla* and SCN pacemaker neurons. Nevertheless, chick pineal cells have been valuable as a model for nonneuronal vertebrate circadian pacemakers at the cellular level. The cells exhibit a robust circadian rhythm of melatonin synthesis and release under constant red light for an extended period of time. Perturbations that act on the melatonin-synthesizing machinery can be distinguished from those acting on the pacemaker driving melatonin rhythmicity. Agents that acutely raise or lower melatonin output without inducing phase shifts act downstream from the pacemaker; agents that induce phase shifts act upon and through the clock mechanism. In contrast to results in neurons, experimental manipulations of membrane potential, voltage-sensitive calcium channels, or cyclic AMP levels do not affect the oscillation of the circadian pacemaker in chick pineal cells. The acute and entraining effects of light appear to be conveyed through completely separate pathways.

Temporal programs are generated via multiple output paths. Circadian pacemakers govern a wide array of rhythms, from biosynthetic to behavioral. The range of waveforms and phases of peripheral overt rhythms may be very different from the central oscillation that drives them. The circadian signal is presumably modified as the pacemaker's outputs are coupled to effector cells via synaptic transmission and hormonal secretion, and indirectly through the regulation of behavior. Because the fidelity of the circadian signal might be altered by such multistep output pathways, overt rhythms could differ substantially from their underlying cellular and molecular oscillations.

SUMMARY

Three animal model systems have been highly useful for studying circadian physiology at the cellular level by means of in vitro culture. The most favorable animals have been the sea hare, *Bulla;* the chick pineal gland; and the mammalian SCN. In *Bulla* and chick pineal, individual clock cells are also the photoreceptors, and they generate the output signals. In contrast, the SCN clock cells of mammals depend on distant photoreceptors in the retinal ganglion cells for light detection. The clock cells of the sea hare and mammals are true neurons, while the chick pinealocyte pacemakers are nonneural. Individual clock cells contain oscillators in all three cases, and these are coupled to produce a stable, precise output rhythm from the multicellular pacemaker.

CONTRIBUTORS

Patricia J. DeCoursey wishes to thank Gene Block, David Earnest, Michael Lehman, Johanna H. Meijer, Terry Page, Rebecca Prosser, Till Roenneberg, William J. Schwartz, Rae Silver, Tony van den Pol, and Martin Zatz for their material contributions to Chapter 6.

STUDY QUESTIONS AND EXERCISES

1. Visit a neurophysiology laboratory to observe one or more of the following techniques:

   - Immunocytochemistry
   - In vitro cell culture
   - Electrophysiological recording

2. With the help of an electrophysiologist, try setting up a pacemaker system in vitro and record signals for 24 h. The experiment will be most successful with a group of at least 12 students who can work in pairs on shifts of about 3 h. An experienced instructor is essential. The site should be as aseptic as possible if mammalian SCN is used. Use either molluscan eye if available or rat SCN. Why should you not use chick pineal? (*Hint*: What is involved in measuring the melatonin output rhythm in the chick cultures?)

   When the preparations are complete, in groups of two or three students, cover 3 h sessions with one half-hour overlap between shifts to serve as a continuity period. Plot the results singly and as means for groups of 3 h covering each team's data. Do the data show a circadian rhythm of the spontaneous firing rate?

3. One very easy, almost foolproof way to observe in vitro culture involves so-called chick-in-a-cup culture. All that is needed is a simple incubator with a clear viewing top, a means of temperature regulation, and a pan of water for providing high humidity. Build one of wood or even use a Styrofoam box with a small light as the heat source. Also needed is a fertilized but non-incubated hen egg from a local hatchery or farmer. Improvise a membranous "cup" of saran wrap as described here.

   Prior to incubation, a simple "cup" should be built to hold the contents of the egg. The cup consists of a 12 cm length of PVC pipe with a diameter of about 7.5 cm. The diameter doesn't matter, as long as a standard petri dish fits over the top as a cover. Nonclinging saran wrap is used as a membrane-like pouch to form the cup, with a sturdy rubber band at the top of the pouch to hold it in place. Let the incubator warm up and stabilize at 35°C for several days before starting. Procure a sleeve of standard petri dishes for covers.

   Incubate the fertilized hen eggs for exactly 72 ± 1 h. Then clean up and organize the work space. When everything is ready, sterilize the egg by wiping it with a weak solution of iodine solution. All of this preparation sounds complicated, but it really isn't, and the actual making of the culture is as simple as cracking an egg carefully with a table knife. Standing close to your incubation cup, gently hit the egg to form a straight crack around the midline. Then carefully press outward to separate the two shell halves and let the entire egg slide, *not drop*, into the saran wrap membrane below. The embryo with its surrounding circle of rapidly expanding blood vessels should be about the size of a dime at this stage, with beating heart clearly visible. If the yolk or blood vessels rupture, start again. If the embryo is underneath, wait a few minutes to see if it floats around to the top. If not, gently paddle the yolk with a sterile teaspoon to encourage the embryo to float around on top. Amazingly, the preparation is extremely robust. It is relatively resistant to bacterial contamination because of the abundant lysozymes in the yolk. Many cultures will continue to develop normally for up to 2 weeks. They may reach even day 19 of 21 incubation days. The chicks, however, are not able to hatch because of calcium deficits without their shells. The chick embryos will be completely transparent or translucent for the first 10 days of culture. Thereafter, feathers develop and the internal organs are no longer visible.

   Now comes the circadian part. Watch for a good overt marker of rhythmicity, such as heartbeat. Try recording for one circadian day using rotating teams of observers. Does the embryo show any circadian rhythms? If so, at what age do they start?

REFERENCES

1. Barinaga, M. 2002. How the brain's clock gets daily enlightenment. Science 295: 955–957.

2. Barrett, R. K., and J. S. Takahashi. 1995. Temperature compensation and temperature entrainment of the chick pineal cell circadian clock. J. Neurosci. 15: 5681–5692.

3. Bellingham, J., and R. G. Foster. 2002. Opsins and mammalian photoentrainment. Cell Tissue Res. 309: 57–71.

4. Bernard, M., D. C. Klein, and M. Zatz. 1997. Chick pineal clock regulates serotonin N-acetyltransferase mRNA rhythm in culture. Proc. Natl. Acad. Sci. USA 94: 304–309.

5. Berson, D. M., F. A. Dunn, and M. Takao. 2002. Phototransduction by retinal ganglion cells that set the circadian clock. Science 295: 1070–1073.

6. Binkley, S. 1990. A Clockwork Sparrow: Time, Clocks, and Calendars in Biological Organisms. Prentice Hall, Englewood Cliffs, NJ.

7. Block, G. D., and S. Wallace. 1982. Localization of a circadian pacemaker in the eye of a mollusk, *Bulla*. Science 217: 155–157.

8. Block, G. D., S. S. Khalsa, D. G. McMahon, S. Michel, and M. Guesz. 1993. Biological clocks in the retina: Cellular mechanism of biological timekeeping. Int. Rev. Cytol. 146: 83–144.

9. Block, G. D., D. G. McMahon, S. Wallace, and W. Friesen. 1984. Cellular analysis of the ocular circadian pacemaker system: A model for retinal organization. J. Comp. Physiol. 155: 365–378.

10. Block, G. D., M. H. Roberts, and A. E. Lusska. 1986. Cellular analysis of circadian pacemaker coupling in *Bulla*. J. Biol. Rhythms 1: 199–217.

11. Blumenthal, E. M., G. D. Block, and A. Eskin. 2001. Cellular and molecular analysis of molluscan circadian pacemakers. In: Handbook of Behavioral Neurobiology Volume 12: Circadian Clocks, J. S. Takahashi, F. W. Turek, and R. Y. Moore (eds.), pp. 371–400. Kluwer Academic/Plenum, New York.

12. Bouskila, Y., G. J. Strecker, and F. E. Dudek. 2001. Cellular mechanisms of circadian function in the suprachiasmatic nucleus. In: Handbook of Behavioral Neurobiology Volume 12: Circadian Clocks, J. S. Takahashi, F. W. Turek, and R. Y. Moore (eds.), pp. 401–432. Kluwer Academic/Plenum, New York.

13. Colwell, C. S. 2000. Rhythmic coupling among cells in the suprachiasmatic nucleus. J. Neurobiol. 43: 379–388.

14. de la Iglesia, H. O., J. Meyer, A. Carpino, Jr., and W. J. Schwartz. 2000. Antiphase oscillation of the left and right suprachiasmatic nuclei. Science 290: 799–801.

15. Ding, J. M., L. E. Faiman, W. J. Hurst, L. R. Kuriashkina, and M. U. Gillette. 1997. Resetting the biological clock: Mediation of nocturnal CREB phosphorylation via light, glutamate, and nitric oxide. J. Neurosci. 17: 667–675.

16. Doyle, A., and J. B. Griffiths. 1997. Mammalian Cell Culture: Essential Techniques. Wiley, New York.

17. Freshney, R. I. 2000. Culture of Animal Cells: A Manual of Basic Techniques, 4th Edition. Wiley, New York.

18. Gillette, M. U. 1991. SCN electrophysiology *in vitro*: Rhythmic activity and endogenous clock properties. In: Suprachiasmatic Nucleus: The Mind's Clock, D. C. Klein, R. Y. Moore, and R. M. Reppert (eds.), pp. 125–143. Oxford University Press, New York.

19. Gillette, M. U., and J. W. Mitchell. 2002. Signaling in the suprachiasmatic nucleus: Selectively responsive and integrative. Cell Tissue Res. 309: 99–107.

20. Gillette, M. U., and S. A. Tischkau. 1999. Suprachiasmatic nucleus: The brain's circadian clock. Recent Prog. Horm. Res. 54: 33–58.

21. Grace, M., V. Alones, M. Menaker, and R. G. Foster. 1996. Light perception in the vertebrate brain: An ultrastructural analysis of opsin and vasoactive intestinal polypeptide immunoreactive neurons in iguanid lizards. J. Comp. Neurol. 367: 575–594.

22. Green, C. B., J. C. Besharse, and M. Zatz. 1996. Tryptophan hydroxylase mRNA levels are regulated by the circadian clock, temperature, and cAMP in chick pineal cells. Brain Res. 738: 1–7.

23. Harrison, N. L., and M. Zatz. 1989. Voltage-dependent calcium channels regulate melatonin output from cultured chick pineal cells. J. Neurosci. 9: 2462–2467.

24. Hattar, S., H-W. Liao, M. Takao, D. M. Berson, and K-W. Yau. 2002. Melanopsin-containing retinal ganglion cells: Architecture, projections, and intrinsic photosensitivity. Science 295: 1065–1070.

25. Herzog, E. D., and W. J. Schwartz. 2002. A neural clockwork for encoding circadian time. J. Appl. Physiol. 92: 401–408.

26. Herzog, E. D., J. S. Takahashi, and G. D. Block. 1998. Clock controls circadian period in isolated suprachiasmatic nucleus neurons. Nature Neurosci. 1: 708–713.

27. Honma, S., T. Shirakawa, Y. Katsuno, M. Namihira, and K. Honma. 1998. Circadian periods of single suprachiasmatic neurons in rats. Neurosci. Lett. 250: 157–160.

28. Jacklet, J. W., and W. Colquhoun. 1983. Ultrastructure of photoreceptors and circadian pacemaker neurons in the eye of a gastropod, *Bulla*. J. Neurocytol. 12: 373–396.

29. Jacklet, J. W., M. Klose, and M. Goldberg. 1987. FMRF-amide-like immunoreactive efferent fibers and FMRF-amide suppression of pacemaker neurons in eyes of *Bulla*. J. Neurobiol. 18: 433–449.

30. Kandel, E. R., J. H. Schwartz, and T. M. Jessell (Eds.) 1991. Principles of Neural Science, 3rd Edition. Elsevier, New York.

31. Klein, D., R. Y. Moore, and S. M. Reppert (Eds.) 1991. Suprachiasmatic Nucleus: The Mind's Clock. Oxford University Press, New York.

32. Korf, H. W. 1994. The pineal organ as a component of the biological clock: Phylogenetic and ontogenetic consideration. Ann. N.Y. Acad. Sci. 719: 13–42.

33. Leeson, T. S. and E. R. Leeson. 1981. Histology, 4th Edition. Saunders, Philadelphia.

34. LeSauter, J., and R. Silver. 1999. Localization of a suprachiasmatic nucleus subregion regulating locomotor rhythmicity. J. Neurosci. 19: 5574–5585.

35. Liu, C., D. R. Weaver, S. H. Strogatz, and S. M. Reppert. 1997. Cellular construction of a circadian clock: Period determination in the suprachiasmatic nuclei. Cell 91: 855–860.

36. Low-Zeddies, S. S., and J. S. Takahashi. 2001. Chimera analysis of the Clock mutation in mice shows that complex cellular integration determines circadian behavior. Cell 105: 25–42.

37. Masters, J. 2000. Animal Cell Culture: A Practical Approach, 3rd Edition. Oxford University Press, Oxford, UK.

38. McMahon, D., S. F. Wallace, and G. D. Block. 1984. Cellular analysis of the *Bulla* ocular circadian pacemaker system II: Neurophysiological analysis of circadian rhythmicity. J. Comp. Physiol. A 155: 379–385.

39. Meijer, J. H., and W. J. Rietveld. 1989. Neurophysiology of the suprachiasmatic circadian pacemaker in rodents. Physiol. Rev. 69: 671–707.

40. Meijer, J. H., J. Schaap, K. Watanabe, and H. Albus. 1997. Multiunit activity recordings in the suprachiasmatic nuclei: in vivo versus in vitro models. Brain Res. 753: 322–327.

41. Meijer, J. H., K. Watanabe, J. Schaap, H. Albus, and L. Détári. 1998. Light responsiveness of the suprachiasmatic nucleus: Long-term multi-unit and single-unit recordings in freely moving rats. J. Neurosci. 18: 9078–9087.

42. Menaker, M., and A. Oksche. 1974. The avian pineal organ. In: Avian Biology Volume 4, D. S. Farner, J. R. King, and K. C. Parkes (eds.), pp. 79–118. Academic Press, New York.

43. Michel, S., M. E. Geusz, J. J. Zaritsky, and G. D. Block. 1993. Circadian rhythm in membrane conductance expressed in isolated neurons. Science 259: 239–241.

44. Moore, R. Y., and R. K. Leak. 2001. Suprachiasmatic nucleus. In: Handbook of Behavioral Neurobiology Volume 12: Circadian Clocks, J. S. Takahashi, F. W. Turek, and R. Y. Moore (eds.), pp. 141–171. Kluwer Academic/Plenum, New York.

45. Nakahara, K., N. Murakami, T. Nasu, H. Kuroda, and T. Murakami. 1997. Individual pineal cells in chick possess photoreceptive, circadian clock and melatonin-synthesizing capacities *in vitro*. Brain Res. 774: 242–245.

46. Natesan, A., L. Geetha, and M. Zatz. 2002. Rhythm and soul in the avian pineal. Cell Tissue Res. 309: 35–45.

47. Nikaido, S. S., and J. S. Takahashi. 1989. Twenty-four hour oscillation of cAMP in chick pineal cells: Role of cAMP in the acute and circadian regulation of melatonin production. Neuron 3: 609–619.

48. Page, T. 2001. Circadian systems of invertebrates. In: Handbook of Behavioral Neurobiology Volume 12: Circadian Clocks, J. S. Takahashi, F. W. Turek, and R. Y. Moore (eds.), pp. 79–110. Kluwer Academic/Plenum, New York.

49. Page, T. L., and K. G. Nalovic. 1992. Properties of mutual coupling between the two circadian pacemakers in the eyes of the mollusk *Bulla gouldiana*. J. Biol. Rhythms 7: 213–226.

50. Pennartz, C. M. A., M. T. G. de Jeu, N. P. A. Bos, J. Schaap, and A. M. S. Geurtsen. 2002. Diurnal modulation of pacemaker potentials and calcium current in the mammalian circadian clock. Nature 416: 286–290.

51. Ralph, M. R., R. G. Foster, F. C. Davis, and M. Menaker. 1990. Transplanted suprachiasmatic nucleus determines circadian period. Science 247: 975–978.

52. Roberts, M. H., and G. D. Block. 1983. Mutual coupling between the ocular circadian pacemakers of *Bulla gouldiana*. Science 221: 87–89.

53. Roberts, M. H., and G. D. Block. 1985. Analysis of mutual circadian pacemaker coupling between the two eyes of *Bulla*. J. Biol. Rhythms 15: 55–75.

54. Ross, M. H., G. I. Kaye, and W. Pawlina. 2002. Histology: A Text and Atlas, 4th Edition. Lippincott Williams & Wilkins, Philadelphia.

55. Schwartz, W. J. 1991. SCN-metabolic activity in vivo. In: Suprachiasmatic Nucleus: The Mind's Clock, D. C. Klein, R. Y. Moore, and S. M Reppert (eds.), pp. 144–156. Oxford University Press, New York.

56. Silver, R., and R. Y. Moore. 1998. Special issue on SCN. Chronobiol. Int. 15: 395–566.

57. Silver, R., J. LeSauter, P. A. Tresco, and M. Lehman. 1996. A diffusible coupling signal from the transplanted suprachiasmatic nucleus controlling circadian locomotor rhythms. Nature 382: 810–813.

58. Takahashi, J. S., P. J. DeCoursey, L. Bauman, and M. Menaker. 1984. Spectral sensitivity of a novel photoreceptive system mediating entrainment of mammalian circadian rhythms. Nature 308: 186–188.

59. Takahashi, J. S., N. Murakami, S. S. Nikaido, B. L. Pratt, and L. M. Robertson. 1989. The avian pineal, a vertebrate model system of the circadian oscillator: Cellular regulation of circadian rhythms by light, second messengers, and macromolecular synthesis. Recent Prog. Horm. Res. 45: 279–352.

60. Underwood, H. 2001. Circadian organization in nonmammalian vertebrates. In: Handbook of Behavioral Neurobiology Volume 12: Circadian Clocks, J. S. Takahashi, F. W. Turek, and R. Y. Moore (eds.), pp. 111–140. Kluwer Academic/Plenum, New York.

61. Underwood, H., R. K. Barrett, and T. Siopes. 1990. The quail's eye: A biological clock. J. Biol. Rhythms 5: 257–265.

62. Underwood, H., C. T. Steele, and B. Zivkovic. 2001. Circadian organization and the role of the pineal in birds. Microsc. Res. Tech. 53: 48–62.

63. van Esseveldt, L. E., M. N. Lehman, and G. J. Boer. 2000. The suprachiasmatic nucleus and the circadian time-keeping system revisited. Brain Res. Brain Res. Rev. 33: 34–77.

64. Vitaterna, M. H., D. P. Ding, A-M. Chang, J. M. Kornhauser, P. L. Lowrey, et al. 1994. Mutagenesis and mapping of a mouse gene, *Clock*, essential for circadian behavior. Science 264: 719–725.

65. Wagner, S., M. Castel, H. Gainer, and Y. Yarom. 1997. GABA in the mammalian suprachiasmatic nucleus and its role in diurnal rhythmicity. Nature 387: 598–603.

66. Welsh, D. K., D. E. Logothetis, M. Meister, and S. M. Reppert. 1995. Individual neurons dissociated from rat suprachiasmatic nucleus express independently phased circadian firing rhythms. Neuron 14: 697–706.

67. Zatz, M. 1991. Light and norepinephrine similarly prevent damping of melatonin rhythm in cultured chick pineal cells: Regulation of coupling between the pacemaker and overt rhythms? J. Biol. Rhythms 6: 137–147.

68. Zatz, M. 1994. Photoendocrine transduction in cultured chick pineal cells. IV. What do vitamin A depletion and retinaldehyde addition do to the effects of light on the melatonin rhythm? J. Neurochem. 62: 2001–2011.

69. Zatz, M. 1996. Melatonin rhythms: Trekking toward the heart of darkness in the chick pineal. Sem. Cell Dev. Biol. 7: 811–820.

70. Zatz, M., J. A. Gastel, J. R. Heath III, and D. C. Klein. 2000. Chick pineal melatonin synthesis: Light and cyclic AMP control abundance of serotonin N-acetyltransferase protein. J. Neurochem. 74: 2315–2321.

71. Zigmond, M. J., F. E. Bloom, S. C. Landis, J. L. Roberts, and L. R. Squire. 1999. Fundamental Neuroscience. Academic Press, New York.

One of the greatest lessons from the molecular biology era is that living cells use a common set of design principles that grow in complexity but retain basic axioms in progressing up the evolutionary tree. In this chapter the circadian tour continues inside the pacemaker cells to the regulatory pacemaker genes and proteins. The pathway to identifying these elements has been circuitous, with many false leads. All successful approaches have applied basic principles of chronobiology to actual regulatory loops.

Genetic and molecular analyses have identified so-called clock genes. Circadian rhythms are based partly on negative feedback loops of processes. Inherent oscillations in the transcript and/or protein levels of specific clock genes play a central role in the generation of the rhythms. The proteins encoded by many of these loci negatively feed back to reduce the level of their own transcripts, thereby providing an autoregulatory system. Chapter 7 first reviews the logic of the analyses, then examines the mechanism for the best-known cases: the fruit fly *Drosophila*, the fungus *Neurospora*, the cyanobacterium *Synechococcus*, and the laboratory mouse *Mus*.

Molecular Biology of Circadian Pacemaker Systems

One ring to rule them all . . . and in the darkness bind them.

—J. R. R. Tolkien

Introduction: Both Conserved and Divergent Elements Are Found in the Basic Clockworks[7,16,48]

Almost as soon as the phenomenon of rhythmicity was described, scientists began theorizing about the basis of the oscillations (Figure 7.1). Early circadian clock models could not be very specific in terms of molecular details because too little was known. The only observable details of the clock were its physiological and behavioral outputs plus the behavior of the whole oscillatory system in response to external resetting cues. All of the four premolecular-era models shown in Figure 7.1 were possible. Early theoretical analysis sought to limit the spectrum of possible models as a way of carrying out discrete biochemical experiments. However, saying that the circadian clock probably represents a feedback loop involving positive and negative elements is not very revealing. The real questions concerned the nature of the elements, as well as where and how they acted. Were they within cells or between cells? Did the elements act in a variety of ways in different organisms? How can the components of the loop be identified? The best approach depends on the way the clock mechanism is visualized. Consequently, the experimental pathways taken to identify the mechanism of the clock have reflected the views of many different scientists.

Despite divergence in approaches, however, the major question has always concerned the biochemical, molecular, and cellular mechanism of an oscillator with the basic characteristics of a circadian clock. Especially challenging have been the means of achieving the necessarily long time constant, the means of entrainment of the oscillator by visible light and various resetting agents, the means of achieving temperature compensation, and the nature of the escapement whereby temporal information generated by the oscillator is used to time the behavior of the cell and organism. Within the past two decades, answers to most of these questions have emerged at the molecular level.

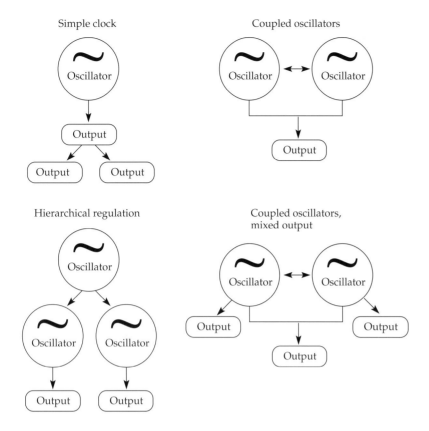

FIGURE 7.1 Early circadian clock models. For further explanation, see the text.

the cellular level; it seems likely that they do not reflect a common mechanism at the physiological and biochemical level in multi-cellular organisms."[48]

To say that the clock in a single-celled organism like *Gonyaulax* or *Tetrahymena* is cellular has always seemed self-evident. In contrast, an almost impossible leap of faith was needed to say the same about a mammalian circadian clock, even after the dominant oscillator was localized to the SCN, the suprachiasmatic nucleus. Moreover, a scientist's view of the intracellular versus intercellular clock controls the choice of system studied, the kinds of experiments done, and clearly the expectation of a common mechanism for rhythms in a variety of phylogenetically diverse organisms. Chapter 6 outlined the key experiments favoring an intracellular oscillator and a relatively similar mechanism in all clocks. Remember, however, that the commonality now seen is based on a very small sample of species studied, and only a very few circadian oscillator components have actually been identified.

An integral part of discussing what is known is a review of what has been done, and why it was done in that way. In the field of clocks, the descriptions of what and why give insight into the resolution of one of the major themes in chronobiology during several past decades: determining at the ultimate level of resolution whether rhythms are the product of intracellular or intercellular feedback loops. The material in previous chapters has already partially resolved the issue by showing that the molecular cycle describing an elemental circadian oscillator can be described within the bounds of single cells without the need to invoke intercellular communication between multiple cells.

The debate surrounding this question encapsulated a major aspect of how scientists thought about rhythms in the years between 1960 and 1990, and therefore colored how they approached the problem of the molecular basis of these rhythms. In 1960, when Erwin Bünning presented the opening address at the seminal Cold Spring Harbor Symposium on Biological Clocks, the whole question of the molecular basis of circadian timing seemed so daunting that he spoke merely of "the clock." A decade later the foreword to a symposium volume stated, "It is still very much an open question whether or not these similarities in physical clock characteristics reflect a common timekeeping mechanism at

For Many Years Theory Dominated Experiment in the Generation of Models for the Feedback Loop Comprising the Clock

Common characteristics of all oscillators provided an intellectual focus to early work[25,51]

All oscillators have three basic requirements: positive input to drive a change, feedback, and a delay in the execution of the feedback. Regulatory mechanisms with these characteristics are prevalent everywhere. Factory assembly lines often have feedback control from end points to earlier points, and an aspect of industrial design is elimination of the resulting undesirable oscillations in the rate of production of whatever is being manufactured. Similarly, the metabolic pathways have feedback control exerted via allosteric enzymes that change their shape and therefore their activity as a result of the binding of ligands.

In the premolecular era from 1950 to 1970, the prospect of assigning functions within input, feedback, or delay to actual molecules to concretely describe the molecular basis of any circadian clock seemed remote. Since 1950 the techniques of molecular biology and molecular genetics have penetrated practically every aspect of the life sciences, including chronobiology. A short list includes the recombinant DNA technologies that allowed scientists for the first time to span the gap between classical genetics and biochemistry in cellular organisms. The new field of molecular genetics eventually provided the answers about how cells could build and operate a circadian oscillator with a period length of a day.

Despite the application of molecular tools, the answers still depend on earlier circadian insights, including the concept and universality of the phase response curve, the identity and speed of action of resetting agents, and the utility of limit-cycle models for the oscillator (see Chapter 3). The body of material on characteristics of circadian oscillators is essential background for understanding their molecular basis.

The glycolytic pathway is an example of a simple feedback oscillator[29,35,59,78]

The well-known glycolytic oscillator is roughly analogous to a circadian pacemaker in its responses to perturbations. The glycolytic pathway provides carbon fragments as metabolic intermediates for biosynthetic reactions and for the production of ATP. The basic biochemistry is simple, consisting of the enzymes responsible for turning glucose and 2 ADP into pyruvate and 2 ATP, while balancing the reducing equivalents (NAD/NADH) in the process (Figure 7.2). Although the system includes at least two potential oscillators surrounding the allosteric enzymes glyceraldehyde-3-phosphate dehydrogenase (GAPDH) and phosphofructokinase (PFK), PFK dominates flux through the pathway and the resulting oscillations.

The point of glycolysis is to make ATP and biosynthetic intermediates. The pathway is regulated accordingly by ATP and by citrate, which is a key intermediate in the citric acid cycle into which glycolysis feeds carbon units. In terms of the glycolytic oscillator, the point of the feedback regulation is to make ATP when the concentration is too low, and to stop the production of ATP when too much accumulates.

PFK's role in glycolysis is to take a phosphate from ATP and add it to fructose-6-phosphate, yielding fructose-1,6-diphosphate plus ADP. Both products of the reaction activate PFK. Consequently, the more products it makes the faster it works—an example of positive feedback creating an oscillation: In the course of each catalytic cycle, an ATP is used and an ADP is produced. The reaction goes even faster, and soon the ATP is used up. When this happens, the rate of production of ADP drops, PFK slows down, and the cycle is complete.

Although the net product of glycolysis is the production of 2 ATPs, these are harvested actually a few steps later in the pathway, providing a lag. The production of ADP and fructose-1,6-diphosphate oscillates both because of the inherent allostery of the enzyme and because of the lag provided by the multiple steps of the pathway. Actually, several things are oscillating, includ-

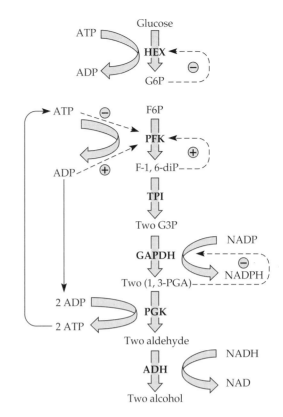

FIGURE 7.2 Negative and positive feedback loops within glycolysis result in the creation of a stable oscillator. Abbreviations for enzyme names are in bold, enzymatic steps are indicated with broad gray arrows, and substrates and products of these processes are abbreviated in plain text. Steps in the pathway where products immediately feed back to regulate enzymes are shown with dotted arrows and the ATP/ADP-mediated glycolytic oscillator feedback loop is shown with black arrows. 1,3-PGA, 1,3-phosphoglyceric acid; ADH, alcohol dehydrogenase; F-1,6-diP, fructose-1,6-diphosphate; F6P, fructose-6-phosphate; G3P, glyceraldehyde-3-phosphate; G6P, glyceraldehyde-6-phosphate; GAPDH, glyceraldehyde-3-phosphate dehydrogenase; HEX, hexokinase; PFK, phosphofructokinase; PGK, phosphoglycerate kinase; TPI, triosephosphate isomerase. See the text for details. (After Chance et al., 1967.)

ing the activity of PFK (which changes on the order of 80-fold), the concentrations of ATP, ADP, fructose-6-phosphate, fructose-1,6-diphosphate, and the amounts of several metabolites such as NAD and NADH that are used and generated later in the pathway. In many experimental situations and with in vivo conditions, concentrations of all glycolytic intermediates oscillate in the range of 10^{-5} to 10^{-3} M with a period length of approximately 7 minutes.

The glycolytic oscillator is a good model for reviewing some of the circadian terms and concepts introduced in Chapters 1 and 3. When all of the components are in a cell-free extract in a test tube and are oscillating, the state of the oscillation surrounding PFK can be fully known simply if the amounts of ADP and ATP that are

FIGURE 7.3 Oscillators and their outputs can be probed by perturbations of elements expected to be within the oscillator (state variables and parameters) or within the output pathway. (A) Output from the glycolytic oscillator measured as the amount of NADH as a function of time. (B) Additions of ADP, a state variable of the oscillation, reset the oscillator in different ways, depending on when in the cycle they are given. (C) Plotting the change in phase resulting from ADP additions versus the phase at which the ADP was given yields a phase response curve of the type seen in Chapter 3. (After Pye, 1971.)

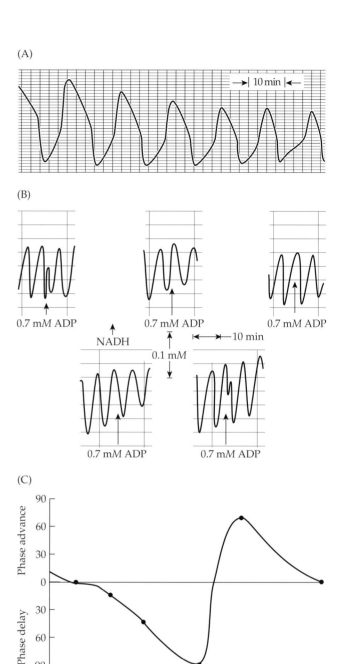

present are known. Therefore, these are the state variables. If they do not oscillate, no glycolytic oscillation occurs. Glucose is used in the reaction, which results in the net production of ATP. Obviously PFK must be present and must act in the manner predicted by its normal biochemistry. These are parameters of the reaction (see Figure 7.3C). As long as they are present or supplied at a constant rate, an oscillation continues.

Several types of perturbations can be carried out on the oscillator (Figure 7.3; Table 7.1). Clearly with no, or too little, glucose (a parameter) added, no oscillation occurs. If more substrate is added through an increase in the rate of input of glucose, the waveform, amplitude, and period of the oscillation will change. Adding a slug of glucose in a stepwise manner will result in a phase shift. If the rate of glucose input is increased sufficiently, the oscillation will cease altogether. The same holds true in all cases with the PFK parameter. Investigators commonly observe the oscillator by following NADH fluorescence as a "hand of the clock"; if a bolus of NAD or NADH is added, the oscillation will appear to change, but it will return later to the previous amplitude, waveform, period, and phase. Finally, the oscillator can be entrained by sinusoidal changes in the input rate of glucose, as long as the changes are within the range $0.7 < T/\tau < 1.2$. The range of entrainment is reminiscent of that observed for circadian rhythms (see Chapter 3). Note that all of these real or imagined changes in the oscillator are due to manipulation of parameters or hands of the oscillator, not state variables.

Clearly manipulations of the state variables ADP or ATP will alter the oscillation: An increase of ADP by 0.7 mM apparently results in weak Type 1 resetting, amounting to a 1.5-minute delay, for instance, when delivered at the minimum of the ADP oscillation, and a 2.5 mM increase in strong Type 0 resetting (see Chapter 3). Likewise, perturbations that indirectly affect the state variables will reset the oscillator. Extensive experiments in vitro and in whole cells have manipulated this oscillator through changes in state variables such as those seen in Figure 7.3. All show dose-dependent strong or weak resetting and can be used to generate a helical resetting surface with an apparent singular point as described in Chapter 3.

No one believes anymore that circadian oscillators are based on the glycolytic oscillator, but most people believe that the circadian oscillator will be a stable limit-cycle type of oscillator, with parameters and state variables analogous to the glycolytic oscillator. An overview of changes wrought by perturbations to oscillator components and outputs is instructive. Table 7.1 compiles the effects of different changes in parameters or state variables on the oscillator. Depending on the effects observed on the period length, phase, and waveform, the true role of the element can be inferred. A defining characteristic of a state variable is that pulses will result in a permanent phase shift but no period effect. Because state variables are by nature oscillatory, a step change cannot be made in them; only a temporary increase or decrease can be made. Table 7.1 presents answers from analyzing a very simple biochemical oscillator with just two state variables, whereas the circadian clock oscillator system will be harder to predict. Learning to think about oscillators, perturbing oscillators, and what the data means will be useful later, when actual potential components of the circadian clock are considered.

Coupling of simple oscillators enhances stability and adjusts the period length of the ensemble[31,55,76,77,78]

Two of the salient characteristics of circadian oscillators are important in their performance. First, they are very robust. Specifically, they are not easily perturbed by changes in the metabolism of the organism while remaining acutely sensitive to appropriate environmental factors. Second, the period length of about 24 h is a very long time in terms of intracellular molecular interactions. One way of introducing these characteristics to a biological oscillator would be to couple more than one simple oscillator so that the output was a function of the ensemble. Moreover, such a process could introduce a frequency reduction whereby separate shorter-period ultradian oscillators might be coupled to produce a circadian output.

In its simplest form, the coupling can be imagined as the familiar "beat frequencies" representing the summed output of two separate sinusoidal oscillators. For individual frequencies of ω_1 and ω_2, w the summed output will be $\frac{1}{2}(\omega_1 - \omega_2)$. For two oscillators cycling every 3 h and 4 h, respectively, the output would be $\frac{1}{2}(\frac{1}{3} - \frac{1}{4})$ = once per day. The result may sound nice, but it's too simple to be true for the biological clock. Consider the case of the same two oscillators with some biological plasticity built into the first to allow its period to drift just 6 minutes to 3.1 hours. Now the beat frequency is 27.5 hours, far outside the normal range of homeostasis for circadian period lengths. In addition,

anomalies are introduced into the shape of the phase response curve. Finally, if the two oscillators are coupled instead of their independent outputs being simply summed, they will mutually synchronize and the long period beat will be lost. It is hard to imagine two oscillators within a single cell remaining completely uncoupled. Problems like these have suggested that circadian clocks are not simply the summed output of ultradian metabolic oscillators, such as the glycolytic oscillator described earlier, and that coupling of oscillators does not provide a plausible mechanism for large-scale frequency reduction of ultradian oscillators to achieve a circadian period length.

The effects of oscillator coupling within an organism should be considered, however, because coupling must clearly exist in some systems, and probably exists in most. A good example can be seen in the growth of a *Neurospora* culture in two dimensions over a petri dish. Because the fungus grows as a syncytium, a plug of the culture taken from any part of the petri dish must have a clock in it. Formally the petri dish culture must consist of a population of coupled oscillators. Another example is the population of neurons in an SCN mammalian circadian pacemaker (see Chapters 5 and 6). Because individual SCN neurons have clocks (see Chapter 6) and the ensemble stays in synchrony, the individual neurons must be coupled to some degree. In each of these cases, the coupled oscillators are to a large degree similar, if not identical. Thus the molecular components of the clocks are the same, and differences would be due to local variations, perhaps in nutrition for *Neurospora* or vascularization in the SCN, or due solely to developmental plasticity.

The first extensive analysis of coupling for oscillators was carried out by Goodwin. Starting with the concept of a metabolic glycolytic oscillator, transcriptional and translational regulatory events were added. Multiple individual feedback oscillators could be effectively coupled because of allosteric effects of metabolites on enzymes, or because of activating/repressing actions at the level of gene expression. If activating or repressing metabolites were made to diffuse from their site of

---

TABLE 7.1   Effects of perturbations on a limit-cycle oscillator and its output

Type of perturbation	Period	Phase	Waveform
State variable, pulsatile change	No effect	Permanent	No effect
Parameter, step change	Permanent	Permanent	Permanent
Parameter, pulsatile change	Transient	Permanent	Transient
Hand of the clock, step change	No effect	?	Permanent
Hand of the clock, pulsatile change	No effect	Transient or permanent	Transient

Sources: Tyson, 1976, and Winfree, 1976.

synthesis to their site of action, useful time delays accrued. Because diffusion is a relatively temperature-independent process, a modicum of temperature compensation of period length could also be developed. The possibilities were endless. Models of this type were popular for some time and emerged in a biochemically discrete form in the chronon model, which will be discussed a little later in this chapter.

Winfree later analyzed a population of similar weakly coupled oscillators. He showed that even though the component oscillators were of slightly different period length, they would mutually entrain to a frequency that was close to one of the individual oscillators. Pavlidis examined the effects of elastic and inhibitory coupling between oscillators and found that a large increase in period could be obtained. For a population of n oscillators of frequency ω, the sum of the outputs of all the coupled units oscillates with a frequency of $[\omega(1 - (n-1)r)^{1/2}]$, where r is the coupling constant. Pavlidis carried out numerical simulations using $(n-1)r$ equal to about 0.9. The ensemble oscillated with a periodicity of about $\omega/4$. The value came close to the circadian value. In addition, simulated pulses of light resulted in a correctly shaped phase response curve.

With sufficient ingenuity, the mathematics of coupled oscillators can be made to closely mimic real observations. The simulations described here provided a sound theoretical, physical, or formal underpinning to a phenomenon that must happen in living rhythmic organisms. In all of these computations, however, the actual components of the oscillators were not specified. The bottom line is not whether biological oscillators can be found; cells are littered with them. Nor is the important issue whether oscillators can theoretically be coupled and have appropriate periodicities; they can. The real need was to determine which of the myriad possible oscillators has evolved as the circadian clock of organisms. Several suggestions have been made.

Three different approaches were used to elucidate the molecular mechanism of circadian clocks[15,23,33,47,69,83]

THREE MAIN QUESTIONS CHALLENGING CHRONOBIOLOGISTS. In reducing the problem of circadian rhythmicity to its simplest form, three foci are evident. These are three long-standing questions of chronobiology. The first asks how a clock works. It addresses the nature of the oscillator or ensemble of tightly coupled oscillators that give rise to a circadian system with its associated properties of stability, robustness, temperature compensation, resettability, and a circa-24 h period length. The second focus concerns input to the core oscillator or oscillatory system and asks how entrainment works. The third asks how the capacity for rhythmicity controls the physiology and behavior of rhythmic cells and eventually the

organism. Because biochemically traceable regulatory relationships must clearly exist among these three aspects of rhythms, three approaches developed for discovering the molecular basis of circadian rhythmicity.

FOLLOWING THE INPUT PATHWAY TO THE CLOCK. Because light is a universal entraining agent, an initially promising route was to determine the nature of the photoreceptor through the use of action spectra for clock resetting, then to isolate the photoreceptor, and finally to follow the regulatory pathway into the oscillator. A variation of this rationale used inhibitors that could block light input without affecting the clock rhythm itself. The possibility was great that different photoreceptors and second messengers might function in different systems. If core mechanisms were conserved, however, then common elements might emerge as the entrainment pathway approached the oscillator. For example, cyclic nucleotides, calcium, and membrane depolarization are on the input pathway in molluscan eye preparations like *Bulla* and *Aplysia* but are involved in output in the chick pineal circadian system (see Chapter 6). However, protein synthesis inhibitors can block light-induced phase shifting in all three systems and in *Neurospora*. A virtue of this approach is that a lot was learned on the journey even if the core was not reached.

TRACKING THE REGULATION OF A RHYTHMIC OUTPUT BACK TO THE CLOCK. A second approach employed the logic of following a regulatory pathway all the way from a rhythmic process back to the clock core. This approach was widely used in many systems, beginning with the study of bioluminescence in *Gonyaulax*. Further studies quickly utilized leaf movements in plants, development in fungi, action potentials in molluscan eyes, and melatonin production in vertebrates. A central clock component in plants, CCA1, was discovered in this way. Analysis of a rhythm in photosynthesis resulted in identification of a rhythmically expressed gene encoding a pigment-binding protein (LHCB). Circadian expression of LHCB is regulated directly by binding of the CCA1 protein to the *lhcb* promoter. Whereas the CCA1 research was a notable success, most other experiments merely described clock regulation of downstream activities (see Chapter 8).

PERTURBING THE OSCILLATION BY PHARMACOLOGICAL MANIPULATION OR GENETIC LESION. The rationale for inhibitor studies is that brief inhibition of critical processes should result in a phase shift of the rhythm, as seen in Figure 7.3. The pharmacological approach was used to examine the involvement of protein synthesis and lipid membranes in the oscillator. It was also applied to the study of chick pineal and molluscan eye preparations (see Chapter 6). Several notable and almost universal results ensued. Short-term inhibition of RNA or protein synthesis reset the clock. Chronic high doses stopped the clock at a unique phase, and chronic long-term partial inhibition changed the period length of the clock. The systems included *Euglena*, *Gonyaulax*, higher plants, fungi, molluscs, birds, and mammals. The results were important, starting in the late 1970s, for they showed the critical nature of periodic translation and transcription. By extension, the results implied the sharing of common elements in some aspects of core oscillators. However, interpretation problems arose with these data. Artifacts associated with unintended targets of inhibitors or unknown cross-regulatory relationships are well-known issues.

Similar in rationale to the use of chemicals to inhibit a function is the use of genetic lesions to inhibit a function by reducing or eliminating the activity of the protein that executes that function. In the twenty-first century a common way to dissect a complex biological problem is the sequential use of the techniques of genetics, molecular genetics, biochemistry, and finally cell biology. At present such an approach seems intuitive, but in 1960 it was not possible. Many believed that attempts to genetically lesion the clock would simply result in a dead organism because clocks would be essential to life. Later, as clock-mutant organisms appeared, genetic approaches were viewed with equal skepticism as being prone to pleiotropic artifact. For instance, the period length of the rhythm in compound action potentials in the isolated *Aplysia* eye could be changed by several hours simply by alteration of the carbon level and ionic content of the bathing medium. An interpretation of these data was that general changes in a cell's metabolism could pleiotropically affect the clock. If this was the case, then perhaps single-gene mutations affecting period length shared a similar pleiotropic origin by simply altering the intracellular milieu. This would mean that clock mutations might never reveal clock components. Ultimately the only answer to such concerns was to analyze the genes, but first some of the intellectual output of these approaches will be considered in the various models for the clock that were considered from the 1960s to 1980s.

Development of Circadian Oscillator Models in the Pre-Molecular Genetics Era Was a Mirror of the Times

A sampling of early models reveals diverse attempts to explain circadian characteristics in biochemical terms[18,20,22,52]

OVERVIEW. A variety of suggestions were made in the 1950s and 1960s as to how "the clock" might work. These suggestions accompanied the rise of biochemistry, the elucidation of metabolic pathways and the importance of feedback in their regulation, and still later the

dawn of molecular biology. All the early models shared common features, including a grounding in the current understanding of cellular metabolism to provide a real or plausible feedback loop with a delay, a means to get the circa-24 h time constant, and a way to explain light resetting and temperature compensation. All the models centered on a solid piece of modeling, or a kernel of circadian physiology such as a rhythmic function of biochemical activities known to be causally interrelated, or an effect of an inhibitor. Then the models drew in conjecture and circumstantial evidence from a broad group of organisms to develop a plausible model.

THE CHRONON MODEL. One of the first and most widely known models was the chronon model (Figure 7.4), which theorized, in every cell capable of circadian rhythmicity, hundreds of genes whose purpose was in timekeeping. Together, all of these genes made up the chronon. The first gene was transcribed, the mRNA translated, and the protein product seen as helping to initiate transcription of the next element in the chronon, with the last element initiating the first one to restart the cycle. The long time constant arose from the length of time required to transcribe the hundreds to thousands of genes in the chronon, to translate their proteins, and to have them move through the cell. Temperature compensation was derived from the temperature-independent rate of diffusion of the mRNA to the cytoplasm and the protein back into the nucleus; we now understand that mRNA is actively exported. Moreover, each time-specific transcript and protein from the chronon provided a ready means for determining changes in gene expression and metabolism that were specific to the time of day.

A long and creative series of models was advanced on the basis of known biochemical feedback loops (see Figure 7.2). Much of the discussion here concerning metabolic coupling of oscillators arose in the course of the development of models for the clock based on such coupling and cross-coupling of intracellular regulatory networks. In some cases the cross-coupling involved ultradian oscillators such as the glycolytic oscillator, mitochondrial calcium fluxes and energy metabolism, and other aspects of regulatory biology. Another model, or set of models, centered on cAMP (cyclic adenosine monophosphate), which would allosterically inhibit adenylate cyclase and activate phosphodiesterase. Through the appropriate choice of parameters in this cAMP feedback loop, a modeler could derive a limit-cycle oscillator with about the right cycle time. Furthermore, the oscillator could be stabilized by mutual coupling with other oscillators within a tissue. Such models drew support from experiments showing circadian period and phase effects resulting from the treatment of cells with inhibitors, such as inhibitors of cAMP phosphodiesterases. Another observation was that in many slow-growing cells and tissues, the cell cycle is

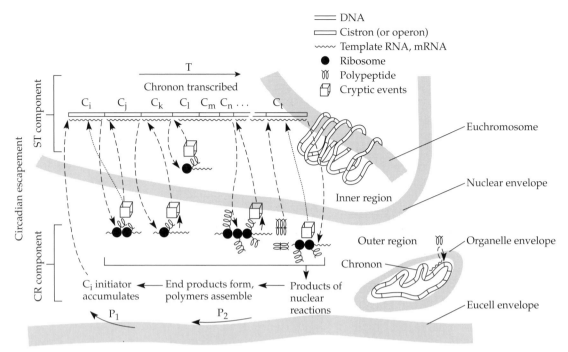

FIGURE 7.4 The chronon model. The chronon model emerged before distinctions were understood between prokaryotic and eukaryotic gene expression. The chronon is a very long piece of DNA that includes many genes, shown as C_i, C_j, C_k, and so on. In the chronon model the circadian cycle consisted of pretranscriptional (P_1), transcriptional (T), and posttranscriptional (P_2) phases. The sequential transcription (ST) component is where each gene in the chronon is transcribed, one by one. Each transcript then diffuses away, is bound by ribosomes and translated into protein. This is the cytoplasmic ribosome (CR) component. The protein diffuses back to the chronon to initiate transcription of the next gene in the chronon. The last gene restarts the cycle. Time of day–specific output through regulation of other genes not in the chronon is readily envisioned. (After Ehret and Truco, 1967.)

gated by a circadian clock. This concept emerged later as a model in which the cell cycle was used as an oscillator. The scientific bases of these models can generally now be traced to genuinely rhythmic clock-controlled output or, in the case of the inhibitors, to pleiotropic effects.

MEMBRANE MODELS. Another early and influential set of models consisted of the membrane models, an example of which is shown in Figure 7.5. The membrane models arose shortly after biological membranes and membrane proteins were "rediscovered" and the fluid mosaic concept of the membrane was articulated in the early 1970s. They were chiefly relaxation oscillators (analogous to pipette washers) in which transport of ions across a membrane was regulated in a such a way that the concentration of ions on one side of a membrane would influence the activity and interaction among the transporters that facilitated their movement to the other side of the membrane.

This concept of a membrane transport-based feedback loop allowed description of a limit-cycle oscillator,

perhaps the first explicit biochemical model using one. Light resetting worked by a light-activated ion channel, the long time constant resulted from the slow rate of pumping, and temperature compensation arose from the known capacity of cells to maintain relatively constant fluidity of their membranes. Thus rates of diffusion of transporters within the fluid mosaic of the membrane would be more or less constant. Even though these models did not provide the description for how circadian oscillators work, the membrane models spawned an enormous amount of interest and research.

All the early models were at some level plausible, but ultimately none were compelling[19]

A common feature shared by these models was that they all started with an explanation of how the clock worked. From there the literature was scanned for consistent data, and eventually experiments were derived that could test the model. In this sense, then, the models were all educated guesses at the final answer, and although none were correct, their rise and fall paralleled the times and are a good mirror on the field as it developed. L. N. Edmunds's book *Cellular and Molecular Bases of Biological Clocks*, which appeared just after the first clock genes, *per* and *frq*, were cloned but before molecular insights into clock mechanisms arose from these efforts, is a good source for detailed discussions of early

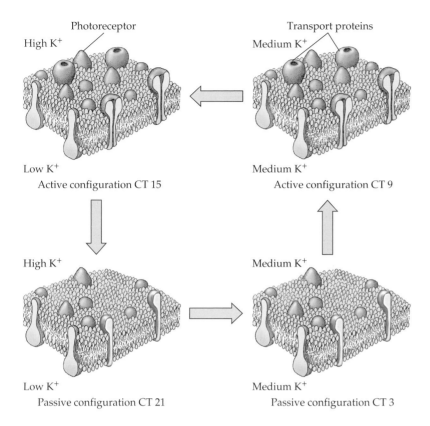

High K⁺   Photoreceptor

Low K⁺
Active configuration CT 15

Medium K⁺   Transport proteins

Medium K⁺
Active configuration CT 9

High K⁺

Low K⁺
Passive configuration CT 21

Medium K⁺

Medium K⁺
Passive configuration CT 3

FIGURE 7.5 A membrane model for the circadian clock. In the fluid mosaic model of biological membranes, proteins were envisioned as floating in a layer of lipid the consistency of olive oil, the fluidity of which was maintained more or less constant. Changes in protein arrangements and associations could cause and affect changes in the direction and activity of ion transporters, thereby affecting ion distribution with respect to the membrane as shown. The photoreceptor was envisioned as a light-regulated potassium channel. In the membrane model, active stages of ion pumping alternated with passive stages of diffusion, as shown, at different circadian times of day. (From Njus et al., 1974.)

clock models and biochemical/pharmacological analyses of the oscillator in the pre-molecular biology era.

A Brief Primer on Genetics, Biochemistry, and Molecular Biology May Help Readers' Understanding of the Molecular Basis of Circadian Rhythmicity[1,44]

Modern molecular biology reflects the fusion of two independent fields: genetics and biochemistry

Questions about how biological clocks work have been around for years, but the ways in which the questions were pursued were fundamentally influenced by the prevailing understanding of how cells work. The second half of the twentieth century was the time during which chronobiology solidified and emerged as a scientific field. That period coincided with the Golden Age

of Biology, an era continuing even now. Two centuries from now, the great flowering of science in the Golden Age will probably be remembered best as the period during which the molecular basis of life came to be understood.

Molecular biology represents the union of aspects of genetics and biochemistry. The impetus for the fusion was the discovery of DNA as the genetic material, as well as the recombinant DNA revolution. To see how circadian oscillators work, the reader must know something about how cells work from a molecular genetics point of view. The brief primer that follows provides the framework of key ideas to help students get started. Keep in mind that entire textbooks cover just single aspects of this topic.

Classical genetics provides a way to organize the observation of heritable characteristics and to track them over generations

CLASSICAL GENETICS. A gene can be defined operationally as the basic unit of inheritance. All life is based on proteins, which carry out biochemical functions, and in all cases the information to encode these proteins is carried in the genes. Thus all organisms have genes, and the science that studies the structure, function, organization, control and transmission of this information from generation to generation is genetics. Single genes are the units of inheritance and are now understood to encode single proteins that execute biochemical functions; this concept is known as "one gene, one polypeptide."

Genes are named, generally, for the phenotype conferred by an allele (a variant) of the gene, or for what the gene encodes. For example, the *gpd* gene encodes the glycolytic enzyme glyceraldehyde-3-phosphate dehydrogenase, and *frq* (short for *frequency*) encodes a protein in the *Neurospora* clockwork that, when mutated, changes the period length of the circadian clock. Gene names are always written in italics but the protein products of genes are never italicized. However, the rules for capitalization and punctuation of gene names, if any, are specific to individual genetic systems—that is, specific to organisms. This can admittedly be confusing for first-time readers, but it is also a level of detail that is not important for the material presented here. Gene and protein names in this book are presented as they appear in the primary literature.

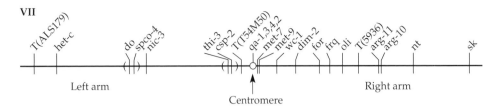

FIGURE 7.6 A stripped-down schematic of the genetic map of linkage group VII from *Neurospora*. The centromere is in the middle, with genes located on the left and right arms with respect to each other. Genes located 5 mm apart on this map will recombine with each other about 6% of the time in every generation. (From Perkins et al., 2001.)

GENETIC MAPPING. Single genes rarely if ever exist by themselves, but rather are found strung together like beads on a string. The strings of genes are called chromosomes, and the same genes always lie in the same place on a particular chromosome (Figure 7.6). Some organisms, including most bacteria, contain a single chromosome, but most eukaryotes contain many chromosomes; humans have 46 chromosomes. Each chromosome is distinct and has its own complement of genes. In the aggregate, all of the genes in the organism are known as the genome.

At some stage during the life cycle of nearly every organism, cells exist that have a single copy of all the genes—that is, a haploid complement of the genes. Bacteria and fungi can spend most of their lives this way as haploid organisms, but in mammals the only haploid stage is the gamete: egg or sperm. When the haploid cells fuse to make a diploid cell (which contains two copies of each chromosome and each gene), blocks of genes lying at the same locations on homologous chromosomes have the possibility of trading places. This process is called recombination. For instance, a diploid cell has two copies of Chromosome 1, each of which contains the same genes, although they may differ in the alleles for some of these genes. As a result of recombination, some of the genes at the end of the right arm of this chromosome could, as a unit, trade places with the same genes at the end of the right arm of the other Chromosome 1. Recombination between homologous chromosomes almost always happens once per generation, with the result that although individual genes last a very long time, individual chromosomes are recombined every generation.

Recombination happens with a finite and predictable frequency, in the same way that you might predict gas station stops on a trip from Boston to Chicago. One cannot know where the stop will happen, but it will happen somewhere, and once it happens somewhere, it won't happen again for a while. In terms of genetic recombination, the result is that genes that lie close together on a chromosome tend to stay together from generation to generation; that is, they are tightly linked. More generally, genes can be mapped along chromo-

somes with respect to their neighbors by the frequency with which they recombine. Genes close together don't recombine very often; genes far apart recombine with greater frequency. Thus, recombination frequency is a measure of how far apart two genes are; in this way the map position of one gene can be determined with respect to other genes on a chromosome. Genes on chromosomes are physically linked, so they tend to stay together (forming a genetic linkage group), and separate chromosomes independently assort each generation. Genetic map location (chromosomal position) and the characteristics (phenotype) that alleles confer on an organism uniquely define a gene. In this way, in clock genetics for instance, one can distinguish (by map location) two different genes, each of which, when mutated, results in a clock that has a period length 3 h longer than normal (the same phenotype). This information would indicate that, in this case, at least two different genes contribute to the determination of period length.

The structure of DNA provided a molecular basis for classical genetics

DNA AND ITS ROLE IN HEREDITY. In the 1930s the connection was made between chromosomes (which can be seen in the microscope) and linkage groups, the groups of genes that tend to stay linked together in a genetic cross. Chromosomes, of course, are biochemically complex things that, when isolated, include lots of proteins, carbohydrates, and nucleic acids, so it was not until the 1940s and early 1950s that the genetic material was confirmed to be DNA.

This nucleic acid polymer had at one point been assumed to be just a boring random series of four nitrogenous bases (adenine, guanine, thymine, and cytosine) linked together by deoxyribose sugars and phosphate groups. By 1950, Chargaff, a biochemist, had determined that the base composition is not wholly random in that the amount of adenine (A) always equals the amount of thymine (T) (and guanine equals cytosine). In 1953, Watson and Crick published the structure of DNA as a right-handed double helix of uniform diameter in which the two strands of DNA wrapped around each other but ran in opposite directions (Figure 7.7). This antiparallel double helix was held together on its inside by hydrogen bonding of the bases such that A bonded exclusively with T and G exclusively with C, thus providing a structural basis for Chargaff's observations.

The bases on the inside were linked together on the outside by the structural backbone of the helix, which con-

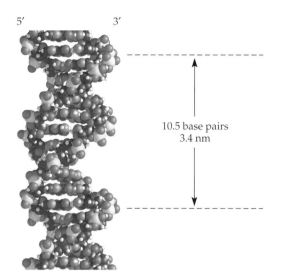

FIGURE 7.7 Space-filling model of the DNA double helix, with bases on the inside and the sugar–phosphate backbone on the outside. The helix is about 2 nm in diameter, and each helix makes one turn around itself every 10.5 base pairs, a distance of about 3.4 nm. Because individual DNA strands run antiparallel, each end of the double helix has both a 3' hydroxyl and a 5' phosphate.

sisted of alternating deoxyribose sugars and phosphates, with the bases attached to the sugars. The phosphates are attached to the 5' position of the sugar, so the DNA backbone can be described with a polarity and ends: a 5' phosphate end and a 3' hydroxyl end. The structure also provided a nice model, later confirmed, for DNA replication: The two strands could be separated, and each would provide a template, via hydrogen bonding, for the assembly of complementary helices. This process is known as semiconservative replication because after replication, the one strand of the double helix that served as the template is from the parent, and the other strand is newly synthesized from individual bases (Figure 7.8).

The cells are now known to have specific enzymes (helicases to unwind the double helix and DNA-dependent DNA polymerases to form the new polymer) whose role it is to facilitate this DNA replication and faithfully preserve the base pairing. Polymerases form DNA by adding preformed units of sugar + phosphate + base (nucleotides) to the 3' end of the polymer. All they require to do this are the four different nucleotide base substrates (ATP, GTP, TTP, and CTP); a single-stranded DNA template; a primer, which is a short piece of DNA whose base sequence is complementary to the template such that the two can hydrogen-bond together; and a buffer of appropriate salts and pH to allow the polymerase to work.

PRACTICAL APPLICATIONS OF DNA REPLICATION. Two widely used applications of this understanding of DNA

replication should be mentioned in passing. One is the ability to determine the sequence of bases along the DNA chain and therefore the information (proteins) encoded therein. DNA sequencing relies on (1) the ability to separate, by electrophoresis, DNA chains of length n from $n + 1$, where $n \lesssim 600$; (2) the ability to make DNA in a test tube using polymerase and the reagents listed above; and (3) the availability of nucleotide analogs that do not contain a 3' hydroxyl (ddATP, ddGTP, ddTTP, ddCTP). When these analogs are incorporated by polymerase into the growing chain, the chain can grow no longer because the necessary 3' end is missing. Thus in a reaction containing

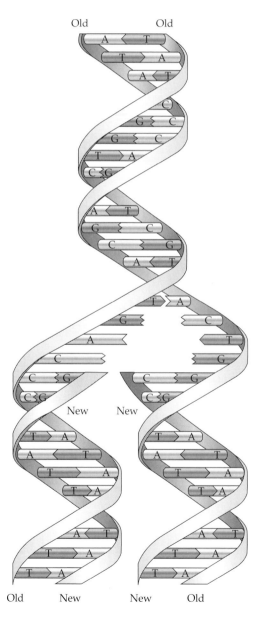

FIGURE 7.8 DNA is replicated in a semiconservative manner, such that each daughter double helix contains one old and one new DNA strand.

a small percentage of ddCTP along with the normal CTP, the chain can sometimes be terminated each time the polymerase needs to add a C to it (the fraction of terminations versus extensions reflects the ratio of normal CTP to ddCTP). Because the longest chains grow to 600 or more nucleotides, the reaction mix will contain a series of shorter chains also, each terminated at a point where a ddC was added in place of a normal C. When this mixture is sized by electrophoresis, the position of each C in the chain can be deduced by the size of the terminated fragment. By extension, the same can be done for A, T, and G, using ddATP, ddTTP, and ddGTP, respectively. The start of the sequence is determined by the primer used. By breaking the DNA from an entire genome up into lots of small pieces and sequencing enough of the small pieces, investigators can determine the DNA sequence of all the genes in a genome.

A second common application of in vitro DNA replication is the polymerase chain reaction (PCR), which takes advantage of the semiconservative manner in which DNA is replicated and the fact that, when heated, the two strands of DNA will separate (Figure 7.9). The PCR mixture includes some template DNA containing the region to be amplified, a heat-stable DNA polymerase, buffer and nucleotide substrates, and high concentrations of two short DNA primers (about 20 nucleotides in length) complementary to the template. The positions at which these primers hybridize to the template determine the boundaries of the region that will be amplified.

Each PCR cycle begins with the mixture being heated to a point where the template DNA strands separate.

On cooling, the complementary DNAs will reassociate in their usual hydrogen bond–driven sequence-specific manner, but because the primers are present at high concentration, each template will bond to a primer rather than to its old full-length template partner. Once the primer hydrogen has bonded to the template, the polymerase will act to make new DNA complementary to the template, lengthening the chain from the 3' end of the primer; in a few minutes up to a few thousand bases are added. The next cycle is initiated again by heating of the mixture, which separates the newly synthesized DNA from its template. Upon cooling, this newly synthesized DNA can itself be hybridized to a primer and act as a template for another round of synthesis. In this way each cycle from here on produces twice as much DNA, so 30 cycles produce up to 2^{30} copies, about a billionfold amplification.

Gene expression involves sequential transcription and translation

Genetic information is contained within the nucleus of a cell and is organized on chromosomes, which are long DNA strands (Figure 7.10). The Central Dogma of molecular biology describes the flow of genetic information within cells, stating that DNA gives rise to RNA, through a process now known as transcription, and RNA provides a template for the synthesis of proteins, through a process known as translation (Figure 7.11). The Dogma was reportedly so named by Crick in the early days of molecular biology, when none of this had yet been proven. Crick, who had been raised as an Anglican, reportedly quipped that everyone knew a dogma was something in which all true believers believed, but for which there was not a shred of evidence.

In the ensuing 40 years, the evidence has become rather solid; the sequence of events described by the Central Dogma is, in fact, how information gets from DNA into the rest of the cell to make things happen. In all living things, transcription occurs when an enzyme (DNA-dependent RNA polymerase, or RNA polymerase for short) binds to DNA at the beginning of a gene, within the region known as the promoter, and uses the DNA to make an RNA copy of the DNA base sequence. RNA is like DNA, except that uracil takes the place of thymine and ribose takes the place of deoxyribose. The details from here on, however, are slightly different in prokaryotes and in eukaryotes.

The structure of a gene and the means through which it is expressed differ in prokaryotes and eukaryotes

PROKARYOTIC GENE STRUCTURE AND EXPRESSION. The gene of a prokaryote such as a cyanobacterium is rela-

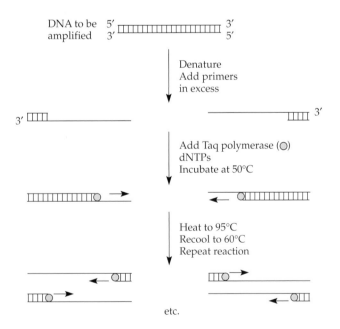

FIGURE 7.9 Schematic of gene amplification by the polymerase chain reaction. dNTPs refer to a mix of all four deoxynucleotide triphosphates.

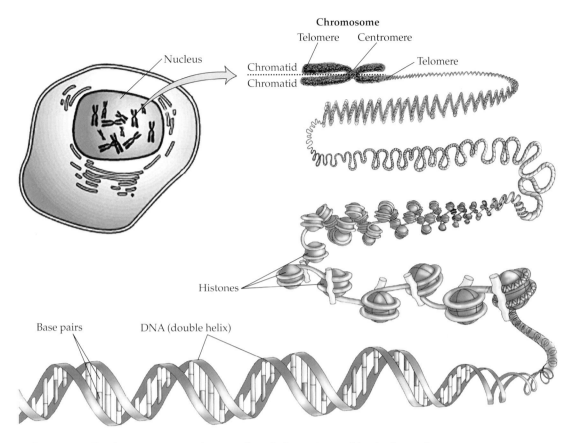

FIGURE 7.10 Putting gene expression together. Information resides in the nucleus in DNA, which is packaged into chromosomes.

tively simple (Figure 7.12). RNA polymerase can recognize and bind to a specific sequence of bases found at the beginning of all prokaryotic genes. This binding is sufficient to initiate gene expression through the synthesis of

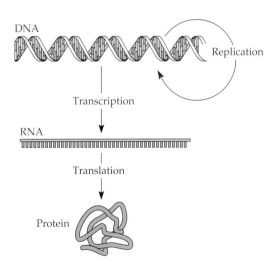

FIGURE 7.11 The Central Dogma of molecular biology. DNA contains genetic information. It replicates to make copies of itself. Information is read out initially by transcription to make RNA, which is then translated to make protein.

an RNA that is complementary to the gene. This RNA transcript is called a messenger RNA or mRNA. Like DNA, RNA is made by extension from the 3′ end of the RNA, so a transcript is made from the 5′ end (the beginning) to the 3′ end (the end). However, the binding of polymerase to the promoter is often not tight, with the result that little RNA is made. Other cell proteins can act as transcriptional activators to help polymerase bind, or as transcriptional repressors to inhibit polymerase from binding. As a first approximation, subject to some additional constraints and regulation, the more RNA there is, the more protein will be made. In prokaryotes, a single promoter can initiate transcription of an mRNA that can encode one or several proteins as will be described here. Transcription continues through the movement of the polymerase along the DNA until a transcriptional termination sequence is encountered and the polymerase exits the DNA helix.

Messenger RNA is translated into protein in a very large molecular machine called the ribosome. The two subunits of the ribosome bind the mRNA starting near the 5′ end of the mRNA and move along it as a unit until the first AUG sequence—that is, an adenine followed by a uracil followed by a guanine—is encountered. This sequence of bases is the signal to begin protein synthesis

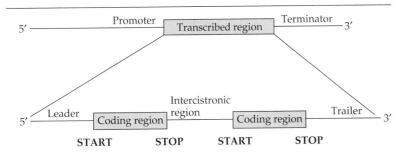

FIGURE 7.12 Basic structure of a polycistronic region containing two genes in a prokaryote. The promoter is where RNA polymerase binds to DNA to make a transcript, which in this case, contains two protein-coding regions separated by an intercistronic region.

using the amino acid methionine; all proteins begin with methionine. Ribosomes scan from the 5′ end of the mRNA, so this end will yield the beginning of the protein. Because amino acids are polymerized through the creation of an amide bond between the COO^- terminating the existing protein chain and the NH_3^+ of the incoming amino acid residue, the start of a protein is the N terminus, arising from the 5′ portion of the mRNA, and the end of a protein chain is the C terminus, arising from a more 3′ portion of the mRNA. From here the sequence of bases on the mRNA is decoded as sequential groups of three bases are read according to the genetic code, in which triplet codons designate specific amino acids (Figure 7.13). These amino acids are linked in sequence into the growing protein bound on the ribosome.

The genetic code is redundant in that more than one distinct triplet codon can encode the same amino acid, but it is not ambiguous; that is, each triplet can be decoded in only one way. Amino acids are added by the ribosome until a STOP codon (UAA, UAG, or UGA) is encountered, at which point the protein is released from the ribosome. The ribosome, however, can remain bound to the mRNA and will move along for a while, and if a second AUG is encountered, the whole process of translation can begin again. The region of transcript between the starting AUG and the STOP codon that the ribosome reads and decodes using the genetic code is called an open reading frame. A series of protein-coding regions under the control of a single promoter is called an operon, and mRNAs that encode more than one protein are called polycistronic mRNAs.

Proteins assume their three-dimensional shape as they are synthesized, in responses to normal physical forces, and generally they are active once they have been released, although in some cases additional processing (such as proteolysis) is required to activate a protein. Activity can be further modified in prokaryotes through posttranslational covalent modifications such as phosphorylation of serine or threonine residues. Phosphorylation introduces a negative charge (from the phosphate) where formerly there was only an uncharged hydroxyl group on the amino acid, and the result can be a change in the structure of the protein, yielding increased or decreased activity, qualitatively changed activities, or altered ability to interact with other proteins. Phosphate groups are added by kinases and removed by phosphatases, and they can be added or removed within minutes in response to a stimulus. Proteins are destroyed by proteases that sever the peptide bonds between amino acids, releasing the individual amino acids for incorporation into other proteins or further metabolism.

EUKARYOTIC STRUCTURE AND EXPRESSION. Several salient differences between eukaryotes and prokaryotes add to the complexity of eukaryotic gene expression (Figure 7.14). First is the fact that eukaryotic cells have more complex internal structures, including a series of membrane-enclosed internal organelles between which proteins must traffic. Second, they generally contain much more DNA than prokaryotes have. Except

Second position in the triplet codon

		U	C	A	G	
First position in the triplet codon	**U**	Phenylalanine	Serine	Tyrosine	Cysteine	U
						C
		Leucine		STOP	STOP	A
					Tryptophan	G
	C	Leucine	Proline	Histidine	Arginine	U
						C
				Glutamine		A
						G
	A	Isoleucine	Threonine	Asparagine	Serine	U
						C
		Methionine and START		Lysine	Arginine	A
						G
	G	Valine	Alanine	Aspartic acid	Glycine	U
						C
				Glutamic acid		A
						G

Third position in the triplet codon

FIGURE 7.13 The genetic code. Information contained in triplet codons in RNA is translated into the language of proteins to specify a single amino acid or to stop translation.

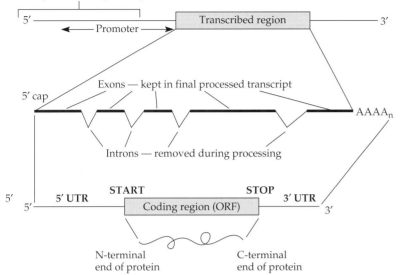

FIGURE 7.14 Basic structure of a gene in a eukaryote. The promoter is where RNA polymerase binds to DNA to make a transcript, an event that requires assistance from dedicated general transcription factors and can be helped or hindered by other transcription factors. The poly-A ($AAAA_n$) tail is added when the end is clipped off, and the primary transcript is processed by removal of introns and shipped to the cytoplasm, where it is translated.

for small amounts in the mitochondrion and chloroplast, all of the DNA is contained in the nucleus. Thus information transfer begins in the nucleus, and in eukaryotes the nuclear process of transcription is physically separated from the cytoplasmic process of translation.

Eukaryotes have three RNA polymerases, but only one is involved in the synthesis of mRNA to be used for making proteins. Eukaryotes don't contain operons or produce polycistronic mRNAs; instead each promoter drives and regulates expression of just one protein-coding region. Unlike the case in prokaryotes, RNA polymerase itself cannot initiate transcription, but instead relies on other proteins in a large complex, the basal transcription apparatus, to prepare a "docking site" on a promoter that allows it to bind and to begin transcription. The assembly of this complex is a major point of regulation of gene expression. This basal complex can recognize a promoter and bind in a rudimentary way, but generally it does so poorly unless helped by still other proteins with high affinity for specific sites in the promoter region. These are the dedicated transcription factors that help assemble the complex with the polymerase on the promoter. As in prokaryotes, there are activators and repressors, and different transcription factors often act together on a promoter in a combinatorial manner to affect gene expression.

Once transcription has been initiated, the primary transcript is made, and it differs in several ways from those in prokaryotes. Each of the following steps is subject to independent regulation:

1. Each primary transcript is capped with a specific structure that will allow export from the nucleus and subsequent translation.

2. Each primary transcript contains regions (introns) that are not used to encode proteins and that must be spliced out to yield the final juxtaposed segments (exons) that make up the processed transcript that is ready for translation. Because the removal of each intron is a separate event, alternate splicing may yield different processed transcripts from a single gene, each capable of directing the synthesis of different but partially overlapping proteins.

3. Whereas in prokaryotes transcriptional terminators signal for polymerase to fall off the template, in eukaryotes transcription continues indefinitely, but the ends of transcripts are determined by another processing event, in which the primary transcript is cut and a series of adenines are polymerized onto the 3′ ends of their transcripts. Thus eukaryotic mRNAs can be distinguished from other RNAs by the 3′ poly-A tails.

4. Final processed transcripts can contain long untranslated regions before the AUG and after the STOP codon. Proteins exist that can bind these untranslated regions (the 5′ and 3′ untranslated regions) and confer regulation on the stability or translation of the mRNAs.

5. Only fully processed transcripts are exported from the nucleus.

Once an mRNA has been exported from the nucleus, it can be bound by ribosomes and translated. The initiation of this process is subject to regulation by proteins that can bind to the 5′ and 3′ untranslated regions. Once translation begins, the nascent protein will begin to assume a shape, and both this shape and the sequence of amino acids in the first part of the protein can carry regulatory information directing the ribosome to halt translation and go to the endoplasmic reticulum prior to resuming synthesis. Proteins translated there are destined for insertion into biological membranes or for export from the cell. If translation continues to the STOP codon, the resulting protein can carry sequences of amino acids that direct transport of the protein back into the nucleus, or into various other cellular compartments. Because this information is contained within the amino acid sequence, it is genetically determined.

Once translation has been completed, proteins are subject to further regulation via covalent modification, and the repertoire of such modification is larger than in prokaryotes. The means by which eukaryotic proteins may be modified include phosphorylation, methylation, and acetylation, all of which can change the local electric charge on the surface of the protein and modify its shape, activities, or interactions with other proteins; glycosylation, the additions of sugars that can influence intracellular targeting and recognition by other proteins; and proteolysis, either to destroy the protein or to modify it into an active form. A common covalent modification of proteins in eukaryotes is the attachment of a short protein called ubiquitin. Ubiquitination marks a protein for transport to and degradation in the proteasome, a huge protein complex whose function is to rapidly degrade proteins.

Proteins are the ultimate products of most genes, and changes in proteins are the results of most mutations[1,70]

PROTEINS AND PROTEIN DOMAINS. Proteins fold and assume shapes that reflect their functions and dictate their activities. They are in a real way small molecular machines, and they often act as if they had parts devoted to specific functions. Such parts of proteins are known as domains. For instance, a protein that acts as a transcriptional activator will have a part (a domain) whose role is to recognize and specifically bind to a certain sequence of bases in DNA, and another domain whose role is to interact with RNA polymerase to promote its activity. Often proteins work together in specific complexes to carry out certain tasks, in which case domains of each protein in the complex are designed to foster the protein–protein interactions that are required to keep the complex together.

In nature, often a variety of different domains can execute the same function; for instance, proteins use at least half a dozen different domains to bind to DNA, several unrelated domains to interact with RNA polymerase, and a variety of domains to facilitate protein–protein interactions. One of the latter, a particular domain that promotes interactions between proteins, is the PAS domain. PAS domains are found in numerous proteins. In some groups they are also involved in sensing light, oxygen, or voltage changes. Some PAS domains also bind small molecules. Interestingly, PAS domain–containing proteins are involved in all eukaryotic circadian clocks, as will be described later.

Domains assume their structures as a result of the sequence of amino acids in that part of the protein. For this reason, the sequence of amino acids can often be used to predict the activity of a protein (by virtue of its domains). Because different domains frequently can be recruited to perform the same function, the conservation of sequences and domain structures among proteins from different organisms is often taken as strong evidence that the proteins evolved from a common ancestor and that they perform the same functions in both systems.

MUTATIONS, THE HERITABLE CHANGES IN GENES. Mutations are heritable changes in DNA; that is, they are permanent changes in the sequence of bases on the chromosome. In some cases mutations can be silent, as for instance, if a single base change were to occur within an intron that is always removed from a spliced transcript. In other cases a single base change can result in a major change in the activity of a protein. Mutations can be as simple as a change in one base to a different one, or the addition or deletion of a single base or multiple bases. Mutations may also be complicated; for example, a chromosome arm may be broken off and attached, with all of its attendant genes, to the end of a different chromosome. In all of these cases, the change in the sequence of bases can result in a change in the characteristics of the organism. Specifically, everything described in this short primer on gene expression, from the structure of individual proteins to the mechanics and regulation of their expression, is directly or indirectly coded within the DNA. As a result, all of these aspects of protein structure, activity, and regulation are subject to mutation.

The effects of some mutations can be very subtle: A change of a single base can result in the substitution of one amino acid for another, yielding a protein that is a little less stable at elevated temperature, so the mutant phenotype appears only when the organism is exposed to elevated temperatures. Alternatively, a single base change that introduced a STOP codon immediately after the AUG sequence would probably result in a complete loss of function of the protein. Mutations can arise spontaneously, or as a result of exposure to chemical mutagens or physical mutagens like UV light.

Regardless of the nature of the mutation, it is a change in the sequence of bases on a chromosome, so it is associated both with a phenotype and with a genetic map location, and therefore the DNA containing it can be isolated and the nature of the mutation described by determination of the DNA sequence. Understanding the nature of the mutation and the resulting phenotype often makes is possible to obtain information about the phenomenon being studied. For instance, if a mutation that gave rise to a long circadian period length were found to have altered a phosphorylation site in a protein, we could infer that phosphorylation of that protein was important in some way for determining period length.

Importantly, the use of genetics and mutations to explore the genes involved in a process need not assume anything about the process. All that is required in an organism is a phenotype to examine (like a circadian rhythm) and sex, the ability to segregate genes in genetic crosses and thus to map them. If these simple requirements are met, the organism will tell the investigator which genes are important for the phenomenon.

An Understanding of the Molecular Bases for Circadian Rhythmicity Is Emerging[16,65]

Common themes among circadian oscillators are evident[8,16,17]

NEGATIVE FEEDBACK LOOPS. One common property of an oscillation is its tendency, in a regular manner, to move away from equilibrium before returning. Negative feedback is one way to achieve this temporary departure from equilibrium. All that is needed is a process whose product feeds back to slow down the rate of the process itself (a negative element), and a delay in the execution of the feedback. Biological oscillators could be built from various different regulatory schemes, as reflected in the wide array of imaginative models described earlier that were proposed in the pre-molecular genetics era as bases for circadian rhythms. A metabolic pathway or an ion flux could in theory work as well as transcription and/or translation to achieve negative feedback. Delays could result from hysteresis, a slowness of response yielding an overshoot when equilibrium is being approached, or from a threshold phenomenon preventing immediate feedback, as in a relaxation oscillator such as a pipette washer. Yet another means of achieving delays is through nonlinearity in responses, as when multiple components must find each other and associate prior to executing feedback. A biological oscillator must also have a positive element, a source of excitation or activation that keeps the oscillator from winding down.

Finally, in theory of course, feedback loops that operate between cells via neuronal connections should work just as well as intracellular loops for building a clock. Such intercellular circuits are very common in the regulation of rhythmic behaviors such as swimming or feeding. Intriguingly however, all clocks have evolved to use very similar molecular solutions to the feedback and delay problem: All known circadian oscillators use feedback loops that close within cells (none require cell–cell interactions), and all rely on positive and negative elements in oscillators. In circadian clock loops, the transcription of clock genes is activated by positive elements and yields clock proteins that act as negative elements. These in turn block the action of positive element(s) whose role is to activate the clock gene(s) (Figure 7.15).

GENERALIZATIONS. In no organism do we understand enough about the general assembly of a circadian system to be able to seriously contemplate an in vitro, or even an in vivo, reconstruction of a circadian clock. However, work to date does allow some generalizations. First, the clock includes, probably at its core, molecular feedback loops that involve genes whose products are not essential to life. Instead the function of these genes is devoted solely to circadian timekeeping, in keeping with the evidence that clocks are adaptive but not essential to life (see Chapter 2). Second, many circadian systems will probably involve the interaction of multiple feedback loops, although at present only a few of these can be described in concrete terms. Third, the clocks in organisms among the eukaryotes that share a most recent common ancestor will share elements and similar components in the assembly and operation of their clocks. These groups are, in order of relatedness, the animals, fungi, and probably higher plants.

The capacity for circadian timekeeping probably arose more than once in evolution, but there is much to suggest that it arose only a few times. Within the eukaryotes studied, a steady stream of functional and sequence-based similarities have emerged as the molecular bases of circadian oscillatory systems have become understood (see Figure 7.15). This deliberately oversimplified model can serve as a reference point for understanding the overall organization of a feedback loop that forms a part of all known clocks.

A generic core circadian feedback loop can be described[16]

The schematic view of a core oscillator model that is presented in Figure 7.15 shows the common elements of the feedback loops that have been described so far, which are generally taken as one of the core oscillatory loops of circadian systems. One might interpret Figure 7.15 as implying that circadian oscillators will be only simple transcription and translation feedback loops, but they will probably not be this simple

The positive element in the loop is the transcriptional activation of one or more clock genes. In the cyanobacterium *Synechococcus*, one model predicts that clock genes will be activated by the KaiA gene product (however, see the discussion that follows for other interpretations). In eukaryotes such as animals and fungi, clock genes are apparently activated through the binding of

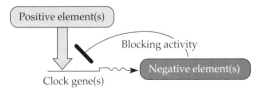

FIGURE 7.15 The basic elements in a circadian feedback loop. (From Dunlap, 1999.)

transcriptional activators, which are paired via interactions of PAS domains within the proteins. These positive elements bind to DNA on the clock gene promoters to activate clock gene transcription. Functionally analogous PAS domain–containing DNA-binding clock elements with similar overall protein sequences have been described in the three eukaryotic organisms—*Neurospora*, *Drosophila*, and mouse—whose clocks have been studied the most thoroughly at the molecular level. All three will be described here in detail. These positive elements drive transcription of clock genes that give rise to mRNAs whose translation generates clock proteins that provide the negative element in the feedback loop. Examples include the KaiC gene product in *Synechococcus*; FRQ in *Neurospora*; PER and TIM in flies; and apparently PER1, PER2, CRY1, and CRY2 in mammals.

The negative element in the loop feeds back to block activation of the clock genes, with the result that the amount of clock gene mRNA, and eventually the amount of clock protein, declines. As the loop cycles, it generates cyclical inhibition of transcription factors (the positive elements). The action of these positive elements on other clock-controlled genes suggests that time information from the oscillator might drive output by regulating target clock-controlled genes, as will be described more in Chapter 8. This robust daily cycling of clock gene mRNA, clock protein, and clock-controlled gene RNA and protein is characteristic of circadian systems. Evidence supporting this loop as a core of circadian oscillators lies in the internal consistency of the underlying genetics: Most genes that have been identified in cyanobacteria, *Neurospora, Drosophila*, mice, and plants as genes that affect circadian clocks can be fit into this framework.

The central importance of these loops is also supported by the fact that environmental effects on these components have been shown to underlie resetting of the clock cycle by light and temperature. Although not all of the details of all systems from cyanobacteria through fungi through humans have yet been described, many of the elements mentioned here are known in all of the systems that have been examined. The threads of similarity among all systems strongly suggest that this feedback loop reflects a common mechanistic core for most, if not all, lineages of circadian oscillators.

The Cyanobacterial Clock Depends on a Negative Feedback Loop Involving Transcription, Translation, and Protein–Protein Interactions[30,36,37,40,49]

Nearly everything known about the molecular basis of cyanobacterial clocks comes from studies on *Synechococcus elongatus*. In cyanobacteria, the clock can regulate a variety of processes, including cell division, amino

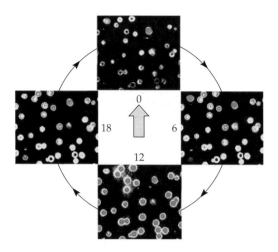

FIGURE 7.16 Colonies of the cyanobacterium *Synechococcus elongatus* containing bacterial luciferase genes driven by the clock-regulated promoter of the *psbAI* gene. The colonies are growing on a petri dish in the dark, and the same area was viewed at four times, each 6 h apart. The size of each colony in this figure reflects the amount of light produced during the time the plate was viewed, and the circadian cycling of light output is clear. Most of the colonies are wild-type colonies, but four different mutants are also shown. (Modified from an original illustration by T. Kondo.)

acid uptake, nitrogen fixation, photosynthesis, carbohydrate synthesis, and respiration. The particular utility of a timing system is manifested in the last two of these, in which the nitrogenase enzyme required for nitrogen fixation is poisoned by the oxygen evolved from photosynthesis. In fact, both processes are circadianly regulated, 180 degrees out of phase (see Chapter 8).

Mutations in genes that affect operation of the *Synechococcus* clock were obtained in a bioluminescence reporter screen in which a clock-regulated photosynthetic gene promoter (*psbAI*) was fused to bacterial luciferase and used to drive rhythmic bioluminescence (Figure 7.16). More than 50 mutants were identified, having period lengths ranging from 14 to 60 h. These genes have been cloned and shown to comprise a cluster of three genes known as the *kai* genes; *kai* is the Japanese word for "cycle" (Figure 7.17). Expression of the *kai* genes is driven from two promoters, one for *kaiA* and one for both *kaiB* and *kaiC*.

The functions of the *kai* genes can be inferred from their regulation and from the phenotypes of alleles. Deletion or overexpression of either *kaiA* or *kaiBC* results in arrhythmicity, but not in the same way. The KaiA protein is constitutively expressed, and loss-of-function mutations of *kaiA* result in very low-level arrhythmic expression from the *kaiBC* promoter. Overexpression of *kaiA* yields constant superelevated expression of *kaiBC* and, again, arrhythmicity. These data suggest that KaiA acts as an activator of transcription. KaiB and C are

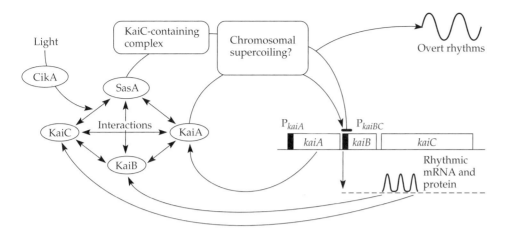

FIGURE 7.17 Interactions among elements in a model of the cyanobacterial circadian oscillator. As described in the text, the *kaiA* gene and the operon containing *kaiB* and *kaiC* are rhythmically transcribed and later translated. The resulting proteins combine with other proteins to form complexes, which may cause global changes in the degree of supercoiling and condensation of the bacterial chromosome. The end result is KaiC-mediated repression of the *kaiBC* operon, thereby constituting a negative feedback loop. P_{kaiA} and P_{kaiBC} refer to the promoters of the *kaiA* gene and the *kaiBC* operon. (After Mori and Johnson, 2001.)

rhythmically expressed with a peak at about circadian time (CT) 15, and three lines of evidence speak to their function. First, inactivation of either or both *kaiB* and *kaiC* yields arrhythmicity and altered expression from the *kaiBC* promoter. Second, overexpression of *kaiC* results in arrhythmicity and in severely dampened expression of *kaiBC* but not of *kaiA*. Third, pulsatile production of KaiC resets the oscillation. These data are consistent with a role for the KaiC gene product as a negative element in the oscillator; no role can currently be assigned to KaiB.

All of the Kai proteins interact, and mutations that alter these interactions affect the operation of the clock. For instance, the interaction between KaiA and KaiB is strengthened by KaiC. Also the strength of the KaiB–KaiC interaction oscillates with the time of day, and a mutation that acts to strengthen the KaiA–KaiB interaction lengthens the period of the clock.

Screens for mutations in genes encoding other proteins that interact with the Kai proteins have revealed additional clock elements, among them SasA (for "*Synechococcus* adaptive sensor"). SasA is a histidine protein kinase of the type usually associated with a DNA-binding, transcription-regulating "response regulator" in a bacterial two-component signaling pair. In such two-component systems, one component senses an environmental change and becomes activated; it then phosphorylates the second component, triggering a response. Loss of SasA does not block rhythmicity, but it does result in a tenfold reduction in *kaiBC* RNA oscillations, as if SasA worked as an amplifier. Transient overexpression of SasA will phase-shift the clock, suggesting a central role. Although the cognate response regulator that would pre-

sumably work with SasA has not yet been identified, this notion fits well with the known effects of KaiC on transcriptional regulation. The KaiC protein has an ATP/GTP nucleotide-binding activity, and KaiC phosphorylates itself as a part of its activity in the clock.

One of the most interesting aspects of the cyanobacterial clock is that it results in rhythmic transcription of all genes in the cell, suggesting a global regulatory mechanism (see Chapter 8). Similarities between the amino acid sequence of KaiC and bacterial helicases has suggested the possibility that KaiC could act as a helicase directly on the chromosomal DNA to change the degree of condensation. Helicases in general act by adding or removing twists in the DNA double helix, and in general the more tightly twisted the DNA is, the more difficult it is for DNA-dependent RNA polymerase to transcribe the genes. Thus, in this model KaiC-mediated circadian regulation of the degree of DNA coiling could globally influence gene expression.

Genetic screens are also beginning to suggest elements of the light input pathway through which the cyanobacterial clock is reset. Among possible input elements is CikA (for "circadian input kinase"), a protein that bears sequence similarity to bacteriophytochromes and histidine kinases. Loss of CikA shortens period, affects phases of some rhythms, and alters entrainment to dark pulses by reducing the phase response curve to one-fourth its normal amplitude, as well as making it antiphasic with the wild type.

Considering that a decade ago nothing was known about the molecular basis of cyanobacterial rhythms, it is clear that great progress has been made, even though much remains to be done. Many of the components in the feedback loop are still to be identified, and the governing principle of the loop is yet to be illuminated. Core regulatory themes, however, include circadian regulation of transcription and translation, and the importance of phosphorylation and protein–protein interactions, which may influence each other.

Generic Core Circadian Feedback Loops in Eukaryotes Involve Interlocking Positive and Negative Regulation of Gene Expression[2,16,17]

The first circadian feedback loops to be understood molecularly were those in the eukaryotic model systems *Neurospora* and *Drosophila*. These systems share some similarities and differences with each other and with cyanobacteria, but they serve well to introduce the additional complexity of eukaryotic clocks. As in the cyanobacteria, rhythmic expression of the transcripts and proteins corresponding to the negative elements is the rule. All known circadian oscillators involve negative regulation of transcription as a part of the loop, but in the cyanobacterial clock it was not clear how that negative regulation was produced. In eukaryotes this negative feedback is known to occur through action of the negative elements in blocking the transcriptional activating ability of the positive elements (Figure 7.18). All positive elements are heterodimers composed of two different proteins that interact via PAS domains. One of these two proteins is constitutively expressed, and the other is rhythmically expressed. In addition to common function and regulation, extended similarity is seen in the amino acid sequences of some of the positive-element proteins. Finally, in all known cases, one of the negative elements has a dual role in the cycle, in that it performs both the negative function of blocking the activity of the heterodimer and also a positive func-

tion in promoting the expression of one of the heterodimer partners. This dual action linking the negative and positive limbs of the cycle creates interlocking regulatory loops, as seen when FRQ promotes the expression of WC-1, PER/TIM the expression of CYC, and PER2 the expression of BMAL1. Circadian output (see Chapter 8) derives in part from the action of the clock-regulated positive elements on the expression of genes whose products do not feed back onto the central clock loop.

A Basic Eukaryotic Clock in *Neurospora* Uses Several Components to Execute Negative and Positive Feedback[16,45]

Both *Neurospora* and *Drosophila* are model systems in which the tools and paradigms necessary for the molecular dissection of circadian timing systems were developed. *Neurospora* normally grows as a syncytium in which many nuclei share a common cytoplasm in a hypha, an individual filament (see Plate 2, Photo G). When cultures are grown on a solid substrate, the *Neurospora* clock controls the pattern of asexual development in the region of the growing front. Aerial hyphae, which can develop into vegetative spores called conidia, arise from mycelia laid down in the late night through early morning. In contrast, mycelia laid down at other times of day do not generally develop further but instead remain largely undifferentiated (Figure 7.19). Although many molecular outputs are known and bioluminescence-based reporters have been developed for monitoring rhythmicity directly in liquid cultures, this rhythmic change in growth habit remains the most obvious manifestation of the *Neurospora* clock. The circadian nature of this developmental switch from surface to aerial growth and differentiation was noted by Pittendrigh over 40 years ago and was used by Feldman in the early 1970s to identify the first *Neurospora* clock-mutant strains, alleles of the *frequency* (*frq*) gene.

Essential components of a circadian oscillator feedback loop in *Neurospora* are known[4,10,11,42]

OVERVIEW. The *Neurospora* circadian oscillator includes interlocked positive and negative feedback loops based on transcription and translation, with a heterodimer of PAS domain–containing proteins acting as a transcriptional activator and a single gene encoding negative elements. Among the elements that constitute the oscillator are the WC-1 ("white collar-1") and WC-2 ("white collar-2") proteins, as well as both *frq* mRNA and the FRQ protein, and several kinases (Figure 7.20). WC-1 and WC-2 heterodimerize to form the white collar complex, or WCC. The interactions between these two different proteins are mediated by PAS domains in each

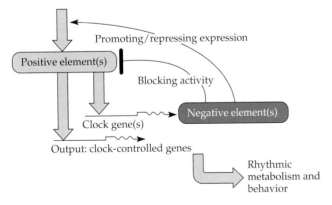

FIGURE 7.18 Features in common among the molecularly transcription/translation-based eukaryotic circadian oscillators. The positive elements are transcription factors that activate expression of the negative elements, proteins that act to block the activity of the positive elements while also promoting their expression. (From Dunlap, 1999.)

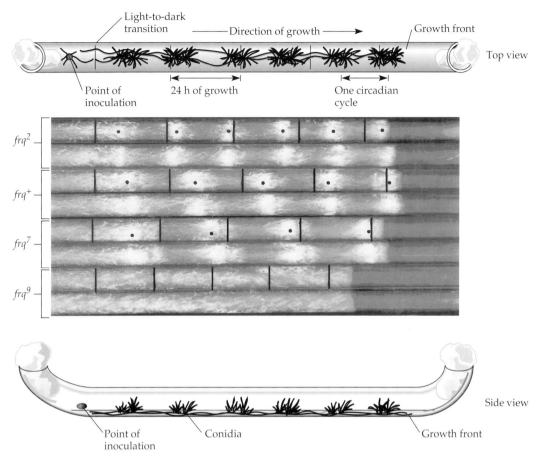

FIGURE 7.19 Analysis of *Neurospora* rhythms on race tubes. Strains are inoculated on the left, grown for a day in the light, and transferred to darkness, which is interpreted as dusk. From there on, the clock runs and controls development in the region of the growth front. For several hours each day the growing front produces aerial hyphae, which eventually differentiate to produce asexual spores called conidia. After a time this developmental switch is reversed, and the aerial hyphae are no longer produced. The growth front thus leaves behind a band of conidiating hyphae morphologically and biochemically distinct from the undifferentiated surface hyphae on either side. This cycle recurs every 21.6 h, and once each region has been laid down the hyphae are developmentally set. Thus after a week of growth in constant darkness, an agar surface is covered with fluffy conidiating bands alternating with undifferentiated surface growth. The period length and phase of the clock are simply read from this pattern of growth, and mutants can be selected by screening for variations in the patterning. *frq²*, *frq⁷*, and *frq⁹* are all alleles of the *frequency* gene; their distinct phenotypes are the results of different mutations in this gene.

protein. The WCC activates expression of the *frq* gene. The *frq* mRNA then encodes two distinct FRQ proteins that feed back to block this activation.

Loss-of-function mutations in *frq*, *wc-1*, or *wc-2* result in loss of normal circadian rhythmicity, although noncircadian oscillations remain in the background of at least the *frq*-null strains. This means that additional feedback loops exist, but whether they are involved in circadian rhythms is still unknown. Mutations in *frq* or *wc-2* that preserve sufficient function to sustain circadian rhythmicity can result in substantial period length defects, yielding periods from 16 to 35 h, as well as resulting in partial loss of temperature and nutritional compensation of the clock. Both *frq* mRNA and FRQ protein are rhythmically expressed in a daily fashion, and FRQ represses the abundance of its own transcript.

GENERATION OF THE CIRCADIAN OSCILLATION. With Figure 7.20 as a guide, the progress of the *Neurospora* clock cycle can be tracked from late at night. At this time most of the FRQ protein in the cell has recently been degraded and *frq* RNA levels are low but are beginning to rise. A period of 10 h is required for *frq* mRNA levels to rise from their trough to peak amounts. The transcription factors WC-1 and WC-2, as the WCC, bind to sites (Clock Boxes) within the promoter of the *frq* gene to drive the circadian rhythm in transcription of *frq*. As the *frq* transcript is produced, the initial primary transcripts are spliced in a temperature-dependent manner. Gradually by early morning, FRQ proteins appear; either a long or a short form of FRQ can be translated, depending on the products of the temperature-regulated alternative splicing of the *frq* transcript.

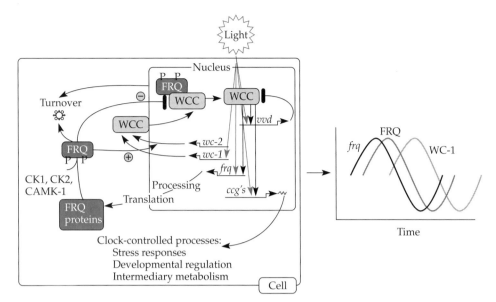

FIGURE 7.20 Molecular components of the *Neurospora* clock. This oscillator incorporates positive feedback from the complex of WC-1 and WC-2 (the WCC) to activate *frq* expression, and from FRQ to enhance WC-1 expression; and negative feedback of FRQ to block the activation of *frq* expression by WCC. The feedback loops are interconnected through the multiple actions of FRQ to block its activators while promoting their synthesis. Specific molecular events are always correlated with certain times of day; in fact, they define that subjective circadian time (CT) for the organism. (CT 18) High levels of WC-1 lead to the formation of WCC dimers that promote transcription of *frq*. (CT 4) Levels of *frq* mRNA peak, and newly translated FRQ appears and begins to move to the nucleus to block WCC. (CT 10) FRQ repression has resulted in lower *frq* transcript levels; FRQ is bound and phosphorylated by kinases; FRQ promotes the synthesis of WC-1 while forming a trimeric complex with the WCC, keeping it inactive. (CT 16–20) Continued phosphorylation makes FRQ unstable, and ultimately it is rapidly degraded, releasing the WCC to activate *frq* expression and start the cycle anew. *ccg*, clock-controlled gene. (From Dunlap, 1999.)

The FRQ proteins made soon form both homodimers and heterodimers, and they participate in three separate actions that together govern aspects of the cycle. First FRQ enters the nucleus, where it interacts in specific ways with the WCC. Because the WCC–DNA complex on the *frq* promoter does not contain FRQ, it is likely that FRQ exerts its negative role in the clock cycle by reducing or preventing the ability of the WCC to bind to its Clock Boxes in the *frq* promoter. By midday, WCC activity is declining to its lowest level just as FRQ levels are rising; together the two turn down the expression of the *frq* gene.

Second, and independently, FRQ begins to be phosphorylated by several kinases, including casein kinases 1 and 2, as soon as it is made. This phosphorylation is important for determining the stability of the FRQ protein, which in turn determines the period length of the clock. At this point, close to midday, FRQ levels are high and continuing to rise; FRQ is becoming phosphorylated and has entered the nucleus to bind to the WCC. In a wild-type cell at room temperature (25°C), typically about 40 molecules of FRQ will exist within the nucleus when FRQ levels are at their peak, compared at the same time to a quarter that many molecules of WC-1 and more than three times that many of WC-2. As a result of FRQ's binding the WCC and reducing its activity, *frq* expression falls and *frq* transcript levels begin to decline. Ongoing translation of the remaining *frq* mRNA, however, allows FRQ protein levels to continue to rise for several hours so that *frq* mRNA levels peak in the midmorning, about 4 to 6 h before the peak of total FRQ in the afternoon.

At about this time the third role of FRQ in the cycle is manifest: FRQ promotes, through an unknown mechanism, the translation of WC-1 from existing *wc-1* mRNA. For this reason, levels of WC-1 begin to rise even as phosphorylation-promoted degradation of FRQ begins. Thus at close to the same time, FRQ is blocking activation of the *frq* promoter by interacting with the WCC while promoting WC-1 synthesis to increase the level of the WCC. FRQ also promotes the synthesis of WC-2, although perhaps as a result of its stability, the levels of WC-2 do not cycle. Because WC-2 is available at constitutively high levels, the enhancement of WC-1 synthesis by FRQ creates an ever increasing amount of WCC, but this WCC is still held inactive when FRQ reduces the ability of the WCC to bind DNA.

Finally, the phosphorylation of FRQ ultimately triggers its precipitous turnover, and the WCC is released. WC-1 levels peak in the night near to when FRQ levels drop to their low point. This wave of WC-1 created by the juxtaposition of FRQ-promoted WC-1 synthesis and the blockage of WCC activation creates a sharp transition, with high WCC activity to initiate the next cycle and to maintain a robust amplitude in the feedback loop.

In summary, the *Neurospora* clock contains positive and negative feedback loops that are interlocked by the multiple activities of FRQ in promoting WC-1 synthesis while blocking WCC activity. The heterodimer of PAS domain–containing proteins, the WCC, acts as a transcriptional activator, and FRQ acts as a dimer as the negative element.

The molecular basis of entrainment to light and temperature changes is understood in Neurospora[12,26,34,43]

The goal of entrainment is to move the day phase of the clock (subjective day) so that it coincides with the day phase of the external world. For entrainment to work, light must advance the clock into the next day when seen late at night but delay the clock into the previous day when seen early in the night. Thus the molecular basis of entrainment by light is that the same photic cue should have opposite effects on the timing mechanism, depending on when light is perceived. In many circadian systems, the clock is not a passive player in this regulation, but instead itself regulates the light response system so that light input to the clock is "gated" (see Chapters 3 and 8). In *Neurospora* the molecular bases for light resetting and gating are coming to be understood.

With clock components that peak in the daytime, such as clock gene mRNAs in *Neurospora* or in mammals, photic induction of the components will reset the clock quite well. In *Neurospora*, light acts rapidly through the WCC to induce high levels of *frq* expression (Figure 7.21A). The WCC mediates the induction of *frq* via light by binding to two sites in the *frq* promoter. WC-1 binds FAD (flavin adenine dinucleotide) and is itself the primary circadian photoreceptor: The FAD cofactor absorbs light, changes its chemistry, and thereby triggers a change in WC-1 and in the shape of the WCC, which ultimately allows the WCC to promote

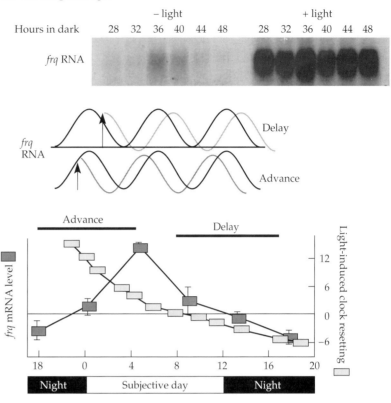

(A) Resetting with light

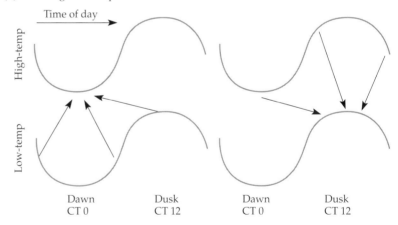

(B) Resetting with temperature

FIGURE 7.21 Resetting the *Neurospora* clock. (A) Light resetting. (Top) The application of light to cultures free-running in the dark (24 to 48 h in darkness) always results in induction of *frq* mRNA. (Middle) This schematic view of light resetting shows how the same molecular event, induction of *frq* by light, can result in opposite clock effects (advances or delays), depending on when the light is seen. If *frq* mRNA levels are low and rising (late night to early morning), light results in a phase advance. If levels are falling (afternoon or evening), light yields a delay. (Bottom) The daily rhythm in *frq* mRNA levels (left axis) is superimposed on the phase response curve to light (right axis). (B) Temperature resetting. The level at about which FRQ protein amounts cycle is higher at higher temperatures, and the amount of FRQ required in the nucleus for effective feedback is interpreted in terms of this level. On the left is a temperature step up: No matter at what phase the step occurs, at the higher temperature there is not enough FRQ to cause feedback, so it is then by definition daytime and more FRQ is made. On the right is a temperature step down: No matter when the step occurs, there is enough FRQ for feedback (subjective nighttime), and FRQ expression is turned off. (From Loros and Dunlap, 2001.)

the transcription of *frq* more effectively. Because light sends *frq* mRNA levels to their peak regardless of when the light is perceived, the clock can always be effectively reset. In the late night, when *frq* mRNA levels are rising to a midmorning peak, rapid induction of *frq* by light to its peak levels quickly advances the clock to a point corresponding to midmorning. Conversely, through the subjective evening and early night, when *frq* mRNA levels are falling, induction rapidly sends the levels back up to peak and sends the clock back to the normal time of the peak, corresponding to midday, yielding a phase delay (see Figure 7.21A). The *Neurospora* light resetting response is also gated by the clock, so at some times of day the same amount of light results in more induction of *frq* than at other times of day. This effect is mediated in part by the VIVID protein (VVD), as will be explained more fully in Chapter 8.

Ambient temperature influences rhythmicity in several ways in *Neurospora*. First, steps up in temperature reset the clock in a manner similar to light pulses. Second, there are physiological temperature limits for operation of the clock, a general feature of circadian rhythms. Third, within these limits the period length is more or less the same. This is the classic clock property known as temperature compensation. Whereas with light, transcriptional regulation is key, temperature effects are mediated through the amount and perhaps the type of FRQ protein made. In fact, temperature regulates both the total amount of FRQ and the relative levels of the long versus short FRQ forms. When the ability of the cell to make either FRQ form is eliminated through mutation, the temperature range permissive for rhythmicity is reduced. This novel adaptive mechanism may serve to extend the physiological temperature range over which the clock functions.

Resetting of the clock by temperature steps also reflects posttranscriptional regulation. At different temperatures *frq* transcript levels oscillate around similar levels, but FRQ amounts oscillate around higher levels at higher temperatures, as Figure 7.21B shows. For instance, at 28°C (the upper curve in Figure 7.21B), the lowest point in the daily cycle of FRQ abundance (late night) is higher than the highest point in the curve (late day) at 21°C (the lower curve in Figure 7.21B). Because the level of FRQ in this oscillator defines the subjective time of day, the "circadian time" associated with a given number of molecules of FRQ is different at different temperatures (see Figure 7.20). That is, at 25°C 40 molecules of FRQ in the nucleus is the peak amount observed, so when 40 molecules are in the nucleus, it is mid- to late subjective night. At 20°C, however, where the peak level of FRQ is lower, only 15 molecules of FRQ in the nucleus might exist at the peak and thus define late subjective night.

In this scenario a shift in temperature corresponds to a shift in the state of the clock (literally a step to a dif-

ferent time), although initially no synthesis or turnover of components occurs. After the temperature step, relative levels of *frq* mRNA and FRQ are self-assessed by the circadian regulatory circuitry in terms of the new temperature, and they respond rapidly: If there is too little FRQ to cause negative feedback, then it must be subjective daytime and more is made. If there is more than enough FRQ to block *frq* mRNA expression, then it must be nighttime and none is made until some FRQ has degraded.

Thus, unlike light, which acts via a photoreceptor outside the loop, temperature changes reset the circadian cycle instantaneously and from within, with the regulatory dynamics of the cycle itself acting as the temperature sensor. In the natural world, temperature changes seen at dusk and dawn can approximate step changes, and surprisingly such nonextreme temperature changes in *Neurospora* and in a variety of other organisms can have a stronger influence on circadian timing than light (see Chapter 3). In all cases, though, light and temperature cues reinforce each other to keep clocks synchronous in the real world.

The Clockwork in *Drosophila,* a Model Animal, Relies on Patterns of Gene Regulation and Feedback Similar to *Neurospora* but Reflects Increased Complexity[2,75,81,82]

Like *Neurospora, Drosophila* is a paradigmatic molecular circadian system whose development has illuminated the details of the operation of clocks at the molecular level. As outlined in Chapter 3, early work by Pittendrigh on *Drosophila pseudoobscura* led to descriptions of the formalisms still used for describing rhythms. The rhythmic characteristic that drove early research was pupal eclosion (emergence, or "hatching"), which takes place in a tightly defined window of time near subjective dawn (Figure 7.22), but in recent decades, studies on *Drosophila melanogaster* have followed the daily crepuscular (dawn and dusk) rhythm in locomotor activity.

Genetic analysis of rhythms in *Drosophila* began with the work of Konopka in the early 1970s; *per* was the first clock gene he identified, and the first one to be cloned a decade later. Much of what is known about rhythms in general derives from the study of this gene, its regulation, and its products. Fruit flies have levels of complexity associated with tissue specialization of function that have offered insights into the circadian systems of multicellular animals beyond those available from *Neurospora* or other microbial systems, demonstrating, for instance, that cell-autonomous clocks are found in most separate parts of the adult fly (see Chapter 11).

(A)

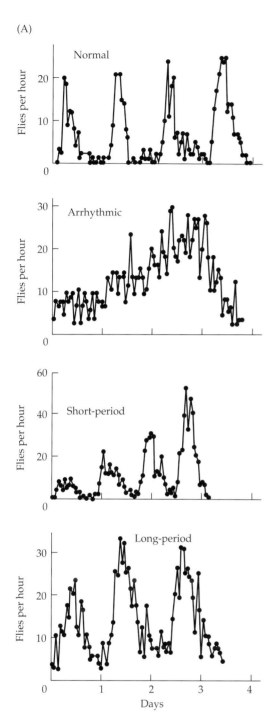

(B)

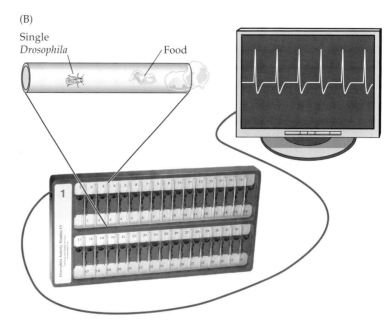

FIGURE 7.22 Rhythmic behavior in wild-type and mutant *Drosophila melanogaster.* (A) Hatching is clock-controlled, and normally about twice as many eggs hatch during the subjective day period as in the subjective night period. When mutagenized adult males were mated to normal females, some of their progeny showed altered timing of hatching. These individuals were collected and examined separately, and the circadian clock controlling the eclosion in these strains was shown to have an altered period length. The arrhythmic, short-period, and long-period mutations all mapped to the same gene, *period*, which was mutated in different ways in the different alleles. (B) Now activity rhythms of individual flies are monitored electronically. Flies are active around dusk and dawn, and their movements break a beam of light. The number of breaks can be tallied with computers. (A from Konopka and Benzer, 1971.)

Many central components of the Drosophila circadian oscillator are known[13,17,28,80]

OVERVIEW. The *Drosophila* circadian oscillator includes interlocked positive and negative feedback loops based on transcription and translation, with a heterodimer of PAS domain–containing proteins acting as a transcriptional activator and two genes encoding negative elements in the feedback loop. The core interlocked feedback loops are defined by the actions of the CLOCK (CLK) and CYCLE (CYC) proteins, *period* (*per*) and *time-*

less (*tim*) mRNAs, and PER and TIM proteins. These are aided by the action of VRILLE (VRI) and three kinases: CASEIN KINASE 2 (CK2), DOUBLE-TIME (DBT), and SHAGGY (SGG) (Figure 7.23). CLK and CYC heterodimerize to form a complex (in Figure 7.23, the CLK–CYC complex, or CCC), the standard PAS–PAS heterodimer of eukaryotic circadian clock loops that is analogous to the WCC in *Neurospora*. CCC activates expression of the *per* and *tim* genes, which then encode proteins that result in feedback to block their own activation.

PER appears to be the protein actually causing repression; TIM is essential for stabilizing PER and regulating its intracellular location. Loss-of-function mutations in *Clk*, *cyc*, *per*, or *tim* result in loss of normal circadian rhythmicity, although noncircadian oscillations have been reported in the background of null strains. DBT is required to phosphorylate PER, leading to its degradation, and phosphorylation of TIM by SGG promotes nuclear entry of TIM–PER; VRI binds to the *Clk* promoter to depress *Clk* expression and thereby to repress *per* and *tim* RNA accumulation. CK2 probably acts to prime phosphorylation of PER and TIM by DBT and SGG. Partial loss-of-function and gain-of-function mutations in

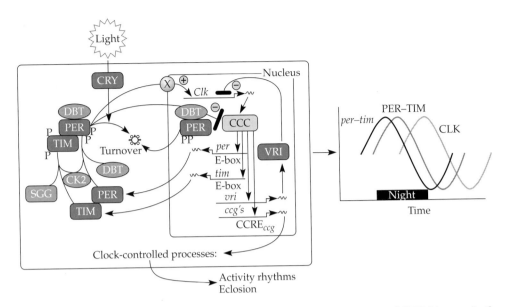

FIGURE 7.23 Molecular components of the *Drosophila* clock. Specific molecular events occur at certain circadian times (CT) of the day. (CT 1) High levels of CLK protein lead to the formation of CLK–CYC dimers that in turn promote transcription of *per*, *tim*, and *vri*. (CT 4) Levels of *per* and *tim* mRNA begin to rise. VRI is made and feeds back to reduce *Clk* expression. (CT 10) PER protein is bound and phosphorylated by CK2 and DBT; phosphorylated PER is degraded, delaying its accumulation. (CT 16) As PER and TIM levels increase, the proteins together form a stable trimeric complex that is capable of moving into the nucleus, a process promoted by the CK2 and SGG kinases. In the nucleus, PER–TIM–DBT represses transcription of *per* and *tim*, and PER–TIM somehow (indirectly) activates *Clk* transcription. (CT 19) PER and TIM proteins are at peak levels, the DBT–PER–TIM heterotrimers are evenly distributed throughout the nucleus and cytoplasm, and activity of the CLK–CYC dimer is inhibited. (CT 24) By morning, phosphorylated PER is no longer bound by TIM, and, unprotected, it is degraded. The breakdown of PER releases the CLK–CYC complex to activate *per* and *tim* expression, starting another cycle. *ccg*, clock-controlled gene; CCRE, clock control regulatory element. (From Dunlap, 1999.)

any of these genes can result in substantial defects in period length (yielding periods from 16 to 35 h) or arrhythmicity. *per*, *tim*, *Clk*, and *vri* mRNA and the corresponding proteins are rhythmically expressed in a daily fashion, and by appropriate phasing of expression, delays, and interlocked loops, the clock cycle proceeds in a robust manner.

GENERATION OF THE OSCILLATION.   The progress of the *Drosophila* clock cycle can be followed from the middle of the day (Figure 7.23). At this time most of the PER and TIM protein in the cell has been degraded, and *per* and *tim* RNA levels are beginning to rise, a process that will take about 10 h to peak. The transcription factors CLK and CYC work together in a heterodimeric complex (the CCC) to bind to sites called E-boxes within the promoters of the *per*, *tim*, and *vri* genes.

Like the WCC in *Neurospora*, the CCC is a transcriptional activator that drives circadian rhythms of transcription of its target genes. The transcripts are produced and are translated into the corresponding proteins. VRI moves to the nucleus and reduces CLK expression. While PER is in the cytoplasm, it is phosphorylated by CK2 and DBT. This phosphorylation makes PER susceptible to degradation thereafter unless it is associated with the TIM protein. Because TIM takes some time to accumulate, PER accumulation is delayed, and this delay

is one of the lags in the feedback loop that give rise to the long (approximately 24h) cycle length.

Also in the cytoplasm, the SGG kinase phosphorylates TIM, probably aided by CK2. Phosphorylation of TIM promotes TIM's access to the nucleus, and thus TIM and PER help each other and DBT into the nucleus. This delay in nuclear accumulation is another lag in the cycle. Gradually, by early evening, PER and TIM proteins appear; they heterodimerize and enter the nucleus. Once in the nucleus, TIM gradually disappears from the complex, and PER brings about a gradual repression of *per* and *tim* transcription. In a manner similar to the repression of WCC activity by FRQ in *Neurospora*, it is likely that PER, or possibly the PER–TIM complex, exerts this negative role in the clock cycle by reducing or preventing the ability of the CCC to bind to the *per* and *tim* promoters. In any case, before midnight, CCC activity has declined to its lowest level just as PER levels rise to their maximum, so together these processes turn down the expression of both the *per* and the *tim* genes.

At this point in the evening, PER is negatively affecting CCC activity in the nucleus. Without this transcriptional activator to drive more mRNA synthesis, *per* and *tim* transcript levels begin to decline, although ongoing translation of the remaining mRNAs allows PER and TIM protein levels to continue to rise for several hours.

Thus *per* and *tim* mRNA levels peak in the early to midevening, about 4 to 6 h before the peak of PER and TIM in the mid- to late evening.

Also at this time, another role of PER in the cycle is seen: PER (or possibly the PER–TIM complex) in some way causes increased expression of *Clk* so that levels of CLK begin to rise even as phosphorylation-promoted turnover of PER begins. Thus at close to the same time in the cell, PER is blocking activation of the *per* promoter by the CCC while promoting CLK synthesis to increase the level of the CCC. Because CYC is available at constitutively high levels, the increasing levels of CLK result in an increasing amount of the CLK–CYC complex, which is still held inactive by PER.

Another protein, VRI, also acts to reinforce this cycle: The CCC binds to E-boxes in the *vri* promoter and promotes its synthesis. Once made, VRI immediately represses expression of *Clk*, so CLK levels fall, and soon *per*, *tim*, and *vri* mRNA levels decline. VRI in effect strengthens the repressive arm of the loop between late day and midnight. In the nucleus just as in the cytoplasm, as soon as TIM leaves, DBT phosphorylates and thereby destabilizes PER, leading to PER's precipitous turnover. As PER is degraded, PER-promoted synthesis of CLK is soon balanced by CLK degradation, and CLK levels peak just after dawn near to when PER levels drop to their lowest point.

The overall effect is a wave of CLK and CCC activity created by the juxtaposition of PER-promoted CLK synthesis and the blockage of CCC activation. This high transcription-activating activity creates a sharp transition to initiate the next cycle and to maintain a robust amplitude in the feedback loop.

The Drosophila *clock is reset through light-promoted protein degradation*[21,50,68,75]

The circadian system in *Drosophila* can be synchronized with the day–night cycle through mechanisms that can detect small changes in the ambient conditions, either brief pulses of light or small steps in temperature. Although temperature resetting mechanisms are still being elucidated, light is understood to reset the clock through a mechanism involving light-promoted turnover of the TIM protein. *Drosophila* uses several rhodopsin- and flavin-based photoreceptors to detect light. Loss of all the opsin-based receptors still permits entrainment, but it takes 1 to 5 days longer to reach a stable phase.

The principal and cell-autonomous photoreceptor for the clock appears to be a cryptochrome (CRY) that uses a flavin as a chromophore to confer light sensitivity on the clock system. In the lateral neurons in the brain of *Drosophila*, this CRY has no role in the core oscillator loop. These brain clock cells are the site of the dominant pacemaker for behavioral rhythms, and they can also receive information about ambient light as detected by

rhodopsin-based photoreceptors in the compound eye and a deep-brain photoreceptor (see Chapter 5). Taken together, these data explain how flies lacking CRY can still be entrained to full-photoperiod light–dark cycles, but not to short-duration light treatments in the manner of normal insects. Specialization in this animal system has separated clock function from light perception; animals lacking CRY (and even lacking eyes altogether) still have functional clocks.

The mechanism through which CRY resets the *Drosophila* clock represents a pleasing symmetry with the mechanism observed in *Neurospora*, in which light induction of *frq* plays a central role. In *Drosophila*, PER and TIM levels peak at night and begin to decline in anticipation of daytime, a period characterized by low levels of both proteins. In this system, light acts acutely through CRY to block the activity of PER–TIM, and eventually this leads to turnover of TIM. Because TIM is required to prevent DBT from phosphorylating PER, the event that leads to its degradation, loss of TIM means loss of PER. CRY is normally present both in the cytoplasm and in the nucleus, and it associates with TIM in a light-dependent manner. When CRY binds to TIM, TIM is covalently modified through attachment of the short protein ubiquitin to it, and this ubiquitination of TIM leads to its subsequent degradation via the proteasome. Because daytime in a fly is characterized by low levels of TIM and PER, light-mediated degradation of the proteins in the early night reduces TIM and PER levels, delaying the phase back to the previous day. Conversely, light-promoted degradation of TIM and PER late at night also reduces their levels but advances the clock into the next day.

The cells of most tissues in the adult fly, including the legs, antennae, gut, and excretory system, have clocks that can run independently (see Chapter 11). All of these peripheral clocks are reset in the same way: through CRY-mediated TIM turnover. Interestingly, in some of these cells CRY may have a role in keeping time beyond just sensing light. Although the organismal clock in most insect cells doesn't run in this way, the possible involvement of CRY in a clock mechanism foreshadows an activity observed in vertebrate systems.

Single-Gene Mutations Have Been Used to Describe a Circadian Oscillator in Mammals[2,60,61,62]

One particularly informative single-gene mutant was identified in a hamster[46,60]

In terms of cellular and molecular complexity, mammalian circadian systems are at a pinnacle. This intricacy provides a canvas upon which to view the genetics of a complex phenotype at its fullest. The mammalian circadian system encompasses multiple layers of coupling. Pacemakers such as the SCN, heart, liver, and kidney are clockshops of coupled oscillators, and the SCN is coupled to all these peripheral oscillators. Coupling is also revealed in the subtle effects through which outputs, such as locomotor activity, can feed back to influence the circadian system itself. These connections are made through the actions of proteins. As a result, genetic mutations can affect the circadian system as a whole. In fact, many mutations are known that affect rhythms in relatively minor ways, but these are still difficult to interpret due to the complexity of the system. Perhaps for this reason, the mutations that have had the greatest influence on the understanding of circadian biology in mammals are single-gene mutations that have robust effects on input or on the core oscillator, as was the case with simpler systems of *Drosophila* and *Neurospora*.

The original mammalian single-gene clock mutant, *tau*, arose as a spontaneous mutation not in the mouse but in a mammal that is a favorite for circadian physiological research, the golden hamster, *Mesocrecitus auratus*. This mutant appeared, unbidden, in a shipment of hamsters from a supply house, and its character was revealed during a preliminary screening on running wheels in preparation for a different experiment (Figure 7.24). The mutation was discovered because of an unusually early onset of locomotor activity under an LD cycle and a short 22 h free-running period in the original heterozygote (*tau*/+) individual. Breeding yielded a homozygote (*tau*/*tau*) that had a very short 20 h period length. The lack of classical genetics information about the hamster made the gene very difficult to clone initially, but *tau* immediately provided an exceptional tool for dissection of physiological aspects of the hamster circadian system (see Chapter 5). Effects of the mutation

on entrainment and activity were noted and used to develop models for entrainment and for circadian gating of the rodent's estrous cycle. Transplantation of SCN tissue from a *tau* mutant to a wild-type donor provided incontrovertible evidence that supported the role of the SCN as the dominant pacemaker in control of locomotor activity rhythms (see Figure 5.19). The relatively rapid recovery of rhythms in transplant recipients revealed a strong role for humoral signals as rhythmic output from the SCN. The data also demonstrated that multiple independent pacemakers could run in the same brain. The mutation was shown to affect not only period length but also temperature compensation in a peripheral clock, thereby connecting these properties in the mammal as they had been connected in model systems. The *tau* mutant has been immensely important in both physiological and genetic approaches. It demonstrated that mammalian circadian systems could be dissected through the isolation and analysis of single-gene mutants, just as in model systems.

The central clockworks in vertebrates is best understood in house mice[2,3,39,74]

Proof of the value of this approach came with the identification of *Clock*, the first gene associated with the molecular mechanism of the clock in a mammal. In a standard forward genetic screen (Figure 7.25), the 25th mouse to be examined displayed a 1 h lengthening of period. In addition to this stroke of luck, the *Clock* mutation was fortuitous for another reason: Because it showed up in a heterozygote, the mutation was not recessive as are the common loss-of-function mutations that typically arise from mutageneses. Instead, the *Clock* mutation was a genetic variant of a much rarer sort that makes an altered but still functional product. Breeding yielded homozygote *Clock/Clock* progeny that displayed an initial period length 3 h longer than wild type, but that soon graded into arrhythmicity in DD conditions (see Figure 7.25B,C). The gene was genetically mapped to chromosome 5. In a molecular tour-de-force that used many novel genetic tools from the mouse genome project, mRNA transcripts arising from the region were identified and the gene was cloned. Confirmation of its identity came in two ways. A very large fragment of DNA containing the wild-type gene was first placed into the genome of the mutant and used to restore the wild-type phenotype. Subsequently, DNA sequencing revealed a single base-pair change that made the CLOCK protein shorter. Sequencing showed that CLOCK was a DNA-binding PAS domain–containing protein. The sequence was similar to the PAS proteins already associated with clocks, *Drosophila* PER protein and WC-1 and WC-2 in *Neurospora*, and CLOCK was later shown to have a function as an activator, analogous to WCC. Conservation of

FIGURE 7.24 The rhythmicity of wheel running in rodents. Rodents love to run on wheels such as this one, and their running is controlled in part by the circadian clock. (Courtesy of C. Cook.)

(A)

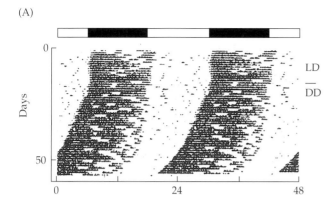

(B)

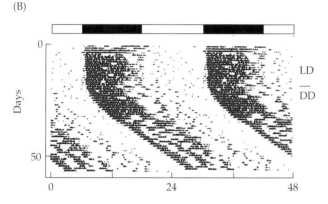

(C)

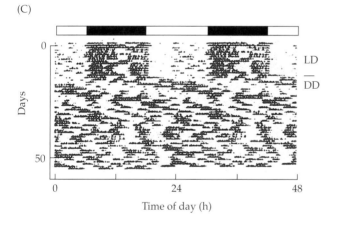

Time of day (h)

FIGURE 7.25 Identification of the *Clock* mutation in mice.
In a genetic screen for genes required to operate the mouse clock, young male mice were mutagenized with the chemical ENU (ethylnitroso-urea). Surviving mice were allowed to breed to unmutagenized females, and the running-wheel activity was examined in first- and second-generation backcrosses carried out to generate mice heterozygous or homozygous for rhythm mutations. Running-wheel activity for one resulting mutant is shown. The mice were initially kept under entrained conditions (LD 12:12) for about 12 days and then kept in darkness to allow their endogenous rhythmicity to be expressed over the subsequent 50 days. (A) The rhythm of a wild-type mouse, showing a typical period length (approximately 23.5 h). (B) The rhythm of a mouse heterozygous for a single-gene mutation in a gene called *Clock*. The nucleus of each cell contains one normal copy and one mutant copy of this gene, and the result is a distinctly long period of about 27 h (27.1 in this individual). (C) The rhythm of a mouse in which each nucleus bears two mutant and no normal copies of the *Clock* gene. The result is a very long period that grades into arrhythmicity. (From Hotz-Vitaterna et al., 1994.)

tive elements in the loop. Other conserved features are the appearance of interlocked loops in which negative elements promote the expression of a positive element and the importance of phosphorylation to regulate clock protein turnover and thereby determine period length.

Rapid progress has been made in identifying components of vertebrate circadian feedback loops[8,27,39,61,62,66,82]

The vertebrate circadian feedback loops are based on transcription and translation, and they employ a heterodimer of PAS domain–containing proteins acting as a transcriptional activator, as well as a family of PER proteins and CRY proteins as negative elements. In addition to conserved design principles, the molecular biology of vertebrate circadian oscillators reveals duplication of several clock factors, different roles for some, and possible loss of roles for others. This duplication and reduplication of components has added greatly to the complexity of the vertebrate clock. Components known to play important roles in vertebrate circadian oscillators are the BMAL1 and CLOCK proteins; two orthologs of the *Drosophila* PER protein (PER1 and PER2); and two cryptochromes, CRY1 and CRY2. Rhythmicity is further promoted by casein kinase 1 (CK1), as seen in *Drosophila* and *Neurospora*, although in mammals two similar forms of this kinase, CK1δ and ε, perform this function. A transcriptional repressor, REV-ERBα, completes the list of known components. A third protein with an amino acid sequence like PER, PER3, exists but may be involved chiefly in output. A protein similar to TIM is found in the SCN, but no role can be ascribed to it yet. Although the CRY proteins have amino acid sequences similar to those found in model systems, their primary functions in the clock appear to be somewhat different (Figure 7.26). The roles of BMAL1

structure and function are important clues for understanding the molecular biology of vertebrate clocks.

Subsequent progress in the molecular dissection of the core mammalian clock has been rapid due to the whole-genome DNA sequences from mouse and human that emerged in the late 1990s. By searching these sequences, chronobiologists have been able to find sequence homologs to many clock proteins known from model systems, especially *Drosophila*. Design principles seen in model systems are conserved in the assembly of vertebrate circadian feedback loops. These include the transcription and translation–based feedback loop, with its resulting rhythms in clock component RNAs and proteins, and the use of PAS–PAS heterodimers as posi-

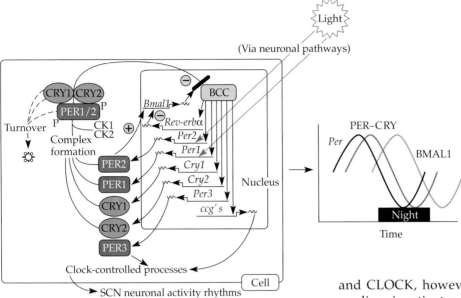

An alternative view:

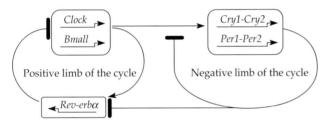

FIGURE 7.26 Molecular components of the mouse clock.
A circadian day can be followed by tracking the circadian times (CT) at which specific molecular events in the clockwork occur. (CT 18) High levels of BMAL1 protein lead to the formation of CLOCK–BMAL1 (BCC) dimers that in turn promote transcription of the *Per*, *Rev-erbα*, and *Cry* genes. (CT 0) Levels of *Per*, *Rev-erbα*, and *Cry* mRNAs are rising, and the corresponding proteins appear. (CT 6) REV-ERBα moves to the nucleus to repress *Bmal1* expression. PER and CRY shuffle to and from the nucleus, and they may be stabilized in the nucleus by mutual association. PER proteins are bound and phosphorylated by CK1, marking them for eventual turnover. Eventually the PER–CRY complex blocks BCC activity. (CT 12) As a result of BCC inhibition, *Per*, *Rev-erbα*, and *Cry* mRNA levels are declining, and likewise from the turnover of PER, REV-ERBα and CRY protein levels are decreasing. BMAL1 levels are rising, phase-lagging the PER2-promoted increase of *Bmal1* mRNA. (CT 18) Turnover of the PER–CRY–CK1 inhibitory complex releases the BCC to reactivate transcription of the *Per*, *Rev-erbα*, and *Cry* genes to reinitiate the cycle. An alternative and deliberately oversimplified view in the lower panel separates the elements of the positive subcycle, happening from midnight to mid-morning, from those in the negative subcycle, which happen at the opposite phase of the cycle. *ccg*, clock-controlled gene. (From Dunlap, 1999; Preitner et al., 2002; and Reppert and Weaver, 2002.)

and CLOCK, however, acting as the PAS–PAS heterodimeric activator of transcription, seem to be fully conserved. In Figure 7.26 they are the BMAL–CLOCK complex, or BCC, the standard PAS–PAS heterodimer of eukaryotic circadian clock loops. BCC activates expression of the *Per, Cry, and Rev-erbα* genes, which then encode proteins that result in feedback to block their own expression. The REV-ERBα protein binds quickly to the *Bmal1* promoter to repress expression. PER and CRY together, as a complex with CK1 proteins, regulate each other's intracellular location while blocking the activity of the BCC. In rodents, mutations showing substantive effects on rhythmicity are known for *Clock, Bmall, Per1, Per2, Cry1, Cry2, Rev-erbα,* and *Ck1ε*, whereas loss of *Per3* has only minor effects on rhythmicity (Figure 7.27).

The products of *Cry1* and *Cry2* appear to be to some extent redundant in that mice missing just one of these have only slightly altered clocks, but mice lacking both have no clock (Figure 7.28). In humans, individuals bearing a point mutation in *hPer2* resulting in the production of an altered PER2 protein display an advanced circadian phase of awakening and activity. This characteristic is what would be expected from the work in model systems for an entrained clock having a short inherent period length (see Chapter 3), and in humans this is manifested as advanced sleep phase syndrome, or ASPS (see Chapter 10).

Ck1δ and CK1ε phosphorylate the PER proteins, leading to their degradation. A spontaneous mutation in this kinase in the hamster (in the *tau* gene) leads to shortening of the rhythm's period. Rhythmic expression of some clock components was seen in model systems and is also seen in vertebrates: mRNAs arising from the *Per* genes, *Bmal1*, *Rev-erbα*, and the *Cry* genes, as well as their corresponding proteins, are rhythmically expressed in a daily fashion (see Figure 7.28). By appropriate phasing of expression, delays, and interlocked loops, the clock

(A)

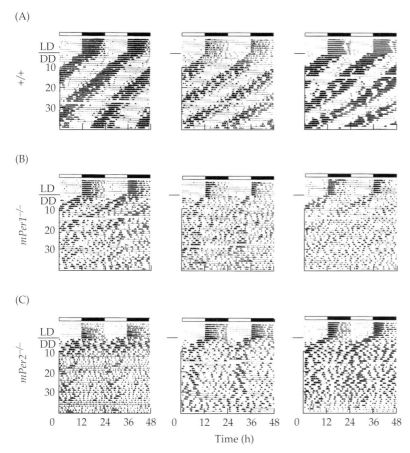

(B)

(C)

Time (h)

FIGURE 7.27 Locomotor activity rhythms are lost in mice lacking *per1* or *per2*. Double-plotted activity records (actograms) of representative wild-type mice, +/+ (A), *mPer1*-deficient mice (B), and homozygous *mPer2*-deficient mice. Vertical bars represent periods of wheel running. Animals kept initially in LD12:12 (the bar above the record) were then transferred to constant darkness (DD) and followed for the number of days shown on the left. All of the wild-type mice maintained robust rhythmicity in DD. Of the *mPer1*-deficient mice, the animal on the left showed weak rhythmicity in DD, while the other two animals were arrhythmic by the third 10-day segment in DD. All three *mPer2*-deficient mutant mice shown were arrhythmic by the end of the record. (From Bae et al., 2001.)

cycle proceeds in a robust manner. However, there appear to be subtle changes in the way the loops are assembled. There are some differences among different vertebrate systems studied, but because the mouse appears to be a good model for humans and most details have been described in the mouse (including the various mutations), the mouse oscillator will be described first.

GENERATION OF THE OSCILLATION. The progress of the clock cycle in a cell within the suprachiasmatic nucleus of a mouse can be followed from midevening to late at night (see Figure 7.26). At this time of day, most of the PER, REV-ERBα, and CRY proteins have been recently degraded, and *Per*, *Rev-erbα*, and *Cry* RNA levels are low but are beginning to rise. It is likely, by the way, that the same processes are happening at these times in the

human brain. The transcription factors BMAL1 (also called MOP3 or, in humans, ARNTL) and CLOCK work together as the heterodimeric BMAL1–CLOCK complex (BCC in Figure 7.26). Together they bind to E-boxes in the promoters of various genes, including *Per1*, *Per2*, *Rev-erbα*, and probably the *Cry* genes, to activate their transcription.

Gradually by early morning the proteins begin to be translated. REV-ERBα moves to the nucleus to repress *Bmal1* expression. PER1–PER2 can move to and from the nucleus and form complexes by physically associating with CRY1, CRY2, and the two forms of CK1; these proteins move to the nucleus. This important interaction appears to mutually stabilize the proteins in a manner reminiscent of that observed with PER and TIM in *Drosophila*. The result of the interaction is that, after a lag probably brought about by the degradation, both PER and CRY proteins accumulate in the nucleus in a codependent manner. Once there, they each participate in several different, but by now perhaps familiar, actions that contribute to robust cycling.

One primary action is, of course, to block the transcriptional activation activity of the BCC, as is required in a negative feedback loop. In a second action, PER2 promotes the expression of *Bmal1*, thereby countering the effect of REV-ERBα repression, and together PER and CRY cause cycling *Bmal1* expression. This interlocked positive feedback loop is functionally, but not mechanistically, similar to the promotion of *Clk* expression by PER–TIM in *Drosophila* and of WC-1 expression by FRQ in *Neurospora*.

In the mouse, the BCC–DNA complex has been reported to contain PER2 and the CRY proteins, so in mammals the negative elements may exert their effect in the clock cycle by reducing the ability of the DNA-bound BCC to activate transcription. This is not the case in *Drosophila* or *Neurospora*, in which the negative elements oppose DNA binding by the positive elements. In any case, by midday, BCC activation is declining to its lowest level as PER2 and CRY levels rise. Together, these two effects turn down the expression of target genes, including *Rev-erbα*. The PER proteins are becoming phosphorylated by the isoforms of casein kinase 1 (CKIδ and CKIε), an action that may influence stability or intracellular location.

Eventually, as a result of inhibited expression, *Per*, *Rev-erbα*, and *Cry* transcript levels begin to decline, fol-

(A)

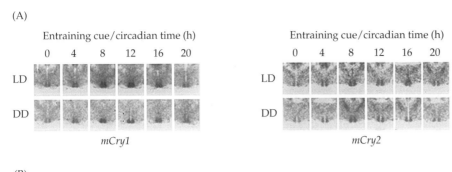

(B)

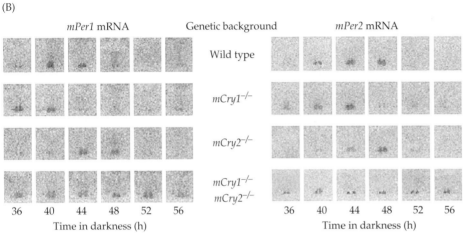

FIGURE 7.28 Rhythmic expression of clock gene mRNAs in the SCN in wild-type and *cry* mutant genetic backgrounds. (A) Expression of *mCry1* and *mCry2* in the suprachiasmatic nucleus of wild-type mice in LD 12:12 and over 20 h in DD. Shown are coronal brain sections through the SCN collected from mice sacrificed at the times noted. (B) Expression of *mPer1* and *mPer2* in the SCN of wild-type and *mCry* mutant mice in constant darkness. The relative RNA abundance of *mPer1* (left) and *mPer2* (right) in the SCN of wild-type (first row), *mCry1$^{-/-}$* (second row), *mCry2$^{-/-}$* (third row), and *mCry1$^{-/-}$mCry2$^{-/-}$* (fourth row) mice in constant darkness was determined by quantitative in situ hybridization. Gene-specific mRNAs appear as dark grains on a light background. Because *mCry1/mCry2* double-mutant mice are arrhythmic, an environmental time scale (hours in DD) was used rather than a circadian time (CT) scale. (From Okamura et al., 1999.)

given what is known about how the oscillator itself works and how resetting works in model systems. In particular, within SCN clock cells are parallels with the mechanism of resetting in *Neurospora*. The overall characteristics of resetting on the whole vertebrate organism have been noted before (see Chapter 3): Behavioral rhythms can be reset by a short-duration light pulse (15 minutes or more), but resetting tends to be slow in that the steady-state phase shift is rarely seen on the first day after the treatment. In addition, the amount of resetting tends to be small, on the order of a few hours at most. Strong Type 0 resetting (discussed in Chapter 3) is never seen in animals after a single light treatment. Light seen early in the night causes a phase delay into the previous day, and light late at night results in phase advances into the next day. Further, the light response is gated such that light at night has effects, whereas light during the day has relatively little effect in terms of resetting the clock.

As noted previously in terms of resetting a clock with

lowed some hours later by a decline in PER, REV-ERBα, and CRY levels. As these negative elements disappear, they leave the BCC to reactivate transcription to initiate the next cycle and to maintain a robust amplitude in the feedback loop. Although this description is lacking some of the mechanistic detail available in model systems that have been studied longer, the molecular analysis of the mammalian oscillatory system is being studied very intensively, so more details will be available in the near future.

Light resets the vertebrate clock by activating transcription of negative elements in the feedback loop[24,54,61,62,67,73]

The general mechanism through which light acts in vertebrates to synchronize internal circadian clocks with the external environment is understood only in broad outline. At that level, however, it can be readily understood,

components that peak in the daytime, such as the oscillator described in *Neurospora*, photic induction of the components serves well to reset the clock. In the mouse and very likely also in people, light is detected by specialized ganglion cells in the retina (see Chapters 6 and 11), probably through the use of melanopsin as a light-sensing protein. Interestingly, the CRY proteins, which are the essential sensors of light in *Drosophila*, apparently share this role in vertebrate light sensing with melanopsin. When light falls on the retina, it is detected by rods and cones of the normal visual system and by specialized chromophores (probably melanopsin and cryptochromes) in ganglion cells. An action potential is triggered and carried via the retinohypothalamic tract to the SCN in the hypothalamus. There it results in release of the neurotransmitters glutamate and pituitary adenylate cyclase–activating peptide, also known as PACAP. These chemicals are detected by several receptors that initiate changes in the ionic environment within the SCN neuron. One of these actions is to release intracellular calcium from stores in the endoplasmic reticulum. SCN cells contain kinases that are activated by calcium, so this transient calcium increase stimulates the activity of several kinases and initiates signaling through a kinase cascade. (In a kinase signaling cascade, phosphorylation of an initial protein kinase is triggered by a stimulatory event, here the glutamate-induced release of intracellular calcium. The proteins representing the initial kinase in the cascade then phosphorylate, and thereby activate, the second [different] protein kinase in the cascade, and all the copies of the second kinase can act on molecules of a distinct third kinase. Because a single enzyme can act very rapidly and catalytically to phosphorylate many target protein kinases at each step, the organization of enzymes and substrates into a cascade clearly will result in a rapid amplification and transmission of a signal to the ultimate step.) Activation of kinase cascades can have multiple effects. Light acts broadly to activate an acute response pathway in SCN cells, thereby promoting the expression of numerous genes, including clock genes. Ultimately the kinase cascades conclude with changes in the structure of the chromatin, the complex of proteins and DNA that make up the chromosomes within the nucleus. One of the phosphorylated target proteins is the cAMP response element–binding protein (CREB), a protein that when activated will bind to specific DNA sequences. CREB binding sites are found in the promoters of the *Per1* and *Per2* genes.

Induction of *Per1* is seen in the SCN of the hypothalamus within 10 minutes of light falling on the eyes (Figure 7.29). This rapid induction of *Per1* by light appears to be the primary stimulus for light resetting of the clock. The *Per2* gene shows delayed light induction, and the levels of *Per3* and *Cry* expression are not affected by light. The *Per1* induction response, as well as the resulting clock-resetting response, is circadianly gated to a modest degree (see Chapter 8). Light seen at night results in a large induction to a high level, whereas light seen in the mid- to late day yields little if any induction. The result is that light seen at night resets the clock in the normal way, yielding delays just after dusk and advances just before dawn. However, light seen during the subjective day has little acute effect on the clock.

The way in which light-induced PER1 acts within the feedback loop to reset the clock is not fully understood. The clock within the SCN cells is reset by a light stimulus within one cycle, so by the next day the new clock phase is achieved, even though it can take many days of transients before behavioral rhythms are completely reset. Thus there is no jet lag within the SCN, and all of the delay in phase shifting of an animal's clock is due to events in the output pathway connecting the clocks in the SCN to the rhythms in the rest of the body.

Many animals have evolved peripheral clocks that are distinct from the central-brain cellular clocks[6,57,61,62]

One of the many dogmas concerning rhythms to fall by the wayside in the last decade was the notion that multicellular organisms had only one central clock that dictated the immediate behavior of all rhythmic cells in the organism (see also Chapter 11). Instead we now appre-

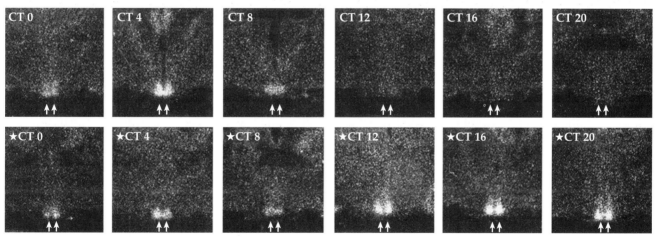

FIGURE 7.29 Light exposure at any time brings the amount of *mPer1* mRNA to subjective daytime levels. At subjective times of day spanning the circadian cycle, mice were exposed to bright light for 30 minutes, returned to the dark for 30 minutes, and then sacrificed. The lower series of micrographs with stars denote light-treated mice; the upper series are from control mice handled but not exposed to light. Coronal brain sections from the top to the bottom of the head through the paired SCN (white arrows) were treated to detect the amount of *Per1* mRNA in the brain, which here shows up as white specks on a dark background. The upper series shows the circadian rhythm in *mPer1* mRNA content, and the lower series shows light induction. (From Shigeyoshi et al., 1997.)

ciate multicellular organisms as a whole collection of circadian oscillatory cells, a true clock shop.

Circadian oscillators in peripheral tissues use many of the same clock molecules as do central brain clocks, and they appear to be assembled into the same feedback loops (see Figure 7.26). However, in addition to driving distinct outputs (see Chapter 8), other differences may exist because these peripheral clocks are not self-sustaining to the extent of central clocks. If the central pacemaker is lesioned or separated from the brain, the peripheral clocks gradually "wind down," so that after a week in constant conditions there is little evidence of rhythmicity. In *Drosophila* and in lower vertebrates such as zebra fish, the peripheral clocks have retained the ability to be reset by light/dark cues; in mammals, however, light reception has been specialized to the retina. The cycling of clock components like the *per* and *cry* genes in mammalian peripheral clocks are phase-lagged by 3 to 9 hours compared to the SCN clock, the exact lag being characteristic of the tissue. Temperature cycles, as well as neuronal and humoral cues from the mammalian brain, may constitute time cues for peripheral oscillators (see Chapter 5).

The Emerging Description of Plant Circadian Oscillators Includes Multiple Coupled Oscillators, Each Apparently Involving Transcription and Translation[37,47]

Elucidation of the molecular mechanism of plant rhythms has been difficult[32,63]

Although deMairan's rhythms in leaf movements were among the first circadian rhythms to be described scientifically, progress in dissecting the molecular mechanism of plant circadian rhythms has been slow. There are several good reasons for this. First, the application of genetics to the dissection of plant circadian feedback loops has been problematic. Only within the past decade have *Arabidopsis* and *Chlamydomonas* emerged as plant genetic and molecular model systems.

Second, the interpretation of clock mutations has been more difficult than in other models. In *Neurospora* and *Drosophila*, the circadian systems run fine in the dark but stop in constant bright light. Therefore, mutations affecting the operation of the clock could be examined in darkness without regard to the light-sensing system. Plants, however, generally require light to live, and the plant circadian system is exquisitely sensitive to the quality (color spectrum) and quantity (fluence) of light. As a result, mutations in genes whose only role is to sense light can appear to affect the circadian system by mimicking the effects of changing light intensity or quality. Thus mutations in genes with no role in the clock can result in changes in circadian period length or phase.

A third factor contributing to slow progress is that plants are more distantly related to either fungi or animals than the latter are to each other. Plant genomes have revealed no proteins like FRQ or PER or WC-1, nor heterodimers consisting of PAS domain–containing proteins that acted as circadian transcriptional activators. This dearth of molecular similarities among clock components means that the conserved hallmarks that were so essential to the dissection of vertebrate clocks have not been available to give plant molecular chronobiologists a leg up.

Finally, in fungal and animal cells a single central circadian oscillator is found. However, in all likelihood plant cells use more than one self-sustained oscillator. Redundancy certainly occurs at the molecular level among clock components. As a result, the loss of individual components is often difficult to interpret.

This overwhelming complexity is highly interesting. Plant clocks regulate both time of day–specific expression of various genes and activities, including pigment production, photosynthesis, leaf movements, and a host of biochemical activities (see Chapter 8). Clocks are also used for photoperiodic time measurement (see Chapter 4). Thus investigators have been able to identify putative oscillator compounds in plants by following output pathways back to the clock.

In one case, control of the photosynthetic rhythm was followed to the nuclear-encoded proteins found in the light-harvesting complex in the chloroplast. From there the path led to rhythmic expression of the *lhcb* gene. Study of the regulation of *lhcb* led to identification of the MYB-transcription factor CCA1, which confers rhythmic expression on *lhcb*. Overexpression of CCA1 results in arrhythmicity. This last surprising observation suggests that CCA1 may be a component of the oscillator, as will be discussed shortly. Similarly, rhythmically expressed plant promoters have been used to drive the rhythmic expression of bioluminescence genes. These rhythmically bioluminescent plants normally make light in the daytime, so clock-mutant plants making light at the "wrong" time could be found. Because the plant clock directs photoperiodic time measurement, mutant genes that cause plants to flower under inappropriate light conditions have also been suggested to encode putative oscillator components.

The organization of plant circadian oscillators suggests the involvement of transcription and translation feedback loops[32,47,63,82]

The overall layout of plant circadian oscillators is most likely a negative feedback loop that incorporates elements of transcription and translation. Early work with metabolic inhibitors in *Chlamydomonas* showed that partial inhibition of protein synthesis would lengthen the period of the circadian cycle. Work in the photosynthetic protist *Gonyaulax* suggested that there were sensitive phases for this inhibition consistent with the existence of a daily cycle of synthesis: Clock-critical proteins were seen as being made at a particular time, acting later, and then being degraded.

Mutations in a variety of genes can affect the period length of the clock. Some of these genes encode genuine clock components, and others affect period by mimicking the effects of light. Because of the likely internal redundancy of the clock mechanism, much work has used dominant mutant effects. These studies include strains that overexpress potential clock components because such dominant effects can still be seen in the presence of another normal component. In contrast, the effects of recessive loss-of-function mutations would normally be masked by a redundant normal gene. Some examples of rhythms arising from mutations in plant genes affecting the clock are shown in Figure 7.30. Although genetic screens targeting clock mutations have turned up several potential components, there is still considerable discussion concerning the identity of the central components of the plant circadian oscillator. It is almost certain that the list is not yet complete.

COMPONENTS OF THE PLANT CLOCK. At the top of the list for likely clock components are two DNA-binding transcription factors: CCA1 and LHY. These proteins are founding members of a family of REVEILLE proteins that have very similar protein sequences and that bind to DNA via MYB domains. MYB-domain proteins are not unique to circadian functions but are used in a variety of regulatory contexts in plants. Phosphorylation of CCA1 by casein kinase 2 (CK2) influences its ability to bind DNA, and LHY can also be phosphorylated by CK2.

The importance of CCA1 and LHY first surfaced in strains overexpressing them one at a time. Overexpres-

sion of either CCA1 or LHY results in arrhythmicity. The fact that loss of either CCA1 or LHY results in a clock that has only a slightly short period length shows that the two proteins are mostly redundant. Because overexpression of either protein results in reduced expression of both *CCA1* and *LHY* mRNAs, probably both CCA1 and LHY act as negative elements in a circadian loop. The expression of both genes cycles with peaks of mRNA and protein near subjective dawn.

Another set of potential clock components consists of a family of *Arabidopsis* pseudoresponse regulators (APRRs) including TOC1 (also known as APRR1) and APRR3, APRR5, APRR7, and APRR9. In a typical two-component signaling system, a sensor histidine kinase is sensitive to and can be stimulated by an environmental signal. The sensor kinase then phosphorylates itself and subsequently transfers the phosphate group to the second component, a response regulator. The response regulator produces the response, often activation of pertinent output genes. Pseudoresponse regulators show extensive amino acid sequence similarity to true response regulators, but they lack the characteristic aspartate amino acid that receives the phosphate group from the first component. Nevertheless, it is likely that such proteins are involved in signal transduction. In the cyanobacterial clock described earlier, the *Synechococcus* gene *sasA* encoded a sensory histidine kinase that interacted with the oscillator component KaiC. All the APRRs examined are rhythmically expressed. TOC1 and APRR3 have peaks near subjective dusk. Loss of TOC1 results in a short period length and in reduced expression of CCA1 and LHY. This is consistent with a role as a positive element in a circadian loop. However, because the clock runs without TOC1, probably TOC1 is redundant with one or more of the other APRRs, including APRR3.

Several plant genes encode proteins involved in circadianly relevant light sensing. Mutations in these genes result in rhythm defects. However, the defects are due to defects in light sensing, which in turn affect the clock. For instance, ELF3 is a protein that cycles with a peak in expression like TOC1. Strains lacking ELF3 are arrhythmic in constant light but are rhythmic in constant darkness, and strains overexpressing ELF3 are still rhythmic. In a similar manner, expression of the *Gigantea* (*GI*) gene cycles, the GI protein interacts with a known light sensor, PHYB, and *GI* mutants have reduced expression of LHY and CCA1 and display diverse period length effects.

(A)

(B)

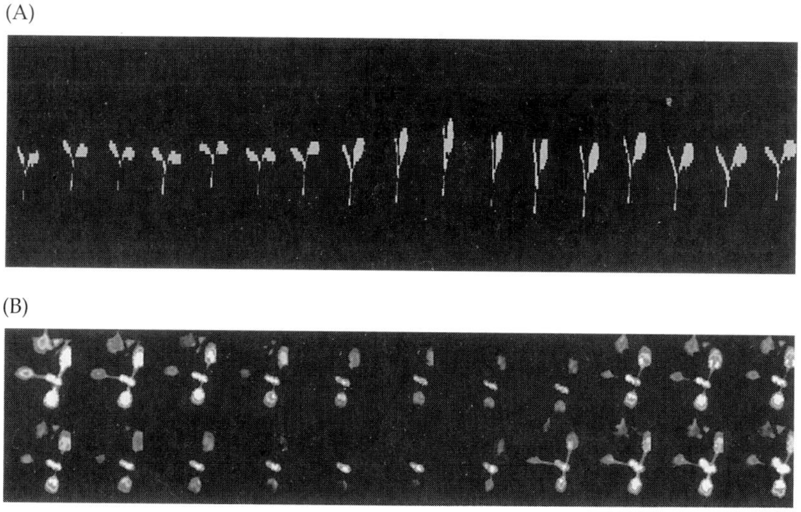

FIGURE 7.30 Rhythms in leaf movements (top) and bioluminescence (bottom) in plants. (Courtesy of R. McClung, T. Michael, and A. Millar.)

GENERATION OF THE OSCILLATION. Not enough is known about plant clocks for the daily molecular events to be described in a convincing manner. However, filling in some temporal gaps with conjecture and extension by analogy to be better-understood systems make it possible to imagine the circadian cycle in *Arabidopsis* (Figure 7.31).

CCA1 and LHY appear to act as negative elements in the feedback loop. Expression of CCA1, LHY, and perhaps other REVEILLE (RVE) proteins rises to a peak shortly after dawn. These proteins use their MYB domains to bind to the DNA in the promoter of *TOC1* and perhaps to other *APRR* genes. This binding blocks expression. The DNA-binding activity is modulated by CK2 phosphorylation and affects the kinetics of the clock cycle. Eventually, expression of TOC1 and perhaps the other APRRs is reduced and their levels decline to a low point in the day. TOC1 acts as a positive element in the feedback loop. Because GI, TOC1, and perhaps the other APRRs are required for expression of CCA1 and LHY, loss of TOC1 or GI means that expression of CCA1 and LHY also falls. By midday the abundance of the MYB factors is declining. When the level of these negative elements has fallen enough, expression of TOC1 can

resume and, in turn, facilitate renewed expression of CCA1 and LHY. In this way the cycle can repeat. Note that the overall effects and action of many of these proteins can only be inferred from their actions on each other. This means that TOC1, for instance, may act in a positive manner directly on CCA1, or it may act in a positive manner on an unknown factor that in turn acts directly to activate expression of CCA1.

Another factor may underlie the extensive redundancy seen among the putative components of the plant clock. Parallel independent coupled circadian oscillators may be operating in the same cells at the same time. Evidence for this is internal desynchronization, the appearance of multiple rhythms having different period lengths in the same organism or from the same cell (see Chapter 3). Such data have been described for gene expression rhythms arising from different parts of the same leaf and from mesophyll cells in *Arabidopsis*, and also for the bioluminescence and motility rhythms in the dinoflagellate *Gonyaulax*. Should multiple oscillators exist within single cells, then they must be normally coupled because special conditions are required to bring about their uncoupling to visualize the separate rhythms. If the oscillators are usually coupled, then loss of one oscillator would not be seen to result in arrhythmicity, but instead would result in a small period effect. Although there is clearly much yet to be learned about plant circadian systems before they can be described with the completeness of fungal or animal clocks, it is also clear that the answers will not be just a simple recapitulation of the insights from the other model systems.

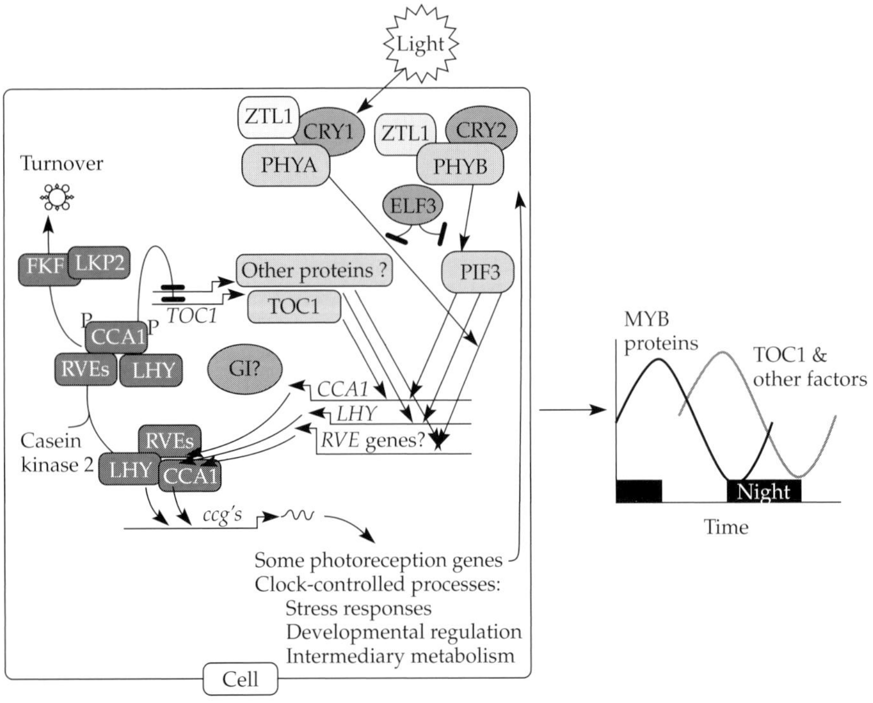

Many light-sensing proteins contribute to light resetting of plant clocks[14,37]

Plant rhythms are as sensitive both to light and temperature cues as fungal rhythms are. Although the mechanisms underlying temperature resetting must await a better explanation of the oscillator itself, light resetting responses are beginning to be understood. Light provides both an environmental cue and the source of energy for plants. Thus plants have developed a broad repertoire of proteins to aid in the detection and utilization of light. Light is sensed

FIGURE 7.31 Molecular components of the plant clock. A schematic view of the regulatory interrelationships among putative components of the higher plant circadian oscillator. During the night, CCA1 and LHY, and possibly additional proteins that bind to DNA using MYB domains, are made. These proteins are phosphorylated and are envisioned as negative elements in the feedback loop, blocking the expression of TOC1 (APRR1) such that the levels of these activators fall. Without the activators, expression of CCA1 and the other MYB proteins will fall. Degradation of CCA1 (and the other MYB proteins?) late in the day is expected to be mediated by FKF, LKP2, and related proteins. As the MYB-class proteins are degraded, TOC1 and related proteins are freed to reactivate transcription of *CCA1*, *LHY*, and the other negative elements. *ccg*, clock-controlled gene. (From Dunlap, 1999.)

through the use of five different phytochromes (PHYA through E) and two cryptochromes (CRY1 and 2). For instance, high-fluence red light is sensed with PHYB, PHYD, and PHYE; low-fluence blue light with CRY1 and PHYA; and high-fluence blue light with CRY1. Expression of PHYB, CRY1, and CRY2 is circadianly regulated, thus providing, along with ELF3, a means for gating light input to the clock.

There are also complex interactions among these receptor proteins. For instance, CRY2 and PHYB interact, and CRY1 is phosphorylated by PHYA in a light-dependent manner. In the aggregate, these proteins and interactions influence gene expression by acting through intermediates. Among these intermediates is PIF3, a bHLH (basic helix-loop-helix) DNA-binding protein. PIF3 binds to the promoters of phytochrome-regulated genes via G-boxes, which have sequences very much like those of the E-boxes bound by the bHLH domains of the BCC (BMAL–CLOCK) complex. The ultimate effect of light is the rapid induction of LHY and CCA1, the likely negative elements in the plant circadian loop. Thus the mechanism by which light phase-shifts (resets) the clock in plants appears to be very reminiscent of how light resetting is understood in *Neurospora* and in vertebrates. Light resets the clock through the acute induction of expression of negative-element genes, whose action then resets the phase of the feedback loop. To continue the similarity, light seems to have no immediate effect on the expression or content of TOC1, a putative positive element in the loop. This result parallels what has been seen in *Neurospora* and in vertebrates.

In summary, even given the current level of understanding, plant clocks appear to share elements of similarity with oscillators understood in fungi and animals. Positive elements peak late in the day, toward dusk. These activate the expression of negative elements that peak early in the day near dawn. Activity of the negative elements may be modulated by CK2 phosphorylation. Light resetting works through the transcriptional induction of the morning-peaking negative elements, with little effect on the positive elements. Distinctions between plant and other circadian systems, however, are clearly going to emerge. A likely distinction will be the apparent relevance of multiple independent coupled oscillators, a precedent which may not exist in either fungi or animals.

A Diversity of Research Approaches Have Led to Similar Mechanisms

A generic core circadian feedback loop can be traced in many eukaryotes[16]

The first circadian feedback loops to be understood molecularly were in the eukaryotic model systems *Neuro-*

spora and *Drosophila*, and these systems have some similarities and differences between themselves and with cyanobacteria, vertebrates, and plants. In all organisms where they are known, rhythmic expression of the transcripts and proteins corresponding to the negative elements is the rule. All known circadian oscillators involve negative regulation of transcription as a part of the loop. However, this negative regulation may be brought about in different ways. In *Neurospora* and in animals, it is brought about by binding of negative element proteins to the PAS–PAS heterodimeric activator. In the cyanobacterial clock it may be brought about through protein–protein interactions or through changes in supercoiling of the bacterial chromosome.

In *Neurospora* and in animals, positive elements are in all cases heterodimers consisting of two different proteins that interact via PAS domains. Of these two proteins, one is constitutively expressed (WC-2, CLK, or CLOCK) and the other is rhythmically expressed (WC-1, CYC, or BMAL1). In addition to common function and regulation, extended similarity in the sequences of amino acids links the positive-element proteins WC-1, CYC, and BMAL1. Finally, in *Neurospora* and in animals, at least one of the negative elements has a dual role in the cycle, performing both the negative function of blocking the activity of the heterodimer and a positive function in promoting the expression of one of the heterodimer partners: FRQ promotes the expression of WC-1, PER–TIM the expression of CYC, and PER2 the expression of BMAL1.

Considerable diversity in mechanistic details is emerging among animal clocks[54]

Most animals use a molecule like PER to assemble a clock. Fungi and animals use PAS–PAS heterodimers as positive elements in circadian feedback loops. However, a striking amount of inventiveness and diversity exists both in molecular clock components and in the roles they play. A well-known nematode, *Ceanorhabditis elegans*, has a circadian clock but does not appear to have strong sequence homologs of either PER or FRQ in its genome. Different animals can have varying numbers of many of the corresponding genes, and comparative studies have revealed a surprising degree of plasticity in how they are used in cellular clocks. In *Drosphila,* for instance, PER and TIM are the central negative elements in the feedback loop. CRY is responsible for sensing light and interacting with TIM to reset the clock. *Drosophila* also contains a second TIM-like protein encoded by a gene called *timeout*. The TIMEOUT protein is essential for development and appears to play a role primarily in that rather than in rhythms. The sequence of the only TIM-like protein to be found in mammals looks more like TIMEOUT than TIM, and this protein is required for development, so its possible role in rhythms cannot be assessed. However, vertebrates may not use TIM in their clocks. *Drosophila* uses CRY primarily as a light sensor. Although this is still an area of active research, the mammalian CRY proteins may instead have primary roles in the circadian feedback loop itself. Among other vertebrates, zebra fish have two different genes encoding BMAL proteins that are differentially regulated, six (!) *Cry* genes, and three *Per* genes. Frogs of the genus *Xenopus* have a similar number. Further, in fish and frogs, the PER1 and PER3 proteins cycle in abundance during constant conditions. PER2, however, the PER family member with the strongest role in the oscillator in mammals, is not circadianly regulated at all, but rather is expressed only in response to light.

The many origins of circadian rhythmicity are still not entirely clear[11,16,63,64,65,82]

Despite the apparent plasticity in the identity and functions of individual components, the logic of the underlying feedback loop seems to be consistent between organisms within the fungal and animal clades. Figure 7.32 expands modestly on earlier figures to detail some familiar aspects of the circadian feedback loop model of circadian oscillatory systems examined in this chapter. Although centered on oscillators from the fungal and animal lineage, from the discussions of cyanobacterial and plant systems it is clear that aspects of this system will also hold true for a broader range of organisms.

A major common theme is the importance of negative feedback loops. As mentioned early in this chapter, it is possible to build biological oscillators in ways that do not require negative feedback, and yet this is how biological oscillators have been built. Further, all the

described oscillators within the fungal and animal lineage use PAS–PAS heterodimers as positive elements. They also share similarities in the way the loops are assembled and interconnected. The fact that the cyanobacterial circadian system had an origin distinct from those in eukaryotes seems likely, and at present too little is known concerning the molecular underpinnings of plant rhythms to make an informed guess about similarities with other clock systems. Given the similarities and differences noted among the circadian systems presently understood in the fungal and animal clades, it can be argued that these similarities all arose from convergent evolution of clocks arising completely independently. An alternative argument based on the same set of similarities predicts that differences have arisen from evolution following a single origin. Over the last 900 million years, fungal and animal clocks have diverged in the details just as the animal clocks have continued to diverge, as described earlier. However, few of the chronobiologists at work today were present for the Cambrian explosion 900 million years ago, so readers will need to decide these questions for themselves.

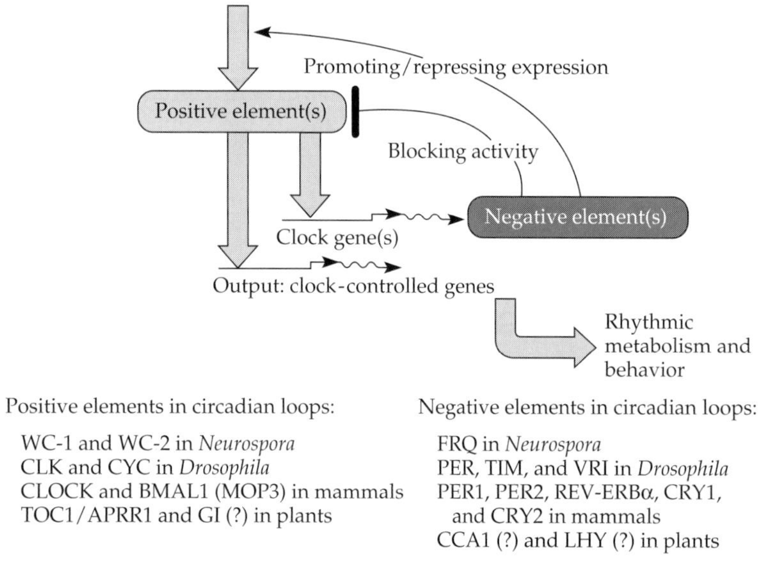

Current models for circadian oscillators continue to be refined[61,62,79,82]

Considering that in 1990 it was not possible to describe the molecular mechanism of any circadian oscillator, it is clear that this aspect of chronobiology is rapidly advancing. However, a great deal remains undescribed, not only in cyanobacteria or in plants' circadian systems, where only the broadest outline of the oscillator is understood, but also in fungi and animals, where at first glance the loop appears to be more or less complete. Early analyses documented the consistent importance of transcriptional regulation to generate the characteristic daily rhythms in clock gene expression that are understood as a hallmark of molecular rhythmicity. More recently, however, data arising from both fungal and animal systems suggests that posttranscriptional regulation of clock gene mRNA stability or translation may also contribute significantly to the determination of period length and phase in circadian systems. Expression rhythms of WC-1 in *Neurospora* or of BMAL1 in mice contribute to but are not essential for rhythmicity, and recent data from *Drosophila* even suggest that a severely weakened but still detectable clock can operate without rhythmic transcription of either *per* or *tim*.

The regulation of metabolism, signaling, and cell–cell interactions has produced in living things a variety of feedback loops. As

Positive elements in circadian loops:

WC-1 and WC-2 in *Neurospora*
CLK and CYC in *Drosophila*
CLOCK and BMAL1 (MOP3) in mammals
TOC1/APRR1 and GI (?) in plants

Negative elements in circadian loops:

FRQ in *Neurospora*
PER, TIM, and VRI in *Drosophila*
PER1, PER2, REV-ERBα, CRY1, and CRY2 in mammals
CCA1 (?) and LHY (?) in plants

FIGURE 7.32 Common themes in descriptions of eukaryotic clocks. (From Dunlap, 1999.)

noted early in the chapter, the first order of business for molecular chronobiologists was to discern which of these feedback loops lay at the core of cellular circadian rhythmicity, and many would argue that that work is done. However, those other noncircadian oscillatory loops remain, and many are probably coupled with the core loops to create circadian systems. In plants, as well as in fungi and animals, additional feedback loops can be coupled to the clock. Some loops are metabolically generated and couple within rhythmic cells; other loops couple via cell–cell communication. In either case these additional loops may influence the action of the principal circadian feedback loops.

When more than one core loop can function independently to generate a rhythm with full circadian characteristics, a cell can contain more than one clock and the clocks can be coupled. *Gonyaulax* and plants provide examples. More often, however, when a core clock loop is lost through genetic lesion, the coupled noncircadian loops can remain oscillatory but lose their circadian characteristics of circa 24-h period length, persistence, compensation, and phase stability. Fungi and *Drosophila* provide examples.

In a similar vein, the existence of robust peripheral circadian oscillators was noted in mammals. At the level of cell-to-cell coupling of feedback loops, the molecular analysis of rhythms begins to merge with the analyses of rhythms at the cell and tissue level described in previous chapters. Elucidating the roles of multiple feedback loops and oscillators in circadian systems will remain a thread of research for some time. Further, the rhythmic daily activation of transcription described here for the generation of rhythmicity also appears to be a major avenue for circadian output, the means through which time information generated by the clock is used to regulate the temporal activity of cells (see Chapter 8). The molecular analysis of circadian rhythmicity will continue to inform and be informed by chronobiology at all levels. The story is far from over.

SUMMARY

An understanding of the molecular bases of circadian rhythmicity is emerging after years of study through genetics and molecular biology. Circadian systems may well involve multiple oscillatory loops, but at present only relatively simple interlocked feedback loops can be described. Negative feedback underlies the regulation of all circadian feedback loops. Transcription and translation are nearly universally involved in the design and operation of these negative feedback loops. In general, circadian feedback loops appear to be assembled from genes and proteins that are not essential for life. They are assembled at least in part from genes and proteins whose sole role in the cell is in circadian timekeeping.

Cyanobacterial and plant clocks are assembled through negative feedback loops that require pro-

tein–protein interactions and that result in daily rhythms in transcription and translation of clock proteins. The eukaryotic fungal clock of *Neurospora* uses a single negative-element gene, *frq*, and two genes encoding positive elements, *wc-1* and *wc-2*, to make components to build a core clock. The positive elements interact via PAS domains to form a heterodimeric protein that acts as a transcriptional activator.

In the animal clock of *Drosophila*, the three negative elements PER, TIM, and VRI regulate the activity of a heterodimeric transcriptional activator. The heterodimer forms via interaction of CLK and CYC through their PAS domains. In the vertebrate clock, two families of PER and CRY proteins, along with REV-ERBα act as negative elements to regulate the expression or activity of a heterodimeric transcriptional activator. The heterodimer forms via interaction of CLOCK and BMAL1 through their PAS domains.

CONTRIBUTORS

Jay C. Dunlap wishes to thank Susan Golden, Carl Hirschie Johnson, Takao Kondo, C. P. Kyriacou, C. Robertson McClung, Hitoshi Okamura, Amita Sehgal, Joseph S. Takahashi, Charles J. Weitz, and Mike Young for their material contributions to Chapter 7.

STUDY QUESTIONS AND EXERCISES

1. Assuming the organism can live long enough, how would blocking protein synthesis for more than a day be expected to affect the clock? At what phase would an animal clock like that of the mouse, or a fungal clock like that of *Neurospora*, be expected to stop? Check your answer in Khalsa and Block, 1992.

2. How many times in evolution do you believe circadian timing systems arose, and why?

3. From which processes is the long (approximately 24 h) time constant understood to arise in the *Neurospora* system, the *Drosophila* system, the mammalian system, and higher plants?

4. If one could mix and match proteins and organisms, which proteins from what different circadian systems could one expect to substitute for one another?

5. In Figure 7.16, try to identify the four colonies that are not wild type, but instead are clock mutants. This was the screen executed by Kondo et al. that resulted in identification of the first cyanobacterial clock mutants.

6. Starting from first principles, design a circadian oscillator that does not involve transcription and translation. How can it be kept insulated from changes in metabolism? How can it be reset by changes in light and temperature?

7. On the basis of the molecular details presented, develop a model for temperature compensation of a circadian feedback loop. See also Chapter 3.

8. A physician has a rural practice. A patient shows up at the clinic door. She is the mother of two teenagers; in addition, she is caring for the teenage niece and nephew of her ne'er-do-well sister. She maintains that she is fine, but she is concerned that none of her charges have a social life. The reason is that they cannot seem to stay up past 8:00 P.M., so dating and parties are out of the question. She wants to know what is causing this. The physician is aware of the work on *hPer2* and advanced sleep phase syndrome,[71] so she asks for a blood sample from the teenagers and arranges to have their *Per2* genes sequenced. Disappointingly the DNA sequence reveals no genetic variation from that of most people. What might the physician do next? What other genes might lie at the root of the sleep problem and how could these possibilities be tested?

REFERENCES

1. Alberts, B., A. Johnson, J. Lewis, M. Raff, K. Roberts, and P. Walter. 2002. The Molecular Biology of the Cell, 4th Edition. Garland, New York.

2. Allada, R., P. Emery, J. S. Takahashi, and M. Rosbash. 2001. Stopping time: The genetics of fly and mouse circadian clocks. Annu. Rev. Neurosci. 24: 1091–1119.

3. Antoch, M. P., E.-J. Song, A.-M. Chang, M. H. Vitaterna, Y. Zhao, L. D. Wilsbacher, A.M. Sangoram, D. P. King, L. H. Pinto, and J. S. Takahashi. 1997. Functional identification of the mouse circadian *Clock* gene by transgenic BAC rescue. Cell 89: 655–667.

4. Aronson, B. D., K. A. Johnson, and J. C. Dunlap. 1994. The circadian clock locus *frequency*: A single ORF defines period length and temperature compensation. Proc. Natl. Acad. Sci. USA 91: 7683–7687.

5. Bae, K., X. Jin, E. S. Maywood, M. Hastings, S. Reppert, and D. R. Weaver. 2001. Differential Functions of *mPer1*, *mPer2*, and *mPer3* in the SCN circadian clock. Neuron 30: 525–536.

6. Balsalobre, A., F. Damiola, and U. Schibler. 1998. A serum shock induces circadian gene expression in mammalian culture cells. Cell 93: 929–937.

7. Bünning, E. 1960. Opening address: Biological clocks. Cold Spring Harb. Symp. Quant. Biol. 25: 1–9.

8. Cermakian, N., and P. Sassone-Corsi. 2000. Multilevel regulation of the circadian clock. Nature Rev. Mol. Cell Biol. 1: 59–67.

9. Chance, B., E. K. Pye, and J. Higgins. 1967. Waveform generation by enzymatic oscillators. IEEE Spectrum 4: 79–86.

10. Cheng, P., Y. Yang, and Y. Liu. 2001. Interlocked feedback loops contribute to the robustness of the *Neurospora* circadian clock. Proc. Natl. Acad. Sci. USA 98: 7408–7413.

11. Crosthwaite, S. C., J. C. Dunlap, and J. J. Loros. 1997. *Neurospora wc-1* and *wc-2*: Transcription, photoresponses, and the origins of circadian rhythmicity. Science 276: 763–769.

12. Crosthwaite, S. C., J. J. Loros, and J. C. Dunlap. 1995. Light-induced resetting of a circadian clock is mediated by a rapid increase in frequency transcript. Cell 81: 1003–1012.

13. Darlington, T. K., K. Wager-Smith, M. F. Ceriani, D. Staknis, N. Gekakis, T. D. L. Steeves, C. J. Weitz, J. S. Takahashi, and S. A. Kay. 1998. Closing the circadian loop: CLOCK-induced transcription of its own inhibitors *per* and *tim*. Science 280:1599–1603.

14. Devlin, P. F., and S. A. Kay. 2001. Circadian photoperception. Annu. Rev. Physiol. 63: 677–694.

15. Dunlap, J. C. 1993. Genetic analysis of circadian clocks. Annu. Rev. Physiol. 55: 683–728.

16. Dunlap, J. C. 1999. Molecular bases for circadian clocks. Cell 96: 271–290.

17. Edery, I. 2000. Circadian rhythms in a nutshell. Physiol. Genomics 3: 59–74.

18. Edmunds, L. N., Jr. 1988a. Biochemical and molecular models for circadian clocks. In: Cellular and Molecular Bases of Biological Clocks, pp. 298–367. Springer, New York.

19. Edmunds, L. N., Jr. 1988b. Cellular and Molecular Bases of Biological Clocks. Springer, New York.

20. Ehret, C. F., and E. Truco. 1967. Molecular models for the circadian clock. I. The chronon concept. J. Theor. Biol. 15: 240–262.

21. Emery, P., V. So, M. Kaneko, J. C. Hall, and M. Rosbash. 1998. CRY, a *Drosophila* clock and light-regulated cryptochrome, is a major contributor to circadian rhythm resetting and photosensitivity. Cell 95: 669–679.

22. Engelmann, W., and M. Schrempf. 1980. Membrane models of circadian rhythms. Photochem. Photobiol. Rev. 5: 49–86.

23. Eskin, A. 1979. Identification and physiology of circadian pacemakers. Fed. Proc. 38: 2570–2572.

24. Foster, R., and C. Helfrich-Forster. 2001. The regulation of circadian clocks by light in fruitflies and mice. Philos. Trans. R. Soc. London [Biol.] 356: 1779–1789.

25. Friesen, O., and G. Block. 1984. What is a biological oscillator? Am. J. Physiol. 246: R847–R851.

26. Froehlich, A. F., J. J. Loros, and J. C. Dunlap. 2002. WHITE COLLAR-1, a circadian blue light photoreceptor, binding to the *frequency* promoter. Science 297: 815–819.

27. Gekakis, N., D. Staknis, H. B. Nguyen, F. C. Davis, L. D. Wilsbacher, D. P. King, J. S. Takahashi, and C. J. Weitz. 1998. Role of the CLOCK protein in the mammalian circadian mechanism. Science. 280: 1564–1569.

28. Glossup, N. J. R., L. C. Lyons, and P. E. Hardin. 1999. Interlocked feedback loops within the *Drosophila* circadian oscillator. Science 286: 766–769.

29. Goldbeter, A., and S. R. Caplan. 1976. Oscillatory enzymes. Annu. Rev. Biophys. Bioeng. 5: 449–476.

30. Golden, S., C. H. Johnson, and T. Kondo. 1998. The cyanobacterial circadian system: A clock apart. Curr. Opin. Microbiol. 1: 669–673.

31. Goodwin, B. C. 1963. Temporal Organization in Cells. Academic Press, London.

32. Green, R. M., and E. M. Tobin. 2002. The role of CCA1 and LHY in the plant circadian clock. Dev. Cell 2: 516–518.

33. Hastings, J. W. 1960. Biochemical aspects of rhythms: Phase shifting by chemicals. Cold Spring Harb. Symp. Quant. Biol. 25: 131–143.

34. He, Q., P. Cheng, Y. Yang, L. Wang, K. Gardner, and Y. Liu. 2002. WHITE COLLAR-1, a DNA binding transcription factor and a light sensor. Science 297: 840–842.

35. Higgins, J. 1964. A chemical mechanism for oscillation of glycolytic intermediates in yeast cells. Proc. Natl. Acad. Sci. USA 51: 989–994.

36. Ishiura, M., S. Kutsuna, S. Aoki, H. Iwasaki, C. R. Andersson, et al. 1998. Expression of a clock gene cluster *kaiABC* as a circadian feedback process in cyanobacteria. Science 281: 1519–1523.

37. Johnson, C. H. 2001. Endogenous timekeepers in photosynthetic organisms. Annu. Rev. Physiol. 63: 695–728.

38. Khalsa, S. B. S., and G. D. Block. 1992. Stopping the biological clock with inhibitors of protein synthesis. Proc. Natl. Acad. Sci. USA 89: 10862–10866.

39. King, D. P., Y. Zhao, A. M. Sangoram, L. D. Wilsbacher, M. Tanaka, M. P. Antoch, T. D. L. Steeves, M. H. Vitaterna, J. M. Kornhauser, P. L. Lowrey, F. W. Turek, and J. S. Takahashi. 1997. Positional cloning of the mouse circadian *Clock* gene. Cell 89: 641–653.

40. Kondo, T., N. Tsinoremas, S. Golden, C. H. Johnson, S. Kutsuna, and M. Ishiura. 1994. Circadian clock mutants of cyanobacteria. Science 266: 1233–1236.

41. Konopka, R. J., and S. Benzer. 1971. Clock mutants of *Drosophila melanogaster*. Proc. Natl. Acad. Sci. USA 68: 2112–2116.

42. Lee, K., J. J. Loros, and J. C. Dunlap. 2000. Interconnected feedback loops in the *Neurospora* circadian system. Science 289: 107–110.

43. Liu, Y., M. Merrow, J. J. Loros, and J. C. Dunlap. 1998. How temperature changes reset a circadian oscillator. Science 281: 825–829.

44. Lodish, H., A. Berk, L. Zipursky, P. Matsudaira, D. Baltimore, and J. Darnell. 2000. Molecular Cell Biology. Freeman, New York.

45. Loros, J. J., and J. C. Dunlap. 2001. Genetic and molecular analysis of circadian rhythms in *Neurospora*. Annu. Rev. Physiol. 63: 757–794.

46. Lowrey, P. L., K. Shimomura, M. P. Antoch, S. Yamazaki, P. D. Zemenides, M. R. Ralph, M. Menaker, and J. S. Takahashi. 2000. Potential syntenic cloning and functional characterization of the mammalian circadian mutation *tau*. Science 288: 483–491.

47. McClung, R. 2001. Circadian rhythms in plants. Annu. Rev. Plant Physiol. Plant Mol. Biol. 52: 139–162.

48. Menaker, M. (Ed.) 1971. Biochronometry. National Academy of Sciences, Washington, DC.

49. Mori, T., and C. H. Johnson. 2001. Circadian programming in cyanobacteria. Sem. Cell. Dev. Biol. 12: 271–278.

50. Naidoo, N., W. Song, M. Hunter-Ensor, and A. Sehgal. 1999. A role for the proteasome in the light response of the timeless clock protein. Science 285: 1737–1741.

51. Nicolis, G., and J. Portnow. 1973. Chemical oscillations. Chem. Rev. 73: 365–384.

52. Njus, D., F. Sulzman, and J. W. Hastings. 1974. A membrane model for the circadian clock. Nature 248: 116–122.

53. Okamura, H., S. Miyake, Y. Sumi, S. Yamaguchi, A. Yasui, et al. 1999. Photic induction of *mPer1* and *mPer2* in cry-deficient mice lacking a biological clock. Science 286: 2531–2534.

54. Pando, M. P., and P. Sassone-Corsi. 2001. Signaling to the mammalian circadian clocks: in pursuit of the primary mammalian circadian photoreceptor. Science's STKE, Nov. 6, http://stke.sciencemag.org/cgi/content/full/OC_sigtrans;2001/107/re16.

55. Pavlidis, T. 1969. Populations of interacting oscillators and circadian rhythms. J. Theor. Biol. 22: 418–436.

56. Perkins, D., A. Radford, and M. Sachs. 2001. The *Neurospora* Compendium. Academic Press, San Diego, CA.

57. Plautz, J. D., M. Kaneko, J. C. Hall, and S. F. Kay. 1997. Independent photoreceptive circadian clocks throughout *Drosophila*. Science 278: 1632–1635.

58. Preitner, N., F. Damiola, L. Lopez-Molina, J. Zakany, D. Duboule, et al. 2002. The orphan nuclear receptor REV-ERBα controls circadian transcription within the positive limb of the mammalian circadian oscillator. Cell 110: 251–260.

59. Pye, E. K. 1971. Periodicities in intermediary metabolism. In: Biochronometry, M. Menaker (ed.), pp. 623–636. National Academy of Sciences, Washington, DC.

60. Ralph, M. R., and M. Hotz-Vitaterna. 2001. Mammalian clock genetics. In: Handbook of Behavioral Neurobiology, Volume 12, J. Takahashi, R. Moore, and F. Turek (eds.), pp. 433–453. Kluwer, NY.

61. Reppert, S. M., and D. R. Weaver. 2001. Molecular analysis of mammalian circadian rhythms. Annu. Rev. Physiol. 63: 647–676.

62. Reppert, S. M., and D. R. Weaver. 2002. Coordination of circadian timing in mammals. Nature 418: 935–941.

63. Roden, L. C., and I. A. Carre. 2001. The molecular genetics of circadian rhythms in *Arabidopsis*. Sem. Cell. Dev. Biol. 12: 305–315.

64. Roenneberg, T., and M. Merrow. 2002. Life before the clock. J. Biol. Rhythms. 17: 495–505.

65. Sehgal, A. (Ed.) 2003. Molecular Biology of Circadian Rhythms. Wiley, New York.

66. Shearman, L., S. Sriram, D. Weaver, E. Maywood, I. Chaves, et al. 2000. Interacting molecular loops in the mammalian circadian clock. Science 288: 1013–1019.

67. Shigeyoshi, Y., K. Taguchi, S. Yamamoto, S. Takeida, L. Yan, et al. 1997. Light-induced resetting of a mammalian circadian clock is associated with rapid induction of the *mPer1* transcript. Cell 91: 1043–1053.

68. Stanewsky, R., M. Kaneko, P. Emery, B. Beretta, K. Wager-Smith, et al. 1998. The *cryb* mutation identifies cryptochrome as a circadian photoreceptor in *Drosophila*. Cell 95: 681–692.

69. Takahashi, J. S., J. M. Kornhauser, C. Koumenis, and A. Eskin. 1993. Molecular approaches to understanding circadian oscillations. Annu. Rev. Physiol. 55: 729–753.

70. Taylor, B. L., and I. B. Zhulin. 1999. PAS domains: Internal sensors of oxygen, redox potential, and light. Microbiol. Mol. Biol. Rev. 63: 479–506.

71. Toh, K. L., C. Jones, Y. He, E. Eide, W. Hinz, et al. 2001. An hPer2 phosphorylation site mutation in familial advanced sleep phase syndrome. Science 291: 1040–1043.

72. Tyson, J. J. 1976. Mathematical background group report. In: The Molecular Basis of Circadian Rhythms, J. W. Hastings and H-G. Schweiger (eds.), pp. 85–108. Abakon, Berlin.

73. Van Gelder, R. N. 2002. Tales from the crypt(ochromes). J. Biol. Rhythms 17: 110–120.

74. Vitaterna, M. H., D. P. King, A.-M. Chang, J. M. Kornhauser, P. L. Lowrey, J. D. McDonald, W. F. Dove, L. H. Pinto, F. W. Turek, and J. S. Takahashi. 1994. Mutagenesis and mapping of a mouse gene, *Clock*, essential for circadian behavior. Science 264: 719–725.

75. Williams, J. A., and A. Sehgal. 2001. Molecular components of the circadian system in *Drosophila*. Annu. Rev. Physiol. 63: 729–755.

76. Winfree, A. 1976. On phase resetting in multicellular clockshops. In: The Molecular Basis of Circadian Rhythms, J. W. Hastings and H-G. Schweiger (eds.), pp. 109–129. Abakon, Berlin.

77. Winfree, A. T. 1967. Biological rhythms and the behavior of populations of coupled oscillators. J. Theor. Biol. 16: 15–42.

78. Winfree, A. T. 1980. The Geometry of Biological Time. Springer, New York.

79. Yang, Z., and A. Sehgal. 2001. Role of molecular oscillations in generating behavioral rhythms in *Drosophila*. Neuron 29: 453–467.

80. Young, M. 1998. The molecular control of circadian behavioral rhythms and their entrainment in *Drosophila*. Annu. Rev. Biochem. 67: 135–152.

81. Young, M. W. 2000. The tick-tock of the biological clock. Sci. Am. 282(Mar): 64–71.

82. Young, M. W., and S. A. Kay. 2001. Time zones: A comparative genetics of circadian clocks. Nature Rev. Genet. 2: 702–715.

83. Zatz, M. 1992. Perturbing the pacemaker in the chick pineal. Discuss. Neurosci. 8: 67–73.

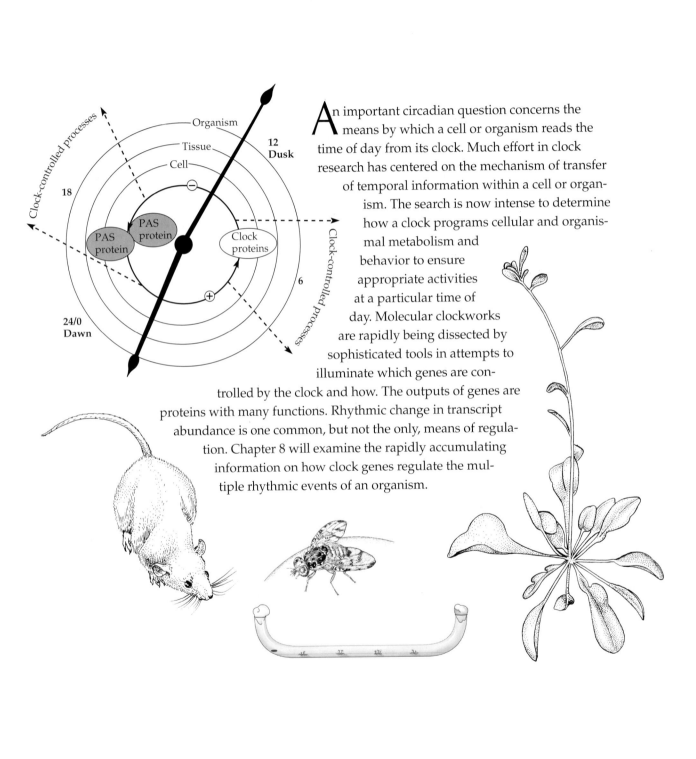

An important circadian question concerns the means by which a cell or organism reads the time of day from its clock. Much effort in clock research has centered on the mechanism of transfer of temporal information within a cell or organism. The search is now intense to determine how a clock programs cellular and organismal metabolism and behavior to ensure appropriate activities at a particular time of day. Molecular clockworks are rapidly being dissected by sophisticated tools in attempts to illuminate which genes are controlled by the clock and how. The outputs of genes are proteins with many functions. Rhythmic change in transcript abundance is one common, but not the only, means of regulation. Chapter 8 will examine the rapidly accumulating information on how clock genes regulate the multiple rhythmic events of an organism.

8

Adapting to Life on a Rotating World at the Gene Expression Level

Time is the substance from which I am made.

—Jorge Luis Borges

Introduction: An Important Aspect of Circadian Regulation of Biological Processes Is Conveying Output Signals at the Gene Expression Level[16]

Many biological processes are timed by the clock in almost all eukaryotes and in some prokaryotes[20,48]

Rhythmic outputs may be manifested at several functional levels in organisms. An example of an intracellular output is the regulation of transcriptional rate for a particular gene. At the tissue level, an example is daily leaf movement (Figure 8.1). The compound leaves of many legumes are capable of movement, folding at night and opening by day, and the cycle is controlled by a circadian clock. The level of the whole organism is exemplified by regulation of sleep–wake patterns. Whatever the process and whatever the level, ultimately all rhythmic behavior patterns have their basis in molecular events at the cellular level.

Programming specific events to occur in an organism at appropriate times of day is by far the most commonly recognized function of endogenous pacemakers. Understanding how this timing works is a basic problem in cellular regulation that has widespread ramifications in many fields of biology. By the time of the Cold Spring Harbor Symposium on Biological Clocks in 1960, a large literature was in place describing clock-controlled phenomena.

The focus of study was recognizing circadian behavioral rhythms and documenting their physical features (see Chapters 1 and 3) and their ecological advantages for organisms (see Chapters 2 and 4). Because clocks had not been localized at the molecular level, little progress had been made in understanding how the clock controls a particular rhythm. Regardless of the complexity of the organism or the

FIGURE 8.1 Clock-regulated leaf movement in the legume *Albizia*. Inset shows nighttime leaf folding.

that periodic changes were being regulated by the clock and were not identical with the clock. To make the distinction clearer, he named these regulated processes the hands of the clock. At the same meeting, Halberg discussed physiological regulation of rhythmic events as transport of the right amount of the right material to the right destination at the right time. He noted that different rhythmic functions could occur at different times of day and therefore have different phases, yet all may be synchronized by a specific timekeeping mechanism. Biochemical and molecular analysis of the hands of the clock was just beginning.

Some of the earliest work on biochemical rhythms involved cycling of carotenoids and chlorophyll molecules in relation to rates of photosynthesis. Before 1960, Hastings was measuring $^{14}CO_2$ incorporation as a measure of photosynthesis. He and others were discussing the idea that nucleic acid metabolism might play a role in clock function. Because rhythms were known in non-dividing, single-celled organisms, he postulated that changes in metabolism of DNA not associated with division or RNA metabolism might play a role.

After examining the incorporation of ^{32}P-labeled inorganic phosphate into both RNA and DNA over time, Hastings concluded that nucleic acid metabolism did not show large changes with circadian activity. Nevertheless, he pointed out the possibility that changes in nucleic acid activity with regard to clocks could have been buried by the "noise" and that techniques to separate individual species of nucleic acids would be useful. The subsequent development of these molecular techniques, as will be described in this chapter, facilitated rapid progress in the understanding of clock hands and how they are coupled to the clock mechanism.

regulatory output of the clock, the circadian clock is universally assembled at the level of the single cell, with biochemical and molecular mechanisms within the cell controlling clock output.

The advent of gene expression research in chronobiology is fairly recent[11,23,56]

Early work on circadian rhythms produced a rich legacy and cataloging of rhythmic biological processes. Targeting research toward understanding how and what the clock regulates had to wait for developments in the field of molecular genetics (see Chapter 7). However, Pittendrigh and colleagues noted in a 1959 report on *Neurospora crassa*, "We do not interpret the presence or absence of zonation as a reflection of the presence or absence of the clock, but rather as a coupling or uncoupling from the clock of the particular processes leading to zonation."[56] They rightly concluded that the process of zonation, asexual spore development in *N. crassa*, was distinct and independent from an underlying clock mechanism.

By the time of the Cold Spring Harbor symposium, scientists were clearly differentiating output from the basic timekeeping mechanism. In the opening talk, Bünning emphasized the importance of understanding

Large-scale molecular screening for clock-regulated output genes was begun in the 1980s[31,39]

By 1990 the issue of how information was being transferred from the clock to the molecular targets to result in rhythmic behavior had become a major question in molecular clock research. Chapter 7 pointed out that conserved, common features of cell-based circadian clocks include the use of negative and positive feedback based on transcription and translation of canonical clock gene products. In fungi, insects, and mammals, a pair of PAS domain–containing transcription factors acting as a heterodimeric complex supplies the primary positive role in the feedback. However, more than one way exists to build output regulation onto this feedback loop.

Although all steps in the process of gene and protein expression can in theory be clock controlled as a means of regulating output, the best-studied and possibly the most widespread means of regulation is clock control of transcription. Clock-controlled genes are called ccg's. In some cases the PAS domain transcription factors themselves are known to be involved in the transcription of output genes. In addition, translational regulation of noncycling messenger RNAs, sometimes by RNA-binding proteins, has also been found in diverse systems.

Knowledge about clocks and clock outputs has grown dramatically in the last decade, and with this growth the questions about clock regulation of output have expanded. One issue is the diversity of clock-controlled processes in all life. Another unknown is how much of the genome, both transcriptome and proteome (collective terms for all of the messenger RNAs and proteins encoded by an organism's genome), is under clock regulation in individual species. Still unresolved is the question of which processes are clock regulated and which are not. The relationship of clock regulation of these processes to fitness of the organism into its environment is another challenging question.

Many Cellular Processes Are Clock Controlled[16,48,69]

Rhythms adapt organisms to temporal niches

A common clock function is programming, the scheduling of cellular and organismal metabolism and behavior such that appropriate activities take place at appropriate times of the day (see Chapters 2 and 4). Rhythmic regulation of a cell's or organism's life may involve diverse activities ranging from basic energy metabolism to cognitive behavior. These rhythmic processes, many occurring across phyla, include phototaxis, enzymatic activities, photosynthetic capacity, insect eclosion, locomotor activity, sleep and many other behavior patterns, electrical or action potentials, respiratory function, cellular morphology, intracellular ionic concentrations, hormonal signaling, and circadian gating of the cell division cycle, to name just a few (see also Plates 1–4). The waveforms of these rhythms take a variety of shapes, from relatively sinusoidal to pulsatile.

Rhythmicity appears to be a widespread feature of normal physiology, and the classic constancy of the internal milieu may emerge largely through the action of rhythmic, mutually opposed underlying control systems. Many more rhythmic processes are well described at the physiological and cellular levels than at the molecular level. This chapter will focus on examples of rhythmic processes whose molecular underpinnings are at least partially understood.

Not all daily rhythms are circadian[3,28,36,46]

Many molecular processes are exogenously driven by the diurnal light–dark cycle. In addition to the entraining role that light plays by resetting the clock's oscillation, photic stimuli may directly drive, modulate, or modify an overt rhythm's waveform. Many processes are acutely inhibited or activated by light and thus are rhythmically driven by the light–dark cycle but show no rhythmicity under conditions of constant environmental light or darkness.

In the marine dinoflagellate *Gonyaulax polyedra*, for instance, the pH of the culture medium increases in the light and decreases in the dark as a result of carbon dioxide uptake and release that is caused by the predominance of photosynthesis and respiration, respectively. Under constant conditions, however, no rhythmic changes are observed. Another good example comes from rhythms of substances in the rat suprachiasmatic nucleus (SCN) (see Chapters 5 and 6). In this site, mRNA and peptide levels of gastrin-releasing peptide (GRP), as well as of vasoactive intestinal polypeptide (VIP), are co-localized in some SCN neurons. However, they exhibit oppositely phased rhythms during the light–dark cycle in vivo, with high levels of GRP during the light and VIP during the dark. Such rhythmicity is not found in constant darkness (Figure 8.2), although a circadian rhythm of VIP secretion is found when brain slices containing the SCN are cultured in vitro.

Interactions between the environment and the clock synchronize an organism to local time[3,16]

The amplitude of a rhythm in constant conditions is often, but not always, smaller than in natural environmental cycles. For instance, a variety of genes from plants, insects, or the filamentous fungus *Neurospora* are independently regulated both by light and by the clock. In constant conditions the abundance of messenger RNA for these genes accumulates during the subjective day and decreases during the subjective night, resulting in a rhythm of mRNA abundance. In an environmental cycle of light and dark, the light during the day phase increases the transcriptional rate of these genes, producing, in combination with the underlying circadian regulation, a higher amplitude of periodic mRNA accumulation. In other examples, however, the amplitude may be decreased in the natural world. The circadian rhythm of tyrosine hydroxylase enzyme activity in a cichlid fish retina is high during subjective night and low during subjective day. However, as the enzyme's activity is also elevated by light in a natural light–dark cycle, exogenous photic and endogenous clock influences are opposed, and rhythm amplitude is decreased.

Rhythms vary among different vertebrate species. The overt manifestation of many circadian molecular rhythms results from the collaborative action of both the endogenous clock and the exogenous natural environment. The same rhythm in different organisms may be regulated by both factors to different degrees. The rhythms of *N*-acetyltransferase (NAT) activity and mela-

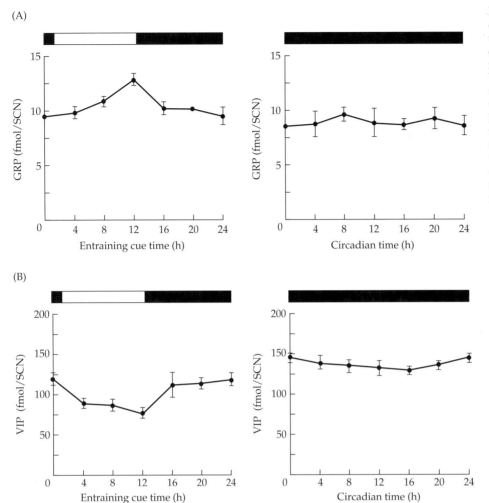

FIGURE 8.2 Diurnal and circadian profiles of two peptides under light–dark and constant dark conditions. (A) Gastrin-releasing peptide (GRP). (B) Vasoactive intestinal peptide (VIP). Rhythms seen in the context of both peptides under light–dark cycles disappear under constant dark conditions, showing that the amount of peptide is not controlled by the clock. Solid and open bars represent dark and light periods, respectively. (From Inouye, 1996.)

tonin levels in vertebrate retina are highest during the dark phase of a light–dark cycle, and this rhythmicity persists in the absence of the external cycle in chicken, quail, pike, and the frog genus *Xenopus*.

However, the light–dark rhythm ceases under constant conditions in the frog genus *Rana*. The primary regulatory role of the light–dark cycle in *Rana* retina extends also to the rhythm of photoreceptor disc shedding, which in this species, unlike most others examined, is not sustained in constant darkness. The biological significance of these interspecies differences in the control of similar rhythms is not understood beyond the fact that they reinforce the notion that circadian rhythmicity is but one method of regulation.

Gating of molecular events is a form of clock regulation[2,16,24,63,64,65,73]

QUANTITATIVE REGULATION.   In the circadian regulation of organismal responses, the clock commonly exerts generally quantitative modulating responses rather than qualitative regulation with complete presence or

absence of a response. In this way the clock and clock output(s) define or enforce temporal windows that promote or reduce the activities of the cellular machinery. This activity is known as circadian gating.

Gating is formally an output property, and a wide variety of processes can be gated in different organisms. One example is cell division. First described in the marine dinoflagellate *Gonyaulax*, the gating of cell division is now seen in a variety of systems, including humans. Other examples are responses to various stimuli. At the organismal level, feedback of output back onto the clock or onto input is common: Examples include locomotor activity in rodents, as will be discussed a little later in the chapter; regulation in the pineal in which the protein CREM regulates ICER, which in turn regulates melatonin synthesis, which can feed back on the mammalian clock in the SCN; circadian control of the *Drosophila* photoreceptor CRY, which thereby gates input; and circadianly controlled sleep and hormone release in humans (which will be discussed in Chapters 9 and 10). Such nested loops are not uncommon in circadian systems, but few of these loops are confined to the circadian oscillatory cells themselves.

A very common aspect of gating is the circadian control of light input, seen in the gating of light induction of the mammalian immediate early genes *c-fos* and *fra-2* (Figure 8.3); in photic resetting, as will be discussed shortly; and in clock regulation of light perception, such as in scorpions (as mentioned in Chapter 5). Such control of light is well known in plants, in the clock-gated "acute" response of light-induced genes, and in the induction of phytochrome B in *Arabidopsis*. The sections that follow will look more closely at two examples of gating.

GATING IN IMMEDIATE EARLY GENES. Photic induction of immediate early genes in the suprachiasmatic nucleus depends on the circadian phase. Light induction of *c-fos* expression in the mammalian SCN provides a good example of gating. In eukaryotes, the information contained in a gene's sequence is expressed as mRNA when auxiliary proteins called transcription factors bind near the start of the gene in the promoter region and act to help DNA-dependent RNA polymerase settle down and begin transcription at the "+1" or "start" site.

The most straightforward regulatory effects on gene expression happen through the alteration of rates of transcription via the binding of transcription factor proteins to promoter DNA. For instance, the nuclear phosphoprotein c-Fos, product of the *c-fos* proto-oncogene in mammals, encodes a transcription factor that can act as part of a heterodimeric protein complex called AP1 to bind specific DNA sequences, thereby altering gene expression by regulating transcription rates.

In the rodent SCN, after a light pulse is administered during the night, levels of *c-fos* mRNA and c-Fos protein are dramatically elevated from essentially undetectable levels. The magnitude of this photic stimulation depends on the circadian phase, showing a phase dependence similar to that already well described for light-induced phase shifts of overt locomotor rhythmicity (see Chapter 2). In constant darkness, light pulses administered during the subjective night will produce delayed or advanced phase shifts of the overt behavioral rhythm, as well as induce *c-fos* mRNA transcription, whereas light pulses delivered during most of the subjective day elicit neither *c-fos* mRNA nor phase shifts.

At what level does this gating occur? The phase dependency of the *c-fos* photoinduction means that *c-fos* is a target of the circadian clock; that is, its induction is clock controlled. The phase-dependent mechanism that gates this photoactivation is unknown and might oper-

(A)

c-fos: light in the subjective night

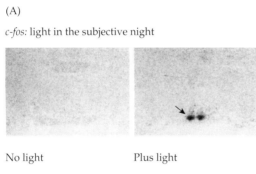

No light   Plus light

fra-2: light in the subjective night

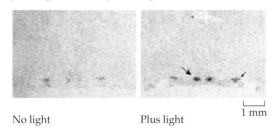

No light   Plus light   1 mm

(B)

FIGURE 8.3 Photic stimulation of *c-fos* and *fra-2* expression in the ventrolateral SCN is gated by the circadian clock. (A) Autoradiographs (in situ hybridization) of representative coronal brain sections from rats killed during subjective night (CT 19), either in darkness or after a single 30-minute light pulse. Large arrows, ventrolateral SCN; small arrow, constitutive *fra-2* signal in the supraoptic nucleus. (B) Messenger RNA levels graphed as the amount of transcript present relative to similar tissue sections collected from rats sacrificed during subjective day (CT 7) or night (CT 19), either in darkness or after a single 30-minute light pulse. CT indicates circadian time. (From Schwartz et al., 2000.)

ate very proximally in the signal transduction pathway, such as in the retina; and/or within SCN cells, such as at the receptor or second-messenger level; and/or on the *cis*-acting regulatory DNA sequences within the *c-fos* gene promoter.

Significantly, the "gate" appears to be specific to the SCN because light stimulation of *c-fos* mRNA and immunoreactive c-Fos protein levels in the intergeniculate leaflet within the rat brain do not depend on time of day of stimulation. This is consistent with observations that *clock* genes and *clock-controlled genes* show little or no cycling in areas of the brain outside of the SCN. The gate may also be specific to photic stimulation because activation of *c-fos* mRNA by D1 dopamine receptors in the fetal rat SCN is also independent of the phase of stimulation. Circadian control of *c-fos* expression persists in the SCN in vitro; electrical stimulation of the optic nerves induces *c-fos* mRNA in SCN tissue explants only during the subjective night. Thus neither the retina nor the rest of the brain is a required constituent of the gating mechanism, although either one may influence it.

The circadian control of c-Fos expression is probably exerted, at least in part, at the level of intracellular transduction pathways. Elevation of both Ca^{2+} and cyclic AMP leads to the phosphorylation of CREB proteins that bind to a cyclic AMP response element in the promoter region of the *c-fos* gene, 60 base pairs before the transcription start site. Note that in the SCN, CREB becomes phosphorylated after a 5-minute light pulse given during the subjective night but not during the subjective day.

Another distinct element in the *c-fos* promoter is the serum response element, 299 to 320 base pairs upstream to the transcription start site. The binding and phosphorylation of protein factors that bind to the serum response element are also of interest, and some data suggest that the mRNA encoding a MAP (mitogen-activated protein) kinase phosphatase activity exhibits phase-dependent photic activation in the SCN. The transcriptional activation of *c-fos* in the SCN probably involves coordinated regulation of cyclic AMP response element sites, serum response element sites, and other sites.

GATING OF PHASE SHIFTING. Light resetting of the circadian oscillator is gated in several systems. As described in Chapter 7, several circadian systems, including *Neurospora* and the mouse, have oscillators with negative elements whose amounts peak in the day phase, and these oscillators are reset through light induction of those negative elements. Specifically, light induces *per1* and *per2* in the mouse and *frq* in *Neurospora*. In each of these cases the degree of light induction of the clock gene is itself modulated by the clock (Figures 8.4 and 8.5). In terms of efficient resetting, this makes good sense for the organism, in that light seen at night needs to have a strong effect in resetting, whereas light seen in the day phase generally should not. This is just what is seen in the mouse in Figure 8.4. Because CREB may play a role in this photic induction, the mechanism for gating of *per1* induction may have elements in common with that described already for the immediate early genes.

(A)

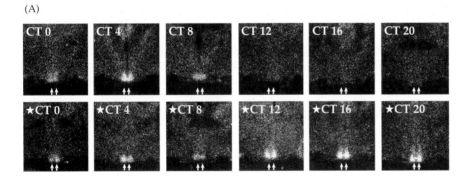

(B)

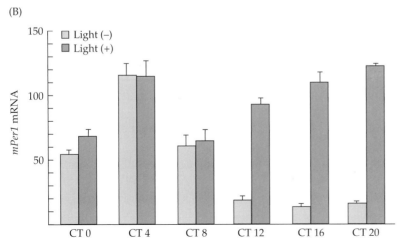

FIGURE 8.4 **Gating of the degree of light induction of the clock gene *mPer1* in the mouse SCN.** Light exposure at any time brings *mPer1* to subjective daytime levels. Mice were given a 30-minute light treatment at different times through the circadian cycle, returned to the dark, and sacrificed 30 minutes later. (A) The top panel shows *mPer1* mRNA in situ hybridization in the SCN (arrows); the bottom (starred) panel shows the light-treated SCN. (B) Densitometric analysis of data such as these from several mice at each time shows that light exposure always induces *mPer1* levels to within twofold of peak levels from the subjective day. Controls ($n = 4$) were not light treated. (From Shigeyoshi et al., 1997.)

A great deal is also known about the mechanism underlying gating of light induction of the clock gene *frq* in *Neurospora*. In terms of gating of the light response, one would anticipate a molecular output from the clock that would feed back to regulate input to the clock, in one or more additional clock-associated feedback loops. In *Neurospora*, both the VIVID (VVD) protein and an antisense transcript that arises from *frq* may provide this output. The first of these outputs has been well described. VVD is a novel member of the family of proteins that contain PAS domains. Its action describes an autoregulatory negative feedback loop that closes outside of the core oscillator but affects all aspects of circadian timing (see Figure 8.5). Expression of *vvd* is controlled by the clock, and VVD in turn feeds back to regulate the expression of several input and output genes, including itself.

One model for its action posits VVD, via its PAS domain, interacting with and transiently down-regulating the transcriptional activator WCC, thereby influencing both input and output. However, the fact that *vvd*-null strains are still robustly rhythmic shows that VVD is not required for circadian rhythmicity. Nonetheless, loss of the VVD protein has far-reaching effects on the perception of light and on the entire circadian system ranging from input as seen in the phase response curve, to oscillator function as measured by period length, to output as manifested in the phasing and expression levels of clock-controlled genes. Clock regulation of the immediate and transient repressor VVD contributes to circadian entrainment by making dark-to-light transitions more discrete; it supplies a molecular explanation for circadian gating of the light response at the level of the cell and oscillator.

The *frq* antisense transcript is transcribed, as its name implies, in the reverse direction of the strand of *frq* mRNA that encodes the FRQ proteins. In fact, although antisense *frq* RNA is several thousand nucleotides in length, it contains no long open reading frames that might encode proteins, so its action is apparently exerted through its complementarity with sense *frq*.

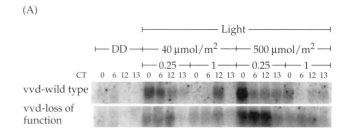

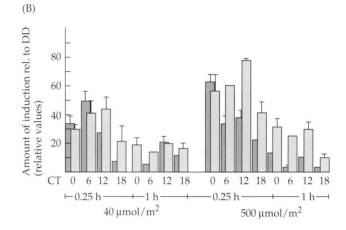

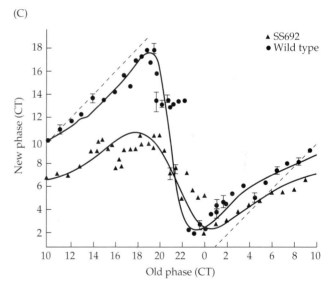

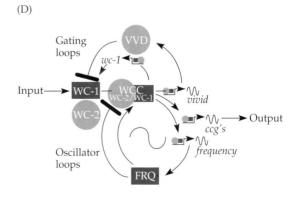

FIGURE 8.5 Gating of light input in *Neurospora* is influenced by the VIVID protein. *Neurospora* cultures kept in the dark were exposed to low light (40 μmol of light per m²) or high light (500 μmol of light per m²) at four different times in the subjective day, and the amount of *frq* induction was determined after 0.25 or 1 h compared to controls. (A) Northern analysis of *frq* light induction from wild-type or *vvd* mutant RNA from harvested cultures. (B) The light-colored bars are from VVD-null strains, and the dark bars are from wild-type strains. (C) Loss of VVD changes the shape of the phase response curve from weak to strong Type 0. The data from a light-induced phase-shifting experiment are plotted as phase transition curves. (D) Model showing how VVD as a clock-regulated output molecule could feed back outside of the core circadian loop to affect input. (From Heintzen et al., 2001.)

Antisense *frq* is light induced and is clock regulated with a peak in the opposite phase from sense *frq*. Loss of antisense *frq* greatly increases the magnitude of phase shifts arising from short light pulses, turning the typical Type 1 phase response curve into a Type 0. Because light-induced phase shifting is minimal at the time of day when antisense *frq* is maximal, and because loss of antisense *frq* potentiates phase shifting, antisense *frq* probably also contributes to circadian gating.

Not all biochemical rhythms necessarily lead to observed rhythms[75]

Observable cellular levels of macromolecules depend on both their synthesis and degradation rates. Rhythms in the levels of molecules with half-lives greater than zero will not exactly match rhythms in the rates of synthesis of these molecules. Obviously these considerations apply to the circadian control of transcription, posttranscriptional processes, and translation. As one example, the markedly reduced amplitude of the rhythm in GAPDH (glyceraldehyde-3-phosphate dehydrogenase) activity and abundance differ when compared to its rate of protein synthesis (measured by reduction of $NADP^+$ to NADPH, Western blot analysis, and [^{35}S]methionine incorporation, respectively, as will be discussed in this chapter). Another example, the rate of transcriptional synthesis of albumin mRNA, shows a very strong circadian rhythm in rat hepatocytes, cells derived from mammalian liver. Nevertheless, the mRNA's long half-life results in a nearly constant level of transcript over the course of the day.

Measurement of the transcription rates of randomly selected liver mRNAs from complementary DNA (cDNA) clones indicates that up to 20% of the genes expressed in hepatocytes may be subject to circadian transcription. However, the percentage of liver mRNAs with significant rhythms in their abundance appears to be much lower. Furthermore, in multicellular organisms the circadian signal will undergo even further modulation as the clock's outputs are conveyed to effector cells via the production of cytokines, the secretion of hormones, transmission by synapses, and even indirectly the rhythmic regulation of behavior.

Differentiating Molecular Outputs from the Clock Machinery Aids in Construction of a Working Clock Model[22,39,56]

Clock outputs can be distinguished from oscillator components[60]

One implication of a simple input→oscillator→output model is the existence of biochemical processes that are controlled by the circadian clock but are not part of the clock mechanism itself (see Chapter 7). Experimental change in the concentration or activity of an output molecule regulated by the clock, including elimination of the output, should not alter the generation of rhythmicity per se. This will hold true only if the output under consideration does not in any way feed back to the clock or onto clock input. A clock-controlled rhythm, whether related to motility, development, or luminescence, has been considered analogous to the hand of a mechanical clock: It can be removed without the progress of the underlying mechanism being altered, as evidenced by the movement of the remaining hand.

In the case of *Gonyaulax*, inhibition of photosynthesis does not alter the bioluminescence rhythm. On the basis of a single clock model, this has been interpreted to mean that photosynthesis represents an output, or hand of the clock. The same result, however, would also be expected if the rhythms of photosynthesis and bioluminescence were driven by two different clocks. The clock hand analogy should be accepted very cautiously for organisms in which known genetic lesions for a feedback loop do not exist.

Knowledge of output regulation may lead back to oscillator components[23]

As discussed in Chapter 7, three main research approaches have been taken with the aim of learning about the molecular footings of circadian rhythmicity. One of these approaches is to track the rhythmic regulation of an output back to the clock. If the biochemistry of a circadian-regulated process is known, then the molecules involved in its control are expected to have some relationship to the oscillatory mechanism itself. It should be possible to follow the output pathway backward to elucidate this mechanism.

The early input–output model specified a strategy to experimentally identify the molecules acting as oscillator components of the clock. The strategy would be effective even if the regulated molecule were also a component of the oscillator itself. In this case, a molecule regulating an output process might be an oscillator component that acts by affecting a rate-limiting step. An illustration of this general principle is the rhythm of glycolysis (see Chapter 7), in which ATP is a product of the glycolytic pathway and also a component of the oscillator.

In both intact yeast and yeast extracts, glycolysis is rhythmic, attributable to feedback on the slowest and rate-limiting step, the activity of the enzyme phosphofructokinase (PFK). When metabolic flux is high, ATP levels rise and PFK is inhibited, shutting down the pathway. The metabolic flux is now reduced, dropping ATP levels and reactivating PFK. In this way, PFK is the focal point of control and a part of the rhythm-generating machinery. Thus by augmenting the activity of the slow-

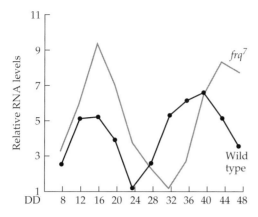

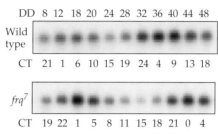

FIGURE 8.6 The *eas* (*ccg-2*) gene is an output gene based on standard criteria. The *Neurospora eas* gene is regulated by the circadian clock. Here transcript abundance is shown over time in a wild-type strain and a 29-h period strain (*frq[7]*). The data yield three notable observations: (1) The endogenous rhythm persists under constant conditions. (2) The period length of the rhythm reflects the genotype of the strain. (3) Inactivation of the gene has no effect on the clock. (Courtesy of D. Bell Pedersen.)

est step, the clock could control the activity of an entire output pathway.

The eas *gene in* Neurospora *helps elucidate circadian gene expression[5]*

A molecular example of a clock hand known not to affect the clock that regulates it consists of the products of the *eas* (also known as *ccg-2*) locus (Figure 8.6). In *Neurospora*, the *eas* (*ccg-2*) gene is rhythmically transcriptionally controlled by the *frq*-based feedback loop. Mutations in FRQ that change clock period length appropriately change the cyclic timing of *eas* (*ccg-2*) transcription and mRNA abundance, confirming that EAS (CCG-2) is regulated by the FRQ-based oscillator (see Figure 8.6).

The EAS (CCG-2)product is a hydrophobin protein, conferring an extracellular, hydrophobic surface of bundled rodlets made from cross-linked EAS (CCG-2)on fungal spores. Various mutations in *eas* exist, including some that confer complete loss of function. As expected, the mutant phenotype is a loss of hydrophobicity resulting in wettability and clumping of *Neurospora* spores. When examined at the overt or molecular levels in these mutants, including the loss-of-function mutant, the clock is found to operate normally. Both the *frq* transcript and the FRQ protein cycle show a wild-type periodicity and amplitude in Northern and Western blot analysis, respectively. Output rhythms, including those of other clock-controlled genes and conidiation, as will be discussed in this chapter, in constant conditions were also indistinguishable from wild-type rhythms, even though the morphology of the spores was altered.

An early and strict definition of a rhythmic output gene or process had three components: (1) The endoge-

nous rhythm persists under constant conditions, (2) the period length of the output rhythm reflects the genotype of the strain, and (3) alteration or inactivation of the gene or process has no effect on the clock. There are caveats to this definition, however. In particular, as the level of sophistication and analysis has increased in circadian systems, examples have emerged in which the product of the output regulation feeds back onto components of the oscillator or input. In this way output affects input, creating nested feedback loops surrounding the core oscillator.

Outputs May Have Effects on Oscillator Components[25,42,50,71]

NEUROSPORA AS A GOOD EXAMPLE. In addition to feeding back to affect input as described in the previous section, outputs can feed back to affect the clock itself, thereby generating a truly nested set of loops in the larger circadian system that surrounds the core oscillator.

The case of the VVD protein in *Neurospora* was described earlier (see Figure 8.5) and is restated here. VVD exerts its effect by interacting with the WC-1 protein, which is one of the PAS–PAS heterodimeric transcriptional activators in the *Neurospora* core feedback loop, in addition to being the blue-light photoreceptor and therefore integral to the light input pathway. Further, the synthesis of VVD is clock regulated and therefore could constitute a part of the oscillatory loop, except that VVD expression also depends on light, so in constant darkness expression falls to near zero. It is not surprising that VVD has little effect on the clock free-running in constant darkness; still, it provides an example of an output that can affect the clock itself, in this case creating a loop that closes within the oscillatory cell.

(A)

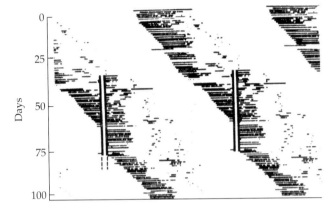

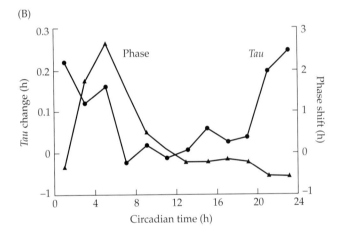

(B)

FIGURE 8.7 **Effects of social interactions on the hamster circadian system.** (A) The phase of the free-running rhythm in running-wheel activity in hamsters can be reset in response to a variety of factors that increase the hamsters' overall activity levels. Tested activities include exploration of a novel environment (including cage changes), social contact or interactions with other hamsters, and confinement to a running wheel. The top shows a single such resetting event at the time indicated by the arrow and the solid bar. The bottom shows the entraining effect of repeated social interactions at the times indicated by the solid black bars. (B) Three hours of novelty-induced wheel running affect the phase of the hamsters' rhythm, as reflected in the phase response curve shown here. Phase-shift response (triangles) and period length (circles) are shown. (A from Mrosovsky, 1988; B from Weisgerber et al., 1997.)

ed out, the surprising result was that the drugs themselves were having little effect, but the change in locomotor activity had a major effect on the clock. In fact, the rodent clock is sensitive to a variety of stimuli that affect locomotor activity, as Figure 8.7 shows.

The molecular basis of this effect has been traced to the oscillatory cells in the SCN. Locomotor activity affects the stability of the PER proteins there, and it can thereby reset the phase of the oscillator. The resetting mechanism, however, is still largely unknown.

Selected Case Studies Show the Molecular and Biochemical Correlations of Physiological and Behavioral Rhythms[31]

An overview indicates that clock regulation of cellular physiology and behavior must relate in some way to changes in cellular biochemistry[23]

A necessary part of understanding physiological or behavioral rhythms is to explain how these rhythms are causally related to underlying oscillations in the cell's biochemistry, such as to changes in the activity of a key enzyme or the levels of a key substrate. Important steps in developing such an understanding are to determine the biochemistry and molecular biology underlying the regulated process and then to identify a focal point or rate-limiting step where the process might be regulated by the clock, as described in Chapter 7 for PFK as a pivotal regulator of glycolysis.

An enzyme's activity might be altered in many ways, ranging from dramatic changes in absolute levels to subtle changes in structure. The powerful techniques of modern molecular biology have made it possible to investigate the mechanisms regulating gene expression, and good examples showing clock control at different levels in many systems have been found. These examples include regulation by transcriptional, translational, and posttranslational mechanisms, all of which affect

HAMSTERS AS ANOTHER GOOD EXAMPLE. Another well-described example of output feeding back on the oscillator comes from work in mammals. Social interactions can affect a broad spectrum of clock properties in hamsters. Clock regulation of locomotor activity in mammals was mentioned in earlier chapters. Locomotor activity is the most commonly used assay for circadian rhythmicity in mammalian systems.

Some years ago a study was being done to screen for clock-resetting effects of hypnotic drugs that would induce sleep in humans, and the outcome seemed to be that some of the drugs had strong phase-specific effects in some rodents but not in others. Of course, hypnotics not only put an animal to sleep, but as a result they also temporarily block the locomotor activity output, so the right control for a hypnotic is to give the drug at the same circadian phase to an animal that is not active then, something that could be done with diurnal and nocturnal animals. When the controls were finally sort-the final amount of gene product produced. In addition, the clock may use different mechanisms at other control points. Cofactors, second messengers, ions, pH, and protein modification can all affect the rate of specific steps in reaction pathways.

In Gonyaulax, behavior patterns and rates of protein synthesis are clock controlled[41,46,47,49]

BIOCHEMICAL CORRELATES. The levels or activities of the biochemical correlates of a rhythm must vary in a specific phase relationship with the rhythm itself. A classic and well-studied example of such a biochemical correlate is the bioluminescence rhythm of the unicell *Gonyaulax polyedra* (Figure 8.8). During the night phase the dinoflagellates acquire the ability to produce light. Between 50 and 100 times more light can be produced by cells during the subjective night phase than during the subjective day phase when cells are placed in constant conditions. Three components are necessary for light production: the substrate luciferin, a luciferin-binding protein (LBP), and the enzyme luciferase (LCF). In the cell, all three components are sequestered in a single specialized organelle called a scintillon.

The *Gonyaulax* bioluminescence rhythm provides an excellent example of a rhythm controlled at the posttranscriptional level—that is, where the circadian regulation happens after the gene has been expressed as mRNA. One correlate of the bioluminescence rhythm is that the number of scintillons is roughly ten times greater in night-phase cells than in day-phase cells.

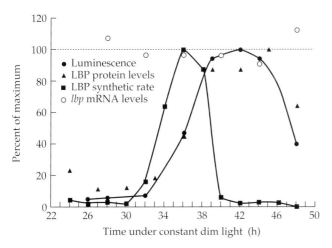

FIGURE 8.8 Temporal activity of LBP. Shown are densitometric scans of *lbp* mRNA (open circles), protein (triangles) and synthetic rate (squares) from Northern and Western blots and pulse-labeled autoradiographs, respectively, as well as the bioluminescent capacity of the cells (filled circles). *lbp* mRNA is not rhythmic. The rates of LBP synthesis peak prior to LBP protein and bioluminescence, which rise concurrently to a maximum and remain high for about 6 h. (From Morse et al., 1989.)

These organelles are formed de novo at the beginning and degraded at the end of each night phase. This activity can be quantified by a count of the number of organelles reacting with antibodies directed against LBP or LCF in cell sections. The disappearance of the organelles and their contents has been confirmed by Western immunoblot analyses of protein extracts taken from algae at different time points within the circadian cycle (see Figure 8.8).

These studies indicate that the light-producing proteins within the scintillon are degraded rather than simply moved to a different subcellular location. In addition, the levels of both LBP and LCF proteins rise and fall about tenfold over the circadian cycle. Although no antibodies are available to measure the enzyme luciferin, this component of the system can be conveniently measured by virtue of its endogenous fluorescence. Luciferin fluorescence of intact cells or their extracts also varies roughly tenfold over the circadian cycle.

The amounts of all three biochemical components of the bioluminescence reaction vary in phase with cellular bioluminescence, indicating that they represent true biochemical correlates to the behavioral rhythm. The molecular basis of these rhythms was further pursued with the expectation that the rhythms in protein content would be the result of underlying rhythms in mRNA coupled with nonrhythmic translation. Surprisingly, however, the reverse was the case (see Figure 8.8): The clock was controlling the rate at which protein was made from a constant pool of *lbp* mRNA, as well as the rate of LCF synthesis. Further research showed that this translational control arises from specific sequences of bases in the 5' untranslated region of the *lbp* mRNA that are bound by proteins capable of blocking translation. The clock apparently controls the synthesis of these proteins. Similar proteins and activities have now been identified in *Chlamydomonas*, another algal model system.

Different rhythms within an organism exhibit different acrophases (peaks) and durations. Clock regulation of cellular processes orchestrates and organizes physiology and behavior along the temporal domain, ensuring that rhythms are appropriately integrated and sequenced for concerted action.

CIRCADIAN REGULATION OF MULTIPLE ACTIVITIES AT DIFFERENT PHASES IN *GONYAULAX*. Several different rhythms have been measured in *Gonyaulax*. In general, the peak (acrophase) of each activity or behavior appears to be appropriate: the middle of the night for bioluminescence, and daytime for photosynthesis (Figure 8.9). *Gonyaulax* also exhibits circadianly regulated motility and aggregation. The rhythm of aggregation is probably associated with daily vertical movement. Cells accumulate near the surface and photosynthesize

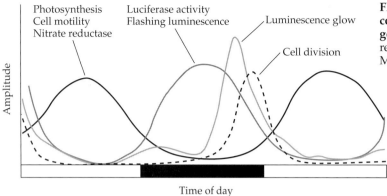

FIGURE 8.9 **Peak phases for some distinct clock-controlled phenomena in the single-celled dinoflagellate *Gonyaulax polyedra*.** Solid and open bars represent day and night phases, respectively. (From Mittag et al., 1995.)

in the daytime, whereas at night they descend vertically to depths at which inorganic nutrients are more abundant. The reason for confining cell divisions to a certain time of day is not immediately apparent, especially because its acrophase differs among different species.

In an extension of the studies already described for LBP and LCF, the rates of synthesis of many individual proteins have been measured at 2 h intervals and found to vary over the circadian cycle. They were seen to have different acrophases, which appear to fall into three major time slots: late day to early night, middle of the night, and late night to early day. Interestingly, durations of increased synthesis rates may also differ; for some proteins, increased synthesis persists only a few hours, but for others it lasts up to 12 h. It is tempting to interpret this variation as an indication of the turnover time of the protein in question, but this point remains to be established.

NITROGEN METABOLISM IN *GONYAULAX* AND THE PHOTOSYNTHETIC BACTERIUM *SYNECHOCOCCUS*. Species of the autotrophic cyanobacterium obtain carbon and energy through photosynthesis (see Figures 2.20 and 5.1). Some are also able to fix N_2 and thus are not dependent on an inorganic nitrogen supply such as nitrates. The biological challenge is that the nitrogen-fixing enzyme nitrogenase is inactivated in the presence of oxygen, a situation clearly incompatible with the O_2 produced by photosynthesis. Some groups of cyanobacteria have solved this problem by cellular specialization in which cells of one type, called heterocysts, fix nitrogen and do not photosynthesize, obtaining nutrients from the other cells, which photosynthesize only. Other unicellular cyanobacteria have used the circadian clock to separate nitrogenase and photosynthesis in time rather than in space; nitrogenase is restricted to the nighttime, when photosynthesis, and thus O_2 production, is low.

A similar behavior can be observed for nitrate metabolism in the eukaryotic unicellular *Gonyaulax* (see Figure 2.4). In this case, however, the main source of nitrogen is not N_2 but nitrate, so the concurrent presence of nitrate-metabolizing enzymes and photosynthesis poses no difficulty. In fact, the simultaneous occurrence of nitrate metabolism and photosynthesis should be advantageous to the algae because the sun's energy can then be funneled directly into reduced nitrogen, as well as into reduced carbon.

The problem is that photosynthesis occurs at the surface of the water column, whereas nitrate, along with most of the other minerals and nutrients that the algae require, is abundant much deeper, between 10 and 20 m below the surface. To allow the organism to retrieve as much nitrate as possible during the daytime, the enzyme that catalyzes the first step in nitrate metabolism, nitrate reductase, is induced in the daytime.

Biochemical studies show that nitrate reductase activity peaks at the same time, midday, as photosynthesis, with an approximately fourfold rhythm. Immunolabeling experiments suggest that the amount of protein may increase from three- to tenfold during the day phase and also is compartmentalized to the site of photosynthesis, the chloroplast. Greater levels of this enzyme during the day than during the night presumably reflect the movement of the algae in the water column. As they descend 10 m or so at night, to an area where nitrate is plentiful, less nitrate reductase activity is required and the levels drop.

In Neurospora, clock-controlled genes play an important role in the regulation of physiology[16,17,27,39,56,58,66]

CLOCK REGULATION OF BIOLOGICAL PROCESSES. Whether metabolism, reproduction, or behavior is considered, clock regulation is a direct reflection of the biology of the particular system being studied. The 1959 Pittendrigh and Bruce paper[55] stated that *Neurospora*'s cyclic production of asexual spores, or conidia, had a periodicity of about 22 h in constant conditions. The 22 h periodicity was said to be maintained across a physiological temperature range and was believed to reflect an underlying biological clock.

Asexual development in *Neurospora* is clock controlled. The clock regulating the asexual developmental cycle of the ascomycete fungus *Neurospora crassa* has

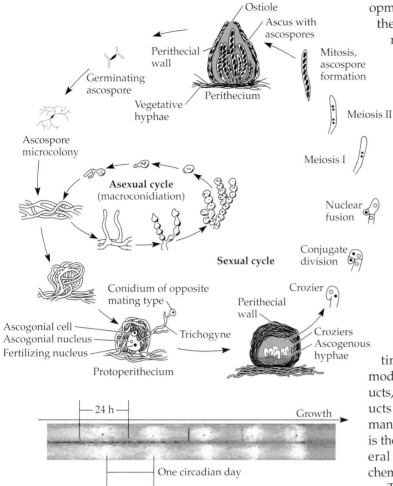

FIGURE 8.10 **Clock regulation in the life cycle of**
Neurospora. Depending on environmental conditions,
Neurospora can adopt either sexual or asexual subcycles, both
of which are clock regulated. In the sexual phase, circadian
regulation is manifest in the number of sexual spores shot
from mature perithecia over time. The most commonly
observed rhythm is that of developmental potential in the
asexual phase as monitored on race tubes (bottom). (From
Davis, 2000.)

opmental switch is also endogenously controlled by
the circadian clock. In constant conditions, clock
regulation of development is initiated in the late
subjective night, producing conidia for a
defined proportion of the day. In the early to
mid subjective circadian day the signal is
switched back and a region of undifferentiated hyphae is laid down until the next
subjective night, resulting in the band-
ing pattern of *Neurospora* asexual development. The clock controls the initiation and periodicity of development
but has no effect on the developmental
process itself. In clockless strains, development is no longer circadianly periodic and the abundance of conidia may be
altered but is otherwise normal.

MULTIPLE MOLECULAR AND BIOCHEMICAL
RHYTHMS. The clock is responsible for
organizing the assembly of the components
involved in asexual development. At specific
times in the subjective day, this large cellular
modification involves many genes and gene products, and many of these genes or their protein products are expected to be expressed in a time-of-day
manner. Although the production of asexual spores
is the best-characterized rhythm in *Neurospora*, several persisting rhythms at the physiological, biochemical, and molecular levels have been described.

The sexual fruiting bodies of the fungus, called
perithecia, contain hundreds of sexual spores
(ascospores) which, when mature, are expelled in a time
of day–specific manner through a perithecial neck. A
clock-controlled gene, *ccg-4*, has been identified as a
Neurospora mating-type pheromone gene, suggesting
that the clock is a direct regulator of sexual mating.
Other known rhythms apply to the production of CO_2,
lipid metabolism, various enzymatic activities, and heat
shock proteins.

The rhythmic production of aerial hyphae and
spores entails major cellular remodeling, with consequent gross changes in protein, carbohydrate, lipid content, and enzyme activities. Many experiments describing these rhythms have employed harvesting of
rhythmically developing cultures; it has been difficult to
know which of these rhythms is independent of conidial production and therefore whether clock regulation of
the rhythms is a direct reflection of circadian control of
development.

IMPORTANCE OF CLOCK REGULATION OF CORE METABOLIC
PROCESSES. In many cases the best-studied rhythmic
output processes that have led to cycling of biochemical
and molecular correlates in organisms have tended to
come from overt rhythms easy to observe and assay. For

been well studied (see Chapter 7; see also Plate 2, Photo
G). This developmental cycle requires just three cell
types (Figure 8.10). Upon germination of an asexual fungal spore, the macroconidiophore or conidia, a branch-
ing mycelium grows as a syncytium with incomplete
cell walls. The cytoplasm and nuclei are able to flow
freely within this syncytium. The aerial hyphae, the second cell type, are produced perpendicular to the substrate, and they elaborate the third cell type, the macroconidiophore, through the formation of complete
cross-walls between these vegetative spores.

Numerous environmental signals are capable of initiating production of conidia, including carbon deprivation in the substrate, desiccation and an increase in O_2
tension, and, notably, light and temperature. This devel-

example, analysis of bioluminescence rhythms in *Gonyaulax* led to the discovery of rhythms in LBP and luciferase activities responsible for the production of light. Dissection of melatonin rhythms in vertebrates led to the discovery of both enzyme activity and transcriptional rhythms for *N*-acetyltransferase (NAT), a key enzyme in the melatonin biosynthetic pathway. Studies of clock regulation of conidiation in *Neurospora* have identified rhythms in conidiation genes like the *eas* (*ccg-2*) gene.

In addition to clock regulation, other levels of regulation were often already known or could be identified when appropriately assayed. From this body of evidence the assumption has been made that clock regulation might be a higher form of cellular and organismal regulation used to control less essential processes, distinct from those imagined for "housekeeping genes" at the cellular level whose expression was expected to be relatively invariant with respect to time.

GLYCERALDEHYDE-3-PHOSPHATE DEHYDROGENASE AS A CORE PROTEIN FOR ALL CELLS. Early studies in *Neurospora* found rhythmic activity of glyceraldehyde-3-phosphate dehydrogenase (GAPHD) in developing cultures. GAPDH catalyzes the first energy-harvesting step in glycolysis, coupling the oxidation of the glyceraldehyde-3-phosphate substrate to its phosphorylation, and it is a pivotal enzyme in the glycolytic oscillator. As discussed earlier, whether the apparent rhythmicity of this core housekeeping protein's activity was generally clock regulated, or was only cycling in response to the cycling of the culture's developmental process was unknown. Screens for mRNAs that cycled in *Neurospora* turned

up a gene, *ccg-7*, which when sequenced, was found to encode GAPDH. Analysis confirmed that both the transcript and the enzyme activities of GAPDH are rhythmic under nondeveloping culture conditions (Figure 8.11),

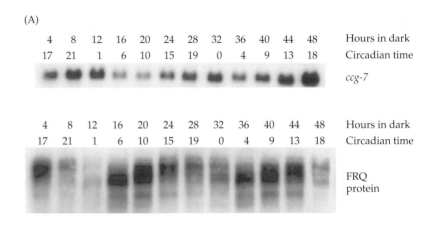

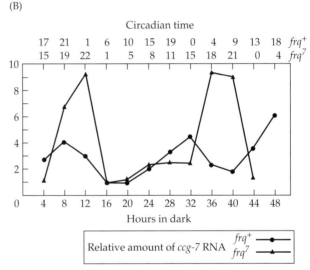

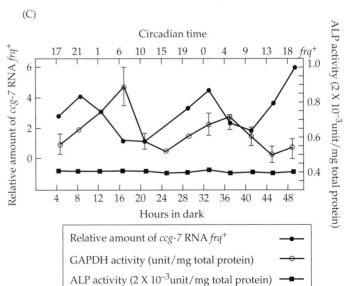

FIGURE 8.11 Circadian regulation of the pivotal glycolytic enzyme glyceraldehyde-3-phosphate dehydrogenase. (A) Northern blots showing rhythmic expression of *ccg-7*, a gene that was identified on the basis of its circadian expression but that turned out to encode the glycolytic enzyme GAPDH. (B) Densitometric analysis of mRNA rhythms from a wild-type clock strain (*frq⁺*) and a long-period mutant (*frq⁷*). (C) The rhythm in mRNA results in rhythmic expression of the GAPDH enzyme. Cycles are shown in both the wild type and a long-period mutant, with noncycling alkaline phosphatase (ALP) activity shown as a control. (From Shinohara, 1998.)

confirming a direct role for the clock in regulating this important enzyme.

Contrary to expectation, *ccg-7* is not regulated by development, and it is also not induced by environmental factors such as light, heat shock, osmotic stress, or carbon or nitrogen starvation. In *Neurospora*, *ccg-7* mRNA levels in general were found to be refractory to change except for the observed modulation by the biological clock, suggesting that this organism is restricting regulation of GAPDH to the clock. In various other organisms GAPDH is regulated by such diverse stimuli as osmolarity, interleukin-2, insulin, and the presence of pathogens or cytotoxic compounds. Although diverse, these stimuli suggest that changes in GAPDH activity could influence many other cellular activities and thereby facilitate adaptation to different challenges or growth conditions.

Intriguingly, a report of clock-regulated GAPDH from the dinoflagellate *Gonyaulax* suggests that the observation in *Neurospora* is not an isolated one and that there may be more global significance to using the clock to regulate GAPDH. Although both *Gonyaulax* and *Neurospora* are eukaryotes, they are evolutionarily extremely distantly related. The unicellular Dinoflagellata belong to the protists and appeared in the fossil record approximately 1.8 billion years ago, during the Precambrian period. Fossil fungi older than the Devonian, 410 to 360 million years ago, are rare; and possibly the earliest Zygomycota, to which the filamentous fungus *Neurospora* belongs, appeared in the Cretaceous period, 146 to 65 million years ago.

Gonyaulax has two nuclear genes encoding GAPDH: one encoding the cytoplasmic isoform, the other a chloroplast-targeted protein. While the plastid form has a high rhythmicity of synthesis, the much lower amplitude of abundance is attributable to the long half-life of the protein. As in *Neurospora*, GAPDH is a day-phase protein that is synthesized late at night or early in the morning.

As distinct from *ccg-7*, levels of *Gonyaulax* GAPDH mRNA do not show rhythmicity, but protein levels do, indicating posttranscriptional clock regulation. Analysis of ribosomes containing GAPDH transcript finds polysome fractions binding transcript at times when the protein is not synthesized, indicating that regulation may occur at the elongation step of protein synthesis. Once translated in the cytoplasm, the protein must still be transported to the chloroplast, and regulation might continue at this step.

Clocks may influence the very core of glycolysis in order to regulate one or more aspects of general metabolism in a circadian fashion, although the exact role for the rhythm of GAPDH is still uncertain in both organisms. If GAPDH oscillations exist to influence cellular activities, other enzymes in fundamental metabolic pathways may also be clock regulated in a manner coordinated with this

GAPDH rhythmicity so as to enhance time of day–specific modulation of metabolism. The question of whether GAPDH is under clock control in other branches of the evolutionary tree of life is an interesting one.

Clock regulation of GAPDH may be particularly interesting with regard to an increasing number of reports concerning nonglycolytic roles for GAPDH polypeptides. Such roles include DNA repair via uracil-DNA glycosylase activity; DNA, mRNA, and protein binding; translational control; protein kinase activity; a possible role in RNA export; interactions with microtubules; and interactions with cell membranes. These activities have been reported mainly for mammalian GAPDH, but such roles for the protein in lower eukaryotes cannot be ruled out a priori, considering the high degree of sequence conservation of GAPDH among all organisms. Of the roles listed here, only assays of uracil-DNA glycosylase activity for the GAPDH protein have been carried out with no detectable activity found in *Neurospora*.

Although numerous clock-controlled genes have been linked to organism-specific behaviors, including aspects of development, animal reproduction, bioluminescence, and photosynthesis, the GAPDH enzyme provides an example of clock regulation of a protein that is common to most living organisms and is important for fundamental aspects of metabolism, suggesting that circadian influences may be more pervasive at the cellular level than was previously anticipated.

Control of behavior in Drosophila *has been studied extensively*[10,22,26,57,73]

REGULATION OF LOCOMOTOR ACTIVITY. Pigment dispersing factor (PDF) is an important regulator of locomotor activity in the fruit fly *Drosophila melanogaster*. The lateral neurons have been genetically, molecularly, and immunologically identified as the anatomical location necessary for the production of rhythmic locomotor activity in the adult fly via the required cycling of *per* expression. A neuropeptide originally named for its ability to disperse pigment granules in the visual system of crabs, pigment dispersing factor is found mainly in a ventral subset of the lateral neurons (LNvs) of the *Drosophila* brain and has been proposed to be an important output signal controlling behavior.

PDF peptide is rhythmically released from the LNvs. Disruption of PDF expression in *pdf*-null mutant (*pdf*[01]) animals results in severe disruption of normal rhythmicity. In LD cycles, rhythms are phase-advanced by 1 h, and importantly, anticipation of lights on is lost, although rhythmicity could still be driven by the light cycle. The *pdf*[01] flies were shown to be weakly rhythmic for the first few days in free-running conditions (DD); then the majority of the *pdf*[01] animals became completely arrhythmic (Figure 8.12). A small percentage of *pdf*[01] animals retained the ability to maintain some

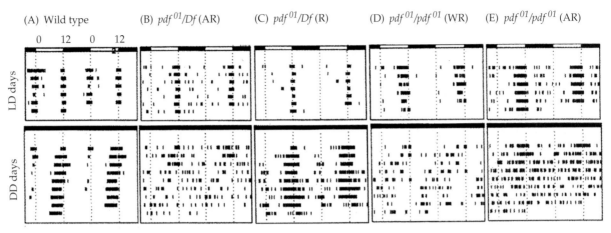

FIGURE 8.12 Representative actograms of flies bearing various *pdf* genotypes. *Df* indicates a deletion of the *pdf* locus. The open horizontal bars across the top indicate day; solid horizontal bars indicate either night (in an LD cycle) or DD free run (in a DD cycle). The activity records are double plotted: the top line tracks days 1–2; the next line, days 2–3; etc. Each tick mark represents an activity greater than 25 events per 30 minute bin. AR, arrhythmic; WR, weakly rhythmic; R, rhythmic by periodogram analysis. For DD days 1–9, the rhythmic wild type (A) displayed power of 195 (actograms with strong rhythmic behavior have power >100). Power values are: (B) 1; (C) 100; (D) 17; and (E) 1. (From Renn et al. 1999.)

rhythmicity over the 9-day course of the experiment, although their period lengths were 1 to 2 h shorter, possibly suggesting that abnormal activity rhythms were somehow feeding back onto the oscillator mechanism.

Another approach to understanding the role of the LNvs as circadian pacemaker cells required for a specific output pathway is to specifically ablate the LNvs cells. Transgenic animals have been made with one of two different cell death genes (*rpr* and *hid*) involved in the apoptotic cell death pathway fused to the *pdf* promoter. LNvs-specific expression of these genes results in cell death of all or most of the LNvs but not of surrounding cells in the fly brain. Ablation of the LNvs resulted in phenotypes very similar to that of the *pdf⁰¹* mutant flies with less than 20% of animals displaying the ability to maintain even weak rhythmicity.

These and other data are consistent with the LNvs as major neuronal pacemaker cells for locomotor activity and the rhythmically released PDF neuropeptides as critical circadian transmitters involved in organizing behavior. A current popular model posits the use of PDF to synchronize rhythms between cells. The ability to retain features of clock-regulated behavior strongly suggests that additional transmitters and possibly additional cells are required to produce completely normal circadian behavior.

VRI AS A CIRCADIAN-REGULATED TRANSCRIPTION FACTOR REGULATING THE PDF NEUROPEPTIDE IN FRUIT FLIES. Differential display has been used as a screen for novel clock-controlled genes in the fruit fly *Drosophila melanogaster*. Differential display techniques are a collection of methods using the polymerase chain reaction

(PCR) to amplify populations of harvested mRNAs into large amounts of cDNA, resulting in the ability to visualize altered expression profiles from the mRNAs using high resolution gel electrophoresis. The gene for a basic leucine zipper (bZIP) transcription factor, *vrille* (*vri*), was found to be down-regulated at ET 20 versus ET 14 in wild-type RNA (ET stands for "entraining cue time"), but expression levels did not change in *per⁰¹* flies (Figure 8.13).

The *vri* mRNA cycles in phase with the *per* and *tim* genes in both DD and LD cycles, and the rhythmicity is lost in *per⁰¹* mutants in LD cycles as well, indicating that the gene is not light responsive. Sequence analysis showed that *vri* has four E-box motifs within its promoter, the recognition site for the dCLK–CYC heterodimer, suggesting that *vri* might be directly regulated by these clock component transcription factors. In *clk* (*jrk*) and *cyc⁰* mutant flies, *vri* expression is no longer rhythmic, as expected, but also lowered.

A 2.8-kilobase-pair fragment of the *vri* promoter containing all four of the E-box motifs was fused to a luciferase reporter gene and transfected into insect cells in culture. In the test cells, but not control cells, an additional vector was transfected with the ability to express dCLK. Only the test cells were able to express luciferase, establishing that dCLK is capable of directly regulating transcription of *vri* in these cells. Expression was severely reduced when the E-box promoter motifs were mutated, reducing the ability of the dCLK–CYC heterodimers to bind.

The extent to which *vri* is a clock-controlled regulator of output pathways in the fly versus a part of the clock mechanism is not completely understood. The VRI transcription factor appears to play an important role in

(A)

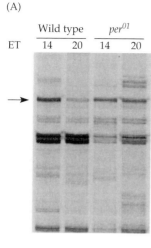

FIGURE 8.13 *vrille* **is a clock-controlled gene.** (A) Identification of a novel clock-controlled gene by differential display. Comparison of RNA from *Drosophila* heads isolated at ET 14 and ET 20 from either wild-type (WT) or *per*[01] flies. ET indicates entraining cue time, time in a 12:12 light–dark (LD) cycle. The arrow shows an amplified fragment expressed more strongly at ET 14 than at ET 20 in wild-type flies, but at an intermediate level at both time points in *per*[01] flies. Sequencing revealed that the fragment is part of the 3' untranslated region of the *vrille* gene. Levels of *vri* RNA oscillate in phase with *tim* transcript levels in wild-type flies (B) but are constantly expressed in *per*[01] flies in LD cycles (C). Results were reproduced at least five times. (D) Levels of *vri* RNA oscillate in constant darkness. Wild-type flies were entrained to LD cycles for 3 days, transferred to constant dark conditions (DD), and collected on the first day in darkness at the circadian time (CT, time in constant darkness) indicated. Quantification of *vri* and *tim* levels, relative to a constantly expressed gene (*rp49*), is shown for each time course in B–D. (From Blau and Young, 1999.)

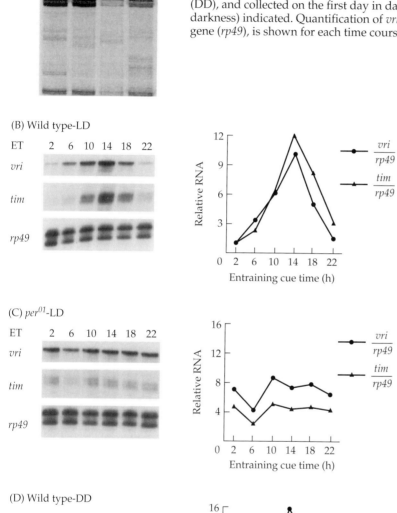

both. Underexpression of *vri* affects circadian behavior and can shorten the circadian period length by half an hour or more. Overexpression results in long period lengths or arrhythmia and severely affects the expression of PER and TIM. Both *per* and *tim* RNA levels are very low, and the resulting low levels of PER and TIM proteins remain cytoplasmic.

These experiments suggest that VRI might be important for repressing the *per* and *tim* promoters, or their activators dCLK and CYC, or possibly even for the nuclear stability of the PER and TIM proteins. In addition, *vri* mRNA expression co-localizes with TIM in the lateral neurons. VRI is required during *Drosophila* embryogenesis. The complete knockout is lethal to embryos, so a null (complete loss of function) phenotype has not been determined for circadian rhythmicity.

The relationship among clock factors, the VRI transcription factor, and the PDF neuropeptide is complex. Numerous experiments have been carried out to develop an understanding of these relationships. At this time many pieces of evidence strongly tie their regulation together, but the mechanisms involved in their interactions are not yet understood. PDF appears to be an indirect, posttranscriptional target of VRI. Levels of *pdf* transcript do not cycle, but levels of the PDF protein do, indicating that some level of posttranscriptional regulation is in place.

Interestingly, the rhythm of PDF is found only in the cell terminals projecting from the small LNvs and not in the cell bodies of the axons. Continuous expression of *vrille* so that it no longer cycles suppresses PDF protein accumulation but has no effect on *pdf* mRNA accumulation.

The *pdf* gene also appears to be a transcriptional target of dCLK and CYC, although not through the binding of an E-box. Mutations in *Clk* and *cyc* reduce *pdf* RNA levels in LNvs. At some level VRI and the clock are affecting the cycling of PDF levels by the rhythmic production of a specific factor necessary for the neuropeptide's expression, maturation, or release after transcription. Possibilities include rhythmic control of translation, or the control of PDF stability or transport or release of the hormone from axon terminals.

Molecular Mechanisms of Circadian Control Illustrate the Importance of Synthesis and Degradation Rates in Rhythms[73,75]

The clock has multiple points of control[74]

At the physiological level the effects of a rhythm in gene expression will be observed only if this rhythmicity is preserved through all the steps of regulation described in Chapter 7: transcript splicing and processing, export from the nucleus, translation into protein, and degradation of both the transcript and the protein. In a case mentioned earlier, rhythmic expression of the albumin gene in the rat liver failed to result in a significant rhythm in albumin mRNA or protein because of the long half-lives of both the mRNA and the protein. In plants, a significant (and generally ubiquitous) rhythm in transcription of the light-harvesting *Lhcb* gene does yield a rhythm in transcript levels, but not a rhythm in protein levels, because of the abundance and stability of the corresponding protein.

Analysis of the most general case is shown through mathematical modeling in Figure 8.14. Even with a transcription rate rhythm that has an infinitely large amplitude, such as for a gene that is absolutely turned off at its "off" phase, in order for a transcript or protein to cycle in abundance, it must be unstable. In addition, the more stable either one of those is, the lower the amplitude of its cycling will be.

The curves in Figure 8.14A show this relationship schematically: With a 1 h mRNA half-life (the shaded curve), the absolute amount of transcript is quite low, but the rhythm amplitude, measured as the ratio of peak to trough, is very high. This ratio drops as the half-life climbs to 5 h and 10 h, and at 20 h the amplitude is only about 1.3-fold, although the absolute amount of transcript at peak has risen by 10-fold. This calculation assumes an absolute cessation of transcription at the minimum, but for most genes this will not be the case.

FIGURE 8.14 The effects of mRNA half-life and transcription rate rhythm amplitude on the observed rhythm in mRNA transcript levels. (A) These curves were calculated with the equation shown, for half-lives ranging from 1 to 20 h. Clearly, as stability (half-life) increases, more mRNA accumulates (as expected) but the rhythm amplitude falls. (B) Similarly, a lower-amplitude rhythm in transcription rate results in a lower-amplitude observed rhythm in mRNA levels. (From Wuarin et al., 1992.)

This problem is approached differently in Figure 8.14B, where the effect of nonzero rates of transcription is factored in at the minimum. Again, as half-life increases, the amplitude decreases (as shown in Figure 8.14A), but in addition, reduction of the rhythm amplitude in the *rate* of transcription itself (from infinite to a more biologically reasonable 5-fold) reduces the amplitude of the observed mRNA rhythm from 40-fold to about 4-fold. As observed rhythms are considered in the following sections, keep in mind the insights provided by these modeling studies.

The fidelity of the circadian signal is affected when output pathways have many steps[35,49]

One consequence of multistep output pathways in which several biochemical components are clock controlled is that the amplitude and waveform of the resulting physiological and behavioral rhythms may differ substantially from the possibly unidentified, underlying molecular rhythms. Oscillations in one step of a pathway do not necessarily result in rhythmicity in the next

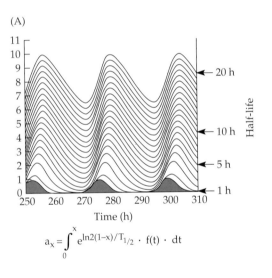

(A)

$$a_x = \int_0^x e^{\ln 2(1-x)/T_{1/2}} \cdot f(t) \cdot dt$$

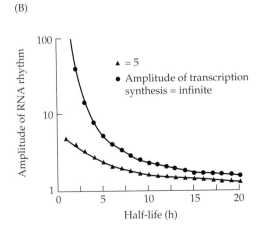

(B)

step or subsequent steps, so the circadian signal may degenerate along the intermediate steps of a multistep biochemical pathway.

For example, a daily oscillation in the levels of the hepatic enzyme cholesterol-7-hydroxylase appears to be driven in rats primarily by a rhythmic transcriptional mechanism employing a *trans*-acting DNA-binding factor, D-box binding protein (DBP), a member of the PAR (proline- and acidic-rich) leucine zipper transcription factor family. A 200- to 400-fold amplitude rhythm of DBP mRNA abundance generates a cholesterol-7-hydroxylase enzymatic oscillation of as little as 10-fold.

On the other hand, if more than one biochemical component in a pathway is clock controlled, the resulting physiological rhythm may have an amplitude much greater than that of any single component. *Gonyaulax* displays a 50- to 100-fold variation in the amplitude of the bioluminescence rhythm, but only a 7- to 10-fold variation in the levels of any of the biochemical components. The reason is that levels of both substrate and enzyme affect the ability of the cell to produce light, so the effect of changing both together is multiplicative. A 7- to 10-fold change in levels of both the enzyme and the substrate can thus produce 50- to 100-fold changes in bioluminescence.

Finally, the waveform of one step may not be faithfully copied in the next step. In *Gonyaulax*, the changes in protein levels produce a waveform that closely resembles a square wave, whereas the bioluminescence rhythm is more sinusoidal. One possible reason is that newly synthesized protein components must be transported and assembled into scintillons before they can function in light production. Thus the lag between the formation of protein and the formation of the organelles may account for the difference in waveform.

Molecular Control Mechanisms at the Cellular Level May Include Transcriptional Processes[5,31,39,41,42,59,73]

Gene transcription is a common target of circadian regulation[6,8,33,40,51]

Two major questions in clock output research are of general interest in the molecular clock field. The first is in understanding how the cell, and thereby the organism, reads the time of day from a functional clock. The second concerns the extent of cell metabolism that is clock regulated in a particular cell type or in an individual whole organism. Both of these questions have been addressed through the identification and examination of clock control of gene activity via the isolation of genes whose abundance is clock regulated over the course of a day, the clock-controlled genes (ccgs).

Clock-controlled genes have turned up in several systems, through both fortuity and targeted searches.

The development of molecular technologies has made possible the examination of whole genomes for rhythmic changes in mRNA abundance or transcription rate. The scope of clock control for metabolism as a whole will be considered later in this chapter; this section will address the underlying molecular mechanisms that result in rhythmic molecules.

Several well-studied examples of circadian rhythms of gene expression have shown that clock control is exerted at the level of mRNA abundance. To achieve this regulation, the clock must be exerting control on transcriptional and/or posttranscriptional processes. In some of these cases the mechanism of regulation is known to be control of the rate of transcription of the clock-controlled genes into mRNAs.

This finding has inspired research to identify the *cis*-acting DNA elements and *trans*-acting protein factors that might mediate these effects on the promoters of regulated genes. The region of DNA upstream, or 5′, to a transcribed region where RNA polymerase and its associated factors bind is called the promoter of a gene. The rates at which RNA polymerase transcribes and new polymerases bind can vary greatly and are determined by chromatin structure, and by the DNA sequences in the gene itself, which determine the auxiliary proteins that will bind to aid the recruitment of polymerase.

Transcription processes are used in circadian regulation in Neurospora[7,40]

SYSTEMATIC SCREENS FOR CLOCK-REGULATED GENES. Several global screens have been carried out in *Neurospora* through a variety of techniques. These screens include morning versus evening mRNA subtractive hybridizations, differential screens involving time of day–specific cDNA libraries, high-throughput sequencing of cDNA clones from two time of day–specific libraries, and cDNA microarray analysis of a *Neurospora* unigene set assembled from two expressed sequence tag libraries. Many clock-controlled genes have been isolated from *Neurospora*. The majority, as well as the best characterized, of these genes are specific to late night or morning, and most but not all of these are also regulated by the environmental cues of light and development.

NUCLEAR RUN-ONS. Transcription is one of the main control points in the regulation of differential gene expression. Transcriptional regulation can be examined with nuclear run-on experiments that determine the transcriptional rate of synthesis for specific genes. Nuclear run-on transcription studies are often used to determine the relative density of nascent mRNA chains as a measure of polymerase activity associated with specific genes at the time of nuclear isolation. Transcriptional synthesis rates can be analyzed through time with run-on analysis, in which rhythmicity in mRNA abundance is usually validated by visualizing the mRNA through Northern blot analysis, RNase protection experiments, or reverse transcriptase PCR.

CIRCADIAN REGULATION OF CCG-1 AND CCG-2. Both *ccg-1* and *ccg-2* are circadianly regulated at the level of transcription. *Neurospora* provides an example. Nuclei were isolated from nondeveloping mycelial cultures, timed to represent samples covering greater than one circadian cycle. The relative transcription rate of each of these timed samples was determined for the *ccg-1* and *ccg-2* genes by addition of necessary factors to the isolated nuclei, allowing transcription to proceed in the presence of [α-^{32}P]uridine triphosphate. Initiation of transcription de novo is not supported; only messages that had already initiated transcription before nuclear isolation will continue to transcribe and be radioactively labeled. The RNA can then be harvested from each sample and used as labeled probes against genes of interest.

Specific rates of transcription for both genes varied with time, and the greatest rates of transcription were seen in the early subjective morning, slightly before or coincident with the peak of mRNA abundance. Between separate experiments and the two genes, the amplitude of the synthesis rhythm varied three- to fivefold, and the maximum rate varied between circadian time (CT) 19 (subjective midnight) and CT 6 (subjective noon), but it occurred generally near CT 0, or dawn, for both genes. The minimum rate is consistent at CT 15, or early night.

Changes in the total amount of *ccg-1* and *ccg-2* mRNA over the circadian cycle indicate that a five- to tenfold change in clock-regulated mRNA abundance is typical for these genes, suggesting that the clock-controlled transcriptional changes here may entirely account for the rhythm in mRNA abundance. Nevertheless, studies like these do not preclude the possibility of other means of control, such as differing RNA stability over time. The results of nuclear run-on experiments do confirm that clock regulation of *ccg-1* and *ccg-2* (*eas*) is mediated, at least in part, by *cis*-acting regulatory elements within the promoter sequences.

CIS-ACTING SEQUENCES AND TRANS-ACTING PROTEIN FACTORS THAT INFLUENCE GENE EXPRESSION. Information from the clock to a particular gene will move as a cascade involving proteins and the nucleic acids to which they bind. At an upstream point in the cascade, the isolated factors must interface with a component or components of the clock mechanism. Clock output pathways are expected to be partly organism specific, depending on individual requirements, but the mechanisms used by the clock to confer temporal rhythmicity

also may be conserved. For transcriptional regulation of a gene to be evaluated, the DNA sequence of potential regulatory regions must be identified.

With the advent of genome-sequencing projects, determining genomic DNA sequences of many organisms is now as trivial as downloading the information from a public database; for examples for *Neurospora*, see http://www-genome.wi.mit.edu/annotation/fungi/ neurospora or http://www.mips.biochem.mpg.de/proj/ neurospora). However, many research organisms do not have genomic database resources. In such cases the genomic sequence of the gene of interest, including the translated and untranslated regions and several hundred to several thousand base pairs (bp) from both the 5' and 3' flanking regions, will need to be determined.

One approach to understanding clock regulation of genes whose cycling mRNA abundance is transcriptionally controlled has been the dissection of *ccg* promoters to identify "clock boxes," often called clock control regulatory elements. Once a specific sequence of DNA has been shown to be involved in the regulation, moving forward to isolate the *trans*-acting transcription factors that bind and control these elements is possible.

Regulation of the *eas ccg-2* gene is complex. The nuclear run-on experiments showed that the *Neurospora eas (ccg-2)* gene is transcriptionally regulated by the circadian clock, but numerous studies have shown additional levels of *eas (ccg-2)* mRNA regulation. The transcript accumulates with nutritional depletion and during asexual development and stress. Furthermore, *eas (ccg-2)* is positively regulated by blue light, and this regulation is independent of the clock. In an *frq* loss-of-function strain (see Chapter 7), light induction is not abolished. To sort out the basis of this complex regulation, deletion analysis of the *eas (ccg-2)* promoter has been carried out to localize the *cis*-acting elements mediating clock, light, and developmental control.

REPORTER CONSTRUCTS FOR DEFINING TRANSCRIPTIONAL REGULATION. By convention, the start site of transcription is termed +1, with each base pair (bp) downstream sequentially numbered: +2, +3, and so on. The region upstream of start is labeled with negative numbers, so 10 bp upstream of start would be –10, and so on. *Cis*-analyses of promoters are possible if the region upstream of start is placed in a cloning vector such that the promoters control the transcription of an easily assayed reporter construct. The transcriptional activity of the promoter is then followed through the activity of the reporter molecule. Many potential ways of assaying a reporter exist, including following the transcript or the protein levels or protein activity of the reporter gene.

Deletion and *cis*-analyses of the transcriptionally controlled *eas (ccg-2)* promoter have separated sequence-specific regions that confer clock, development, and light regulation. Promoter DNA was cut with restriction enzymes into discrete blocks and cloned into plasmid vectors to control the expression of the bacterial hygromycin phosphotransferase (*hph*) gene. A quantity of about 2 kilobase pairs of DNA 5' to the start of transcription was found to be sufficient to recapitulate normal regulation. A 68-bp sequence, the activating clock element (ACE), located at −50 to −118 bp, close to the start site of transcription, when deleted, disrupted rhythmicity of the reporter gene (Figure 8.15).

When the activating clock element was placed in the context of a heterologous promoter driving the *hph* gene and containing no other sequences from *Neurospora*, it was found sufficient to drive cycling of the construct. In addition, the region of DNA sequence lying between −625 and −1900 bp was found to confer repression of expression during mycelial growth that is relieved upon initiation of asexual development. A strong activator of transcription exists at about −1200 within this region. Sequences necessary for positive light regulation were determined to lie between −625 and −460 bp, and another region between −102 and −218 bp was found to be involved in the kinetics of light induction. In *Neurospora*, the regulatory information for clock, development, and light induction is found in separable motifs within the promoter of the *eas (ccg-2)* gene.

Transcriptional control in plants is widespread[4,21,22,25,67]

PLANTS HARVEST SUNLIGHT. It is easy to imagine reasons why plants have evolved the ability to tell time in order to respond appropriately to their environment. As sessile organisms, plants depend on daily changes in their surroundings that cannot simply be taken advantage of or avoided by locomotion. Importantly, photosynthetic organisms are uniquely dependent on the availability of sunlight for nutrition. The ability to anticipate daily changes in light quality and abundance might readily confer adaptive advantage for occupying a particular ecological niche.

Many clock-regulated processes in higher plants are known, including aspects of photosynthesis; stomatal opening; leaf, cotyledon, and flower movements; cell and hypocotyl elongation rates; hormone production; CO_2 uptake; Calvin cycle reactions; and photoperiodic timing of flower induction (Figure 8.16). Many clock-controlled genes have fortuitously been discovered and documented in the literature, most displaying rhythmic levels of mRNA, but also examples of posttranslational modifications (phosphorylation) of outputs have been documented.

PHOTOSYNTHESIS. Several genes involved in the control or biochemistry of photosynthesis have been particularly well characterized, with circadian expression usually peaking shortly after dawn. The chlorophyll *a/b*–binding genes, which encode proteins involved in light harvesting (*cab/Lhcb* gene family), display clock-regulated rates of promoter activation. Extensive resection and mutagenesis studies of the *Lhcb* promoters have shown multiple influences controlling gene expression. In addition to the clock, environmental light, mediated through the phytochrome red/far-red photoreceptors and blue-light photoreceptors, including cryptochromes, has strong regulatory influence.

Light regulation is intricate, with an acute and transient activation of gene expression after the onset of light and additional long-term control of both amplitude and period in the presence of light that is modulated by both fluence and wavelength. These regulators also interact with each other: The clock modulates the response to light, and light resets the clock. Resection analysis of a promoter indicates that many of these responses may occur within a short region of the promoter with overlapping binding sites for several transcriptional regulators. One of these regulators is CCA1, a MYB-related transcription factor necessary for promoter regulation

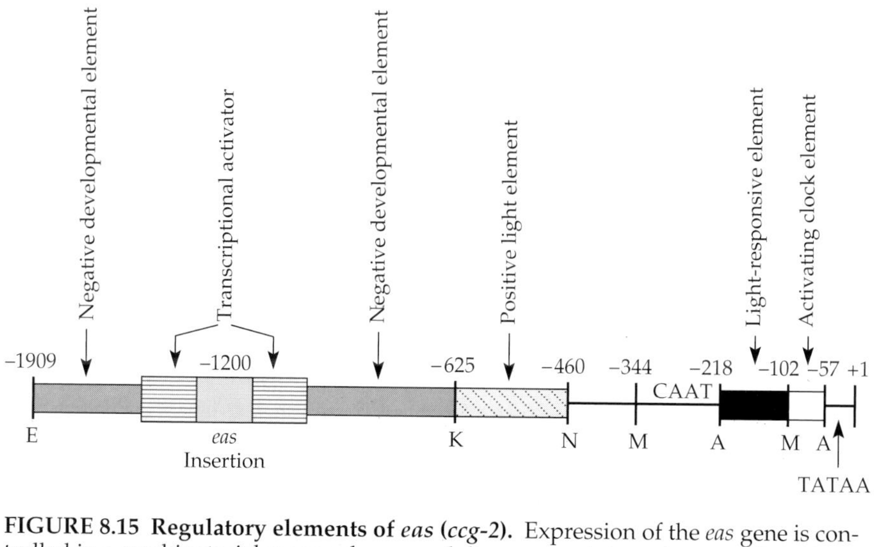

FIGURE 8.15 Regulatory elements of *eas* (*ccg-2*). Expression of the *eas* gene is controlled in a combinatorial manner by several diverse regulatory elements. Specific transcriptional regulators can, under appropriate conditions, bind to each of these elements to cause or block transcription, and the overall rate of transcription reflects their integrated actions. (From Bell-Pedersen, Dunlap, and Loros, 1996.)

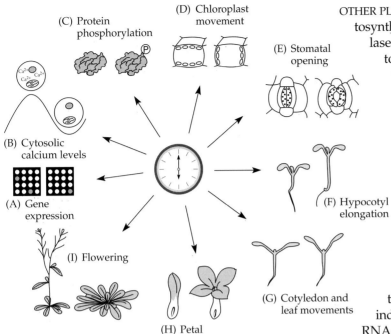

(C) Protein phosphorylation

(D) Chloroplast movement

(E) Stomatal opening

(B) Cytosolic calcium levels

(A) Gene expression

(F) Hypocotyl elongation

(I) Flowering

(G) Cotyledon and leaf movements

(H) Petal opening

FIGURE 8.16 Plant clocks control numerous biological processes. (A) The expression of several genes shows circadian rhythms. Two examples are the genes encoding the light-harvesting chlorophyll *a/b*–binding proteins (LHCB or CAB) and nitrate reductase (NIA2). Many of the proteins encoded by these genes are associated with photosynthesis and its related biochemical and physiological activities. The use of fluorescent differential display and high-density DNA arrays (shown here) to monitor global expression profiles should give an indication of the range of genes showing circadian control. (B) Cytosolic concentrations of free calcium have been shown to oscillate with a circadian rhythm in *Arabidopsis*. Considering the importance of calcium as both a secondary messenger and a cofactor for many enzymes, this might be a means by which the clock regulates a variety of cellular processes. (C) The clock regulates the phosphorylation of some proteins. *Kalanchoe fedtschenkoi* offers the best-studied example, exhibiting circadian activity of a kinase that phosphorylates phosphoenolpyruvate carboxylase. At a higher level of organization, chloroplast movement (D), stomatal opening (E), hypocotyl elongation (F), and cotyledon and leaf movements (G) in *Arabidopsis* all show circadian rhythms. In *Kalanchoe*, petal opening shows a circadian rhythm (H). The clock is also vital for synchronizing developmental processes such as flowering time (I). Mutations in all the putative clock-associated genes cause altered photoperiodic control of flowering. (From Barak et al., 2000.)

by phytochrome. CCA1 is also a strong candidate for an oscillator component in the plant clock. Many clock-controlled genes of plants may be controlled directly through binding sites on their promoters by *trans*-acting factors that are part of the circadian oscillator, either CCA1 or the related LHY transcription factor (see Chapter 7).

OTHER PLANT GENES. Many genes not involved in photosynthesis are also rhythmically expressed. The catalase genes *CAT2* and *CAT3*, involved in removing toxic H_2O_2 from the cell, cycle 12 h out of phase with each other, peaking in the subjective morning and evening, respectively. The mechanism of opposite phase coupling to an oscillator is not understood, and in the case of the *CAT* genes it is not known if they are coupled to the same or different oscillators.

Two closely related genes in the cress *Arabidopsis thaliana* have been isolated by homology to a cycling gene arising from a subtractive hybridization experiment using the white mustard, *Sinapsis alba*. The *Arabidopsis* genes cycle with a peak of expression around dusk, and the DNA sequences of the *CCR1 (Atgrp8)* and *CCR2 (Atgrp7)* genes indicate that they are highly similar to glycine-rich RNA-binding proteins. The proteins are capable of interacting with RNA and are localized within the nucleus, indicating regulatory roles for these proteins. Transcript cycling of *CCR2* is robust, with a greater than 40-fold amplitude; the protein also cycles with a healthy amplitude of about 8- to 10-fold.

The peaks in steady-state protein are about 4 h delayed relative to the transcript peaks, and transcript levels do not begin to rise until protein levels have reached their trough. The delayed oscillation of the RNA-binding protein relative to its transcript could be, but is not necessarily, the result of an autoregulatory feedback loop similar to those seen in oscillator mechanisms in fungi and flies. Interestingly, when *CCR2* is overexpressed, the cycling of its own promoter is strongly suppressed, although overall expression is not abolished.

When the cycling of other plant clock-controlled genes was examined in either free-running conditions (constant light) or LD cycles in the *CCR2* overexpression strain (Figure 8.17), no significant differences were found between the control plants and the overexpressing line for *Lhcb*, *CAT3*, and other cycling genes. However, cycling of the related *CCR1 (Atgrp8)* gene was fully repressed, with a concomitant reduction in overall expression levels.

Periodicity of the *CCR* transcripts is under control of *TOC1* (see Chapter 7). In *TOC1* short-period mutants the *CCR* gene expression rhythms are also shortened. The *CCR2* circadian feedback acts as both a target for the clock and a modulator of other downstream clock-controlled genes. An attractive model postulates this autoregulatory loop to be a "slave" oscillator receiving input from a "master" oscillator capable of controlling or modulating only a limited set of clock-regulated outputs.

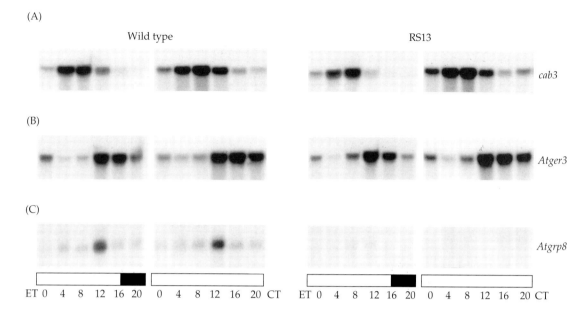

FIGURE 8.17 Influence of *CCR2* (*Atgrp7*) overexpression on other circadian-regulated transcripts. RNA was isolated from wild-type plants and the representative transgenic line RS13, which were harvested at 4 h intervals in a light–dark cycle (LD 16:8) and on the second day after transfer to continuous illumination. The Northern blot with 10 µg of RNA was hybridized with a *cab3* probe, which recognizes all the *cab* transcripts (A), a probe for a germinlike protein, *Atger3* (B), and a gene-specific probe derived from the 5′ untranslated region of the *CCR2* (*Atgrp8*) gene (C), respectively. Solid and open bars represent dark and light periods, respectively. (From Heintzen et al., 1997.)

GLOBAL GENE ANALYSIS. New technologies for examining gene expression are revolutionizing all fields of biology. Experiments examining clock regulation of individual genes, their transcripts, and protein products have already been considered in this chapter. The experiments are labor intensive, and many years of analysis by many experimenters have gone into the characterization of a relatively few genes. Recent advancements in high-throughput DNA sequencing and the resulting availability of genomic and cDNA sequence information has allowed the development of technologies to measure the RNA expression of thousands of genes in a single set of experiments.

We now have the capability to begin to understand clock control of a significant proportion of an organism's entire transcriptome by simultaneously examining the temporal regulation of large sets of genes. One technique, microarray-based expression analysis, is now in common use. In DNA microarrays, spots of DNA, either oligonucleotides or PCR fragments, are placed in a tightly ordered pattern on a glass microscope slide. Several hundred to thousands of genes can be represented on a single slide, and the position and sequence identity of each gene is known. Just as when a single DNA sequence is used to probe a Southern or Northern blot, these DNA sequences that are bound to the solid surface represent the probe against an RNA target.

Test and reference mRNA samples are harvested, converted to cDNA with reverse transcriptase, and then labeled, each with a distinct fluorescent probe. The microarray and cDNA target are incubated together. The level of gene expression relative to the reference sample for an individual spot can be monitored by examination of the amount of fluorescence from one dye or the other. Taking a timed series of RNA extracts and incubating them with duplicate microarrays allows a temporal profile of expression from the genes represented on the array. These methods produce enormous data sets, and much scientific analysis is involved in the analysis-of-variance statistical methods used to determine what constitutes a significant change in gene expression level. Cluster analysis is a method used to group genes according to the similarity of their expression patterns from microarrray experiments.

MANY GENES IN HIGHER PLANTS ARE CLOCK CONTROLLED. Microarray experiments in clock-regulated gene expression in *Arabidopsis* have shown approximately 6% of the genome to be expressed rhythmically, 453 genes out of a total of 8,200 genes examined. Many previously characterized clock-controlled genes were among the 453, lending validation to the statistical analysis used. When the cycling genes were clustered by the timing of peak expression, all phases were found to be well represent-

(A)

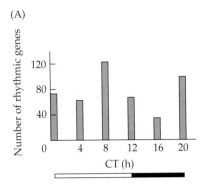

(B)

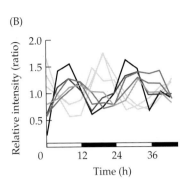

FIGURE 8.18 Regulation of clock-controlled genes in *Arabidopsis*. (A) The expression of different clock-controlled genes peaks at different times throughout the day. (B) Genes implicated in cell wall modification are clock controlled. (C) Speculative model for function of genes depicted in (B). (D) All these evening-phase genes possess a common promoter motif. Solid and open bars represent dark and light periods, respectively. (From Harmer et al., 2001.)

(C)

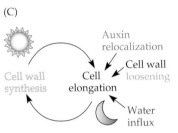

(D)

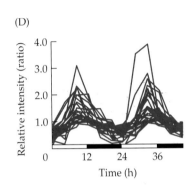

elements was fused to a luciferase reporter and shown capable of driving rhythmic luciferase. This potential evening *cis*-acting sequence was then mutated to change the sequence, resulting in substantial loss of the ability of the 130 bp *cis*-element to drive rhythmic transcription.

Several microarray studies have recently been published and are ongoing in many of the well-characterized model systems for clock study (Table 8.1 and Figure 8.19). Clearly this and related technologies in the near future will have tremendous impact on our knowledge of what the clock programs in specific organisms, as well as our understanding of the mechanisms used to couple the clock to output.

ed, with some underrepresentation in the middle of the night (Figure 8.18).

Numerous biochemical pathways were identified as having clock-regulated components, including genes involved in cell wall remodeling and water uptake. Sequences of promoters that peaked late in the subjective day were computationally scanned, and a highly conserved nine-nucleotide "evening element" was identified 46 times in 31 evening genes. The *CCR2* gene promoter contains four of the evening elements. A 130 bp region of *CCR2* promoter DNA containing one of these

Transcriptional regulation contributes to rhythmic gene expression in vertebrates[3,8,9,13,18,19,30,38,54,55,70,72,74]

OVERVIEW. Many studies have shown a broad array of functions under circadian regulation in the vertebrate eye (see Chapter 5). Retinal adaptation to daily light–dark transitions is complex and involves a massive

---

TABLE 8.1	The extent of circadian control of gene expression in different organisms.

Organism or part(s) studied	Phylogenetic group	Percentage of genome that is clock controlled	Data source
Synechococcus	Cyanobacteria	100	Liu et al., 1995
Neurospora	Filamentous fungi	2–4	Nowrousian et al. 2003
Drosophila	Insects	1–4	Claridge-Chang et al., 2001; McDonald and Rosbash, 2001
Rat/mouse fibroblasts in culture	Mammals	2	Duffield et al., 2002
Mouse liver	Mammals	9	Akhtar et al., 2002
Mouse liver and kidney	Mammals	7	Kita et al., 2002
Mouse SCN and liver	Mammals	6–7	Panda et al., 2002
Mouse liver and heart	Mammals	8–10	Storch et al., 2002
Arabidopsis	Plants	2–6	Harmer et al., 2000; Schaffer et al., 2001
Gonyaulax	Marine algae	3	Unpublished data from O. K. Okamoto and J. W. Hastings

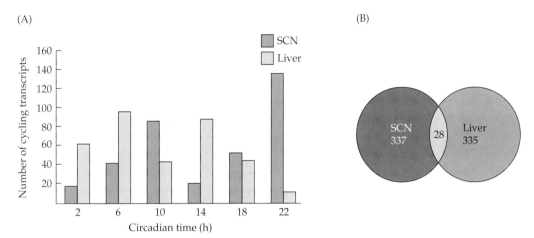

(A)

(B)

FIGURE 8.19 Circadianly regulated genes in the SCN and liver. Transcripts arising from 365 SCN genes and 363 liver genes were found to be circadianly regulated. (A) Distribution of the SCN and liver transcripts among circadian phases in each tissue. The evidence clearly indicates time of day–specific expression. (B) Venn diagram showing the degree of overlap among the cycling genes from the two tissues. Although a few genes (including those encoding cycling components of the clock in the SCN and peripheral oscillatory cells) were found to be circadianly expressed in both tissues, most cycling genes were tissue specific. (From Panda et al., 2002.)

amount of cellular remodeling with attendant changes in many genes and proteins. Much of this change is transcriptionally regulated. In the early to mid-1990s, several reports demonstrated that intact eyes, tissues, and isolated cells cultured from vertebrate eyes contained an independent circadian oscillator. In many cases individual molecular rhythms could be followed.

Circadian oscillators in the retinal photoreceptor cells of chick and the African clawed frog, *Xenopus*, have been studied in vitro. The mRNA levels for red-sensitive iodopsin, the predominant cone pigment in chicken retina, are regulated by a circadian clock in embryonic cell culture, with a two- to threefold increase in mRNA abundance just before subjective dusk. This increase corresponds to a phase just before the peak of cone photoreceptor disc shedding in vivo. Nuclear run-on assays show that iodopsin mRNA synthesis increases 3 to 6 h before expected light offset, demonstrating that the rhythm is generated, at least in part, at the level of transcriptional synthesis.

XENOPUS RETINA. One of the best-studied and best-understood models for circadian biology of photoreceptors is the *Xenopus* retina. Retinas containing the photoreceptive layer of cells can be isolated from either tadpoles or adults, and these eyecups (see Chapter 5) can be maintained in flow-through culture for several days. Numerous molecular output rhythms have been characterized in *Xenopus* retina, including the neurohormone melatonin, detectable in the flow-through medium. Melatonin (see Chapters 4 and 6), which is found to be high at night and low during the day in all vertebrate classes, is a major pineal gland hormone, but it is also found in retina. The best-described function for melatonin is its role in seasonally periodic behavior, but it is also known to function as a circadian modulator.

At least two and maybe more steps in the production of melatonin are clock regulated, and several lines of evidence indicate redundancy in clock control of melatonin synthesis. The mRNA for the rate-limiting, as well as first, enzyme in melatonin synthesis, tryptophan hydroxylase (TPH), is rhythmic and peaks during the early night, coincident with the production of melatonin. Nuclear run-on assays demonstrate transcriptional regulation, with *TPH* mRNA synthesis rate increasing over the course of the subjective day.

Early studies had shown circadian activity of a later enzyme in the pathway of retinal melatonin synthesis, *N*-acetyltransferase (NAT). Much of the rhythmic pattern and amplitude of melatonin synthesis is due to the activity of this enzyme. It is now known that the overall regulation of NAT is highly complex. NAT is strongly regulated by numerous factors, many of which are organism dependent. Two highly conserved forms of NAT regulation include the circadian clock (Figure 8.20) and light.

Photic stimulation has two effects on NAT. The first is a rapid negative effect on abundance that is regulated in part through a cAMP-dependent pathway and second by acting through the clock to entrain NAT activity rhythms in what is believed to be a cAMP-independent mechanism. An additional level of clock regulation may come through the rhythms in melatonin receptors. Two subtypes of melatonin receptors are expressed in the *Xenopus* retina: Mel (1b) and Mel (1c). The mRNAs for both of these proteins are known to

(A)

Entraining cue time (h)

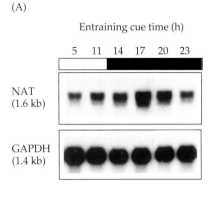

(B)

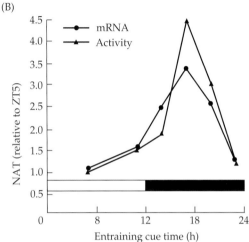

FIGURE 8.20 Daily rhythm in NAT mRNA and activity. Dissociated pineal cells from 1-day-old chicks were cultured for 5 days in a light cycle (LD 12:12, white light on at ET 0) and were harvested at the indicated times. (A) Representative Northern blot analysis of NAT and GAPDH mRNAs (20 μg total RNA per lane). (B) Quantitative analysis of the Northern blot, after normalization to the GAPDH signal. The filled-circle curve shows the levels of NAT mRNA in cells cultured with 5 μg/ml actinomycin D between ET 11 and ET 14. NAT activity was measured at each time point, shown by open circles. All values (mRNA and activity) represent the mean of duplicate determinations and are expressed relative to the ET 5 time point. Similar results were obtained in four separate experiments. Solid and open bars represent dark and light periods, respectively. kb, kilobases. (From Bernard et al., 1997.)

vary in a diurnal rhythm of expression in light–dark cycles, suggesting a further rhythmic level of regulation on melatonin action. The multiple points of rhythmic regulation within the melatonin biosynthetic and metabolic pathways suggest strong selection for biological rhythmicity of this hormone.

CLOCK PROTEINS REGULATE OUTPUT. Clock-driven rhythms in transcription factors can regulate gene expression in target tissues. Albumin DBP, the first identified member of the PAR leucine zipper transcription

factor family, is a regulatory protein enriched in liver but found in several other tissues, including the SCN. Two other members of this family, hepatocyte leukemia factor (HLF) and thyroid embryonic factor (TEF), share extended sequence similarities with DBP in the bZIP (basic leucine zipper) DNA-binding domains and in the PAR domain, a region rich in prolines and acidic amino acids.

DBP, TEF, and HLF are all circadianly expressed in various tissues, although DBP shows the strongest rhythms (Figure 8.21). Levels of *dbp* mRNA and DBP protein both exhibit robust circadian rhythms, with their peaks phased about 3 and 6 h, respectively, after maximal DBP transcription. In hepatocytes, amplitudes of *DBP* mRNA rhythms are high and may reach 160-fold with a peak in late evening to early night and then drop to undetectable levels by morning.

In nocturnal rodents, cholesterol synthesis and catabolism in the liver are higher during the dark feeding and activity phase, when the animals require more bile acids to absorb ingested lipids. The rate-limiting enzyme for cholesterol synthesis and conversion to bile acids is made by the *CYP7* gene, encoding cholesterol-7α-hydroxylase, a member of the P450 superfamily. Levels of this enzyme also exhibit a robust circadian rhythm in mRNA abundance, peaking at about midnight. To a large extent, this rhythm is regulated by a transcriptional mechanism.

In the case of cholesterol-7α-hydroxylase, circadian transcription is driven by the rhythmic binding of DBP to *cis*-acting elements within the promoter of the *CYP7* gene. In addition, TEF and HLF, whose bZIP domains have the ability to bind to the same cognate DNA sequences that the DBP protein binds to, are able to bind *CYP7* DNA and may also play a significant role in the rhythmic production of *CYP7* mRNA. Supporting evidence comes from DBP-knockout mutant mice, which are viable with a functioning pacemaker and also show a reasonably normal amplitude of circadian *CYP7* mRNA, although the phase of the rhythm is advanced 4 h. The extent to which, in the presence of DBP, these other PAR family members participate in the oscillating transcriptional regulation of *CYP7* versus the ability to substitute when the DBP protein is absent is not known.

VASOPRESSIN IN THE SCN. Clock component transcription factors can regulate gene expression in an SCN output. Arginine vasopressin is a neuropeptide found in the brain and involved in peripheral effects of salt and water balance. Vasopressin mRNA is rhythmic in the SCN, regulated by the CLOCK–BMAL1 feedback oscillator (see Chapter 7). In homozygous *Clock/Clock* mutant animals the rhythmicity is abolished, and overall levels of transcript remain at low nighttime levels (Figure 8.22), although overall levels of gene expression

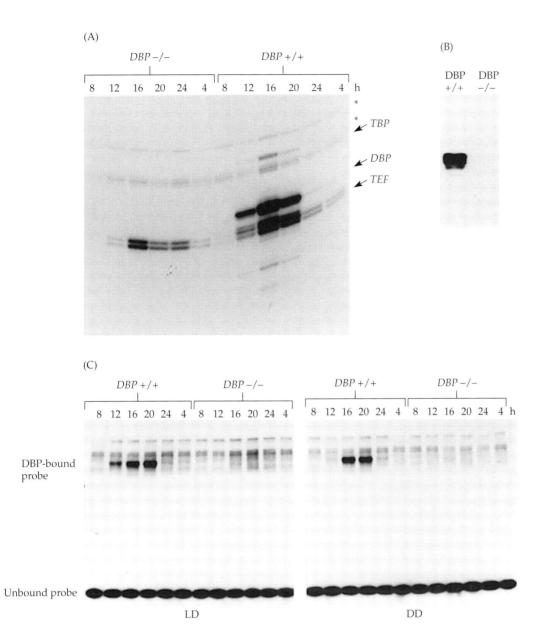

FIGURE 8.21 Expression of *DBP, TBP,* and *TEF* mRNA and DBP protein in wild-type and *DBP* mutant mice. (A) The normal rhythm of *DBP, TBP,* and *TEF* mRNA; and the loss of *DBP* rhythmicity in mice lacking this gene. (B) Corresponding protein data for DBP. (C) Electromobility shift assays in which nuclear extracts from rat liver cells harvested at different times around the clock were incubated with a radiolabeled double-stranded oligonucleotide containing the DBP-binding recognition sequence (GTTACGTAAT). DNA cells containing nuclei that contain DBP will bind the DNA, so it migrates as a band far up in the gel away from the unbound probe at the bottom. Wild-type mice show a clear rhythm for DNA-binding ability. In both light–dark cycles (left) and constant darkness (right), nuclear extracts from mutant mice lacking the *DBP* gene show no binding activity. (From Lopez-Molina et al., 1997.)

are not reduced in the SCN of these mutants. Vasopressin is also expressed in regions of the brain outside of the SCN in a nonrhythmic fashion. Noncycling expression in the nearby supraoptic nuclei (SON) was largely unaffected by mutations in the clock gene.

Examination of 1.5 kilobases of the 5′ promoter sequence of the vasopressin gene sequence revealed that it contains the 6 bp E-box motif, CACGTG, at positions −154 to −149. The phase of the mRNA rhythm for vasopressin is similar to *mPer* mRNA rhythms, and *mPer* expression is known to be mediated by the CLOCK transcription factor through E-boxes within the *mPer* regulatory sequences. A 200 bp region of the promoter, including the CACGTG motif, was assayed for binding of CLOCK, in a standard reporter assay, showing that the CLOCK–BMAL1 heterodimer could activate tran-

(A)

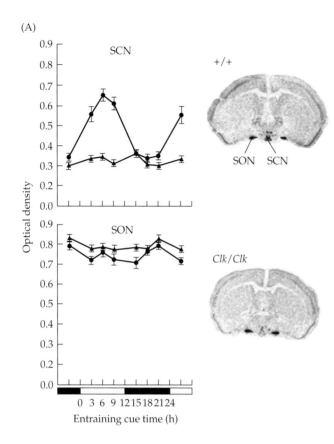

FIGURE 8.22 A model for direct modulation of output by core clock components. (A) Temporal profiles (left) of vasopressin RNA levels in the SCN (top) and SON (supraoptic nucleus) (bottom) of wild-type mice (solid lines) and *Clock/Clock* mice (dashed lines), and coronal sections (right) through the SCN and SON stained to detect vasopressin mRNA. The mRNA rhythm seen in wild-type mice is lost in the SCN in *Clock* mutant mice. Levels are reduced in the SCN, but not in the SON of double-mutant *Clock/Clock* mice. This means that the vasopressin gene can be expressed in these animals but that the necessary transcriptional activator(s) for abundance and rhythmicity are missing in the SCN. Because the vasopressin promoter contains an E-box, one of those could be the CLOCK–BMAL heterodimer, which is known to be reduced in activity in the SCN (see Chapter 7). For both graphs, solid and open bars represent dark and light periods, respectively. (B) A model for the idea that CLOCK expression in the SCN contributes not only to the operation of the circadian oscillator itself, but also directly to the transcriptional regulation of an output gene. (From Jin et al., 1999.)

AFFINITY OF DNA BINDING. Many genes that bind rhythmic transcription factors may not be rhythmically expressed. The nuclear concentration of hepatic DBP oscillates between 10 nM and 2 μM from morning to evening. Consider that a target gene whose promoter binds DBP with an equilibrium dissociation constant (Kd) considerably less than 10 nM would not exhibit circadian transcription because the *cis*-acting element would be occupied by DBP throughout the circadian cycle. In contrast, target genes with low-affinity sites for DBP (Kd in the micromolar range) would be stimulated by the transcription factor only for the short time interval when nuclear DBP concentrations are close to maximal. For example, the promoter of the cytochrome P450 gene harbors a high-affinity DBP-binding site, yet P450 gene transcription does not exhibit significant circadian oscillation. Transcription of the cholesterol-7α-hydroxylase gene, whose DBP recognition sequence binds DBP with a relatively low affinity, is strongly circadian.

The entire Synechococcus *transcriptome is clock controlled*[29,31,34,37,46,52]

A RHYTHMIC GENOME. As noted earlier in this chapter, some species of cyanobacteria have employed their internal circadian clocks to temporally separate the incompatible metabolic activities of photosynthesis and nitrogen metabolism. The mechanism of control was thought to be, at least in part, regulated gene expression because it was known that nitrogenase mRNA abundance was higher at night than during the day. In fact, cyanobacteria such as the model molecular system of *Synechococcus elongatus* use their clocks to regulate a variety of physiological processes.

(B)

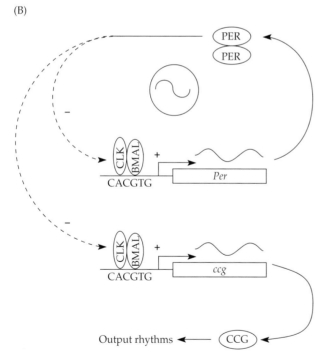

scription from the vasopressin E-box. Therefore, the circadian feedback mechanism is directly linked to a clock-controlled gene by the same positive and negative components that drive the core oscillator.

A sophisticated reporting system to monitor promoter activity in *S. elongatus* was developed by a consortium of collaborating laboratories. A gene cassette of the bacterial luciferase genes *luxA* and *luxB* from *Vibrio harveyi* was placed behind a well-characterized promoter, *psbAI*, from *S. elongatus*. The *psbAI* gene encodes the D1 protein, a major component of the photosystem II reaction center in photosynthesis. When *S. elongatus* was transformed with the DNA construct containing the luciferase genes under control of the *psbAI* promoter, the bacteria were found to glow rhythmically. This bioluminescence could be monitored in vivo, a powerful tool in analyzing molecular rhythmicity of promoter activity. Not surprisingly, the mRNA abundance for the *psbAI* gene was also found to be cyclic.

Through a variation of a "promoter trap" experiment, the technology made it possible to ask how widespread circadian regulation of transcription in *S. elongatus* was. A DNA library of *S. elongatus* genomic fragments fused upstream of (in front of) a promoterless *luxAB* gene cassette was constructed. The library was then transferred into the *S. elongatus* genome, where DNA sequences homologous to those in the library allowed recombination at numerous sites throughout the genome with high efficiency. Bacteria were kept in constant low light, and luciferase expression could be monitored from multiple colonies spread on a single petri dish. When the *luxAB* gene cassette is inserted into the genome behind a transcriptionally active promoter region of DNA, the bacterial cells will express luciferase and glow. Changes in bioluminescence over time for each transformed colony reflected gene expression activity from individual promoters "trapped" upstream of the *luxAB* genes.

More than 3,000 colonies have now been individually examined for temporally regulated expression patterns. The astonishing result is that all the promoters showed circadianly rhythmic activity. Although the amplitudes, waveforms, and phasing of the rhythmic patterns showed wide variation (Figure 8.23), the rhythms all shared a single period length. Peak expression levels for 80% of the genes in this organism is near subjective dusk.

TRANSCRIPTION MECHANISM IN *SYNECHOCOCCUS*. Only rhythmic expression patterns have been found, leading to the hypothesis that the clock is mediating rates of transcription in cyanobacteria in a global fashion, possibly in conjunction with distinct *trans*-acting factors. Because a single type of multisubunit RNA polymerase is responsible for all bacterial transcription, one idea is that some part or parts of the core transcription machinery itself might be clock regulated. Sigma factors (σ) are detachable subunits of the core polymerase that play a major role in recognizing promoter elements directing RNA polymerase to initiate transcription.

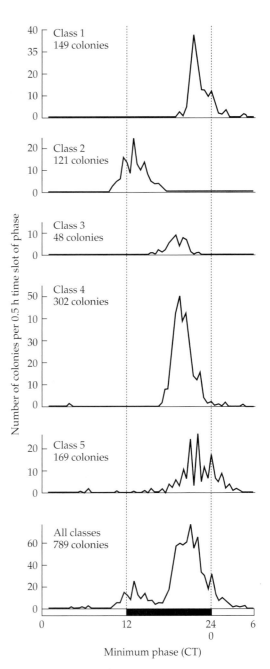

FIGURE 8.23 The cyanobacterial circadian system regulates gene expression throughout its entire genome, but not all genes are expressed at the same phase. Distribution of phases observed for 789 colonies corresponding to reporters for randomly chosen genes within the genome. Whereas most genes tend to be expressed during the late subjective day, the presence of subgroups within this time is apparent. Note that phases are plotted at the minima. Solid and open bars represent dark and light periods, respectively. (From Liu et al., 1995.)

Sigma factor promoters are expressed with a strong circadian cycle in *S. elongatus*. Examination of the RpoD4 sigma factor by immunoblot analysis showed rhythmic abundance of the protein. Elimination of specific group

2 sigma factors can result in changes of amplitude, phasing, and even period length of specific reporter genes. Overexpression of RpoD3, resulting in high levels of this sigma factor at all circadian times, could abolish rhythmic expression from some, but not all, promoters, indicating functional timekeeping even in the presence of abundant RpoD3.

Changes in the mix of sigma factor species available to compete for association with the core RNA polymerase over circadian time are thought to be at least partially responsible for global gene rhythmicity, but they are not the whole story. Other models have been proposed, including global temporal remodeling of the bacterial chromosome, but they have yet to be tested.

Circadian Regulation of Translation Is Another Mechanism for Rhythmic Protein Synthesis[31,46]

Translational regulation is responsible for the rhythm of Gonyaulax LBP synthesis[47,49]

PROTEIN ABUNDANCE IS RHYTHMIC IN *GONYAULAX*. In *Gonyaulax*, the rhythm of bioluminescence is accounted for by changes in the levels of luciferase and luciferin-binding protein (LBP), discussed earlier in this chapter, both of which are degraded at the end of the night phase. Clearly, these proteins must be newly made each evening if their levels are to rise in the cell. The rate of synthesis can be measured directly by the incorporation of radioactively labeled amino acids. When incorporation rates are high, protein synthesis rates are high; the incorporation rate is highest when protein levels are rising fastest in the cell. Conversely, when little of the radiolabel is incorporated, the synthesis rate is low. In *Gonyaulax*, [^{35}S]methionine radiolabel is incorporated into LBP protein only at the start of the night phase, with the observed changes in the rate of LBP synthesis accounting perfectly for the rise in measured protein levels (see Figure 8.8).

The initial regulation that results in changes in the total accumulation of protein over the course of the cycle can result from various different mechanisms, including transcriptional control, as discussed earlier. However, the changes in LBP protein synthesis rate occur despite constant levels of *lbp* mRNA. Analysis of the *lbp* mRNA from timed poly-A RNA samples harvested from rhythmic *Gonyaulax* cultures shows no significant changes of either message abundance or size on Northern blots.

One interpretation is that the *lbp* mRNA, even though at similar abundance at different times of day, might carry structural differences allowing differential translational capability. Timed poly-A RNA samples can be translated in vitro in translationally competent cell-free extracts called rabbit reticulocyte lysates and the

resulting amount of LBP protein examined. In these experiments LBP protein was produced in equal amounts. The dinoflagellate cells must be controlling the translational efficiency of the *lbp* mRNA in order to generate bursts of protein synthesis.

Protein synthesis is a key point of regulatory control for many cellular processes, and this type of control has the advantage of allowing cells to alter protein levels rapidly without requiring transcript synthesis, processing, or export into the cytoplasm. Translational control can be global or gene specific. For example, heat shock and other forms of environmental stress can inhibit translation of most cellular mRNAs while promoting the selective initiation of translation of specific messages needed for damage control. In any organism, including *Gonyaulax*, no evidence supports circadian control of globally regulated translation, suggesting that LBP control is probably gene specific. Gene-specific translational regulation, though not well understood, is thought to occur by a variety of mechanisms. Specialized mRNA-binding proteins have been found to be an important means of mRNA-specific translational control.

UNTRANSLATED REGIONS OF MRNAS PLAY A ROLE. For a translational control mechanism to function, two components are required. First, a specific sequence of ribonucleotides must be present in the mRNA, a *cis*-acting element, to serve as a tag to differentiate this mRNA from all other (nonregulated) mRNAs and allow its translation only at specific times. In many cases the untranslated regions of a transcript, on either the 5′ end or the 3′ end, play an important role in translational regulation. Second, a diffusible protein factor, a *trans*-acting element, must be present in the cell to recognize and bind the nucleotide sequence tag, resulting in either temporal regulation of RNA processing or the inhibition or stimulation of translation. This factor must vary in either amount or activity over the course of the circadian cycle.

In the case of the intronless *lbp* mRNA, time of day–specific splicing was not a possibility. Both the 5′ and the 3′ untranslated regions were examined in an electromobility shift assay for binding of a *trans*-acting factor. A short sequence of the *lbp* mRNA untranslated region was synthesized in the laboratory and radiolabeled, incubated with soluble cellular extracts, and analyzed by gel electrophoresis. When a component of the soluble fraction binds to the labeled mRNA probe, its movement though the gel is slowed (retarded). Protein extracts made from *Gonyaulax* day-phase cells contain a factor, named the circadian-controlled translational regulator, or CCTR, that retards the movement of an *lbp* mRNA probe corresponding to a novel, 22-nucleotide imperfect UG repeat region within the *lbp* 3′ untranslated region. Night-phase extracts actively translating *lbp* mRNA are unable to cause retardation of the labeled RNA in the gel.

Further experiments indicate that this 22-nucleotide region is single-stranded, forms a type of secondary structure called a hairpin loop, and is required to allow binding of the CCTR. The temporal relationship of the binding activity of the CCTR in relation to LBP expression, scintillon number, and bioluminescence predicts that the CCTR functions as a repressor of translation, possibly through steric hindrance of the translational machinery. A probable role for the CCTR factor is to recognize and inhibit translation of the *lbp* mRNA during the day. This inhibition is relieved only at the start of the night phase, and LBP protein can then be synthesized.

An RNA-binding protein may mediate circadian regulation of the Drosophila eclosion rhythm[44,45]

A GENE AFFECTING FRUIT FLY EMERGENCE. In an attempt to identify potential clock-controlled regulatory proteins, a genetic screen for transposon-induced mutations that would specifically affect clock output but not the oscillator mechanism was carried out in *Drosophila*. The screen was designed to specifically detect mutations affecting the daily timing of adult eclosion (hatching of the adult fly, the final ecdysis in holometabolous insect development), but not rhythms in locomotor activity.

A dominant mutation called *lark* was found with eclosion peaks unusually early in the day (phase-advanced) when pupal populations were synchronized to either light–dark or temperature cycles. The free-running period of the animal was wild type, and there was no change in the phasing or periodicity of activity rhythms. The *lark* mutant allele has pleiotropic phenotypes, and homozygosity at the *lark* locus results in embryonic death. Further genetic analysis indicates that the LARK protein may play essential roles in diverse developmental processes, including fertility, wing morphogenesis, and embryonic development. The products of *lark* are first required probably during embryogenesis, and then again at the pupal stage for the regulation of adult eclosion.

Studies in the embryo show the *lark* mRNA to be widely expressed within the nervous system, consistent with a role in neural development, but the transcripts do not cycle either in late pupae or in adult fly heads. Manipulation of gene dosage can result in either early (decreased expression) or late (increased expression) eclosion, indicating that the lark gene encodes a repressor protein that regulates the daily timing of eclosion. Such a repressor might inhibit the eclosion process at certain times of day, and by this mechanism ensure the normal temporal gating of development.

When the LARK protein was examined for cell location, it was found in the nucleus of most neurons in the late pupal brain, with the exception of the neurons of the ventral nervous system that contain crustacean car-

dioactive peptide (CCAP). In CCAP cells, LARK has a cytoplasmic location. Both these neurons and their CCAP product are thought to play an important role in regulation of insect ecdysis, consistent with a role for LARK in circadian regulation of adult eclosion in fruit flies.

MUTATIONS IN THE LARK PROTEIN. Temporal regulation of LARK protein was examined via Western blot analysis in both total pupal extracts and adult head protein extracts, as well as by antibody detection of LARK protein in pupal brain sections. In all cases protein abundance was rhythmic, allowing a model for the circadian control of adult eclosion by the LARK repressor protein (Figure 8.24). Furthermore, in the CCAP-containing neurons, LARK is cytoplasmically located, possibly indicating its specific functional role. DNA sequence analysis reveals that the *lark* gene has striking similarity to the RNA recognition motif (RRM) class of RNA-binding proteins. Members of this large family of RNA-binding proteins are known to have roles in translation control of messenger RNAs, RNA processing, RNA transport, and regulation of RNA stability.

The cytoplasmic localization of the protein suggests that LARK is a repressor molecule that regulates an aspect of the translation of mRNAs encoding downstream components of the eclosion output pathway. Alternatively, LARK might negatively regulate the processing (splicing) of target RNAs. The LARK protein has three regions that are potentially capable of binding RNA: RRM1, RRM2, and a retroviral-type zinc finger (RTZF) motif (Figure 8.25), each with consensus RNA-binding motifs.

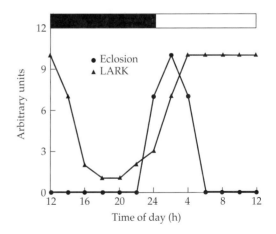

FIGURE 8.24 Model for the control of adult eclosion by the LARK repressor. Note that the numbers on the y-axis are in arbitrary units and are not meant to represent LARK abundance or number of adults. Solid and open bars represent dark and light periods, respectively. (From McNeil et al., 1998.)

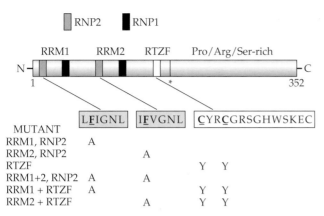

MUTANT	LF̲IGNL	IF̲VGNL	C̲YR̲CGRSGHWSKEC	
RRM1, RNP2	A			
RRM2, RNP2		A		
RTZF			Y	Y
RRM1+2, RNP2	A	A		
RRM1 + RTZF	A		Y	Y
RRM2 + RTZF		A	Y	Y

FIGURE 8.25 Schematic representation of LARK protein showing the amino acid substitutions present in single and double mutants. The locations of the two RRM domains and the RTZF (retroviral-type zinc finger) are indicated on the shaded horizontal box representing LARK. The asterisk indicates residue 178, which marks the C-terminus of a truncated LARK protein. The sequences of the two RNP2 motifs and the RTZF are shown below the schematic of LARK (the RNP1 and RNP2 regions are part of the RRMs. RNPs are highly conserved regions of the protein allowing recognition of the RNA). The underlined residues of the RNP2 and RTZF sequences represent the amino acid substitutions present in the various mutants. The positions and types of substitutions are indicated at the bottom. (From McNeil et al., 2001.)

A genetic approach to structure and function analysis of LARK was taken to determine if these motifs have function in vivo. The sequence of each site was altered by site-directed mutagenesis. Highly conserved amino acids, previously shown to be critical for RNA–protein interactions (the RNP regions), were altered. The mutant *lark* genes were then placed into *lark* loss-of-function host fly strains. Not surprisingly, the mutants displayed various developmental defects in both embryonic and maternal functions. Remarkably, the circadian gating of eclosion behavior was completely normal in the *lark* mutations, whether with RRM1, RRM2, or RTZF alone, or when double mutants of either RRM1 or RRM2 occur in combination with a mutation in RTZF.

There are several possible interpretations. The results are evidence that normal RTZF function may not be necessary for the circadian regulation of adult eclosion. In addition, single mutations in either RRM1 or RRM2 have no effect. One explanation might be that the RRM-binding regions contain the ability to act redundantly such that mutations in only one site are fully functional. Alternatively, the mutations may not have completely abolished RNA–protein binding activity, and clock regulation of eclosion is less sensitive to incomplete loss of LARK function than are other LARK phenotypes.

SUMMARY

Many different molecular mechanisms are used in different organisms for the circadian regulation of clock outputs. The business end of a clock is to serve as an internal regulatory network. The most common use of internal pacemakers is to organize biological events to occur at the most favorable times of day for the life of a specific organism. Regulation by the clock covers a broad range of important life functions, ranging from a cell's basic biochemistry to an organism's pattern of behavior.

In some species the clock may regulate expression of the entire genome. In others, clock regulation appears to be more modest. Understanding how this timing works at the level of genes and proteins has become an important area of research that affects many fields of biology. The biochemical and molecular mechanisms involved in pacemaker function at the level of the cell controls clock output through the control of gene expression and activity. Rhythmic transcription is a common level of regulation, sometimes but not always resulting in rhythmic mRNAs. Cycling mRNAs do not always lead to rhythms in protein abundance.

Cycling transcripts are called clock-controlled genes, or ccg's. Changes in transcript abundance may occur by daily changes in the rate of messenger RNA synthesis or by timed changes in RNA stability. Rhythmically expressed transcription factors bind the promoters of many genes, only some of which may cycle. Clock-controlled gene transcripts may be directly regulated by rhythmic components of the central oscillator or by clock-regulated transcription factors that are themselves ccg's. In some cases, mRNA levels are constant but changes in the efficiency of translation can produce daily bursts of protein synthesis. RNA binding proteins can act as regulators of translation.

Another means of controlling cellular events is the gating of gene expression. Gated genes can be turned on by a stimulus at some times of day, but not others. Outputs can be differentiated from the cell-based pacemaker yet may be able to affect pacemaker function.

CONTRIBUTORS

Jennifer J. Loros, J. Woodland Hastings, and Ueli Schibler wish to thank Carla Green, Susan Golden, F. Rob Jackson, David Morse, and William J. Schwartz for their material contributions to Chapter 8.

STUDY QUESTIONS AND EXERCISES

1. Daily circadian regulation of programs of gene transcription is a universal aspect of circadian regulation of output. Given the discussion of the molecular mechanism of the clock in Chapter 7, why might this have evolved in this way?

2. What advantages does an organism gain when input to its clock is gated? Are there any disadvantages?

3. An investigator working in a biotech company is studying genetic variants of a new organism from Amazonia that produces a novel antibiotic in a circadianly regulated manner. The production is controlled through circadian regulation of a single enzyme that is the rate-limiting step in the biosynthetic metabolic pathway. One strain produces more of the antibiotic but in a nearly constitutive manner; another strain produces less of the enzyme but in a distinctly circadianly regulated manner. A genetic cross reveals that this difference is due to a difference in a single gene.

   a. What are some possible causes for this difference? That is, what might be the identity or function of this single gene?

   b. The investigator is instructed to cross the two strains anyway and to try to isolate an organism that produces as much antibiotic as the high producer but with a strong rhythm like the low producer. Is this possible? Why or why not?

4. What is meant by the phrase "combinatorial regulation of gene expression"?

5. Jet lag may be considered a problem in output regulation. Microorganisms don't get jet lag in that output rhythms are typically reset within a day of a phase change; vertebrate organisms do get jet lag. Follow the course of regulation between the central clock and (a) a rhythm in liver function that peaks at subjective noon and (b) a rhythm in cardiac function that peaks at subjective midnight. Now follow the rhythms after a 5 h advance or delay phase shift such as that encountered in travels between the eastern United States and the United Kingdom. What roles might gating and regulation via peripheral oscillators play in the creation of jet lag?

REFERENCES

1. Akhtar, R. A., A. B. Reddy, E. S. Maywood, J. D. Clayton, V. M. King, et al. 2002. Circadian cycling of the mouse liver transcriptome, as revealed by cDNA microarray, is driven by the suprachiasmatic nucleus. Curr. Biol. 12: 540–550.

2. Albrecht, U., Z. Sun, G. Eichele, and C. Lee. 1997. A differential response of two putative mammalian circadian regulators, *mper1* and *mper2*, to light. Cell 91: 1055–1064.

3. Anderson, F. E., and C. B. Green. 2000. Symphony of rhythms in the *Xenopus laevis* retina. Microsc. Res. Tech. 50: 360–372.

4. Barak, S., E. M. Tobin, C. Andronis, S. Sugano, and R. M. Green. 2000. All in good time: The *Arabidopsis* circadian clock. Trends Plant Sci. 5: 517–522.

5. Bell-Pedersen, D., J. C. Dunlap, and J. J. Loros. 1992. The *Neurospora* circadian clock-controlled gene, *ccg-2*, is allelic to *eas* and encodes a fungal hydrophobin required for formation of the conidial rodlet layer. Genes Dev. 6: 2382–2394.

6. Bell-Pedersen, D., J. C. Dunlap, and J. J. Loros. 1996. Distinct *cis*-acting elements mediate clock, light, and developmental regulation of the *Neurospora crassa eas* (*ccg-2*) gene. Mol. Cell. Biol. 16: 513–521.

7. Bell-Pedersen, D., M. Shinohara, J. J. Loros, and J. C. Dunlap. 1996. Circadian clock-controlled genes isolated from *Neurospora crassa* are late night to early morning specific. Proc. Natl. Acad. Sci. USA 93: 13096–13101.

8. Bernard, M., D. C. Klein, and M. Zatz. 1997. Chick pineal clock regulates serotonin N-acetyltransferase mRNA rhythm in culture. Proc. Natl. Acad. Sci. USA 94: 304–309.

9. Besharse, J. C., and P. M. Iuvone. 1983. Circadian clock in *Xenopus* eye controlling retinal serotonin N-acetyltransferase. Nature 305: 133–135.

10. Blau, J., and M. W. Young. 1999. Cycling *vrille* expression is required for a functional *Drosophila* clock. Cell 99: 661–671.

11. Bünning, E. 1960. Biological clocks. Cold Spring Harb. Symp. Quant. Biol. 25: 1–9.

12. Claridge-Chang, A., H. Wijnen, F. Naef, C. Boothroyd, N. Rajewsky, et al. 2001. Circadian regulation of gene expression systems in the *Drosophila* head. Neuron 32: 657–671.

13. Coon, S. L., P. H. Roseboom, R. Baler, J. L. Weller, A. Namboodiri, et al. 1995. Pineal serotonin N-acetyltransferase: Expression cloning and molecular analysis. Science 270: 1681–1683.

14. Davis, R. H. 2000. *Neurospora*. Oxford University Press, New York.

15. Duffield, G. E., J. D. Best, B. H. Meurers, A. Bittner, J. J. Loros, et al. 2002. Circadian programs of transcriptional activation, signaling, and protein turnover revealed by microarray analysis of mammalian cells. Curr. Biol. 12: 551–557.

16. Edmunds, L. N., Jr. 1988. Cellular and Molecular Bases of Biological Clocks. Springer, New York.

17. Fagan, T., D. Morse, and J. W. Hastings. 1999. Circadian synthesis of a nuclear-encoded chloroplast glyceraldehyde-3-phosphate dehydrogenase in the dinoflagellate *Gonyaulax polyedra* is translationally controlled. Biochemistry 38: 7689–7695.

18. Fonjallaz, P., V. Ossipow, G. Wanner, and U. Schibler. 1996. The two PAR leucine zipper proteins, TEF and DBP, display similar circadian and tissue-specific expression, but have different target promoter preferences. EMBO J. 15: 351–362.

19. Green, C. B., and J. C. Besharse. 1994. Tryptophan hydroxylase expression is regulated by a circadian clock in *Xenopus laevis* retina. J. Neurochem. 62: 2420–2428.

20. Halberg, F. 1960. Temporal coordination of physiologic function. Cold Spring Harb. Symp. Quant. Biol. 25: 289–310.

21. Harmer, S. L., J. B. Hogenesch, M. Straume, H. S. Chang, B. Han, et al. 2000. Orchestrated transcription of key pathways in *Arabidopsis* by the circadian clock. Science 290: 2110–2113.

22. Harmer, S. L., S. Panda, and S. A. Kay. 2001. Molecular bases of circadian rhythms. Annu. Rev. Cell Dev. Biol. 17: 215–253.

23. Hastings, J. W. 1960. Biochemical aspects of rhythms: Phase shifting by chemicals. Cold Spring Harb. Symp. Quant. Biol. 25: 131–143.

24. Heintzen, C., J. J. Loros, and J. C. Dunlap. 2001. VIVID, gating and the circadian clock: The PAS protein VVD defines a feedback loop that represses light input pathways and regulates clock resetting. Cell 104: 453–464.

25. Heintzen, C., M. Nater, K. Apel, and D. Staiger. 1997. AtGRP7, a nuclear RNA-binding protein as a component of a circadian-regulated negative feedback loop in *Arabidopsis thaliana*. Proc. Natl. Acad. Sci. USA 94: 8515–8520.

26. Helfrich-Forster, C., M. Tauber, J. H. Park, M. Muhlig-Versen, S. Schneuwly, et al. 2000. Ectopic expression of the neuropeptide pigment-dispersing factor alters behavioral rhythms in *Drosophila melanogaster*. J. Neurosci. 20: 3339–3353.

27. Hochberg, M. L., and M. L. Sargent. 1974. Rhythms of enzyme activity associated with circadian conidiation in *Neurospora crassa*. J. Bacteriol. 120: 1164–1175.

28. Inouye, S. T. 1996. Circadian rhythms of neuropeptides in the suprachiasmatic nucleus. Prog. Brain Res. 111: 75–90.

29. Iwasaki, H., and T. Kondo. 2000. The current state and problems of circadian clock studies in cyanobacteria. Plant Cell Physiol. 41: 1013–1020.

30. Jin, X., L. Shearman, D. Weaver, M. Zylka, G. DeVries, et al. 1999. A molecular mechanism regulating output from the suprachiasmatic circadian clock. Cell 96: 57–68.

31. Johnson, C. H. 2001. Endogenous timekeepers in photosynthetic organisms. Annu. Rev. Physiol. 63: 695–728.

32. Kita, Y., M. Shiozawa, W. Jin, R. R. Majewski, J. C. Besharse, et al. 2002. Implications of circadian gene expression in kidney, liver and the effects of fasting on pharmacogenomic studies. Pharmacogenetics 12: 55–65.

33. Kloppstech, K. 1985. Diurnal and circadian rhythmicity in the expression of light-induced plant nuclear messenger RNAs. Planta 165: 502–506.

34. Kondo, T., C. Strayer, R. Kulkarni, W. R. Taylor, M. Ishiura, et al. 1993. Circadian rhythms in prokaryotes: Luciferase as a reporter of circadian gene expression in cyanobacteria. Proc. Natl. Acad. Sci. USA 90: 5672–5676.

35. Lavery, D., and U. Schibler. 1993. Circadian transcription of the cholesterol 7α hydroxylase gene may involve the liver-enriched bZIP protein DBP. Genes Dev. 7: 1871–1884.

36. LeSauter, J., and R. Silver. 1998. Output signals of the SCN. Chronobiol. Int. 15: 535–550.

37. Liu, Y., N. Tsinoremas, C. Johnson, N. Lebdeva, S. Golden, et al. 1995. Circadian orchestration of gene expression in cyanobacteria. Genes Dev. 9: 1469–1478.

38. Lopez-Molina, L., F. Conquet, M. Dubois-Dauphin, and U. Schibler. 1997. The DBP gene is expressed according to a circadian rhythm in the SCN and influences circadian behavior. EMBO J. 16: 6762–6771.

39. Loros J. J., and J. C. Dunlap. 2001. Genetic and molecular analysis of circadian rhythms in *Neurospora*. Annu. Rev. Physiol. 63: 757–794.

40. Loros, J .J., S. A. Denome, and J. C. Dunlap. 1989. Molecular cloning of genes under the control of the circadian clock in *Neurospora*. Science 243: 385–388.

41. Markovic, P., T. Rönneberg, and D. Morse. 1996. Phased protein synthesis at several circadian times does not change protein levels in *Gonyaulax*. J. Biol. Rhythms 11: 57–67.

42. Maywood, E. S., N. Mrosovsky, M. D. Field, and M. H. Hastings. 1999. Rapid down-regulation of mammalian period genes during behavioral resetting of the circadian clock. Proc. Natl. Acad. Sci. USA 96: 15211–15216.

43. McDonald, M. J., and M. Rosbash. 2001. Microarray analysis and organization of circadian gene expression in *Drosophila*. Cell 107: 567–578.

44. McNeil, G. P., A. J. Schroeder, M. A. Roberts, and F. R. Jackson. 2001. Genetic analysis of functional domains within the *Drosophila* LARK RNA-binding protein. Genetics 159: 229–240.

45. McNeil, G. P., X. Zhang, G. Genova, and F. R. Jackson. 1998. A molecular rhythm mediating circadian clock output in *Drosophila*. Neuron 20: 297–303.

46. Mittag, M. 2001. Circadian rhythms in microalgae. Int. Rev. Cytol. 206: 213–247.

47. Mittag, M., D. Lee, and J. W. Hastings. 1995. Are mRNA binding proteins involved in the translational control of the circadian regulated synthesis of luminescence proteins in *Gonyaulax*? In: Evolution of Circadian Clock, T. Hiroshige and K-I. Honma (eds.), pp. 97–114. Hokkaido University Press, Sapporo, Japan.

48. Moore-Ede, M. C., F. M. Sulzman, and C. A. Fuller. 1982. The Clocks That Time Us. Harvard University Press, Cambridge, MA.

49. Morse, D., P. M. Milos, E. Roux, and J. W. Hastings. 1989. Circadian regulation of bioluminescence in *Gonyaulax* involves translational control. Proc. Natl. Acad. Sci. USA 86: 172–176.

50. Mrosovsky, N. 1988. Phase response curves for social entrainment. J. Comp. Physiol. 162: 35–46.

51. Nagy, F., S. A. Kay, and N-H. Chua. 1988. A circadian clock regulates transcription of the wheat *Cab-1* gene. Genes Dev. 2: 376–382.

52. Nair, U., J. S. Ditty, H. Min, and S. S. Golden. 2002. Roles for sigma factors in global circadian regulation of the cyanobacterial genome. J. Bacteriol. 184: 3530–3538.

53. Nowrousian, M., G. E. Duffield, J. J. Loros, and J. C. Dunlap. 2003. The *frequency* gene is required for temperature-dependent expression of many clock-controlled genes in *Neurospora crassa*. Genetics. *In press*.

54. Panda, S., M. P. Antoch, B. Miller, A. Su, A. Schook, et al. 2002. Coordinated transcription of key pathways in the mouse by the circadian clock. Cell 109: 307–320.

55. Pierce, M. E., H. Sheshberadaran, Z. Zhang, L. E. Fox, M. L. Applebury, et al. 1993. Circadian regulation of iodopsin gene expression in embryonic photoreceptors in retinal cell culture. Neuron 10: 579–584.

56. Pittendrigh, C. S., V. G. Bruce, N. S. Rosenzweig, and M. L. Rubin. 1959. A biological clock in *Neurospora*. Nature 184: 169–170.

57. Renn, S. C., J. H. Park, M. Rosbash, J. C. Hall, and P. H. Taghert. 1999. A PDF neuropeptide gene mutation and ablation of PDF neurons each cause severe abnormalities of behavioral circadian rhythms in *Drosophila*. Cell 99: 791–802.

58. Rensing, L., A. Bos, J. Kroeger, and G. Cornelius. 1987. Possible link between circadian rhythm and heat shock response in *Neurospora crassa*. Chronobiol. Int. 4: 543–549.

59. Ripperger, J. A, and U. Schibler. 2001. Circadian regulation of gene expression in animals. Curr. Opin. Cell Biol. 13: 357–362.

60. Roenneberg, T., and D. Morse. 1993. Two circadian oscillators in one cell. Nature 362: 362–364.

61. Schaffer, R., J. Landgraf, M. Accerbi, V. V. Simon, M. Larson, et al. 2001. Microarray analysis of diurnal and circadian-regulated genes in *Arabidopsis*. Plant Cell 13: 113–123.

62. Schwartz, W. J. 1997. Understanding circadian clocks: From *c-fos* to fly balls. Ann. Neurol. 41: 289–297.

63. Schwartz, W. J., A. J. Carpino, H. O. de la Iglesia, R. Baler, D. C. Klein, et al. 2000. Differential regulation of *fos* family genes in the ventrolateral and dorsomedial subdivisions of the rat suprachiasmatic nucleus. Neuroscience 98: 535–547.

64. Shearman, L., M. Zylka, D. Weaver, L. Kolakowski, and S. Reppert. 1997. Two period homologs: Circadian expression and photic regulation in the suprachiasmatic nuclei. Neuron 19: 1261–1269.

65. Shigeyoshi, Y., K. Taguchi, S. Yamamoto, S. Takeida, L. Yan, et al. 1997. Light-induced resetting of a mammalian circadian clock is associated with rapid induction of the mPer1 transcript. Cell 91: 1043–1053.

66. Shinohara, M., J. J. Loros, and J. C Dunlap. 1998. Glyceraldehyde-3-phosphate dehydrogenase is regulated on a daily basis by the circadian clock. J. Biol. Chem. 273: 446–452.

67. Somers, D. E. 1999. The physiology and molecular bases of the plant circadian clock. Plant Physiol. 121: 9–20.

68. Storch, K. F., O. Lipan, I. Leykin, N. Viswanathan, F. C. Davis, et al. 2002. Extensive and divergent circadian gene expression in liver and heart. Nature 417: 78–83.

69. Sweeney, B. M. 1987. Rhythmic Phenomena in Plants. Academic Press, San Diego, CA.

70. Tosini, G., and M. Menaker. 1996. Circadian rhythms in cultured mammalian retina. Science 272: 419–421.

71. Weisgerber, D., U. Redlin, and N. Mrosovsky. 1997. Lengthening of circadian period in hamsters by novelty-induced wheel running. Physiol. Behav. 62: 759–765.

72. Wiechmann, A. F., and A. R. Smith. 2001. Melatonin receptor RNA is expressed in photoreceptors and displays a diurnal rhythm in *Xenopus* retina. Brain Res. Mol. Brain Res. 91: 104–111.

73. Williams, J. A., and A. Sehgal. 2001. Molecular Components of the Circadian System in *Drosophila*. Annu. Rev. Physiol. 63: 729–755.

74. Wuarin, J., and U. Schibler. 1990. Expression of the liver-enriched transcriptional activator protein DBP follows a stringent circadian rhythm. Cell 63: 1257–1266.

75. Wuarin, J., E. Falvey, D. Lavery, D. Talbot, E. Schmidt, et al. 1992. The role of the transcriptional activator protein DBP in circadian liver gene expression. J. Cell Sci. Suppl. 16: 123–127.

Circadian rhythmicity permeates practically every aspect of the behavior and physiology of mammals, driven by the SCN circadian pacemaker as primary regulator. Nevertheless, animals, including humans, live in a real world that is constrained by climatological events and changes, scarce food supplies, dangerous predators, shelter needs, the necessity of finding a mate, and the demands of dependent offspring. With all of these contingencies subject to the relentless scythe of natural selection, the often conflicting interactions of so many multiple drives makes some type of prioritization essential for survival.

Nowhere is this masking of biological clock signals seen more clearly than in humans. Not only do humans disregard clock signals, but they create their own environments, in places as bizarre as underground research caves in the antarctic ice sheet or gigantic workstations in outer space. The resultant freedom from the usual strictures of ecological niches and tropic pyramids makes the story of human chronobiology unique. For these reasons, like the burst of the most dazzling fireworks at the end of a celebratory display, consideration of human rhythms is the finale to this book, beginning with normal rhythmic function in Chapter 9, and concluding with their relevance in Chapter 10.

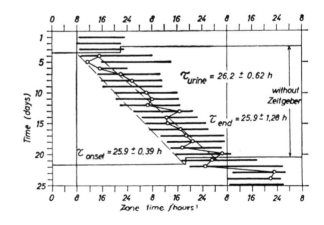

Human Circadian Organization

The clock upbraids me for wasting time.
— William Shakespeare, *Twelfth Night*

Introduction: Humans Are Intensely Curious about Their Own Daily Rhythms[60,65,71]

Most humans in Western industrialized nations today live in a highly managed environment, driven by the clock more than by any other factor. Weekdays are dominated by early alarm clock jolts, a dash to school or the office, and a weary return home in evening for recuperation and sleep. Although weekends are more relaxed, scheduling is still highly important. Natural clues about the time of day play a much less important role for urbanites than in the lifestyle of rural cultures or of earlier civilizations. Relax now for a few moments and imagine living for several weeks away from the bustle of normal city life in a very deep cave, or in an underground bomb shelter. Think about the lifestyle in such constant environments, especially the lack of a wristwatch or clock.

In a deep cave, complete silence and darkness reigns, and the temperature remains at a chilly 50°F for the entire year. Water condenses on the walls and drips monotonously. The fact that a headlamp beam penetrates the gloom only a few feet restricts many activities (Figure 9.1). The assignment is to live on a normal daily schedule without time clues and estimate the passage of time. Life in such a timeless world is hard to imagine. Judging the passage of days, or even determining when to go to bed is difficult. Estimating the number of hours slept and deciding when to get up and start the day's activity time are challenging tasks.

Now try a journey through time and space to a tiny apartment seven stories underground beneath the main hospital of Munich, Germany. Called simply the Bunker, from the German word for "bomb shelter," this structure dates back to World War II. The unused facility is accessed via a series of dimly lit, steep, musty

FIGURE 9.1 The environment in a cave is constant.

staircases, and then a seemingly endless walk along winding corridors suggestive of a catacomb. High above in the busy city, trolley cars rumble along the tracks, full of passengers. They hurry to work early each morning and return home each evening just as sunset announces evening mealtime and approaching bedtime. No clues about the outside world penetrate the Bunker to hint at the time of day. The assignment is to live here for a month in order to estimate perception of time without a watch or a radio or a telephone. The Bunker is sparsely furnished, but plentiful food is provided daily by unseen caretakers who deposit groceries and other necessities in an outer chamber. The question of interest is how someone might react to these circumstances, both psychologically and physiologically.

Perhaps surprisingly, these scenarios are real, and they will be described in this chapter as part of the story of human chronobiology research. Previous chapters have made clear the importance of biological timekeeping in the physiology and behavior of many living organisms. The aim of this chapter is to explore human circadian timekeeping systems. In contrast to the wealth of information on circadian regulation in humans, very little is known about the longer infradian rhythmicities, such as menstrual and circannual rhythms, or about possible photoperiodic regulation of reproduction. For that reason, this chapter will deal almost exclusively with circadian rhythms.

After starting with the early history of human interest in timekeeping, the chapter will progress to modern study methods, then give an overview of free-running periods and entrainment in humans, followed by a systematic consideration of the major circadian rhythms in humans. The final section will consider the special developmental aspects of human chronobiology in

infants and in the elderly. The stage will thus be set for Chapter 10 to survey malfunctions of this timekeeping system, not only in a medical context but also in healthy individuals during the abnormal pursuits of time-zone transitions or working at night.

Interest in Time Measurement Has a Long History[15]

Human awareness of daily, monthly, and annual cycles stretches back to prehistoric civilizations[19]

Earlier chapters have emphasized that all living clocks display the same basic timekeeping properties, although the structural components may differ widely. The major reasons for considering humans separately will emerge as the chapter proceeds. Time structuring exists universally on our planet, measured in days as Earth turns on its axis, months as the moon orbits Earth, or years as Earth orbits the sun. In the past, humans must have responded to these cycles when it came to hunting, to seeking safety during hours of darkness, and to choosing times for irrigating land and planting food.

Records of astrological and calendrical tables, together with the alignment of some of the stone henges in northern Europe (see Chapter 1) and the pyramids in Egypt, all point to a human awareness in early civilizations of these external cycles (Figure 9.2). Ancient peoples, however, had little appreciation for internal timing. Some of the earliest indications of internal timing in humans came from measurements of human body temperature. A British surgeon in 1836 noted daily temperature fluctuations not associated with fever. However, his ideas and other data strongly supporting temperature rhythms in humans were negated for many decades by the ideas of two very prominent European physiologists, as will be discussed shortly.

The mind's time sense has important implications for time perception and use by humans[3,24,36,66]

Subjective time and time memory are unique aspects of human time perception. The great majority of material in this book concerns physiological time measurement. In humans, however, time awareness is a vital aspect of time use. The capacity for concept formation, self-awareness, and other higher learning processes tends to greatly modify the application of physiological time measurement to behavioral output in humans. With an increase in the pace of modern industrialized societies, humans in many cultures equate passage of time not

(A)

(B)

FIGURE 9.2 Environmental rhythms. (A) Lunar rhythm with full moon rising over the Grand Canyon. (B) Solar rhythm with sunset over Hawaii. The lunar and solar cycles were important events in the lives of early humans.

When the basal forebrain is injured, memory of the event remains, but the chronology of events may be lost. If the crucial hippocampus is damaged, serious amnesia may ensue, with loss of ability to store new memories. Damage to the temporal lobe adjacent to the hippocampus partially impairs memory of past events. The cortex of the temporal lobe is especially sensitive, and injury may permanently delete years or even decades of self-memories. Very little is known about time-stamping by the brain. This process, by which the brain assigns a specific time and place to an event, stores it, and then retrieves it on command, is still poorly understood.

Interest in the physiological functions of humans grew in the nineteenth century[6,7]

THE IMPACT OF PAVLOV AND HIS THEORY OF CONDITIONED BEHAVIOR. A world-renowned Russian scientist, Ivan Pavlov (1849–1936), was indirectly associated with the slow acceptance of circadian rhythms in humans. Pavlov was awarded the Nobel Prize in physiology and medicine in 1904 for his work on neural regulation of digestion in mammals. His clearly formulated theory of conditioned learning had an immense impact on the scientific world of his day and continues to influence learning theory in the field of psychology even today. His theory of so-called classical conditioning stated that certain innate responses called unconditioned reflexes could be modified by repeated association with an artificial stimulus to achieve a reliable new response. For example, the salivation of a dog to the savory smell of meat could be conditioned to occur in response to a new stimulus, such as the ringing of a bell.

The atmosphere of late-nineteenth-century physiology was charged with dissension over innateness versus learning as the basis of vertebrate behavioral responses. Pavlov tipped the balance strongly in favor of learning. With regard to innate circadian regulation in humans, Pavlov's ideas attributed any temperature rhythmicity to acquired conditioned responses and profoundly delayed acceptance of these temperature rhythms as manifestations of true endogenous circadian timing.

THE IMPACT OF BERNARD AND HIS CONCEPT OF THE INTERNAL MILIEU. A second important factor in the

with a quantitative measurement but rather with productivity. Time is worth money. As a result, the standard is the amount of work accomplished, not the absolute time spent on a task. Humans are more often concerned, therefore, with subjective awareness of the passage of time than with real time.

Relatively little is known about brain regulation of the mind's time. In some fashion the human brain is uniquely endowed to organize and prioritize experiences into a chronology of memory events and to recall them at will. The intact brain acts as a sieve to magnify or delete certain events and then relegate these to either short-term or long-term memory. The dichotomy is particularly clear in elderly individuals. A World War II veteran may recall with vivid details the minute-by-minute events of D-Day 50 years ago, but fail to recall what he ate for breakfast just an hour earlier.

The basal forebrain, hippocampus, and temporal lobe are important processing centers for time memory. Most current information comes from brain-damaged patients.

ignoring of marked daily body temperature rhythms in humans was the influence of the most prominent French physiologist of his day, Claude Bernard (1813–1878). Bernard noted that land mammals must deal with large-scale environmental changes in temperature, water availability, gravity, and wind. He documented many physiological mechanisms that protect animals from extreme physical and biotic forces. Later the concept was expanded by American physiologist Walter Cannon, who coined the term *homeostasis* for this monumental concept concerning physiological protective mechanisms.

Homeostasis is the maintenance of the internal milieu of the body within very narrow limits in spite of the tendency for an individual's surroundings and activities to change. For example, reflex responses precisely regulate body temperature, osmotic pressure and composition of the blood, and other responses. The concept of internal constancy of warm-blooded vertebrates closed the minds of scientists to accepting changes in body temperature as a normal, endogenously programmed response in healthy humans.

CRACKS IN THE HOMEOSTASIS THEORY. Acceptance of temperature rhythms in humans as a normal event was very slow. The medical field tended to view temperature deviations from a fixed norm, such as fever or hypothermia, as a sign of illness. Even today most home thermometers have a red arrow at 98.6°F or 37°C designating so-called normal temperature. The outstanding research of Jürgen Aschoff on human temperature regulation was very effective in gaining acceptance first of endogenous temperature rhythmicity, and eventually of human rhythmicity in general.

Study of human circadian rhythms accelerated very quickly after 1950[6,45]

A FEW EXAMPLES OF HUMAN RHYTHMS. Many daily behavioral, physiological, and biochemical rhythms in humans are in synchrony with the daily light–dark cycles (Figure 9.3). A few representative cases suffice here to introduce human rhythms. The observed rhythms tend to fall into one of two groups. The first group contains rhythms that peak during the day and are associated with the activity phase of the individual. Core temperature, mental performance, physical efficiency, gastrointestinal activities, blood pressure and heart rate, and hormone adrenaline secretion all fit into this group. The second group concerns rhythms that show a peak during nocturnal sleep. This group is smaller and includes growth hormone and melatonin. Details of human rhythmic functions will be considered later in the chapter.

EARLY CAVE AND BUNKER EXPERIMENTS. An endogenous component of rhythmic variations in physiological and

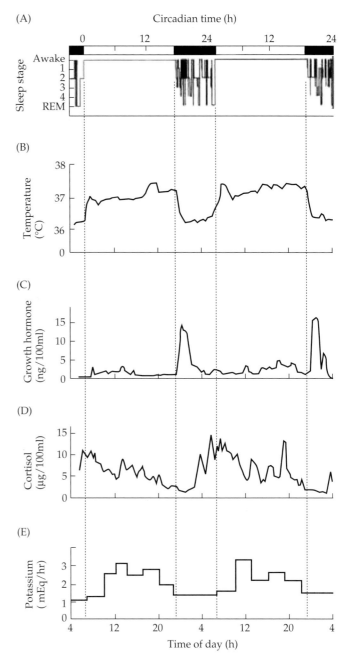

FIGURE 9.3 Representative daily rhythms in an LD 16:8 schedule for 48 h. Subjects were asleep at night and awake during the day. (A) EEG recording of sleep stages. (B) Core body temperature. (C) Growth hormone secretion. (D) Cortisol secretion. (E) Urinary potassium. The time scale for (A) is circadian hours, while the scale for (B–E) is real-time hours. (From Moore-Ede and Sulzman, 1981.)

behavioral variables in animals such as hamsters and rats has been demonstrated by studies of activity patterns under conditions of constant darkness and constant ambient temperature. The best way to provide convincing evidence for humans was to perform similar experiments. Because no time cues are present in such experiments, the method is described as a free-running protocol. Before early work on the endogenous component of human circadian rhythms became feasible, special environments and experimental protocols had to be created to eliminate the effects of surroundings or lifestyle of the subjects on the results.

Early experimenters used some very unusual work sites as laboratories. Underground caves, isolated soundproofed chambers, or the Arctic in summer provided relatively constant conditions. In such locations, much of the complexity due to human interactions and unknown lighting conditions was reduced or eliminated. In the 1930s, the famous American sleep physiologist, Nathaniel Kleitman, and his colleagues descended into a cave and lived on a non-24 h sleep–wake cycle for up to a week in order to measure their body temperature rhythms. The main question was whether the rhythm was innate or learned. The data were somewhat ambiguous but provided evidence for an innate nature because body temperature cycles showed independence from sleep–wake cycles.

Another cave experiment was carried out in 1962 by a young cave enthusiast named Michel Siffre. He lived in a 375-foot-deep cave in the French Alps without a watch and recorded his sleep–wake rhythms for 62 days. The cavern was totally dark except for the light from his headlamp, and the environment was dank and bone chilling during his entire stay. By means of an improvised telephone, Siffre informed his friends every time he went to bed or awoke. The result was an astonishing graph of his 24.5 h free-running sleep–wake rhythm (Figure 9.4). Remarkably, he lost track of 25 days because he didn't count presumed napping periods and these turned out to be 8 h nights.

In 1957, Mary Lobban and her colleagues lived in Spitsbergen, a Norwegian archipelago in the Arctic Ocean, during the time of the summer midnight sun. Photic time cues in the continuous light were minimal during the subjective day, and at night the subjects slept in blacked-out tents. Without the test subjects' knowledge, their watches had been doctored to provide either 21 h or 27 h days instead of 24 h days. Nine solar (24 h) days were equivalent to only 8 days with a 27 h period.

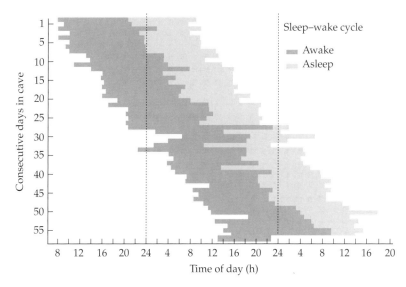

FIGURE 9.4 **Body temperature rhythm of an individual living in a deep, lightless cave for 62 days.** (From Palmer, 2002.)

Such a protocol, first employed by Kleitman in his cave studies, is now called forced desynchronization, and it is still widely used in human experiments. The variables measured showed differences in the relative importance of exogenous and endogenous constituents (Figure 9.5). The sleep–wake cycle showed a 27 h

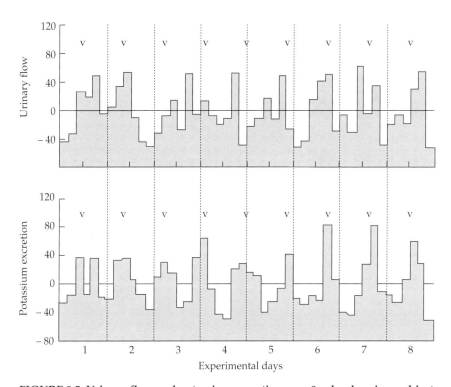

FIGURE 9.5 **Urinary flow and potassium excretion over 9 solar days in a subject living on a forced-desynchronization schedule.** Subject lived on experimental 27 h days in Spitsbergen during the continuous daylight of summer. v, noon by solar time. Dotted vertical lines indicate middle of subjective night. (From Lewis and Lobban, 1957.)

rhythm. Urine flow followed the sleep–wake cycle close-ly, with eight peaks when subjects were awake, and eight troughs when they were asleep. In contrast, the urinary potassium rhythm showed nine peaks and troughs, coinciding with the 24 h solar day. The desyn-chronization between the rhythms is particularly marked in the middle of the beat cycle, days 4 and 5, when the two rhythms are opposite in phase.

Another very important early experiment, directed by Aschoff, took place in a converted former bomb shelter deep underground in Munich, Germany (Figure 9.6). In 1965 the results were published in the journal *Science* in a paper entitled: "Circadian Rhythms in Man: A Self-sus-tained Oscillator with an Inherent Frequency Underlies Human 24-Hour Periodicity."[6] Multiple variables were recorded, including urinary electrolytes, core body tem-perature, and mealtimes. The subjects lived in an envi-ronment without time cues or direct social interactions, and they could select their own day–night LD schedules.

Key findings were an approximately 25 h rest–activ-ity cycle that continued to oscillate in synchrony with the other physiological variables. The ratio of sleep to waking activity remained close to 1:2. The paper drew attention to the multitude of variables that were under apparent circadian control and to the endogenous nature of these cycles. Because the period of the meas-ured rhythms did not match the 24 h period of the solar day, the rhythms were described as circadian or free-running. Aschoff's article marked the beginning of the modern study of circadian physiology in humans.

These pioneering studies under challenging condi-tions provided evidence for the endogenous nature of rhythmicity in human physiology and behavior. However, the study sites were unpleasant or inconven-ient laboratories, and taxing logistics often prevented detailed or complex measurements. Perhaps even the subjects were atypical humans, since they were willing to withstand the discomforts of cold caves or the isolation of the Bunker. Furthermore, the nature of human lifestyle introduced artifacts that obscured the true underlying rhythm. Clearly, new protocols carried out in comfortable, convenient facilities were a very high priority.

INTEREST IN EXOGENOUS VERSUS ENDOGENOUS CAUSA-TION OF HUMAN RHYTHMS. Some aspects of human daily rhythms probably result from behavioral changes associated with the light–dark and sleep–activity cycles. After all, humans are physically and mentally active during the daytime. The environment is light, busy, and

(A)

(B)

(C)

FIGURE 9.6 Munich bunker experiment in 1961. (A) Sign at the entrance to the Bunker, signed by Professor Aschoff. Translated from the German, the sign states, "Entry forbidden! Experiment! Quiet!" (B) Early bunker stu-dent subject working on an assignment during a 28-day bunker sojourn. (C) Leaving the bunker in good spirits.

stimulating. At night, in contrast, humans rest and recuperate during sleep. During this time, growth hormone and cortisol promote the growth of tissues, and fat rather than glucose is metabolized. Conceivably the body temperature falls at night merely because individuals are less active. Sleep is also accompanied by a change from the upright to the prone posture, resulting in cutaneous vasodilation and increased heat loss.

If human daily rhythms were evoked merely by external conditions, they would be described as exogenous. Then they would closely parallel the reflex responses of baroreceptors, chemoreceptors, or osmoreceptors. All would be responses to environmental and behavioral changes, turned on and off directly by external switches. However, the cave experiments showed that this was not exclusively the case. Exogenous rhythms contrast strikingly with endogenous rhythms, which have an internal cause based on a self-sustained circadian pacemaker system.

One of the main aims of chronobiology, particularly in its early years, was to resolve the controversy over endogenous versus exogenous causation of daily rhythms. Scientists were eager to determine to what extent external changes were sufficient to account for the observed daily rhythms in humans, and how endogenous factors contribute to the rhythm. Two lines of indirect evidence indicated that exogenous causes could not be a full explanation of daily rhythms in humans.

After long-haul flights to the east or west crossing several time zones, individuals suffered temporarily from jet lag. The symptoms included fatigue during the new daytime combined with an inability to sleep well in the new nighttime. Additional symptoms included a reduced ability to concentrate or to perform complex mental and physical tasks, and a loss of appetite accompanied by a bloated feeling after meals. The second line of evidence was the occurrence of symptoms in jet lag resembling the so-called shift worker's malaise common in night workers (see Chapter 10). For the moment, just remember that individuals could not fully and immediately change their habits to a new sleep–wake schedule. Another endogenous factor independent of the environment appeared to be involved.

The initial protocols of free-running and forced-desynchronization experiments established that the measured rhythms had endogenous and exogenous components. The endogenous component is generally of more interest than the exogenous component. However, the exogenous effects may alter the indirect output rhythm that is measured, resulting in so-called masking effects. Calling them such does no more than draw attention to a desire to assess what is really happening to the endogenous component. In most other biological studies, the direct exogenous effects produced by behavioral and environmental changes are of prime interest. Masking has an important practical implication for chronobiology research when indirect measurements are being used. Output "hands of the clock" such as body temperature, heart rate, or hormones will not provide accurate information on the endogenous phase, period, or amplitude of circadian rhythms in humans if masking is present.

THE NEED FOR BETTER FACILITIES AND METHODS. The pilot experiments in caves, in the Arctic in summer, and in the Bunker generated much enthusiasm for research on human rhythms and set the stage for the modern scientific study of human circadian physiology that began in the late 1960s. Soon special isolation units were created (Figure 9.7). The first was an underground labora-

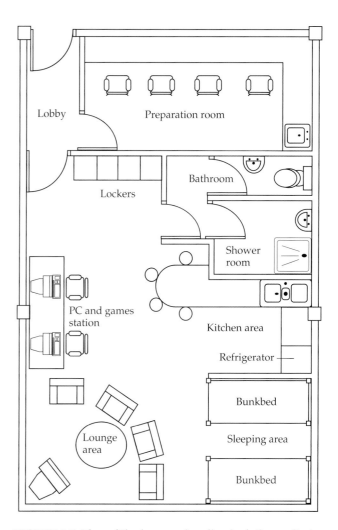

FIGURE 9.7 **Plan of the human circadian isolation unit at Liverpool John Moores University.** The unit provides areas for sleeping, cooking, relaxing, and studying. An LD cycle is programmed with artificial light, and the whole facility is sound attenuated to reduce any time cues. Monitoring equipment is stationed in an adjacent room. (Courtesy of J. Waterhouse.)

tory constructed at Andechs, Germany. A later version consisted of the converted top floor of a hospital in New York. Currently eight human chronobiology observation facilities exist in Europe, Japan, and the United States. Some have kitchens; in others, meals are provided. Some are designed for individuals, others for groups of volunteers. Modern units enable programmed control of lighting and elimination of disturbances from noise and other people. They also provide safe, efficient, comfortable facilities for the volunteers.

Isolation units offer great advantages for chronobiological studies, but a difference still remains between human subjects studied in such circumstances and other animals. Biological timing may well play a role in the organization of the physiology and behavior of humans. However, humans have conscious knowledge of time throughout the day. Such a conscious knowledge of clock time or even the free will of humans adds to variability in behavior and could override circadian signals coming from an internal biological clock.

The term *isolation*, as used in the current context, refers purely to separation from the complexities associated with living normally or to isolation from time cues. It does not refer to sensory deprivation or physical deprivation of any kind. The subjects residing in these units are monitored carefully and cared for by the research staff, technicians, nursing staff, and other qualified staff. In some experiments, volunteers live individually in single rooms, and meals are provided by staff members. In other cases, subjects can spend their waking time in communal rooms together with other volunteers, and they may cook their own food. The arrangement depends on the detailed aims of the experiment. In most cases the subjects describe their experiences as peaceful, productive, and even euphoric.

New Methods Have Been Developed in the Past 25 Years

Methods must accommodate the unique parameters of humans as research subjects[60,70]

INTERDISCIPLINARY INTEREST IN HUMAN CIRCADIAN CLOCKS. Human chronobiology issues have received growing attention in the past few decades in disciplines such as psychology, physiology, the neurosciences, and medicine. As the direct applicability of chronobiological principles to human welfare becomes more and more apparent (see Chapter 10), interest in applied fields of human services grows. However, humans are not rats, mice, or hamsters, and the positive and negative aspects of using humans as research subjects must be carefully evaluated. The uniqueness of humans as subjects for chronobiological studies is important to consider (see Figure 11.7).

ADVANTAGES OF HUMANS IN CIRCADIAN RESEARCH. In terms of size, intelligence, self-awareness, and moral and ethical concerns, as well as communication capacity, humans are uniquely different from all other animal species. These attributes impart several advantages over other animal or plant models for the study of chronobiology. Because of the large blood volume in humans and its relatively easy accessibility, many hormones and other blood constituents can be measured. Alternatives to blood, such as saliva and urine, collected from volunteers can also be studied. In addition, mood and mental performance can be measured because humans can communicate their feelings orally or in writing. Human volunteers can also be instructed to give samples and attempt certain actions, such as exercise or sleep, at particular times.

DISADVANTAGES OF HUMANS IN CIRCADIAN RESEARCH. Several disadvantages and constraints also apply to studies of human circadian rhythms. Humans are highly intelligent social animals who, in these experiments, live complex lives in an artificial environment heavily modified by domestic lighting. They tend to be choosy about living details, and often they must have particular kinds of food. They eat canned or frozen food rather than natural foods. They may have very particular, individualistic lifestyle preferences about recreation, and other habits. The results obtained from highly urbanized humans may differ greatly from those obtained from primitive tribes that continue to live in close contact with nature, but these tribes have rarely been studied from a circadian point of view.

One very funny story about the difficulties of keeping human volunteers reasonably happy in chronobiology isolation experiments comes from J. D. Palmer's delightful book *The Living Clock*.[60] He relates the experience of a sculptor who was eager to live alone for several weeks in an isolation chamber in total darkness, wired to recording instruments:

> In return for serving as a human guinea pig, he negotiated to have an enormous mound of wet clay stacked in the darkness with him so he could try sculpting by feel only. Mick had to practice feeling his way around his new quarters, preparing meals in the pitch darkness, bathing safely and passing need-requests as notes slipped under the door. Fortunately he was a touch typist. Everyone, sculptor and investigators alike, were anxious to see how the sculpture and isolation experiment would turn out. Hopes were high. But after only a few days in isolation the subject began pounding on the wall and yelling to be let out. 'You never gave me anything I asked for in my notes,' said an enraged Mick. 'But you only shoved blank pages out under the door' was the reply. The problem: After weeks of practice runs and training, and every major contingency planned for, someone had forgotten to put a ribbon in the typewriter![60]

Even one of the most fundamental aspects of the environment for humans, the light–dark cycle, is not well defined because humans generally control their light–dark cycle through the light switch. In addition, the natural light–dark cycle varies with location. Cities differ greatly in light profile from rural areas, and light varies with season. Light exposures may be very different in apartments with south-facing windows compared with those that have north-facing windows. Isolation of individuals from these environments is not easy. Such complexity casts doubt on the interpretation of experimental results.

Cost considerations for isolation unit experiments are very important. The facilities are expensive not only to build, but also to maintain. The experiments are labor-intensive and often last several weeks or even months. These factors might well have contributed to the common practice of collecting multiple variables in such experiments. Experiments are often designed by an interdisciplinary team of investigators with multiple missions. Data may be collected simultaneously about a subject's mood, performance, polysomnographically recorded sleep, waking EEG, heart rate, respiration, and blood hormone levels.

A second issue confronting human research is that of ethical constraints. Ethical considerations have always been paramount for investigators working with human subjects. Because experimental interventions can be carried out only with the full cooperation of completely informed volunteers, the range of studies is limited. Both psychological and physiological consequences of any protocol must be carefully considered by the volunteers themselves, by the physicians and scientists involved, and independently by an institutional review board.

Several distinct laboratory protocols have been developed [2,11,22,27,43,48,50,52,69]

CONSTANT ROUTINE. The constant routine has become a cornerstone of studies of chronobiology. The prototype method was first described by Mills's group in 1978. Since that time the protocol has been modified and improved by many groups, particularly Czeisler's at Harvard. The constant routine is a concerted attempt to remove the complexities that accompany a human subject's environment and lifestyle and thereby remove or minimize the masking effect of the exogenous component of a rhythm. In a constant routine the subject is required to stay awake and sedentary, preferably lying down and relaxed for 24 to 40 h. The environment is maintained with constant temperature, humidity, and lighting. The subject must engage in similar activities throughout the routine, generally reading or listening to music, and must take identical meals at regularly spaced intervals. In an alternative protocol, these factors can be distributed uniformly across the 24 h day.

The constant-routine protocol has often been used with measurements of core body temperature. Under these conditions, the rhythm of core temperature does not disappear, even though its amplitude decreases (Figure 9.8). The component of the temperature rhythm that remains must arise from within the body and therefore is the endogenous component of the rhythm. Effects of the environment and lifestyle do exist, however, because the temperature rhythm observed in the constant routine differs from that measured under normal day–night conditions. The difference between the two curves is the exogenous component. For core temperature, the exogenous effects come mainly from physical activity that raises core temperature, and from sleep and lying down, which lower it. These two components are in phase in subjects living normally, for physical activity and wakefulness raise core temperature in the daytime, while sleep and inactivity lower it at night.

Many variables have been studied under constant-routine protocols. Heart rate exhibits marked variation in the presence of a sleep–wake cycle. It shows circadian variation during a constant routine, although with a much smaller amplitude. In contrast, blood pressure has higher values during the day than at night in normal living but does not show any discernible circadian variation during the constant-routine protocol. In other words, the rhythm of heart rate has a weak endogenous component, but blood pressure is almost entirely exogenous in origin.

Melatonin and its metabolites offer good possibilities for human circadian studies under constant routines. Administering the urinary metabolite 6-sulfa-

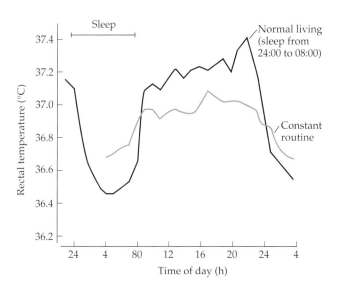

FIGURE 9.8 Mean core body temperature of a group of eight individuals in a constant routine. For further explanation see text. (From Minors and Waterhouse, 1981.)

toxymelatonin is a useful, noninvasive method. For the most accurate assessments, light must be maintained at sufficiently low levels to avoid suppression of melatonin production. Plasma melatonin is, at present, the least masked and most accurate method for measuring circadian phase in constant routines.

Several very important disadvantages are always present in the constant-routine protocol. Because volunteers must stay awake for at least 24 h, they will be sleep deprived. The resultant fatigue will have a definite influence on some variables, such as mental performance and subjective alertness. The conventional 40 h limit on duration of a routine also greatly restricts the quantity of data that can be collected. Removing the masking effects of sleep and activity introduces other undesirable effects, and limiting the duration means that less than two cycles of data can be collected. To avoid sleep deprivation problems, a variant of the constant routine is sometimes used in which volunteers are kept in a constant posture and fed regular, equal meals, but sleep is allowed. The addition of sleep to the routine removes the difficulties due to sleep loss, but it does nothing to remove the masking effects of sleep and its associated changes.

VARIATIONS ON CONSTANT ROUTINES. The aim of the constant routine is to minimize the exogenous component of a rhythm experimentally. Exactly how this is done depends on the variable being measured. The preceding description applies to variables that are affected mainly by activity and posture. The protocol has been modified for variables in which the masking effects are different in nature or importance.

Melatonin secretion is suppressed by bright light. As a result, blood samples must be collected in dim light (Figure 9.9). In general, variations in the amount of activity and food intake are relatively unimportant in melatonin assays. The time at which melatonin secretion starts its nocturnal rise can be used as a phase marker. This convenient marker point is called the dim-light melatonin onset, or DLMO. The time of cessation of melatonin secretion, referred to as DLMOff, has also been used. Some protocols dictate measurement of the excretion rate of melatonin in the saliva or measurement of its main urinary metabolite, sulfatoxymelatonin. A cosine curve is fitted to values obtained over the course of 24 h in order to estimate the time of the peak of the fitted curve.

Studies of the circadian rhythms of the hormones associated with the control of carbohydrate and fat

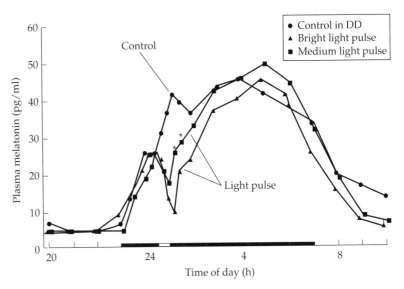

FIGURE 9.9 **Circadian changes in the concentration of plasma melatonin of three subjects during the night** in control conditions of continuous darkness, and in two LD schedules, one with a medium bright light, and the other with a medium bright light pulse between 00:30 and 01:00. Light bar at bottom indicates the LD schedule and pulse time. (From Bojkowski et al., 1987A.)

metabolism have other problems that must be taken into consideration. Van Cauter's group has used a constant intravenous infusion of glucose in place of having subjects eat identical, evenly spaced meals. Supposedly, pulsed plasma concentrations of glucose from meals can act as a masking factor and consequently must be controlled, but eating identical meals does not guarantee a constant uptake of nutrients into the bloodstream. Digestion and absorption from the gut are important, and both processes show circadian variation. A continuous intravenous infusion of glucose can be used to circumvent some of the problems and maintain a constant level of glucose in the plasma. It cannot, however, guarantee a constant uptake of glucose into the cells of the body because uptake is modified by the influence of insulin and growth hormone acting on cell membranes.

FORCED DESYNCHRONIZATION. The forced-desynchronization protocol is based on the fact that the human circadian clock is not able to adjust to an imposed lifestyle whose period differs substantially from 24 h. In its earliest form, the protocol was used in the Lobban experiments in the constant daylight of a high-arctic summer. In more recent experiments, individuals lived in an environmentally controlled facility. If a subject lives on 28 h days with 9.3 h of sleep and 18.7 h of activity each day, or 21 h days with 7 h of sleep and 14 h activity, the endogenous component of a measured rhythm continues to run with its circadian period close to 24 h. As a result, the endogenous and exogenous components of a rhythm can be separated because they

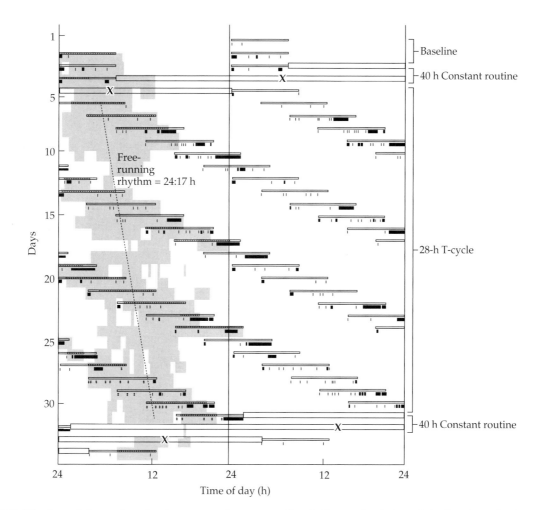

FIGURE 9.10 The forced-desynchronization protocol.
Double-plotted actograph of a forced-desynchronization protocol for a 25-year-old man in an environmental scheduling facility for 34 days. After three baseline sleep episodes (narrow open horizontal bars), a 40 h constant routine (wide open horizontal bars) was performed, followed by a 28 h sleep–wake schedule, and finally a constant routine. Core body temperature was recorded continuously. The X indicates the timing of the core body temperature nadir during the constant routines. The gray shading in left plot indicates when core body temperature was below the mean, and the dashed line indicates the timing of the fitted core body temperature minimum during the forced-desynchronization segment of the protocol. The estimated period of the core body temperature rhythm was 24:17 h. All sleep episodes were recorded electronically. Wakefulness within scheduled sleep episodes is indicated by black bars below the scheduled sleep episodes. Note that the subject experiences far less wakefulness when the sleep episodes occur close to the circadian temperature nadir and that most wakefulness occurs at the end of scheduled sleep episodes. (From Dijk and Czeisler, 1994.)

move progressively out of phase and back into phase again (Figure 9.10). The endogenous component loses or gains about 4 h per cycle of the imposed day. The length of time taken for the two rhythms to regain their original phasing relative to each other is termed the beat cycle. In the case of 28 h days, beats will occur every 7 solar days because 7×24 h equals 6×28 h. In the case of 21 h days, only 7 solar days are required because 7×24 h equals 8×21 h.

If temperatures are measured throughout a beat cycle, they can be averaged in one of two ways. A protocol using artificial days of 27 h will serve for illustration. In one method, results from nine 24 h cycles are averaged into a single 24 h rhythm. Any phase of this averaged rhythm will then have been mixed with all phases of the imposed 27 h sleep–wake cycle. Provided that the sleep–wake cycle is similar from day to day, any effects due to it will be canceled out. The averaged rhythm observed will represent the mean of the endogenous component of the measured rhythm over the course of the beat cycle (Figure 9.11A). In the second method, rhythms are averaged over a period equal to the imposed 27 h sleep–wake cycle. Each phase of this averaged 27 h cycle will be mixed with all phases of the endogenous cycle, thereby canceling out the effects of the endogenous component (Figure 9.11B). Masking

(A)

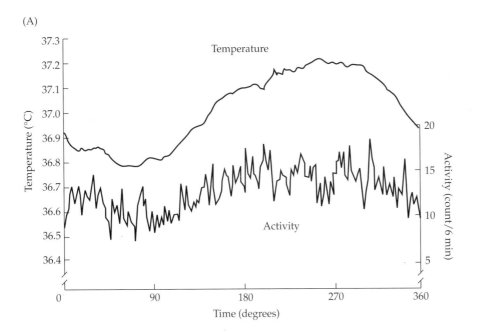

(B)

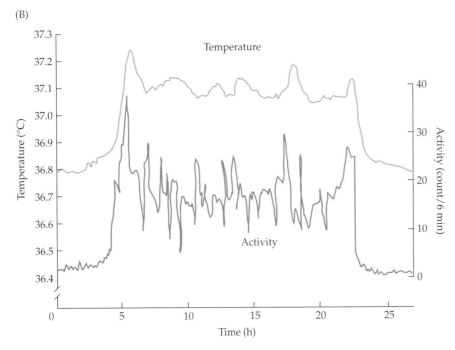

FIGURE 9.11 Analysis of data from a forced desynchronization protocol. (A) The endogenous component of mean core body temperature and activity rhythms from a group of subjects living on 27 h days. The rhythms have been averaged over a period equal to the free-running period, where 360 degrees equals one cycle. Averaged in this way, the effects of the sleep–wake cycle have been canceled out, and the remaining component is endogenous in origin. (B) The exogenous component associated with mean core body temperature and activity rhythms from a group of subjects living on 27 h days. The rhythms have been averaged over a period equal to the imposed 27 h sleep–wake cycle. For further explanation, see the text. (From Waterhouse et al., 1999.)

effects in the form of bursts of activity and transient increases in temperature are evident just after waking, at mealtimes, and just before bedtime.

A more recent protocol places study subjects on high-frequency sleep–wake schedules, with periods ranging from 20 minutes to 3 h. The protocol can be repeated on different volunteers with different timing of the sleep component of the sleep–wake cycle. In a protocol characterized by a 3 h sleep–wake cycle of 1 h sleep and 2 h wake time per cycle, for example, the following schedule for three different volunteers results:

Experimental schedule 1: Sleep at 24:00–01:00, 03:00–04:00, and continuing every 3 h

Experimental schedule 2: Sleep at 01:00–02:00, 04:00–05:00, and continuing every 3 h

Experimental schedule 3: Sleep at 02:00–03:00, 05:00–06:00, and continuing every 3 h

A variable is measured throughout the 3 experimental days. Then combining the results from all 3 days makes it possible to produce separate "sleeping" and "waking" curves. Each curve allows correction for the immediate effects due to alternation between sleeping and waking. In this way, the underlying effects of the circadian sleep–wake cycle can be discerned. The protocol is clearly just an alternative design to separate the exogenous and endogenous contributions.

VALUE OF CONSTANT ROUTINES AND FORCED DESYNCHRONIZATION. The constant-routine and forced-desynchronization protocols have confirmed the presence of an endogenous, clock-driven component, as well as an exogenous component, in human rhythms. The nature of the exogenous masking component depends on the variable under consideration. Two further examples will emphasize the point. First, airway caliber is affected similarly to core temperature. It increases with physical and mental activity in response to adrenaline release, and it drops during sleep in response to parasympathetic nervous system activity.

Second, the two protocols have shown that the relative weighting of the endogenous and exogenous components of a rhythm depend on the function considered. For example, the rate of excretion of water by the kidneys and the rate of secretion of growth hormone by the pituitary gland both have dominant exogenous components. Water intake affects kidney excretion, and sleep affects hormone production. In contrast, the excretion of potassium in the urine and the secretion of cortisol into the plasma both have smaller exogenous than endogenous components. The two components are similar in magnitude in the case of core temperature (see Figure 9.8).

Although the constant-routine protocol is often regarded as providing the best standard for describing a circadian rhythm and for inferring the output of the circadian clock, it suffers limitations. By its very nature, the protocol is too demanding for field conditions and for individuals who cannot give informed consent. The protocol is also inappropriate for use over extended periods of time. It is unusable for multiple consecutive days after a time-zone transition or during night work. In addition, because sleep is prohibited, variables affected by the amount of time spent awake cannot be assessed satisfactorily.

Unlike the constant routine, the forced-desynchronization protocol does not have the problem of loss of complete sleep periods. However, forced desynchronization can give only an average value over the course of a beat cycle, and the other problems associated with constant routines apply.

A different strategy must be developed for studies under field conditions[30,39,40,63,73]

CHOICE OF A FIELD MARKER. The methods described in the preceding section are laboratory bound; other protocols must be adopted for living under natural field conditions. In practice, the subjects behave much more normally with regard to their environment and sleep–wake cycle. Subjects are likely to be physically active and to show a variability in their patterns of activity day by day. They must carry with them any necessary measurement devices. In all likelihood the subjects will not be able to tolerate shortened sleep periods, or forgo them completely, because of family or job considerations. In addition, the circadian marker chosen must be acceptable to both subject and research staff. Finally, demasking techniques must be devised to correct the data for any exogenous environmental and sleep–wake cycle effects. Because of masking effects, field studies in chronobiology present a real challenge.

Many laboratory procedures are unacceptable when subjects are living at home and working at jobs. Methods that require subjects to be awakened at night are unsuitable. Sampling of saliva and urine, for example, falls into this category. Similarly, continual drawing of blood or the use of an intravenous catheter is not feasible. Practical markers include core temperature, movement, heart rate, and blood pressure, but none of these is ideal. The measurement of heart rate and blood pressure can be automated, but inflation of the pneumatic cuff to measure blood pressure may disturb sleep. Heart rate is dominated by exogenous effects, making demasking procedures essential. Rectal temperature is reliable for measurements of core temperature, but wearing a rectal probe is unacceptable to many, particularly during activity or at social events. Tympanic, oral, and midstream urine temperatures have also been used, but they all require the subject to be wakened from sleep. If recordings are not made during the period of sleep, to avoid excess fatigue, then a third of the potential data is missed.

Even for usable techniques, interpretational and other problems arise. For example, several methods of recording core temperature have been tried. A temperature-measuring device consisting of a thermistor pill can be swallowed and later retrieved from feces. Alternatively, an insulated skin temperature can be taken from the armpit. Rectal, gut, and insulated armpit temperatures in subjects living normally have been compared (Figure 9.12). The records of rectal and gut temperature are very similar, but insulated skin temperatures were very different. One interpretation of the skin data is that the insulation of the armpit probe is satisfactory only if the arms are not raised above the head. However, blood reaching the armpit from adjacent, non-insulated areas of skin, such as the arm, might have an effect. If other areas of the skin are considered as potential sites for measurement of core temperature, another possible problem arises. Changes in some parts of the cutaneous vasculature are the means by which heat loss is regulated to produce the circadian rhythm of core temperature. As a result, the area of skin will show a cir-

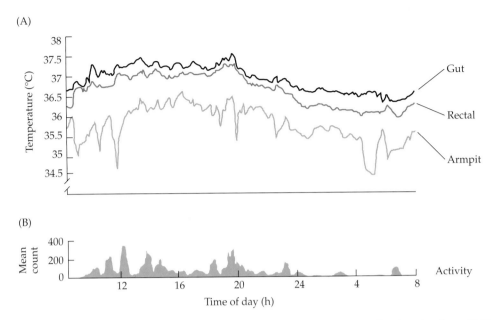

FIGURE 9.12 Comparison of gut, rectal, and insulated armpit estimates of core temperature relative to activity. (A) Rhythms of gut, rectal, and insulated armpit temperatures for a subject living normally, sleeping from 00:30 to 09:00. (B) Mean activity count. For further explanation, see the text. (From Edwards et al., 2002.)

cadian rhythm in blood flow and blood temperature when the subject is at rest in a thermally neutral environment. The armpit temperature will peak when core temperature is falling most rapidly and show a trough in the early morning when core temperature is rising most quickly. The armpit temperature data agree with this interpretation: the temperature effects are delayed appropriately about 6 h compared with core rectal or gut temperature (see Figure 9.12). Although the use of gut temperature appears to be a satisfactory alternative to rectal temperature, the pills are expensive and recycling them is unacceptable. For now, then, experimental investigations of core body temperature must continue to rely on rectal or armpit measurements instead.

ANALYSIS AND INTERPRETATION OF FIELD DATA. Demasking methods use the raw data obtained in the field and then attempt to correct them mathematically to account for masking effects of exogenous influences. Most are comparatively new methods, and they are not universally accepted. More details of the methods and of the criticisms leveled against them can be found in the references cited at the start of this section. The aim here is to show that attempts are being made to unravel the complex mixture of endogenous and exogenous components found in most raw data. The methods have generally been used primarily with temperature data, but some have also been applied to studies of blood pressure, heart rate, and some urinary constituents.

Some of the simplest models have added 0.3°C to temperatures associated with sleep. Other models have removed an exogenous temperature profile from raw data. The exogenous profile was obtained by comparison of raw temperature data with normative endogenous temperature data. Such a profile is based on estimates of the mean temperature rhythm of a group of subjects who had remained sedentary and awake throughout a 24 h period (see Figure 9.8). The profile amounts to the difference between the two curves. The exogenous profile can then be subtracted from other sets of raw data, thereby exposing or demasking the endogenous component of the temperature rhythm. These methods are limited, however, for they assume that the exogenous effect is the same for all subjects and experimental conditions, and that it does not vary between successive days. The last assumption would be especially questionable if the subject's activity or sleep–wake profile were altered by factors such as night work.

More recently, an improved method called "demasking by categories" has been devised. A diary is kept by every subject to document exogenous effects for each day. The subject records different categories of activity, such as sitting, standing, or walking. Activity is measured by an actimeter on the nondominant wrist or as a function of heart rate. The method chosen can thus be tailored to the subject's habits. The sizes of the masking effects due to sleep and to different amounts of activity are then estimated. The aim is to produce a set of corrected temperatures that approximate a sinusoidal curve more closely than the original masked data do. The assumption that the demasked temperatures follow a sinusoidal shape is based on the profiles of the endogenous temperature rhythm observed in constant routines (see Figure 9.8) and calculated from forced-desynchronization studies (see Figure 9.11A).

Additional highly technical mathematical methods to correct for masking have been developed but are outside the scope of this textbook. However, the key issue is to assess how accurately these demasking methods estimate the output from the human circadian clock. The answer is not known due to the fact that direct measurement of output is not possible in humans. Because core temperature rhythm during constant routines is considered the most accurate measure, the issue becomes one of comparing constant-routine data and demasked data. The evidence suggests that the more recent correction methods remove at least some of the masking effects of activity and sleep. These methods provide better estimates of the circadian clock, therefore, than the use of raw data does. Clearly the development of such methods is necessary if field studies are to be pursued and the chronobiological investigation of humans is not to become laboratory bound.

Fundamental Circadian Principles Also Apply in Humans[18,21,34,53,57]

In humans, the primary circadian pacemaker is the suprachiasmatic nucleus[56]

The site of the endogenous component, the suprachiasmatic nucleus (SCN), cannot be directly investigated physiologically in humans for ethical reasons, nor can its molecular chemistry and genetics be examined by gene manipulation. A small amount of histological data is available from cadavers. Most conclusions, however, must be inferred from other animals. Histological evidence indicates that an SCN is present in humans in the ventral hypothalamus just above the optic chasm (Figure 9.13; see also Plate 3, Photo B). Each half of the bilobed nucleus contains about 50,000 cells. The SCN has obvious regional and cellular spe-

cialization. Histochemical studies show that the neurons of the SCN have receptors on their surfaces for several substances, including GABA (gamma-aminobutyric acid) and melatonin.

Free-running rhythms have been documented[8,22,23,46,50]

The value of free-running experiments in establishing an internal component to the circadian timing system has already been described. Results from such measurements in humans show that the great majority of individuals have a free-running period in excess of 24 h, usually about 25 h. A question arises whether the value of this free-running period is really the intrinsic period of the human clock. Even though such a value was accepted for some time, current opinion has changed. Two lines of evidence can be used to evaluate the issue. The first involves important methodological differences between measurements of free-running experiments in humans and other animals. The second line of evidence concerns the effect of light of even very low intensity on the human circadian clock.

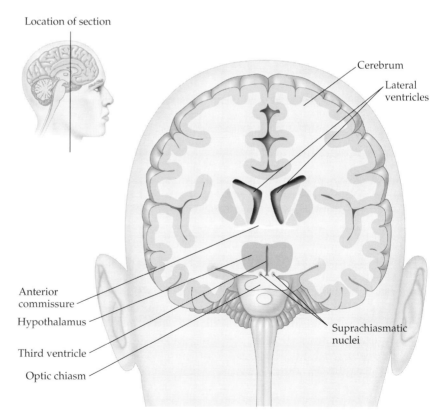

Location of section

Cerebrum

Lateral ventricles

Anterior commissure

Hypothalamus

Third ventricle

Optic chiasm

Suprachiasmatic nuclei

FIGURE 9.13 The site of the human SCN. The human suprachiasmatic nucleus is located at the base of the hypothalamus and just above the optic chiasm. (From Waterhouse and Åkerstedt, 1997.)

The original free-running experiments in humans were not conducted in constant darkness. Subjects could switch on the lights when they woke up and turn them off when they went to bed. Even when the lights were kept on continuously, the sleep–wake cycle generated a light–dark cycle because of eye closure associated with sleep. In view of the sensitivity of the human circadian pacemaker to light, the observed 25 h period did not reflect the intrinsic period of the human circadian clock because the period had been modified by light exposure.

During a free-running experiment, the phase relationship between the sleep–wake cycle and the core body temperature rhythm always changes. Under entrained conditions, humans typically go to sleep 6 h before the temperature minimum. In contrast, under free-running conditions, subjects select to go to sleep soon before this minimum. If a cyclic sensitivity to light is phase-locked to the core body temperature rhythm, the change in phase relationship between sleep and core temperature implies that light is falling on different parts of the phase response curve, or PRC. When subjects are entrained, light at bedtime has little effect compared with light at waking, soon after the temperature minimum. Under entrained conditions, therefore, light falls primarily on the phase-advance portion of the PRC. In contrast, during the free-running condition, evening light falls on the phase-delay portion of the PRC, and morning light is less effective in causing a phase advance because it is too late after the temperature minimum. The net result is that the observed period in free-running conditions will have lengthened.

In nonhuman experiments, investigators can circumvent these problems by placing animals in constant darkness or by blinding them and studying them while they are exposed to a normal light–dark cycle. These options are obviously very difficult or impossible to apply to humans, and indirect methods must be considered.

The preceding theoretical considerations lead to three predictions. First, under conditions of continuous dim light, the free-running period should be closer to the intrinsic circadian period. Second, the period in free-running individuals who are blind should be a close approximation to the intrinsic period if the individuals are normal with regard to their circadian system except for the lack of light input. Third, uniform distribution of light across the whole circadian cycle should lead to an observed period close to the free-running period.

All three predictions have been confirmed. When subjects selected their own sleep–wake cycles and associated dim light–dark cycles, the average period was 24.3 h. For individuals who were blind, the observed period while they were living in society was significantly less than 25 h. When subjects underwent a forced-desynchronization study with the light–dark cycle progressively impinging equally on all phases of the

endogenous oscillator, the observed period of the temperature, melatonin, and cortisol rhythms averaged 24.2 to 24.3 h.

The sleep–wake schedule and the endogenous core temperature rhythm become desynchronized in a forced-desynchronization protocol, and such desynchronization also occurs spontaneously in some individuals during free-running studies. This phenomenon, called spontaneous desynchronization, has important implications for an understanding of the human circadian timing system. During spontaneous desynchronization, the sleep–wake schedule and associated light–dark cycle occur at many phases of the core temperature rhythm. Under these conditions the observed period of body temperature is significantly shorter than during free-running conditions, and it generally equals 24.5 h or less.

All these results indicate that the intrinsic period of the human circadian clock is closer to 24 h than was originally believed, as long as confounding effects of light are eliminated. The free-running period shows remarkable stability over time. Differences may exist between individuals, but repeated assessments of the period in individuals who are blind, several years apart, have yielded estimates differing by a few minutes only.

Entrainment is an important aspect of human circadian behavior[10,12,16,17,18,21,28,33,34,47,49,53,54,62]

THE HISTORY OF LIGHT AS A SYNCHRONIZER. The light–dark cycle is the major environmental cycle for entraining circadian rhythms of many animals. In humans, the role of light has been controversial in the past, mainly because the sensitivity of the human clock to light had been underestimated. The miscalculation led to experiments in which the light–dark cycle was not carefully controlled. In some cases dim room light was considered darkness.

One important early study applied approaches previously used in animal studies, and the results appeared to confirm the role of nonphotic signals in the synchronization of human circadian rhythms. The sleep–wake cycle and other circadian rhythms reentrained to an abruptly shifted light–dark cycle. In these experiments, however, the light–dark cycle was not the only periodic stimulus available. Regular gong signals were used to summon the subjects to perform various tasks and led the investigators to propose that social signals act as synchronizers in humans. Unfortunately, the social synchronizer view held sway for some time. Many chronobiologists believed that social entrainment was the major mode in humans and that light was ineffective. Endocrinologists, however, had studied the rhythm of the hormone cortisol and had obtained evidence that the light–dark cycle was a powerful stimulus for entraining the cortisol cycle. In about 1980, bright light was also shown to elicit a powerful neuroendocrine response in humans: the suppression of melatonin. As a result of the latter two important findings, the role of the light–dark cycle was reinvestigated.

Photic entrainment was assessed by studies of volunteers under conditions similar to those used in animal studies. A strict light–dark cycle was applied, with dark meaning absolute darkness. During the free-running sections, the subject controlled the lighting by sleeping in the dark and by switching on the lights when awake. During the scheduled LD cycle section, however, a schedule of darkness was imposed from 22:00 to 06:00. In this part of the protocol, no auxiliary lighting was available. Under these conditions the sleep–wake cycle and the rhythm of core body temperature appeared to be synchronized to a shifted light–dark cycle (Figure 9.14). However, this interpretation was criticized on the grounds that darkness would induce people to sleep, thereby affecting the sleep–wake cycle and the entrainment of the circadian oscillator.

The turning point in the acceptance of the light–dark cycle as a major synchronizer in humans came with carefully executed experiments that cleared away all doubts. Scheduled exposure to pulses of bright light having an intensity of 10,000 lux and lasting 3 to 7 h was repeated on several consecutive days. These light pulses shifted circadian rhythms of body temperature, melatonin, and sleep propensity, even when the sleep–wake cycle was kept constant. When multiple variables were assessed, the phase shifts of all of them were of similar magnitude, suggesting that all were driven by a single circadian pacemaker.

PHASE RESPONSE CURVES TO LIGHT. The effect of light on the internal clock of humans depends on the time at which it is presented. The relationship between the time of light exposure and the magnitude of phase shift of the clock is called a phase response curve, or PRC (see Chapter 3). Various phase response curves to light have been published for the human circadian timing system. The methods used differ slightly in the different studies. Some approaches used single isolated light pulses during free-running conditions. In others, light pulses were presented on 3 consecutive days, and assessments were made during the initial and final days.

Phase advances were consistently obtained when light was given after the body temperature minimum, and phase delays when light was administered before the temperature minimum. In young adults with a

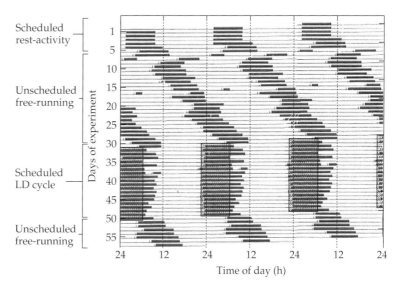

FIGURE 9.14 Triple-plotted actograph showing the sleep–wake cycle of a single individual under imposed LD schedules and self-selected light schedules. The protocol is shown on the left. Gray areas, emphasized by box outlines, indicate scheduled LD cycle. For further explanation, see the text. (Modified from Czeisler et al., 1981.)

habitual sleep episode from midnight to 08:00, the temperature minimum will be located at approximately 06:00. Humans thus resemble other diurnal animals, which show phase advances from light late in the rest phase and delays from light early in the rest phase. The human pacemaker is relatively insensitive to light during the middle of the day, but small phase shifts can be induced by light exposure in the daytime. Apparently a true dead zone, where no phase shifts are produced, does not exist in humans, but the issue should be tested further.

Several other facts about human PRCs are important, including resolution, background lighting, and effects of light amplitude. Most light PRCs have been produced by the administration of light pulses lasting several hours. As a result, a high-resolution PRC allowing precise description of shifts at all circadian phases is not available. In addition, the light pulses were delivered against a light background as a convenience for the human subjects. So-called dark varied from a few lux to room light intensity of about 150 lux. Such a background light cycle contributes to the observed phase shifts and thereby complicates the interpretation of human PRCs.

The effects of light on phase have been investigated quite extensively, but effects on amplitude have received much less attention. Theoretically, the amplitude of circadian oscillation can be reduced or enhanced in response to light stimuli. Available data indicate that carefully timed administration of light in laboratory experiments can reduce the amplitude of the body temperature, melatonin, and alertness rhythms.

EFFECTS OF DURATION, INTENSITY, AND SPECTRAL COMPOSITION OF LIGHT.   Little is known about the effect of light duration on phase shifting. The most detailed data available come from intermittent light exposure studies using either LL or DD schedules. Surprisingly, the effectiveness of a light pulse is only marginally reduced when interrupted by dark episodes. Apparently the first minutes of a light pulse are most effective in causing phase shifts.

The relationship between light intensity and magnitude of phase shifts is logarithmic rather than linear (Figure 9.15). Dose response studies indicate that ordinary room light of 150 to 300 lux exerts approximately half the effect of a saturating bright light of 10,000-lux intensity in phase shifting. Dose response curves have been obtained for the effects of light on melatonin suppression and also for mental alertness. The remarkable sensitivity of the human circadian pacemaker to dim light is also illustrated by entrainment data. Dim light of 1.5 lux is barely sufficient for reading but can maintain entrainment in most individuals on a 24 h LD cycle.

Theories have come full circle. Once the human circadian pacemaker was considered insensitive to light, but now data show it responding to candlelight. However, entrainment in very dim light drifts in many cases without scheduled sleep and total darkness. Sensitivity to low-intensity light is important for the many humans in the present age who work indoors and have very little exposure to strong, natural daylight (see Chapter 10). The discipline of going to bed at a regular time each day and sleeping in darkness is undoubtedly very important for entrainment in many individuals.

The spectral composition of ordinary fluorescent light sources is very different from that of outdoor light. Although the disjunct waveform is perceived by the human visual system as white light, questions arise about the relationship of the spectral composition of light to the magnitude of phase shifts. At the moment, however, the issue is not completely resolved because recent research has implicated specialized retinal ganglion cell photoreceptors in circadian photoreception (see Chapter 6).

Spectral sensitivity measurements of the effect of light on melatonin suppression show that shorter wavelengths are most effective. Such a result is very different from visual system sensitivity. In addition, mutant mice lacking all rods and cones entrain to light–dark cycles; they also show both phase shifting and suppression of melatonin in response to light pulses. These findings imply that the circadian responses to light are at least partly mediated by a photoreceptor system distinct from the retinal rods and cones.

NONPHOTIC ENTRAINING AGENTS.   The preceding sections summarized the evidence that light exposures affect the human circadian pacemaker, but they did not directly ask whether the light–dark cycle is the sole or most important synchronizer for humans. Evidence for the preeminence of the light–dark cycle in entrainment of the human circadian pacemaker comes from the study of individuals who are blind.

A variety of causes and degrees of blindness exist, and a clear association is evident between the severity of blindness and sleep problems. Many blind individuals experience sleep problems, in spite of living in a highly social 24 h environment. Monitoring a reliable marker of the circadian system, such as 6-sulfatoxymelatonin, has demonstrated that most individuals who are totally blind free-run. If social cues or sleep–wake schedules imposed by work schedules or other social constraints were the most important synchronizers of the human circadian clock, blind individuals would not free-run while living in society.

A few blind individuals appear synchronized to the 24 h day in spite of the absence of light perception. In some totally blind individuals, ocular light exposure can still elicit melatonin suppression. The effect is probably mediated by a circadian photoreceptive system distinct from the visual system, as described earlier in this discussion and in Chapter 6. In these individuals, light may still be acting as an

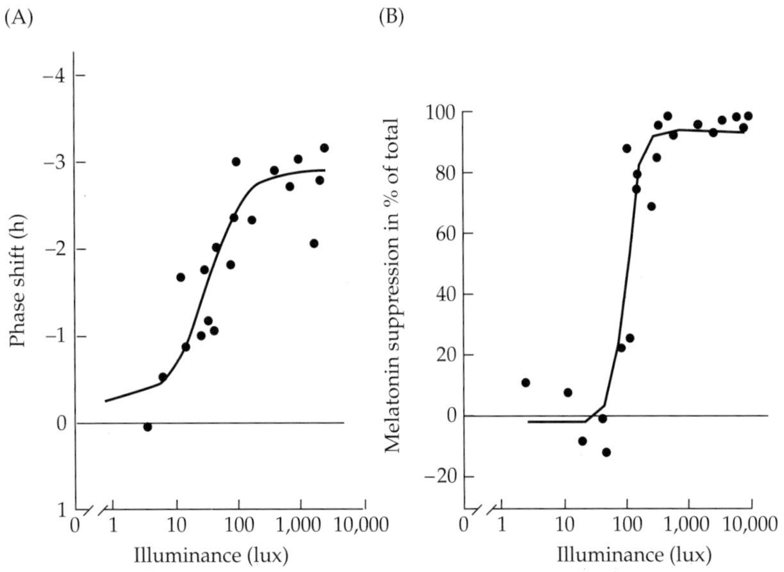

FIGURE 9.15 **Illuminance response curve of the human circadian pacemaker.** (A) The shift in the phase of the circadian rhythm in plasma melatonin was assessed on the day following exposure to a 6.5 h light stimulus. (B) The light treatment also resulted in acute suppression of plasma melatonin during the light exposure. Degree of suppression is expressed as a percentage of total. Individual subjects are represented by solid circles. (From Zeitzer et al., 2000.)

entraining agent indirectly through melatonin suppression. Exclusion of all photic stimuli is therefore critical before a conclusion of nonphotic synchronization in blind individuals can be accepted.

If any nonphotic entraining agents exist in humans, their nature remains unclear. In sighted subjects but not in blind individuals, naps taken during various times of day cause phase shifts opposite to light pulse effects at the same time of day, and these shifts are probably due to a change in light exposure associated with shutting of the eyes. The postural changes associated with the sleep–wake cycle are accompanied by major changes in sympathetic activation and could play a role. The effect of motor activity, particularly physical exercise, has been investigated by several groups. A few of the studies suggest very small phase shifting due to exercise.

One of the most extensively studied nonphotic synchronizers is periodic administration of the hormone melatonin. The SCN contains melatonin receptors, and animal studies have demonstrated that exogenous melatonin can phase-shift circadian rhythms in vivo and in vitro. Administration of melatonin to free-running blind individuals can synchronize the sleep–wake cycle and the endogenous melatonin rhythms in some but not all individuals. In sighted individuals, melatonin administration can facilitate reentrainment of endogenous circadian rhythms to an altered rest–activity and light–dark cycle. A PRC has been measured for melatonin ingestion. The shifts tend to be the inverse of those produced by light. Melatonin ingestion in the afternoon and early evening advances the circadian clock, and in the night and early morning delays it. Because of these phase-shifting effects, melatonin has been called an internal synchronizer.

The role of the endogenously generated rhythm of melatonin in the human circadian cycle is uncertain, but the phase-shifting effects of light and melatonin appear to reinforce each other. Melatonin is normally secreted overnight in the hours of darkness, and light suppresses melatonin secretion by an amount proportional to light intensity. Bright light directly advances the phase of the circadian clock in the early morning just after the temperature minimum via the light PRC; indirectly, light suppresses melatonin secretion and prevents the phase-delaying effect that melatonin would have exerted at this time via its own PRC.

Temporal Regulation of Several Physiological Functions Illustrates the Importance of Circadian Rhythms in Humans[21,30,58]

Humans are more aware of the sleep–wake rhythm than any other circadian rhythm[1,20,25,26,44]

GENERAL PATTERN. Humans are acutely aware of their sleep–wake rhythms. A typical person goes to sleep at 23:00 to 24:00 local time and rises at 07:00 to 08:00.

Marked variations in bedtimes and wake times exist both within and between age groups. For most individuals living in highly industrialized societies, factors like shift-work schedules, classes, and school schedules of children are major factors in the sleep–wake schedule. On weekends and holidays, sleep schedules may change, and some people may take a nap in the afternoon in addition to the major nocturnal sleep episode. Thus the social environment is a major determinant of sleep–wake schedules, even though it is not necessarily in synchrony with the external natural light–dark cycle. Some individuals experience considerable difficulty synchronizing their sleep–wake schedules to these social schedules.

The sleep–wake cycle can be studied and described by various techniques. Subjective estimates of sleep and wake times can be recorded in sleep diaries. Subjects can document the timing of naps, as well as estimates of sleep quality. All these facts provide useful information on the regulation of sleep. Recording by actimeters can be conducted unobtrusively for prolonged periods of time. The data indicate sleep duration and sleep efficiency. Actimeters are being used increasingly in field studies and laboratory studies of human sleep–wake regulation.

The standard for the assessment of sleep is polysomnographic recording of sleep. Data collected include electroencephalograms (EEGs) to record brain waves, electro-oculograms to record eye movements, and electromyograms to record muscle tone. On the basis of these variables, the records can be scored as wakefulness, non-rapid eye movement (non-REM) sleep, and rapid eye movement (REM) sleep. Non-REM sleep is usually further subdivided into stages 1 through 4 on the basis of the amplitude and frequency of the EEG waves, which can be assessed visually or quantified by spectral analysis. Stages 3 and 4 are collectively known as slow-wave sleep.

Such methods are used to quantify the time course of typical features of the sleep EEG, such as the low-frequency slow waves of slow-wave sleep (0.5–4 Hz), and faster phenomena, such as the 12 to 14 Hz sleep spindles. A plot of these sleep stages is called a hypnogram. A typical alternation of the sleep stages can be observed during normal nocturnal sleep (Figure 9.16). Non-REM and REM sleep alternate with an ultradian periodicity of approximately 70 to 90 minutes. The duration of REM episodes, as well as the density of rapid eye movements in REM sleep, increases in the course of nocturnal sleep. Slow-wave sleep is prominent in the first part of the sleep period and almost absent in the second half. Considerable stage 2 sleep is present in the second half of a sleep period. Sleep and the ultradian alternation between non-REM and REM sleep are accompanied by many changes in physiology, ranging from heart rate and heart rate variability to plasma renin concentrations.

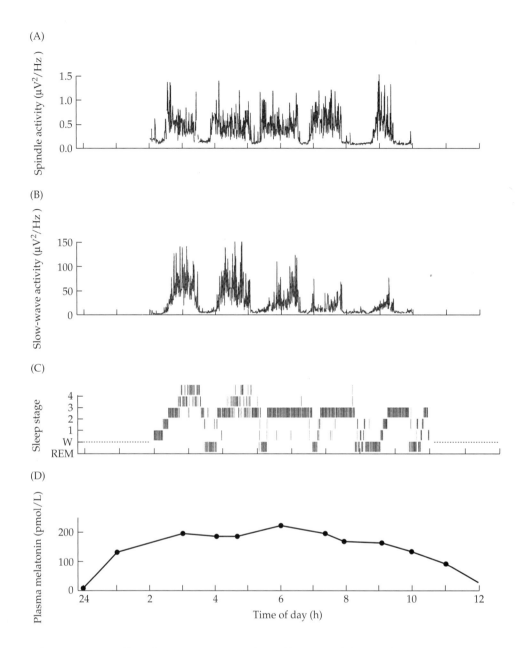

FIGURE 9.16 Baseline sleep. Time course of spindle activity (A); EEG slow-wave activity (B); sleep stage (C); and plasma melatonin concentrations (D) during a nocturnal sleep episode in a 22-year-old male. (Courtesy of D-J. Dijk, J. F. Duffy, and C. A. Czeisler.)

RHYTHMIC FACTORS AFFECTING THE SLEEP–WAKE CYCLE. An entrained young adult living in society goes to sleep about 6 h before the body temperature minimum, which occurs 1 to 2 h after the onset of melatonin secretion. During free-running conditions, in contrast, sleep is initiated close to the temperature minimum and terminated about 6 to 8 h into the rising limb of the temperature rhythm. The change in the internal phase relationship between sleep and core temperature provides some clues about the circadian regulation of sleep structure. The time course of slow-wave sleep remains unaltered, but the time course of REM sleep changes during free-running conditions. The normal increase in the duration of successive REM episodes over the course of a night's sleep is altered, and REM sleep shifts toward the beginning of sleep.

During prolonged temporal isolation, a change in the phase relationship between the sleep–wake and the body temperature cycles may precede spontaneous internal desynchronization. The core temperature rhythm continues with a period slightly greater than 24 h, while the sleep–wake cycle might be about 16 h or, more frequently, about 32 h (Figure 9.17). As Figure 9.17 indicates, internal synchrony prevailed during days 1 to 14, with activity and temperature periods both equal to 25.7 h. After the start of desynchronization on day 15, the activity cycle had a mean period of 33.4 h, and the temperature period was 25.1 h.

The simplest interpretation of these data is that the sleep–wake cycle and the core temperature rhythm are governed by separate oscillators. Normally the two are coupled in the presence of an external light–dark cycle but become uncoupled in the absence of an LD cycle. The period of the oscillator driving the temperature cycle is close to 24 h and stable; the period of the sleep–wake oscillator is labile and can deviate considerably from the circadian range. Not all chronobiologists agree. Some researchers have argued that the apparent spontaneous desynchronization is an artifact of how subjects designate sleep episodes: as either major sleep episodes or naps. However, spontaneous internal desynchronization between the sleep–wake and temperature cycles has been reported in studies in which all major and minor sleep episodes were recorded and reported. Others have pointed out that spontaneous desynchronization is not observed in nonhuman animals, but few data are available to support that contention. Sleep is rarely recorded in animals, and the recording of multiple variables is even rarer.

Spontaneous internal desynchronization may be an artifact in humans of the profound influence of conscious thought during wakefulness about when to go to sleep. Evidently cortical functions change during spontaneous internal desynchronization. When individuals are asked to estimate the passage of hourly time intervals, a strong correlation between duration of the hourly estimates and duration of the wake episode is seen. Possibly a long estimate of hourly time goes along with a long time to initiate sleep. However, time estimation of minute intervals during spontaneous internal desynchronization continues to covary with the core temperature rhythm. The neurobiological basis of these intriguing phenomena remains unknown.

Spontaneous internal desynchronization means that sleep episodes are initiated at many phases of the core body temperature cycle. The distribution of phases at which sleep is initiated, however, is not uniform. Very few sleep episodes are initiated during the interval of about 4 to 6 h before the temperature minimum. This phase has been called, therefore, the wake maintenance zone. In contrast, most sleep episodes are initiated approximately 3 to 5 h after the temperature minimum.

Under entrained conditions these two phases of the core temperature rhythm occur close to habitual bedtime and wake time, respectively. The duration of a sleep episode during desynchronization generally varies from approximately 4 h to about 16 h. The longest sleep episodes are initiated in the hours before the temperature minimum; the shortest episodes, after minimum. Wake durations must also vary greatly, but they do not appear to have a major influence on subsequent sleep duration. Instead, the phase of the circadian pacemaker, as indexed by the core temperature rhythm, appears to be the major determinant of sleep duration.

Several features of sleep structure are also affected by the circadian phase at which sleep occurs. Slow-wave sleep and REM sleep are affected very differently. Slow-wave sleep declines during sleep episodes initiated at all circadian phases. In contrast, the time course of REM sleep is modulated by the circadian phase at which sleep occurs. The peak of the circadian rhythm of REM propensity is located at, or shortly after, the minimum of the core temperature rhythm (Figure 9.18). These data provide compelling evidence that the circadian clock

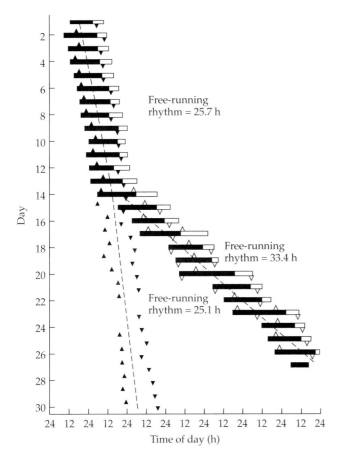

FIGURE 9.17 Spontaneous internal desynchronization during a free run in isolation, plotted as an actogram. Black bars indicate wake times; open bars, sleep times; triangles, temperature maxima and minima pointing up and down, respectively. Temperature maxima and minima are plotted twice to show the circadian rhythm as solid triangles and their relation to the sleep–wake cycle as open triangles. For further explanation, see the text. (From Wever, 1975.)

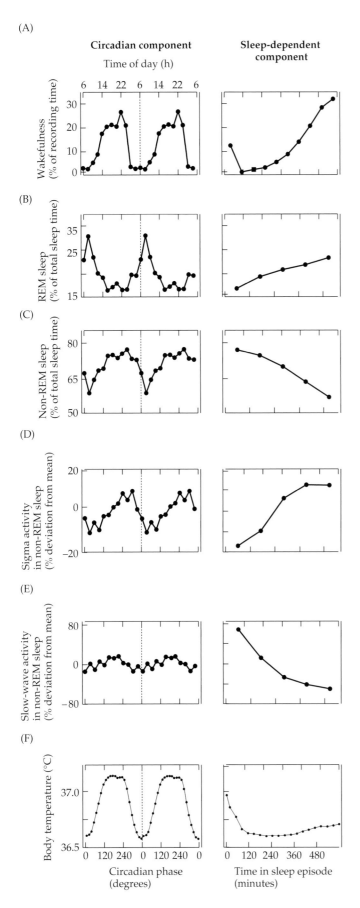

driving the temperature rhythm exerts a major influence on sleep duration and REM sleep.

Sleep length depends also on the ease of initiating sleep and maintaining sleep. Experiments have been performed to investigate how easy it is to fall asleep at different times of the day. One protocol consisted of dividing the 24 h of the day into 72 segments of 20 minutes each. In each of these segments, the volunteer was allowed 7 minutes to fall asleep in the comfort of a sleep laboratory, and a record of electrical activity in the brain indicated the amount of sleep obtained. The subject was awakened if asleep at the end of the 7 minutes and was required to remain awake for the next 13 minutes. The next 20-minute segment was treated in the same way, and the protocol continued until the end of the last segment. The results indicate that falling asleep was easier at night and more difficult during the day. The ability to fall asleep mirrored the circadian rhythm of body temperature. Low body temperature during the latter part of the night was associated with greater ease of getting to sleep. The higher the temperature, the harder it was to fall asleep and the less sleep was obtained in a 7-minute segment.

NONRHYTHMIC FACTORS INFLUENCING THE SLEEP–WAKE CYCLE. The sleep–wake cycle and the composition of sleep are not determined solely by rhythmic factors. Sleep timing depends to some extent also on the degree of stimulation. Sleep will be started earlier if the subject is bored, or later if exciting or interesting circumstances prevail.

Sleep physiologists have known for many decades that sleep structure and sleepiness itself depend in part on the duration of wakefulness. The results have provided evidence that homeostatic processes contribute to sleep regulation and operate with relatively slow time constants. A subject can be deprived of slow-wave sleep by acoustic stimulation that does not induce wakefulness. After deprivation, slow-wave sleep rebounds in the undisturbed part of the night. If the selective slow-wave sleep deprivation lasts the entire sleep episode, then rebound occurs during the subsequent sleep episode. Ap-

FIGURE 9.18 Influence of circadian phase (left) and time elapsed since start of sleep episode (right) on various parameters describing sleep. Wakefulness during sleep episode (A), REM sleep (B), non-REM sleep (C), sigma (sleep spindle) activity (D), slow-wave activity (E), and body temperature (F). Note that the circadian data are double-plotted. Zero degrees in (F) represents the timing of the fitted nadir of the circadian rhythm of core body temperature. Under entrained conditions, the nadir occurs at approximately 06:00; time of day is shown in (A). (After Dijk and Czeisler, 1995.)

parently slow-wave sleep is regulated very accurately. Quantitative analysis of the sleep EEG shows that slow-wave sleep provides a quantitative measure of sleep homeostasis.

Slow-wave sleep seems to be very intense sleep. It is normally concentrated in the first part of the sleep period. As sleep drive decreases, the amount of slow-wave sleep also declines. As a result, extensions of the amount of time awake should lead to more intense sleep, which in turn should obliterate the need for a major increase in sleep duration.

Support for the notion of an intensity component to sleep comes from dose response studies of the effect of acute curtailment of sleep duration on subsequent alertness and performance. Results indicate that the first 2 to 4 h of sleep contribute the most to the recovery of alertness and efficiency. Conversely, a nap in the late afternoon will temporarily improve performance, as well as reduce slow-wave sleep in the subsequent nocturnal sleep episode. The homeostatic facet of REM sleep regulation is not as easily quantified, but the presence of such a mechanism is undisputed.

All these results imply that the sleep–wake oscillator tracks how long a person has been awake or asleep. This characteristic makes the oscillator suitable for fulfilling the function of sleep homeostasis—that is, controlling the average level of sleep debt. Sleep–wake behavior itself is a major determinant of sleep propensity and sleep structure. It is part of the causal loop that generates the sleep–wake oscillation, and a change of behavioral state will immediately reset the oscillation. The sleep homeostasis function, of course, is very different from the role of sleep–wake behavior in contributing to the oscillations of body temperature or melatonin, for these rhythms will persist in the absence of sleep.

THE INTERACTION BETWEEN RHYTHMIC AND HOMEOSTATIC FACTORS. The sleep–wake cycle is regulated by the interaction between oscillatory and homeostatic processes. This concept has been very successfully incorporated into mathematical models of the sleep–wake cycle. These models have led to many predictions related to sleep duration and sleep structure in laboratory experiments and field studies in which the sleep–wake cycle was desynchronized from endogenous circadian rhythmicity. They have also provided insights into the regulation of waking performance. Finally, integration of the concept of sleep homeostasis and circadian rhythmicity offers a functional explanation for the interactions between the two processes and their phase relationship during entrainment.

Under normal conditions, circadian time and sleep–wake histories change in synchrony. Humans wake up in the morning a few hours after the core temperature minimum and stay awake until approximate-ly 6 h before the minimum. During the waking day, time awake increases progressively and, during sleep, homeostatic sleep pressure dissipates in a monotonic manner.

Sleep propensity is consistently highest at the circadian phase that would normally occur in the late evening hours and lowest at the circadian phase normally occurring in the morning hours. For example, sleep latency generally becomes longer in the course of the daytime and then shortens after the onset of nocturnal melatonin secretion, to reach a minimum near the time of the minimum of the core temperature rhythm. Thus humans normally go to sleep shortly after the peak in the circadian drive for wakefulness and wake up shortly after the peak in the circadian drive for sleep (see Figure 9.18).

The factors influencing the sleep–wake cycle have been considered in some detail, not only because they are well documented but also because they exert such a profound influence on behavior and the body as a whole. The sleep–wake cycle produces important changes in posture, activity, light exposure, and food intake, all of which are important from a chronobiological point of view. For many variables, the effects of the sleep–wake cycle contribute to the exogenous masking effects of the circadian rhythm. The variables that have a strong exogenous component appear to adjust in phase after the sleep–wake cycle has been phase-shifted. Blood pressure, heart rate, the secretion of insulin and growth hormone, and many components of urinary excretion are examples.

Indeed, the phases of these rhythms are often expressed in terms of their relationships to midsleep. Separating the endogenous component that adjusts in phase with the clock markers, such as plasma melatonin and cortisol, from core temperature requires a constant routine, forced desynchronization, or demasking of the raw data.

Core body temperature has been studied extensively in humans[42,59,61]

Core body temperature reflects the balance between heat loss and heat production mechanisms. The most reliable measurement comes from a rectal thermistor. Core temperature has been the subject of many circadian studies, primarily because it is relatively easy to measure and has a clear endogenous component. Temperature rises from early daylight and then starts to decline a few hours before habitual bedtime (see Figure 9.18). Typically it falls rapidly when a person goes to bed and reaches a minimum in the latter half of sleep. Brain temperature shows a similar time course.

The range of body temperature is typically 36 to 37.5°C, though deviations outside this range can be observed in response to strenuous activity, hot or cold showers, and changes in posture. Core temperature is protected by a variety of thermoregulatory reflexes, including the tone of cutaneous blood vessels, sweating, and shivering. The set point of thermoregulation is higher during the day and lower at night, and thermoregulatory reflexes appear to be more effective by day than at night. Such an arrangement might be advantageous in preventing core temperature from rising too high in the daytime, when activity is highest.

From a chronobiological viewpoint, core temperature is valuable because it shows an endogenous component that is comparable in size to changes produced by other factors. Sleep lowers body temperature; activity and food intake raise it. A brightly lit and dynamic environment also raises body temperature. Under normal circumstances, these factors combine with the circadian clock to raise body temperature in the daytime. During the evening, the combined effects of clock, decreasing activity and food intake, and a quieter environment all aid in promoting sleep. The change in posture with sleep and the evening onset of melatonin secretion also lower core temperature.

The mechanism by which the circadian rhythm of core temperature is produced has been known for some time. The fact that the rhythm persists during constant routines shows that it originates from heat loss rather than heat gain mechanisms. Furthermore, because the rhythm is almost identical in temperate and tropical regions of the world, it must reflect a changing set point rather than passive heat loss from the body.

Alertness and mental performance show circadian variation[31,55]

The rhythms of alertness and its opposing sleepiness vary during wakefulness. Alertness is normally lowest immediately after waking because of sleep inertia, but it rises rapidly in the next few minutes as an individual wakes up. Alertness then rises slowly throughout much of the day (see Plate 4, Photo G). A transitory decrease known as the postlunch dip may occur during the early and midafternoon, but not until the evening does alertness begin to fall quite sharply. The drop signals the time to go to sleep.

Assessments of many aspects of mental performance have revealed some variation across the waking day (Figure 9.19). The evaluation can be made with tasks involving different amounts of vigilance, sensory input, central processing, or motor output. For example, complex tasks such as logical reasoning yield the best scores in the late morning, and tests involving simple reaction time or visual search show optimal performance in the late afternoon and early evening. Short-term memory,

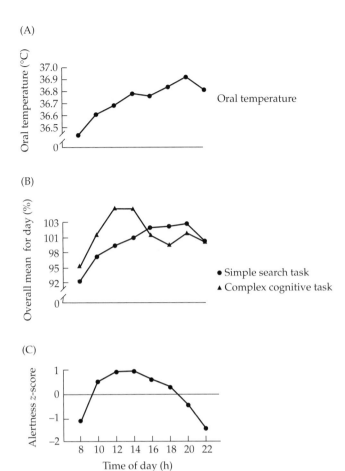

FIGURE 9.19 Daytime temperature, mental performance, and mood rhythms. (A) Daytime oral temperature. (B) Two mental performance tasks, measured as a percentage of the overall mean. (C) Mood (alertness), measured as a z-transformed score. z-scores indicate how different from normal the associated recorded values are. The mean score for the set is 0, and the variation is expressed as multiples of the SD. Thus, a score of −1 is 1 SD below the overall mean. (From Folkard, 1990.)

such as remembering a hand of cards or a telephone number long enough to dial, shows a different diurnal profile. After a high at wake-up time, it apparently deteriorates during the day.

Tests of mental performance are particularly susceptible to the conditions under which they are given. Environmental noise and lighting may induce artifact, and the results may even be influenced by a person's attitude at the time of a test. Because of the effects of fatigue on performance, evaluations are also rarely performed at night.

The concordance of rhythms of alertness and mental performance with core temperature have given rise to a hypothesis that the daytime rise in body temperature promotes neuronal activity and central nervous system function. Fatigue also plays a role. More complex tasks peak earlier in the day than simple ones. Both the sleep–wake cycle and mental performance also clearly have rhythmic functions in phase with core temperature, and both have a buildup of a particular factor, such as sleep drive or fatigue, during awake periods. For all of these reasons, reliance on mental performance tests in a quantitative sense is risky. At best, the tests should be repeated several times, and a concerted effort should be made to standardize conditions under which tests are run.

Melatonin concentration is low in the daytime and high at night[4,5,48,64]

In the presence of a light–dark cycle, melatonin concentrations in plasma are low during the day and high at night (Figure 9.20). These rhythmic variations in melatonin are also observed in blind individuals and persist in sighted individuals kept in very dim light conditions for up to 3 weeks. The rhythm of melatonin is driven by the circadian pacemaker and is not dependent on the external light–dark cycle. Apart from the effect of light, the rhythm is relatively immune from masking effects. Because of its stability, the melatonin rhythm is a favorite marker of the circadian clock in human laboratory studies.

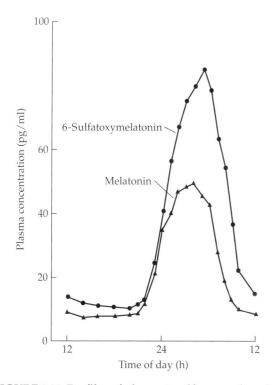

FIGURE 9.20 Profiles of plasma 6-sulfatoxymelatonin and melatonin from 22 normal individuals. Values are plotted as mean. (From Bojkowski et al., 1987B.)

Approximately 20 years ago the abrupt, dose-dependent drop in plasma melatonin concentrations was noted when volunteers were exposed to bright light during the biological night. Office light intensity of 100 to 300 lux can elicit about half as much suppression of plasma melatonin as outdoor dawn or dusk light. The maximum effect occurs at the two twilight times, when light intensity is about 10,000 lux. The drop in plasma concentrations reflects suppression of melatonin synthesis in the pineal. Light intensity is measured by retinal photoreceptors that activate GABA-ergic neurons in the SCN via the retinohypothalamic tract. The neurons modulate the autonomic nervous system outflow through the superior cervical ganglion and thereby regulate the pineal gland (see also Chapters 4, 5, and 6).

Plasma cortisol peaks at wake time and declines to a minimum at sleep onset[41,68]

The hormone cortisol is secreted by the adrenal cortex in response to adrenocorticotropic hormone (ACTH) secreted by the pituitary, which in turn is under control of corticotropic releasing hormone (CRH) released by the hypothalamus. Plasma cortisol concentrations reach a peak at about habitual wake-up time and decline in the course of the waking day to reach a minimum shortly after sleep onset (see Figures 9.3D and 10.2A). Superimposed on this 24 h rhythm are pulsatile variations, originating mainly from the hypothalamic nuclei that secrete CRH. The internal clock imposes a circadian modulation on the frequency and size of the secretory pulses.

Because of the combination of pulsatile release and circadian modulation, the profile of cortisol concentration in the blood is complex, resulting in difficulties in describing cortisol rhythms quantitatively. One solution has been to describe the rhythm as a mixture of cosine curves. The fundamental period is 24 h, with harmonics such as 12 h, 8 h, 6 h, and 4 h superimposed. Although theoretically practical from a mathematical point of view, the solution is not biologically very meaningful. Measurement of cortisol concentrations has other problems. The concentration of cortisol in the plasma is determined by the rate of its secretion, and in addition to the clock, both light and sleep themselves influence cortisol secretion. Elaborate mathematical conversion is required to relate secretion rate to plasma level.

Plasma cortisol has a strong endogenous component. The weak exogenous component arises from food intake and the sleep–wake cycle, possibly through changes in posture. Cortisol is occasionally used as a marker of the circadian clock in research studies. However, the greater interest in cortisol concentrations lies in its role as an immunosuppressive hormone. It has potential for clinical investigations of asthma and rheumatoid arthritis.

Rhythm parameters vary considerably among individuals[9,29,38]

Plasma cortisol, melatonin, and core body temperature are three commonly used markers of the circadian clock of humans. When care is taken to control any masking factors, these markers appear to give similar results for phase changes. Day-to-day variance in phase is slightly greater for cortisol and core temperature than for melatonin. On the surface, melatonin is apparently the best marker of the three. The issue will remain unresolved, however, until the endogenous clock can be assessed directly rather than through inference of its activities by a marker.

Individuals show differences. Even subjects who are day-active and sleep at night exhibit great variation in choice of sleep times. Similarly, the distribution of wake-time activities varies. At one extreme are people who prefer to get up early, to do the most mentally and physically demanding activities early in the day, and then to spend a quiet and relaxing time in the evening before an early bedtime. At the other extreme are individuals who prefer to rise late in the day, to delay the most demanding activities to later in the day, and to retire late at night.

Such differences in habits become the basis of questionnaires that quantify activities as a score for chronotype or degree of "morningness" of the individual. When such questionnaires are applied to a population, most people are positioned near the mean score, and these individuals are referred to as intermediate types. In contrast, the two tails of the normal distribution are described as morning types or "larks," and evening types or "owls" (see the Study Questions and Exercises section of this chapter for an example of such a questionnaire).

Investigations of the rhythm phases for core temperature, sleep propensity, and other variables show differences between morning and evening types. For example, the peak of the melatonin rhythm and the minimum of core body temperature occur at an earlier clock time in morning types than in evening types. However, different variables are changed by different amounts, reflecting differences in entrainment, as well as in the size of masking effects. The constant-routine protocol shows a significant difference between morning and evening types in the profiles of circadian core temperature (Figure 9.21).

Being a morning type or evening type does not normally cause any difficulties for the individuals concerned. They merely differ slightly from each other and from intermediate types in their chosen sleep times and in the way they prefer to organize their waking days. However, problems can arise under certain circumstances. Evening types may have difficulties with early shifts, and intermediate types with late shifts, if the amount of phase advance or delay becomes excessive.

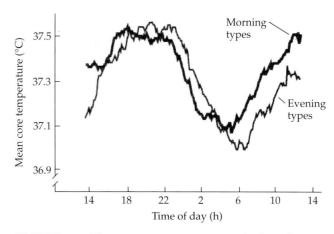

FIGURE 9.21 The mean core temperature rhythm of seven morning types and seven evening types during a constant routine. (From Kerkhof and Van Dongen, 1996.)

Circadian Rhythmicity Shows Marked but Typical Changes from Birth to Old Age[18,74]

Circadian rhythms develop gradually in the first year of life[32,37,67,74]

In the unborn fetus, circadian as well as ultradian rhythms are present because the mother provides a rhythmic environment for her fetus. Blood flow to the fetus is affected by the mother's posture. The maternal hormonal profile also varies. These rhythms undoubtedly affect the fetus.

Short-term ultradian rhythms at birth transform in the first year into circadian patterns (Figure 9.22). Immediately after birth, full-term babies are ultradian in their sleep–wake rhythms, and they show periodicities of 4 h or less. As the neonate develops, the amplitude of its circadian rhythms increases and ultradian rhythms decrease in importance. Ultradian rhythms are more marked in premature neonates than in full-term babies. These changes take place at the same time as the infant is developing neurologically and is beginning to become more responsive to its environment.

Two issues immediately come to mind. The relative importance of exogenous and endogenous components in babies during their early development is important. A second question concerns the relationship between the endogenous components of the ultradian and circadian rhythms during infancy. The ideal method would be to study the baby on a constant-routine protocol. Such protocols are usually impossible with babies, however, except for infants studied in intensive care incubators.

The available data suggest that the circadian rhythm of core temperature in a newborn infant is independent of environmental rhythms and of the sleep–wake cycle. Apparently the circadian rhythm of core temperature arises before the sleep–wake cycle has been established.

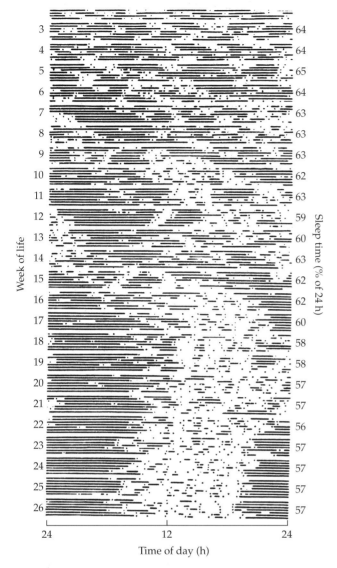

Week of life (left axis) · Sleep time (% of 24 h) (right axis)

Right-axis values (top to bottom): 64, 64, 65, 64, 63, 63, 63, 62, 63, 59, 60, 63, 62, 62, 60, 58, 58, 57, 57, 56, 57, 57, 57, 57

Left labels (weeks): 3, 4, 5, 6, 7, 8, 9, 10, 11, 12, 13, 14, 15, 16, 17, 18, 19, 20, 21, 22, 23, 24, 25, 26

X-axis: 24   12   24   Time of day (h)

FIGURE 9.22 Daily incidence of sleep, wakefulness, and feeding in an infant observed from days 11 to 182 of life. Black bars indicate sleep; white bars, wakefulness. (From Helbrügge, 1960.)

The phases of the temperature rhythm differ widely among babies, indicating that they cannot be produced only by a direct effect of the light–dark cycle.

Evidently the sleep–wake cycle of some babies even in a natural light–dark cycle seems to free-run, thus pointing toward an internal cause. The 3 to 4 h ultradian rhythms are present soon after birth, with a timing that matches the rhythm of feeding and caregiving. These rhythms appear to be mainly exogenous.

Many of the preceding observations were recorded in two recent studies of ultradian and circadian rhythms in a group of full-term, healthy neonates, assessed on two occasions: 2 days and 4 weeks after birth. The sleep–wake cycles and rhythms of heart rate, systolic blood pressure, and core temperature were measured for 24 h on both occasions. During this time the babies were fed on demand about every 4 h and lived alone in a room exposed to natural daylight. In these studies the records were demasked for the direct effects of the sleep–wake and light–dark cycles. The sleep–wake cycle was split into five behavioral categories based on observation: (1) deeply asleep, (2) lightly asleep, (3) awake and inactive, (4) awake and moving, (5) awake and crying. The demasking process showed that core temperature was directly raised when the baby was awake, particularly if it was moving or crying, and directly lowered if it was asleep, particularly deeply asleep.

Two days after birth, a low-amplitude circadian rhythm in core temperature was present, but the peaks of these circadian rhythms showed considerable interindividual variation. In contrast, heart rate, systolic blood pressure, and activity did not show significant circadian rhythms. Instead they exhibited ultradian rhythms with a period of about 4 h (Figure 9.23). By the fourth week, the amplitudes of the circadian rhythms of core temperature had risen, and the peaks indicated less interindividual variation. Most fell in the first half of the light period. Circadian rhythms in activity, heart rate, and systolic blood pressure remained nonsignificant. Ultradian rhythms had increased in amplitude in all variables by 4 weeks of age. Activity, heart rate, and systolic blood pressure rhythms were synchronous with the pattern of feeding and caregiving, and the core temperature rhythm also appeared related to exogenous caregiving influences.

In these two studies a weak circadian rhythm in core temperature was present immediately after birth. Its development was endogenous and not a consequence of the development of circadian rhythms in the sleep–wake cycle or the cardiovascular system. However, when rhythms are measured some days after birth, the babies have been exposed to the alternation of light and dark and other external circadian rhythmicity. Ideal subjects for such studies are babies that have been kept for the entire period of the study on a constant routine with minimal exposure to either daily or ultradian influences. The closest routine is approximated in intensive care nurseries for very premature neonates. For many of these infants, respiratory distress and jaundice are common, requiring maintenance in an incubator for extended periods of time.

One such study dealt with neonatal babies that had been born just 24 to 30 weeks after conception. Both circadian and ultradian influences from the environment were minimal because the ward lights were left on continuously. Feeding was intravenous and continuous, and medical and hygiene care were given when needed, with no discernible rhythmicity. Circadian and ultradian rhythms with periods ranging from 3 to 20 h

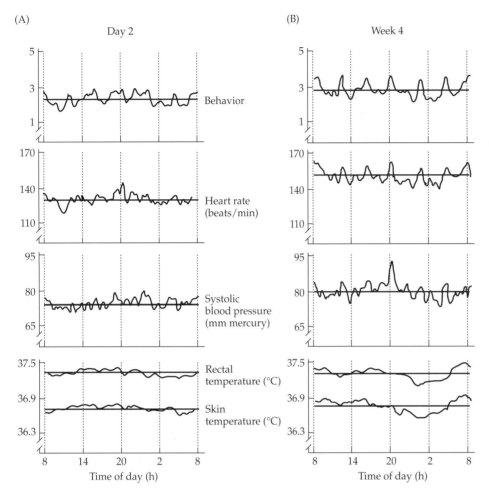

(A) Day 2

(B) Week 4

Time of day (h)

FIGURE 9.23 Mean daily rhythms in a group of full-term healthy babies studied separately in a normal lighting environment on day 2 (A) and in the fourth week (B) after birth. Horizontal lines indicate daily means. "Behavior" measured wakefulness in relative units on a scale of 1 to 5: 1, deeply asleep; 2, lightly asleep; 3, awake but inactive; 4, awake and active; 5, awake and crying. (From Weinert et al., 1994.)

were simultaneously present. The circadian rhythms tended to increase in strength. However, circadian rhythms were more reliable than ultradian rhythms with regard to their presence and their phasing. Circadian rhythms, with a period just over 24 h, were almost always present, but ultradian rhythms, with a wide range of periods, were present only transiently.

For one premature male born 26 weeks after conception, data were collected each hour for 14 weeks. For the first 13 weeks the baby was in constant conditions of light and feeding. Although he suffered from neonatal jaundice and required occasional oxygen, he was otherwise healthy and developed normally. During week 14 he was placed in the ward with normal LD lighting and fed every 3 h (Figure 9.24). In most of the first 13 weeks, a circadian period in the range of 23 to 26 h was present, implying endogenous origin. In week 14, the rhythm was exactly 24 h, possibly due to masking or entrainment. The results confirm that a weak circadian

output is present even in very young babies. The fact that ultradian rhythms were erratic in their presence and period suggests that, in healthy babies, these rhythms are normally produced by rhythmic inputs from caregiving and feeding. These findings also make it unnecessary to postulate the presence of ultradian oscillators.

The obvious candidate for the origin of circadian rhythmicity is the SCN, even though it is believed not to be fully developed at birth. Recent studies of mammalian SCN cells cultured in vitro and investigated in vivo indicate that circadian rather than ultradian rhythms are the main ones generated. Electrical coupling between these elements may be responsible for generating a robust circadian output. Supposedly the coupling between these cells is initially weak. Coupling develops in the weeks after birth, when interconnections between neurons increase in the brain as a whole. Until such couplings have been established, the circadian output from the SCN is weaker and more labile than in the adult.

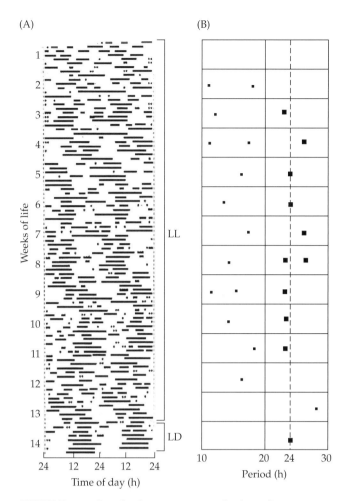

FIGURE 9.24 Core body temperature rhythm of a premature baby in constant conditions of an intensive care incubator for 13 weeks in LL, and in LD in week 14. (A) Double-plotted actogram showing an hourly record of core temperature, with lines indicating temperature for that hour above the daily mean. (B) Main frequencies in the temperature record calculated by maximal entropy spectral analysis. (From Tenreiro et al., 1991.)

With neurological development of the neonate, such a circadian output becomes more robust. In addition, as the baby becomes more responsive to the daily environment, its behavior begins to change. It is more active and takes in more food in the daytime, and it is more likely to sleep at night. All these factors promote entrainment of the internal clock and superimpose upon it an exogenous component. The result of these changes is the development of circadian rhythmicity, with endogenous and exogenous components appropriately phased for a diurnal individual.

Changes in timing and amplitude of circadian rhythms are seen in the elderly[26]

The timing of sleep, its duration, and its consolidation all change significantly through the human life span. Teen-agers and young adults often sleep until late morning or early afternoon; many older people wake up in the early morning hours, and are unable to go back to sleep. Such age-related changes in sleep timing could reflect a chronobiological change, or they might be merely a consequence of age-related changes in scheduling. Retired older people, for example, have more opportunity to take a nap in the afternoon. Constant-routine studies have demonstrated an earlier timing of the endogenous component of circadian rhythms in the elderly. The circadian clock appears to be set to an earlier solar time.

The response of the circadian pacemaker to light does not appear to differ between elderly and young subjects in a way that could account for the observed change in phase of the rhythms. Assessment of the intrinsic period of the internal clock has not revealed a significant age-related reduction in the intrinsic period. However, a difference in intrinsic period of as little as 6 minutes could lead to a difference in the entrained phase of as much as 1 h. Too few data are available to detect such a small difference; more data are needed. The demonstration that the period of the free-running rhythm of melatonin in a group of blind men did not differ significantly between two assessments made about 10 years apart also shows that the intrinsic period does not change greatly. In this case also, more subjects need to be studied.

Another age-dependent circadian parameter is amplitude. Many endocrine rhythms fall in amplitude with age. Declines could result from endogenous or exogenous factors. An important finding is that the SCN becomes smaller in the eighth decade of life. In terms of specific functions, agreement has not been reached on the core temperature rhythm. Data from a constant routine has indicated a fall in the amplitude of the temperature rhythm in some studies but not in others. The difference might arise from the selection of subjects. Volunteers are naturally active and healthy, and these individuals might be unrepresentative of elderly people in general. Possibly the healthy volunteers are less likely to show a reduced amplitude of their rhythms than the older population in general.

The decrease in amplitude of the circadian rhythm of core temperature will be due partly to exogenous factors. Decreased physical activity, taking naps in the daytime, and the decreased ability to have uninterrupted sleep at night may play a role. Declining vision may reduce an individual to LD entraining influences. Decreased social and physical activities in the daytime might also lead to a weakening of entrainment. A decline in amplitude of the temperature and melatonin rhythms would reduce the likelihood of consolidated sleep during the night. However, the earlier timing of waking relative to endogenous circadian rhythms persists when the sleep of older people is studied during the forced-desynchronization protocol. In addition, older people sleep less than young people at all circadian phases, and they cannot maintain high levels of sleep consolidation after the temperature minimum. These results lead to the conclusion that an age-related change exists also in the interaction of circadian rhythmicity and the sleep homeostat. Taken together, these factors explain why insomnia is a common complaint in the aged.

The Circadian System Plays a Critical Role in Human Daily Performance[6,70]

The circadian clock has two main functions in humans. First, it promotes daytime responses such as physical and mental activity and their associated biochemical and cardiovascular changes (see Plate 4, Photo G). Clock regulation separates these responses from nighttime responses involving recovery and restitution during a period of inactivity. The second role is to enable preparations to be made for the dramatic switch between the active and resting phases. Such profound neurophysiological changes cannot be brought about rapidly but require ordered changes in the rates of many biochemical and physiological functions to prepare in advance of actual need.

The responsiveness of living organisms to their environment is in part the result of reflex mechanisms activated purely exogenously. For example, baroreceptor, chemoreceptor, and osmoreceptor reflexes are called forth to combat changes produced by the environment. Reflex reactions are direct responses through the nervous system to environmental stimuli acting as immediate switches. In contrast, circadian timing allows preparation in advance based on a daily temporal framework and program. Anticipation is a vital property and benefit of endogenous biological timing on a revolving planet that has very regular, predictable day–night changes. A circadian pacemaker system enables organisms, including humans, to anticipate environmental changes rather than follow them. The difference has been described as predictive homeostasis to distinguish it from the more conventional reactive homeostasis.

SUMMARY

Since earliest recorded history, humans have shown interest in their own daily rhythms. However, an appreciation of the endogenous circadian oscillator governing a great array of rhythmic functions in humans was not widely accepted in until the mid-twentieth century. Gradually the principles of circadian timing established from research in nonhuman organisms were accepted for human physiology. Overt rhythmicity was recognized as the interaction of a free-running, internal self-sustained pacemaker with synchronizers of the external world. The chief of these synchronizers is the light–dark cycle of the solar day.

Early research used sites such as caves, underground rooms, or arctic continuous daylight in summer to provide constant conditions for studying the endogenous element of circadian rhythms. Later, modern hi-tech human facilities provided convenience, comfort for the research volunteers, avoidance of masking in data recording, and the skilled staff needed to carry out human studies successfully. Humans have unique advantages and disadvantages for circadian research.

Among the innovative protocols developed for the new facilities were the constant routine and forced desynchronization for laboratory studies, as well as various methods for subjects in free-living field studies. These methods confirmed the fundamental properties of circadian rhythmicity in humans that had been almost universally documented for nonhuman organisms.

Humans free-run in constant conditions with a period of approximately 24.3 h. Many endogenous and exogenous factors play a role in the overt rhythmic outputs of humans. The environment acts not only as a masking influence but also as an entraining agent. Entrainment is brought about almost exclusively by the light–dark cycle. Humans are very sensitive to phase shifting even by very dim light. The photic phase response curve documented for humans resembles that of other diurnal animals but tends to show slight responsiveness in the subjective daytime. Very few other entraining agents are known for humans; one notable exception is exogenously administered melatonin.

The primary pacemaker is the suprachiasmatic nucleus of the ventral hypothalamus. The effects of the SCN permeate the body because they act on core temperature, plasma hormone concentrations, outflow of the sympathetic nervous system to the pineal gland, and the sleep–wake cycle.

The most widely studied human rhythms are core body temperature rhythm and the sleep–wake cycle. The temperature cycle has a strong endogenous component but is masked by activity. The sleep–wake cycle is modified by many exogenous factors, and it appears to act in concert with a mechanism that maintains sleep homeostasis. Other important circadian functions in humans include the alertness and mental performance cycle, the physical activity cycle, the plasma cortisol cycle, and the endogenous melatonin cycle. The expression of each of these rhythms shows considerable interindividual variation.

Developmental progression in the expression of rhythmicity is seen from birth to old age. Rhythms in newborn humans are largely ultradian, with periods of wakefulness every 3 to 4 h, but this rhythm consolidates in the first year into a circadian pattern. In the elderly, the amplitude of circadian rhythms is commonly reduced.

CONTRIBUTORS

James M. Waterhouse and Patricia J. DeCoursey wish to thank Torbjorn Åkerstedt, Josephine Arendt, Charles A. Czeisler, Derk-Jan Dijk, David F. Dinges, and Kenneth P. Wright, Jr. for their material contributions to Chapter 9.

STUDY QUESTIONS AND EXERCISES

1. The following test (adapted from Horne and Ostberg, 1976) assesses if an individual is a morning type or an evening type. For each of question, choose the most appropriate answer. Then check the answer key below.

   1) What time would you choose to get up if you were free to plan your day?
   a. 05:00–06:00   b. 06:00–07:30   c. 07:30–10:00   d. 10:00–11:00

   2) For important business meetings, scheduling at the peak of mental powers is important. When would you prefer to schedule this meeting?
   a. 08:00–10:00   b. 11:00–13:00   c. 15:00–17:00   d. 19:00–21:00

   3) What time would you choose to go to bed if you were entirely free to plan?
   a. 20:00–21:00   b. 21:00–22:15   c. 22:15–00:30   d. 00:30–01:45

   4) A friend wishes to go jogging and suggests starting at 07:00–08:00. How would you describe this time for you?
   a. Good   b. Reasonable   c. Difficult   d. Very difficult

   5) Every individual performs physical work best at certain times of day. What time would be best for you?
   a. 08:00–10:00   b. 11:00–13:00   c. 15:00–17:00   d. 19:00–21:00

   6) If you have to go to bed at 23:00, how would you feel?
   a. Not at all tired, and unable to get to sleep quickly
   b. A little tired, but unlikely to get to sleep quickly
   c. Fairly tired, and likely to get to sleep quickly
   d. Very tired, and very likely to get to sleep quickly

   7) When you have been up for half an hour on a normal working day, how do you feel?
   a. Very tired   b. Fairly tired   c. Fairly refreshed   d. Very refreshed

   8) At what time of the day do you feel best?
   a. 08:00–10:00   b. 11:00–13:00   c. 15:00–17:00   d. 19:00–21:00

   9) Another friend suggests jogging at 22:00–23:00. How would you describe this time for you?
   a. Good   b. Reasonable   c. Difficult   d. Very difficult

Answer Key: Score each question with the following points, then total the points. For questions 1 through 5 and 8: a = 1, b = 2, c = 3, d = 4. For questions 6, 7, and 9: a = 4, b = 3, c = 2, d = 1. Your score can range from 9 to 36. Here are the suggested scores for determining activity types: 9–15, definitely a morning type; 15–19, moderately a morning type; 20–25, an intermediate type; 26–30: moderately an evening type; 31–36: definitely an evening type.

Most individuals score between 20 and 25; only about 5% of the population scores 14 or less, or 31 or more. This type of questionnaire is widely used experimentally, and it has been translated into several languages.

One very important relationship pointed out in this chapter concerns alertness relative to the body temperature cycle. Design a simple experiment to demonstrate this relationship. Shown here is an alertness scale that has been widely used in research:

Alertness scale	
Alertness score	**Condition**
7	Full alertness, no fatigue
6	Very slight fall in alertness, very slight fatigue
5	Slight fall in alertness, slight fatigue
4	Moderate fall in alertness, moderate fatigue
3	Fairly marked fall in alertness, fairly marked fatigue
2	Very marked fall in alertness, very marked fatigue
1	Extremely marked fall in alertness, extremely marked fatigue

Plan a protocol for measuring your temperature at regular 2 h intervals throughout your wake period. Many ways are available for measuring temperature, but the most available for students is probably a sublingual so-called fever thermometer from the local drugstore. To gain an accurate reading, keep your mouth closed and breathe through your nose for the full 5 minutes of the reading, and refrain from talking, drinking, or eating for 15 minutes before the start of the temperature reading.

3. What effect do naps and activity have on alertness during the day? Have a subject take a nap, and then make measurements after the subject awakens. Compare results with those taken at the same time of day on a day without a nap.

4. In exercise 2, what are some possible masking factors in your daily routine that could interfere with the temperature reading? How could these factors be reduced during the experiment?

5. This chapter has emphasized repeatedly the problems of masking in human research, particularly in field studies. How well do constant routine and forced desynchronization account for masking effects? Quickly review Chapters 2 and 3, and summarize the kinds of masking that are seen in nonhuman circadian studies. Why has so much less effort been devoted to demasking of nonhuman animal data than to demasking of human data?

6. Exercise increases body temperature, but does it do so equally at different circadian times of day? Plan a short laboratory exercise to test this question, remembering what this chapter has reported about exogenous effects of activity on body temperature.

REFERENCES

1. Åkerstedt, T., and T. Gillberg. 1981. The circadian variation of experimentally displaced sleep. Sleep 4: 159–169.

2. Åkerstedt, T., K. Hume, D. Minors, and J. Waterhouse. 1998. Experimental separation of time of day and homeostatic influences on sleep. Am. J. Physiol. 274: R1162–R1168.

3. Angrilli, A., P. Cherubini, A. Pavese, and S. Manfredini. 1997. The influence of affective factors on time perception. Percept. Psychophys. 59: 972–982.

4. Arendt, J. 1995. Melatonin and the mammalian pineal gland. Chapman Hall, New York.

5. Arendt, J., L. Wetterberg, L. Paunier, P. C. Sizonenko, and T. Heyden. 1977. Melatonin radioimmunoassay: Human serum and cerebrospinal fluid. Horm. Res. 8: 65–75.

6. Aschoff, J. 1965. Circadian rhythms in man: A self-sustained oscillator with an inherent frequency underlies human 24-hour periodicity. Science 148: 1427–1432.

7. Aschoff, J. 1990. Sources of thoughts from temperature regulation to rhythms research. Chronobiol. Int. 7: 179–186.

8. Aschoff, J. 1998. Circadian parameters as individual characteristics. J. Biol. Rhythms 13: 123–131.

9. Baehr, E., W. Revelle, and C. Eastman. 2000. Individual differences in the phase and amplitude of the human circadian temperature rhythm: With an emphasis on morningness-eveningness. J. Sleep Res. 9: 117–127.

10. Beersma, D., and A. Hiddinga. 1998. No impact of physical activity on the period of the circadian pacemaker in humans. Chronobiol. Int. 15: 49–57.

11. Boivin, D., C. A. Czeisler, D-J. Dijk, J. F. Duffy, S. Folkard, et al. 1997. Complex interaction of the sleep-wake cycle and circadian phase modulates mood in healthy subjects. Arch. Gen. Psychiatry 54: 145–152.

12. Boivin, D., J. F. Duffy, R. Kronauer, and C. A. Czeisler. 1996. Dose-response relationships for resetting of human circadian clock by light. Nature 379: 540–542.

13. Bojkowski, C. J., M. Aldhous, J. English, C. Franey, A. L. Poulton, et al. 1987A. Suppression of nocturnal plasma melatonin and 6-sulphatoxymelatonin by bright dim light in man. Horm. Metab. Res. 19: 437–438.

14. Bojkowski, C. J., J. Arendt, M. C. Shih, and S. P. Markey. 1987B. Melatonin secretion in humans assessed by measuring its metabolite, 6-sulfatoxymelatonin. Clin. Chem. 33: 1343–1348.

15. Brearley, H. C. 1919. Telling Time through the Ages. Ingersoll, New York.

16. Cagnacci, A., K. Krauchi, A. Wirz-Justice, and A. Volpe. 1997. Homeostatic versus circadian effects of melatonin on core body temperature in humans. J. Biol. Rhythms 12: 509–517.

17. Campbell, S., P. Murphy, and A. Suhner. 2001. Extraocular phototransduction and circadian timing systems in vertebrates. Chronobiol. Int. 18: 137–172.

18. Carskadon, M. A., S. Labyak, C. Acebo, and R. Seifer. 1999. Intrinsic circadian period of adolescent humans measured in conditions of forced desynchrony. Neurosci. Lett. 260: 129–132.

19. Castleden, R. 1987. The Stonehenge People: An Exploration of Life in Neolithic Britain, 4700–2000 BC. Routledge & Kegan Paul, London.

20. Chandrashekaran, M., G. Marimuthu, and L. Geetha. 1997. Correlations between sleep and wake in internally synchronized and desynchronized circadian rhythms in humans under prolonged isolation. J. Biol. Rhythms 12: 26–33.

21. Czeisler, C. A. 1995. The effect of light on the human circadian pacemaker. In: Circadian Clocks and Their Adjustment (Ciba Foundation Symposium 183), D. Chadwick and K. Ackrill (eds.), pp. 254–290. Wiley, Chichester, UK.

22. Czeisler, C. A., E. N. Brown, J. Ronda, R. Kronauer, G. Richardson, and W. Freitag. 1985. A clinical method to assess the endogenous circadian phase (ECP) of the deep circadian oscillator in man. Sleep Res. 14: 295.

23. Czeisler, C. A., G. S. Richardson, J. C. Zimmerman, M. C. Moore-Ede, and E. D. Weitzman. 1981. Entrainment of human circadian rhythms. Light-dark cycles: A reassessment. Photochem. Photobiol. 34: 239–247.

24. Damasio, A. R. 2002. Remembering when. Sci. Am. 287 (Sept): 66–73.

25. Dijk, D-J., and C. A. Czeisler. 1994. Paradoxical timing of the circadian rhythms of sleep propensity serves to consolidate sleep and wakefulness in humans. Neurosci. Lett. 166: 63–68.

26. Dijk, D-J., and J. F. Duffy. 1999. Circadian regulation of human sleep and age-related changes in its timing, consolidation and EEG characteristics. Ann. Med. 31: 130–140.

27. Duffy, J. F., and D-J. Dijk. 2002. Getting through to circadian oscillators: Why use constant routines? J. Biol. Rhythms 17: 4–13.

28. Duffy, J. F., R. Kronauer, and C. A. Czeisler. 1996. Phase-shifting human circadian rhythms: Influence of sleep timing, social contact and light exposure. J. Physiol. (Lond.) 495: 289–297.

29. Duffy, J. F., D. Rimmer, and C. A. Czeisler. 2001. Association of intrinsic period with morningness-eveningness, usual wake time, and circadian phase. Behav. Neurosci. 115: 895–899.

30. Edwards, B., J. Waterhouse, T. Reilly, and G. Atkinson. 2002. A comparison of the suitabilities of rectal, gut, and insulated axilla temperatures for measurements of core temperature in field studies. Chronobiol. Int. 19: 579–597.

31. Folkard, S. 1990. Circadian performance rhythms: Some practical and theoretical implications. Philos. Trans. R. Soc. London [Biol.] 327: 543–553.

32. Helbrügge, T. 1960. The development of circadian rhythms in infants. Cold Spring Harb. Symp. Quant. Biol. 25: 311–323.

33. Honma, K-I., S. Honma, K. Nakamura, M. Sasaki, T. Endo, and T. Takahashi. 1995. Differential effects of bright light and social cues on reentrainment of human circadian rhythm. Am. J. Physiol. 268: R528–R535.

34. Honma, K-I., S. Honma, and T. Wada. 1987. Phase-dependent shift of free-running human circadian rhythms in response to a single bright light pulse. Experientia 43: 1205–1207.

35. Horne, J., and O. Ostberg. 1976. A self-assessment questionnaire to determine morningness in human circadian rhythms. Int. J. Chronobiol. 4: 97–100.

36. Johnston, A., and S. Nishida. 2001. Time perception: Brain time or event time? Curr. Biol. 11: R427–R430.

37. Kennaway, D. J., G. E. Stamp, and F. C. Goble. 1992. Development of melatonin production in infants and the impact of prematurity. J. Clin. Endocrinol. Metab. 75: 367–369.

38. Kerkhof, G., and H. Van Dongen. 1996. Morning-type and evening-type individuals differ in the phase position of their endogenous circadian oscillator. Neurosci. Lett. 218: 153–156.

39. Khalsa, S., M. Jewett, J. F. Duffy, and C. A. Czeisler. 2000. The timing of the human circadian clock is accurately represented by the core body temperature rhythm following phase shifts to a three-cycle light stimulus near the critical zone. J. Biol. Rhythms 15: 524–530.

40. Klerman, E., H. Gershengorn, J. F. Duffy, and R. Kronauer. 2002. Comparisons of the variability of three markers of the human circadian pacemaker. J. Biol. Rhythms 17: 181–193.

41. Krieger, D., W. Allen, and F. Rizzo. 1971. Characterization of the normal pattern of plasma corticosteroid levels. J. Clin. Endocrinol. Metab. 32: 266–284.

42. Lack, L., and K. Lushington. 1996. The rhythms of sleep propensity and core body-temperature. J. Sleep Res. 5: 1–11.

43. Lavie, P. 1986. Ultrashort sleep-waking schedule: III. "Gates and forbidden zones" for sleep. Electroencephalogr. Clin. Neurophysiol. 63: 414–425.

44. Lavie, P. 2001. Sleep-wake as a biological rhythm. Annu. Rev. Physiol. 59: 327–329.

45. Lewis P., and M. Lobban. 1957. Dissociation of diurnal rhythms in human subjects living on abnormal time routines. Q. J. Exp. Physiol. 42: 371–386.

46. Lewy, A. J., and D. A. Newsome. 1983. Different types of melatonin circadian secretory rhythms in some blind subjects. J. Clin. Endocrinol. Metab. 56: 1103–1107.

47. Lewy, A., V. Bauer, S. Ahmed, K. Thomas, N. Cutler, et al. 1998. The human phase response curve (PRC) to melatonin is about 12 hours out of phase with the PRC to light. Chronobiol. Int. 15: 71–83.

48. Lewy, A., T. Wehr, F. Goodwin, D. Newsome, and S. Markey. 1980. Light suppresses melatonin secretion in humans. Science 210: 1267–1269.

49. Lockley, S. W., D. J. Skene, K. James, K. Thapan, J. Wright, and J. Arendt. 2000. Melatonin administration can entrain the free-running circadian system of blind subjects. J. Endocrinol. 164: R1–R6.

50. Middleton, B., J. Arendt, and B. Stone. 1996. Human circadian rhythms in constant dim light (<8 lux) with knowledge of clock time. J. Sleep Res. 5: 69–76.

51. Minors, D., and J. Waterhouse. 1981. Circadian Rhythms and the Human. Wright, Bristol, UK.

52. Minors, D., J. Waterhouse, and W. Rietveld. 1996. Constant routines and "purification" methods: Do they measure the same thing? Biol. Rhythm Res. 27: 166–174.

53. Minors, D., J. Waterhouse, and A. Wirz-Justice. 1991. A human phase-response curve to light. Neurosci. Lett. 133: 36–40.

54. Miyamoto, Y., and A. Sancar. 1998. Vitamin B2-based blue-light photoreceptors in the retinohypothalamic tract as the photoactive pigments for setting the circadian clock in mammals. Proc. Natl. Acad. Sci. USA 95: 6097–6102.

55. Monk, T., D. Buysse, J. Carrier, B. Billy, and L. Rose. 2001. Effects of afternoon "siesta" naps on sleep, alertness, performance, and circadian rhythms in the elderly. Sleep 24: 680–687.

56. Moore, R. Y., and R. Leak. 2001. Suprachiasmatic nucleus. In: Handbook of Behavioral Neurobiology Volume 12: Circadian Clocks, J. S. Takahashi, F. W. Turek, and R. Y. Moore (eds.), pp. 141–170. Kluwer Academic/Plenum, New York.

57. Moore-Ede, M., and F. Sulzman. 1981. Internal temporal order. In: Handbook of Behavioral Neurobiology Volume 4: Biological Rhythms, J. Aschoff (ed.), pp. 215–241. Plenum, New York.

58. Moore-Ede, M., F. Sulzman, and C. Fuller. 1982. The Clocks That Time Us. Harvard University Press, Cambridge, MA.

59. Murphy, P., and S. Campbell. 1997. Nighttime drop in body temperature: A physiological trigger for sleep onset? Sleep 20: 505–511.

60. Palmer, J. D. 2002. The Living Clock: The Orchestrator of Biological Rhythms. Oxford University Press, Oxford, UK.

61. Refinetti, R., and M. Menaker. 1991. The circadian rhythm of body temperature. Physiol. Behav. 51: 613–637.

62. Rimmer, D., D. Boivin, T. Shanahan, R. Kronauer, J. F. Duffy, and C. A. Czeisler. 2000. Dynamic resetting of the human circadian pacemaker by intermittent bright light. Am. J. Physiol. 279: R1574–R1579.

63. Ross, J. K., J. Arendt, J. Horne, and W. Haston. 1995. Night-shift work in Antarctica: Sleep characteristics and bright light treatment. Physiol. Behav. 57: 1169–1174.

64. Shochat, T., R. Luboshitzky, and P. Lavie. 1997. Nocturnal melatonin onset is phase locked to the primary sleep gate. Comp. Physiol. 42: R364–R370.

65. Siffre, M. 1964. Beyond Time. McGraw Hill, New York.

66. Stix, G. 2002. Real time. Sci. Am. 287 (Sept): 36–39.

67. Tenreiro, S., H. Dowse, S. D'Souza, D. Minors, M. Chiswick, et al. 1991. The development of ultradian and circadian rhythms in premature babies maintained in constant conditions. Early Hum. Dev. 27: 33–52.

68. Van Cauter, E. 1989. Endocrine rhythms. In: Biological Rhythms in Clinical Practice, J. Arendt, D. Minors, and J. Waterhouse (eds.), pp. 23–50. Butterworth, London.

69. Van Cauter, E., D. Desir, C. Decoster, F. Fery, and E. Balasse. 1989. Nocturnal decrease in glucose tolerance during constant glucose infusion. J. Clin. Endocrinol. Metab. 69: 604–611.

70. Waterhouse, J., and T. Åkerstedt. 1997. The body synchronic. In: 1998 Medical and Health Annual, pp. 80–91. Encyclopedia Britannica, Chicago.

71. Waterhouse, J., D. Minors, M. Waterhouse, T. Reilly, and G. Atkinson. 2002. Keeping in Time with Your Body Clock. Oxford University, Oxford, UK.

72. Waterhouse, J., D. Weinert, D. Minors, G. Atkinson, T. Reilly, et al. 1999. The effect of activity on the waking rhythm in humans. Chronobiol. Int. 16: 343–357.

73. Waterhouse, J., D. Weinert, D. Minors, S. Folkard, D. Owens, et al. 2000. A comparison of some different methods for purifying core temperature data from humans. Chronobiol. Int. 17: 539–566.

74. Weinert, D., U. Sitka, D. Minors, and J. Waterhouse. 1994. The development of circadian rhythmicity in neonates. Early Hum. Dev. 36: 117–126.

75. Wever, R. 1975. The circadian multi-oscillator system of man. Int. J. Chronobiol. 3: 19–55.

76. Zeitzer, J. M., D-J. Dijk, R. Kronauer, E. Brown, and C. A. Czeisler. 2000. Sensitivity of the human circadian pacemaker to nocturnal light: Melatonin phase resetting and suppression. J. Physiol. 526: 695–702.

Early human cultures were fascinated by the phenomenon of daily and seasonal aggravation of symptoms or prevalence of certain illnesses. The first known Egyptian medical text, dating back almost to the time of the construction of the Great Pyramid in 2500 B.C., dealt with a few items that would be considered today to be circadian malfunction. Through the ages, the importance of biological rhythms in pathophysiology have been clearly recognized. The early epoch of psychiatry was characterized by treatments that today are considered entraining pulses and sleep–wake cycle manipulations: sun baths, daily walks, strenuous exercise, selective diets at specific times of day, cold baths, and sleep deprivation. Chapter 10 takes a broad view of the relevance of circadian rhythmicity in society today. The many aspects of temporal disorder introduced by human diversity of lifestyle are considered, and ideas are advanced for improving the quality of life in a time sense.

10

The Relevance of Circadian Rhythms for Human Welfare

The clock, not the steam engine, is the key-machine of the modern industrial age.

—Lewis Mumford, U.S. social philosopher

Introduction: Circadian Rhythmicity Has Vital Implications for Many Aspects of Human Lifestyle in the World Today[37,41,49]

Circadian rhythmicity in humans has received growing attention in the past few decades in such diverse fields as psychology, physiology, medicine, psychiatry, and neuroscience. The reasons are clear. Humans live almost worldwide in a round-the-clock economy. Caravans of 18-wheel tractor-trailer trucks roar down the interstates chiefly at night in order to make better time. Nurses and interns defy circadian constraints and work extended shifts up to 72 h long in order to provide continuity of care to patients. Factories faced with profit margin concerns use shifts that expose workers to night hours and constantly swinging shift schedules. A 24 h demand keeps many gas stations, grocery stores, and restaurants open at all hours. Urbanites expect emergency services from hospitals, police, and utilities around the clock. Power plant operators must ceaselessly monitor their behemoths for possible malfunction. Compressed workweeks demand 12 h workdays. The litany of circadian stresses could continue much further.

A chronobiological prescription is available. Ignoring it might bring disaster; implementing it might mean health. The price of a mistake can be appalling. One short example will illustrate the point. At 1:23 A.M. on April 26, 1986, fatigued operators at the Chernobyl nuclear power plant in rural Ukraine were carrying out a safety check operation and inadvertently initiated a malfunction of Reactor 4 (Figure 10.1). The operators were at the low point of their circadian efficiency and temperature cycle, and in the reactor meltdown crisis that immediately ensued, they made some poor judgments about stopping the world's worst nuclear disaster.

FIGURE 10.1 Chernobyl nuclear power plant after the reactor meltdown.

Over 190 tons of highly radioactive uranium and pulverized graphite cooler rods were expelled in a cloud that traveled around the entire Northern Hemisphere. Local pilots and firefighters tried valiantly to quench the explosion and impound all reactors in a massive sarcophagus of concrete dumped by relays of airplanes. First reports were optimistic: only 13 people dead. The actual truth is devastating. Residents of the nearby village of Chernobyl were exposed to deadly radiation 90 times greater than that from the Hiroshima bomb. Estimates suggest that 9 million human beings were affected by the explosion and its fallout, including 3 to 4 million children.

More than 400,000 people were evacuated from their homes in approximately 2,000 towns and villages in Belarus and the Ukraine. Another 1.8 million continue to live in radioactive zones in Belarus, and 3 to 5 million still reside in radioactively contaminated areas in the Ukraine. One radiation specialist judges that the Chernobyl accident will cause 475,000 fatal cancers worldwide and an equal number of nonfatal cancers. The indirect effects are equally frightening. In Belarus alone, 25% of prime farmland and forest is heavily contaminated by radioactivity, and 10% is unusable for the indefinite future.

The cost of ignorance about human circadian physiology is very high. A brief summary of areas of human activity affected by circadian rhythmicity tells the tale. The many aspects of circadian impact on human physiology can be divided for convenience into about 6 categories (Table 10.1). Many of the categories overlap. Some effects, such as jet lag or reduced efficiency, are merely uncomfortable. Others are deadly. Some of the effects,

such as traffic mortality, are very dramatic, acute, and terminal. Others may be equally lethal but accrue insidiously over years or decades of time.

Chapter 9 pointed out that the fundamental principles of circadian physiology seen throughout the plant and animal kingdoms apply also to humans. A mesh of moral and ethical codes protects human dignity and well-being from some invasive experimental methods that are feasible in lower animals, but through the translucent veil of restraint, humans appear to resemble other diurnal mammals. The unique difference, of course, is human intelligence. Curiosity, self-awareness, and great inventiveness allow humans to understand complex ideas and apply them to daily life. On the darker side, human ingenuity allows us to degrade our environment, to create unusual lifestyles, and to achieve a geographic distribution on the planet that no other species can match.

This chapter will consider the stresses on circadian organization and the malfunctions of this time-keeping system. Not only must the clinical contexts

TABLE 10.1	Areas of human behavior and physiology affected by circadian rhythmicity

1. Acute disasters
     Accident rates in traffic and industry
2. Occupational and travel stresses
     Round-the-clock lifestyles: 24 h commercial services
     General performance efficiency in the workplace and school
     Nurses' and residents' schedules
     Nuclear and nonnuclear power plant operators
     Factory shift work
     Interstate trucking
     Military operations
     Jet lag
     Space flight
3. Medical diagnosis and treatment
     Medical diagnostics
     Circadian onset of certain serious illnesses
     Medical treatment timetables
4. Sleep–wake syndromes and mood disorders
     Adjustment to extreme photic conditions: arctic and antarctic inhabitants
     Rhythmicity in the blind
     Sleep–wake problems in teenagers and students
     Rhythmicity in the elderly
     Sleep and mood disorders
5. Treatment of circadian dysfunction
     Chronopharmacology
6. Legal measures for regulating work hours and scheduling
     Federal legislation for work schedules

be addressed, but also effects in healthy individuals who undertake the biologically abnormal pursuits of jet-setting across time zones or working at night. The aim will be to minimize circadian problems by describing mechanisms that are effective in healthy individuals living in normal circadian frameworks.

The chapter started with a shocking example of circadian malfunction in work efficiency at Chernobyl. The discussion turns now to the fundamental circadian properties that are effective in both recognition and treatment of chronobiological malfunctions. A survey of the approximately 6 areas of circadian relevance for humans will follow (see Table 10.1). One category centers on circadian-related acute disasters. A second group concerns occupational and travel stresses. The third group focuses on circadian timing of medical diagnosis and treatment. A fourth group considers the chronopathology of about 10 loosely related sleep–wake syndromes and mood disorders. A fifth section evaluates ways of treating circadian dysfunction by light schedules and drugs such as melatonin. The final category deals with legal measures for regulating work hours and scheduling in industries that have particularly vulnerable circadian considerations.

Several Fundamental Properties of Circadian Systems Are Directly Applicable to Improvement of Human Performance[11,21,42]

The structure of the human SCN appears similar to that of other mammals[40]

The structural details of the human SCN (suprachiasmatic nucleus) pacemaker system were considered in Chapter 5 and briefly in Chapter 9. Only a sketch will be given here as review. The clockworks reside in the SCN pacemake (see Plate 3, Photo B). Input for entrainment comes primarily through specialized retinal ganglion cell photoreceptors, which connect via the retinohypothalamic tract to the SCN. Neural output projections are both intrinsic within the SCN and extrinsic to nearby nuclei, with further neural relays to every part of the body (see Chapter 5).

The general picture of circadian physiology in humans is similar to that of other diurnal mammals[9,26,34]

SCN REGULATION. The SCN rhythmically regulates a multitude of functions in humans. Sleep–wake cycle, levels of specific types of cells in the blood, body temperature, alertness and efficiency, and hormone levels in the blood are just a few examples. One example of three hormones will suffice at the start (Figure 10.2). The free-running circadian period of plasma ACTH, cortisol, and

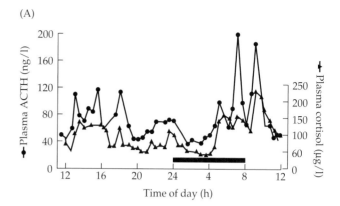

(A)

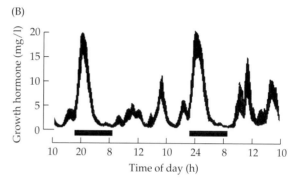

(B)

FIGURE 10.2 Circadian rhythms in healthy subjects of hormones relative to sleep. (A) Plasma adrenocorticotropic hormone (ACTH) and cortisol. (B) Growth hormone. Bars indicate times of sleep. (From Krieger et al., 1971, and Parker et al., 1979.)

growth hormone in humans is about 24.3 h. In subjects living normally entrained, however, a period of exactly 24 h prevails, adjusted by rhythmic environmental entraining agents (see Chapters 1–3).

ENTRAINMENT. Most important are the light–dark cycle and the rhythmic secretion of melatonin. The effect of light depends on the dose, but even in subjects living indoors, rhythmic exposure to domestic and office lighting of 150 to 300 lux is sufficient to produce entrainment with a normal late-evening bedtime and darkness for the sleep period. The effects of light in producing adjustment are described by the phase response curve (PRC) to light (see Figure 3.5 in Chapter 3). Light in the hours before the minimum core body temperature produces a delay of the clock, and light after the temperature minimum, produce a phase advance. Because the period of the human internal clock is greater than 24 h, the advance portion of the phase response curve is what is normally used.

Melatonin, the darkness hormone, is also an important synchronizer. Sometimes it is called an internal synchronizer because it is produced within the body. However, the fact that its production normally depends on the light cycle means that it must have a modulatory rather than a primary role. Melatonin also shows effects on the internal clock that are phase dependent. Its PRC is the inverse of a light PRC.

Although normally produced within the pineal gland, melatonin can also be taken orally as a pill. Ingestion of melatonin in the evening promotes a phase advance; in the morning, a phase delay. Because light suppresses melatonin secretion in a dose-dependent manner, light and melatonin normally act in concert. Morning light advances the clock both directly by causing advance phase shifts and indirectly by suppressing melatonin secretion. At night, melatonin reinforces sleepiness and the nighttime decline in core temperature. Exogenous melatonin during the day induces transient sleepiness and a decline in core temperature.

FIGURE 10.3 Vehicular accidents are very common, and the casualty rate is high.

Other synchronizers have been proposed, including rhythmic food intake and exercise, although the evidence for their importance is not strong. However, the concept that lifestyle might act indirectly, as a synchronizer, is valuable. Humans eat and are physically, socially, and mentally active in a bright stimulating environment in the daytime; at night they usually stop eating and seek rest and sleep in a dark, quiet environment. Such a lifestyle exposes people to the light–dark cycle and enables the normal phases of melatonin secretion to take place. As a result of entrainment of their clock to a 24 h day, individuals can show a dichotomy in their behavior, physiology, and biochemistry between day and night. Deviations from diurnality are the cause of most circadian malfunctions.

Ignoring Circadian Rhythmicity May Increase Accident Rates[37,41,49]

Traffic accident rates appear related to circadian timing[20]

The toll from car collisions and crashes is great (Figure 10.3). Vehicular accidents clearly have many causes, ranging from driving under the influence of alcohol or drugs or recklessness of teenage drivers, to heart attacks or falling asleep at the wheel. However, the evidence is very strong for involvement of a circadian element.

In a study of 6,052 traffic accidents judged to be fatigue related, the highest rate occurred at an individual's body temperature and efficiency low in the hours shortly after midnight. Values catapulted from about 200 at 6:00 P.M., to 900 at midnight, 1,100 at 2:00 A.M., 900 at

4:00 A.M., and 700 at 6:00 A.M., then dropped to a daytime low. Of course, darkness plays an important role in general road visibility, and the accident rate for night drivers is undoubtedly affected by several factors in addition to the circadian component.

The Exxon Valdez supertanker accident had circadian connotations[34]

Shortly after midnight on March 24, 1989, the supertanker *Exxon Valdez* started to thread the narrow channel across Bligh Reef in the pristine wilderness of Prince William Sound, Alaska. She was carrying 1.3 million barrels of crude oil when she struck the reef and tore a gaping hole in her side. About 258,000 barrels of oil spilled into the sound and soon spread in an ugly blackening oil slick seaward for several hundred kilometers. Beaches, verdant forest fringe, benthic substrate, and myriad creatures in the sound were coated with the toxic oily ooze. Vessel damage was $25 million, and the cost of cleanup during 1989 alone was $1.85 billion. The damage to wildlife was inestimable. Even today, 13 years later, effects are apparent in spite of the intense efforts of volunteer and government crews to contain and clean up the spill.

The causes of the *Exxon Valdez*'s grounding have been examined and debated at many levels. In the opinion of the U.S. National Transportation Safety Board, the third mate failed to properly maneuver the vessel because of fatigue and excessive workload. The issue of late-night hours, fatigue, excessive continuous work hours, and circadian efficiency low point again come to the fore in this disastrous acute circadian dysfunction.

The accident at Three Mile Island was the most serious in U.S. commercial nuclear power plant history[49]

At 4:00 A.M. on March 28, 1979, a chain of incidents began that resulted in the melt-down of fuel rods in the second reactor at the Three Mile Island nuclear power plant in southeastern Pennsylvania, and low-level radioactivity gaseous bursts were released for several days. Catastrophe was averted in the nick of time without deaths or injuries to plant workers or to members of the nearby bustling community of Middletown, Pennsylvania. Eventually, however, the crisis led to evacuation of many local residents, massive cleanup efforts at the plant, and decommissioning of the reactor (Figure 10.4). According to the official legal report, it also led to "large scale, sweeping changes in reactor opera-tor training, human factors engineering, and changes throughout the US for emer-gency response planning, and changes of many other management policies for nuclear power plants."[34]

FIGURE 10.4 Three Mile Island nuclear power plant. The facility is shown here after the accident. (Courtesy of the U. S. National Archives and Records Administration.)

One of the key points in understanding the chain of events of Three Mile Island is the bureaucratic phrase "human factors engineering." The accident began at 4:00 A.M. Details are technical, but basically the plant under-went a failure in the water coolant equipment located in the nonnuclear part of the plant. With frightening speed, steam built up in the generators, until the turbine and reactor shut down. The crisis was aggravated by prob-lems in the emergency backup-feed water system in another part of the plant. Without adequate cooling, the nuclear fuel overheated and started to melt the zirconi-um cladding of the radioactive fuel rods, which then reacted with water. As a result, a potentially explosive hydrogen bubble built up above the reactor core.

Although the event was not a true meltdown, some of the fuel rods did melt and contaminate the water coolant. Radioactivity was released both in water that flowed into the reactor building basement during the accident, and in contaminated air released into the atmosphere during the later degassing procedure. The plant instruction manuals contained no advice for these events, which had never happened before, but reduc-tion of the dangerous hydrogen bubble was essential to reduce the potential for an explosion.

The important point to remember is that many com-plicated steps were immediately implemented by the operators trying to solve problems in an unprecedented series of events at 4:00 in the morning. Some had been working continuously on extended shifts and were at a circadian low point of their internal temperature rhythm. Fatigue was at a maximum, and efficiency in solving new problems with great speed was at a mini-mum. A conclusion of circadian efficiency malfunction is almost unavoidable.

Occupational or Travel Stresses May Affect Performance[2,28,50,69]

Modern expectations place 24-hour demands on society[41]

At present a 24 h supply of services is demanded by the public in technologically advanced nations. Access to emergency medical services, availability of electric, gas, and water utilities, a food supply, and transportation services are deemed essential round the clock. The hours during which shops are open and banking facilities are available have also increased, and these demands extend work hours. Many industrial processes must run continuously, and news-gathering media must work on a 24 h basis. All of these needs give rise to nonstandard working hours.

Factory shift work is necessary for industrialized countries[13,49,59]

DEFINITION OF A SHIFT WORKER. For the purpose of this chapter, a shift worker is defined as a nonfarm worker who does not regularly work a standard daytime sched-ule. Needless to say, the work hours of nonstandard schedules vary considerably. This discussion will consider a standard daytime schedule to be approximately 8 h per day, 5 consecutive days a week, during daytime hours over the span from roughly dawn until dusk in Temperate Zone countries. A variation of the standard week is the compressed workweek, consisting of four 10 h units per week, as will be described shortly.

In shift work, unlike day work, the majority of work hours occur at night, when the strongly diurnal human runs into many circadian problems. Often shift work involves a swinging schedule that rotates either between day and night shifts, or among morning, evening, and night shifts. The change from one shift to another is usually made at 3-week intervals in the United States, but more frequent changes, every week or even every day, are quite common in Europe and other countries.

Still another variant of shift work is the extended work schedule. In the medical profession, the extended schedule was formerly a 72 h shift, intended to provide continuity of care to a patient by the same hospital staff. The ultimate extended work schedule is found in wartime military assignments for troops in combat zones. Here the demands of active fighting may call for nearly continuous duty over periods of many days or weeks.

Shift work and extended work schedules are defended by industry and the medical professions on four grounds. In some cases an extended period of time is required to complete a particular job, as in hospital environments. In other cases the basis is a continuous demand for services in occupations with limited employees, as described already. Another defense of the practice is based on economic factors, such as the need to fully use expensive capital equipment in factories around the clock. Finally, technologically advanced equipment such as nuclear power plants or space shuttles requires 24 h monitoring.

The most comprehensive data on the use of shift work in the United States is the census, which is conducted at 10-year intervals to cover demographics of every household. According to a recent census, about 16% of the working population is involved in full-time night work or shift work in the United States (Figure 10.5). The percentage is substantially higher for part-time workers. For the average daytime employee, the social implications of altered work hours for those nearly one in five full-time shift workers are hard to imagine.

PUBLIC INTEREST ISSUES. Almost as important as the number of hours or the time of day worked are the career demands and stresses imposed by a specific job. A crucial element is the degree of responsibility of the job. The effects of inefficiency of a night worker tightening bolts on a refrigerator cannot compare with those of a surgeon undertaking brain surgery or heart transplant on a work schedule extended into the circadian critical

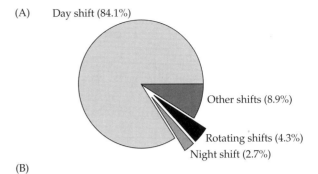

FIGURE 10.5 Shift work in the United States. (A) Percentages in various categories of shift work for full-time employees. (B) A modern factory assembly line. (A from Liskowsky, 1991.)

zone. Nor does the responsibility associated with bolt tightening compare with the responsibility of a nuclear power plant operator. The latter individual's duty is to regulate the power plant and avoid disasters such as Chernobyl. Because of the element of career responsibility, some specialized shift work will be considered a little later in the chapter in terms of specific careers.

THE COMPRESSED WORKWEEK. Before the discussion turns to shift work, the compressed workweek schedule will be considered briefly because of its growing prevalence. The work is usually performed 10 h per day for 4 days or 12 h per day for 3 days, leaving 3 or 4 days free each week, but a variety of other schedules exist. Widely different occupations champion the use of compressed workweeks. Among the most common industries using the compressed workweek are the chemical industry, the oil industry (especially on offshore drilling rigs), and the steel industry. Some nuclear power plants also use a compressed work schedule. Some airline pilots and airborne support staff are scheduled for long international hauls in a compressed week because of their need to return to home base after a long flight.

A major problem from the workers' point of view is the fatigue of consecutive 12 h days. However, workers often state that this downside is offset by more rest days and by not having to experience the hassle associated with travel to and from the workplace. From the employer's point of view, moonlighting concerns are important, as are the decreased quality of work produced, particularly toward the end of the shift. With some types of employment, public concern may arise over safety issues.

TYPICAL WORK SHIFT SCHEDULES. In night work, circadian entrainment may depend on many factors, including the type of schedule and exposure to natural time cues. Shift schedules vary enormously. Some workers must rotate shifts every 2 to 3 days. Other workers are on non-24 h shift schedules, such as those working 18 h days on nuclear submarines. These workers generally do not adapt their circadian systems to a great extent and can become chronically sleep deprived. Permanent night workers can show some adjustment, but individuals vary greatly, and a tendency persists to live normal diurnal schedules on rest days. In some unusual environments, such as offshore oil rigs or Antarctica, complete adaptation to a night-shift schedule can occur within a few days and may be attributed to quiet, dark sleeping quarters, together with lack of conflicting social factors or light exposure.

Short of actually working on a shift schedule, one can best appreciate the impact on the mental and physical state of the participant by examining the rotation schedules for workers of a plant on a traditional weekly shift schedule. One way to visualize the demands, for example, is to plot a work actogram of an individual worker on an 8 h workday in a swing-shift assignment (Figure 10.6). Notice the irregularity of hours and the violation of reasonable circadian dictates.

The mental and physiological stresses on the individual worker are great. These strains are magnified for married workers, especially if they work on unmatched shifts. Even greater is the extension of stress in families with young children in which both parents work on shifts. Night workers in particular may find it nearly impossible to find quiet time for essential sleep when young, noisy children are present. Clearly the quality of family time deteriorates for many shift workers.

Proponents of rapidly rotating shift systems, in which the shift worked changes every 2 to 3 days, argue that the rapid rotation enables some normal days to be worked each week, thereby decreasing the amount of sleep loss that might accumulate. However, such schedules do not include sufficient time for any useful amount of adjustment of the circadian system. The key point is the fundamental clash among the requirement of society (round-the-clock coverage), the workers (a normal social life), and chronobiology (stable sleep–wake schedules that would enable adjustment).

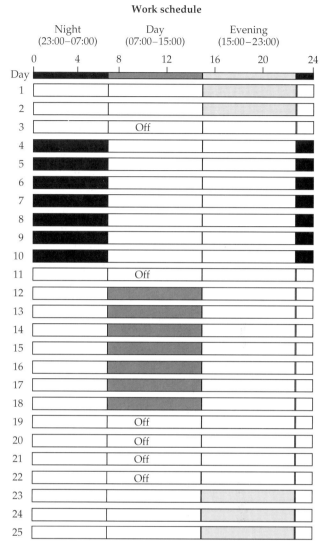

FIGURE 10.6 Actogram of work hours for an individual worker on a three-unit rotating shift. Black bars denote night shift; dark gray, day shift; light gray, evening shift; white bars indicate sleep or leisure time; and "Off" indicates no shift on a particular day. (Courtesy of P. DeCoursey.)

MAJOR PROBLEMS FOR THE SHIFT WORKER. Two of the main concerns about a 24 h society are the quality of performance on the night shift and the loss of sleep. In addition, shift workers have an increased risk of heart disease and other major diseases. Even though accidents in the workplace are comparatively rare, performance is often worse and errors more frequent on night shifts or swing shifts than on regular permanent daytime assignments.

Errors leading to industrial accidents arise from several causes. Environmental conditions of lighting, temperature, noise, and weather must be considered, but circadian time-of-day factors are certainly critically important. The amount of time since awakening and particularly the amount of time on duty cannot be ignored. Workload, other aspects of stress, sleep loss, and sleeping in the workplace also come into the equation. Three of these factors apply directly to the night worker: circadian factors, sleep loss, and time awake.

For those who work at night, circadian rhythms adjust poorly. Therefore, the shift includes the time of worst performance, coincident with the minima of the temperature and adrenaline rhythms and the peak of the melatonin rhythm. A comparison of urinary excretion of adrenaline of a worker on a swing shift of 3 weeks day work to 3 weeks night work indicates that the adjustment to night work is poor and slow. During the daytime shift, values are high during work and low during sleep (Figure 10.7A). In contrast, during the first week of night work, high levels of adrenaline are excreted at about the time of sleep (Figure 10.7B) and would be associated with difficulties both in initiating sleep and in having consolidated sleep. Low values during the work period would be associated with decreased alertness and mental performance. Even by the third week (Figure 10.7C), adjustment is incomplete, with decreased adrenaline levels during work time. Moreover, when daytime work is resumed (Figure 10.7D), the loss of any adjustment is very rapid (compare Figure 10.7A and D).

In addition, subjects might be sleep deprived, and they are also working at the end of the waking phase. Independent research has already used measurements of eye movements and mental performance to assess risk factors and relate them to the feeling of alertness. In one study, if alertness fell below 7 on a subjective scale of 1 to 15, the individual was at risk with regard to performing tasks adequately and safely.

Newer and more comprehensive models based on the concept of circadian and homeostatic factors controlling sleep (see Chapter 9) have also been developed to predict alertness at different circadian phases. A worker's alertness is directly related to the ability to perform tasks without risk. Alertness is determined by $S + C$, the sum of a circadian component C that is in phase with body temperature, and a homeostatic component S that falls exponentially with time awake (Figure 10.8). During sleep, this latter factor dissipates, probably in association with slow-wave sleep, in a process designated S'. If $S + C$ falls below 7, alertness is sufficiently compromised for an individual's performance to be at risk. The right side of the figure shows a specific case of a shift worker on his first night shift of a swing shift. After sleeping from 07:20–12:00 and then working the night shift, his alertness falls into the risk zone. (See also Figure 10.7 for a typical swing shift schedule.)

Many of these negative consequences of night work result from a mismatch between the sleep–wake cycle

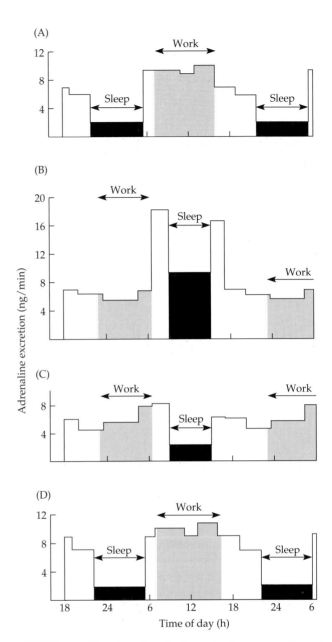

FIGURE 10.7 The circadian rhythm of urinary excretion of adrenaline for a worker during various parts of a swing-shift schedule. (A) During the last week of day work. (B) During the first week of night work. (C) During the third week of night work. (D) During the first week back on day work. (From Åkerstedt and Froberg, 1975.)

and the circadian timing system. Phase adjustment may be poor, not only with swing shifts, but also with assignments of extended schedules of night work. The mismatch occurs because the natural light–dark cycle continues in spite of the night worker's personal sleep–wake and artificial light–dark schedules. The resultant clash prevents the individual's circadian rhythms from aligning fully with the schedule of night work and day sleep. Equally bad is the rapid loss of any

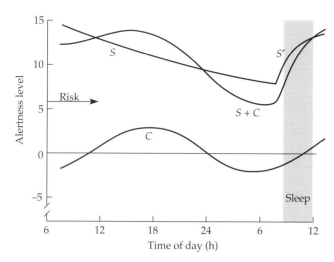

FIGURE 10.8 Schematic model of the interaction between circadian and homeostatic factors in determining alertness. C is the circadian component and S is the homeostatic sleep factor that decreases with time awake. The sum of $S + C$ allows prediction of the degree of risk in performance. Sleep restores S in a process designated S'. The shaded area at right portrays a specific case of a shift worker's sleep regime after changing to night shift work. For further explanation, see text and Figure 10.7. (From Folkard and Åkerstedt, 1991.)

hard-won adjustment due to rotation to a new shift schedule or to intervening rest days (see Figure 10.7D).

Being sleep deprived and staying awake longer should make sleep easier to initiate and maintain, through the actions of the sleep homeostat. However, daytime sleep sessions are shorter and less consolidated than sleep sessions at night, indicating that the effect of the sleep homeostat is outweighed at least partially by environmentally mediated disturbances. Circadian factors include attempting to sleep when the core temperature and adrenaline rhythms are on their rising phase.

Recently it has become clear that eating meals during an unadapted night shift leads to higher blood fat (triglycerol) than an identical meal during the daytime does. Triglycerol is a risk factor for heart disease, and shift workers have a higher fasting level of triglycerol and a greater risk of heart disease than the general population. Thus the long-term consequences of shift work need to be taken into account along with the short-term effects on sleep and the accident rate.

REMEDIAL MEASURES FOR FACTORY SHIFT-WORK MALADJUSTMENT. Unless too few consecutive nights are worked to make it worthwhile, adjustment of the individual's circadian system to the work schedule is highly recommended. Circadian principles clearly dictate that shift schedules should be planned to move later with each shift because the period of the human clock is

greater than 24 h. Timed exposure to bright light can also be used to help adjustment to night work. The direction of adaptation is determined by the timing of light. It should be given before the body temperature minimum to delay the internal clock or after the temperature minimum to advance it. Sleep problems should abate as the body temperature rhythm shifts to align with sleep time.

Because the two major treatments for all maladjustments in circadian phase are phase shifting by light and by melatonin, remedies for almost all the phase disorders considered for factory shift work and for the circumstances to follow will be similar. To avoid tedious repetition, only a brief specific overview will be given in each occupation/travel section. The section on chronopharmacology later in the chapter will cover more detailed general material common to all the conditions.

In laboratory studies that simulate factory shift-work operations, circadian rhythms generally shift about 1 h per day in dim light conditions, and as much as 2 to 3 h per day in bright light conditions. In field conditions, duration of bright light from 2 to 7.5 h and intensities ranging from 2,000 to 12,000 lux have been tested for their capacity to reentrain. Probably 3 h of moderately bright light suffices.

A variety of remedial schedules for night-shift workers have been tested. In Figure 10.9, for example, day 1 is the last occasion for sleeping at night; days 2–6 and 9–10 are days with night shifts. The light is given before the temperature minimum, which will produce a phase delay according to the light phase response curve. The pulse of light is given later on successive night shifts. During rest days in the middle of a set of night shifts, sleep is recommended from 4:00 A.M. to noon

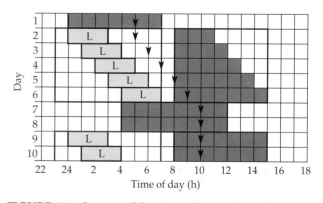

FIGURE 10.9 One possible protocol to produce adjustment of the circadian system to night-shift work. Sleep times are shown as hatched areas. L represents the timing of the 3 h pulse of bright light on each night shift; arrowheads indicate the supposed time of the temperature minimum on successive days. For further explanation, see the text. (From Eastman and Martin, 1999.)

(04:00–12:00), not at the normal time from midnight to 7:00 a.m. (24:00–07:00). The recommended time is intermediate between normal time and time of sleeping during night shifts.

The amount of favorable phase shift in night workers improves with reduction in their exposure to early-morning light on the way home. Recognition of the importance of darkness has grown recently. For a light-induced phase shift to be successful, light must also be avoided at specific times of day. Wearing dark sunglasses or welder's goggles during the trip home from the night shift and carefully darkening bedroom windows both contribute to the efficacy of the regimen of using a light pulse during night work.

PROMOTING ADJUSTMENT TO NIGHT WORK BY MELATONIN AND DRUGS. The use of melatonin as an alternate phase-shifting agent to light has been considered to some extent. Ingesting a pill is much easier and less time-consuming than sitting in front of a light box. Melatonin may enhance adaptation to simulated night-work conditions, but its use has been studied much less than that of light. Melatonin causes general sleepiness after ingestion and should not be taken if driving is required. Because melatonin is relatively unknown in its physiological effects, its use remains extremely controversial (see the section on chronopharmacology later in this chapter).

Schedules of nurses, hospital interns, resident physicians, and hospital surgeries often involve extended work hours[60]

REGISTERED NURSES. The largest health care profession is registered nursing. Unlike factory shift work, the profession is dominated by women. Health care is a round-the-clock profession. Injured and sick patients in hospitals need 24 h care by highly qualified health care professionals. Although physicians make most of the health care decisions and take care of patient interviewing, surgery, and prescribing during hospital care, nurses are the ones who diligently carry out patient care from hour to hour.

Nurses have always been considered invaluable in hospital care, but in the scheduling of their training and work hours, their lives in a circadian sense have often been of little concern. Historically, a nurse's training and career have involved circadian disruption and the high emotional stress of working with patients for the duration of her professional life. Many anecdotes describe the workload, stress, and circadian malfunction of hospital nurses working on night shifts or extended duty, but few data are available about prevalence of shift work. Little is known about the consequences of shift work on a nurse's health, work performance, family and

social life, or quality of patient care. The available information indicates that lack of control of scheduling is an important factor in job dissatisfaction and job turnover. These data show that rotating or irregular shift work produces more digestive problems, more tension and stress, higher injury rates, and more disruptions in family and social life than regular shift work does. Almost nothing is known about the connection between shift-work hours and patient care.

Nursing schedules are notorious for their rapid rotation. Phase adjustment is often not practical, but light can play a helpful role. Two field studies have been published about nurses who worked at night in rapidly rotating shift systems. In the first, light boxes were placed in a room that was used by the nurses at the start and end of the night shift, as well as during rest breaks. In the second study, the nurses wore light visors throughout the shift. The phase-shifting effects in both studies were small.

HOSPITAL INTERNS, RESIDENT PHYSICIANS, AND SURGEONS. Graduate medical training is called internship or residency. Historically, this training has involved long, intensive hours for several years at night, on weekends, and on holidays. The work schedules of these physicians have been grueling. The policy of extended hours for resident training dates back to about 1900, when residents lived on the hospital grounds. The student doctors were on permanent call as a substitute for established physicians who might be tending their practices in their offices, carrying out surgery in the hospitals, or making visitation rounds.

Prior to about 1950, the standard procedure was day duty for 5 or 6 days a week and on-call duty for 5 nights after a day's work. On a busy night the resident worked continuously without sleep for 24 h or more. If no calls arose, even the best of hospitals simply provided a cot in a small dorm room for catching sleep until a call arrived. The most astounding part of the story is that these young interns were mostly in their mid-20s and often were married with young families, but they were not given a salary or compensation of any other kind.

Conditions have improved in 50 years, but only partially. Now most residents are paid a small living wage. As recently as 1990, however, the U.S. Congress stepped in to legislate work hours for hospital interns. The reasons for that need were compelling. Many interns in the largest U.S. hospitals were working 72 h extended shifts. Many were offered a substantial bonus to encourage them to accept a long shift at least once a month. The official reason was that patient care requires continuity. Little heed was paid to the effect of physician fatigue on patient care.

Fortunately, the work schedule of hospital interns has been shortened in the past decade as a result of federal legislation regulating work hours for interns and nurses in hospitals on the basis of circadian principles. Currently, residents' working hours are much more stringently controlled in most hospitals, with work assignments based on circadian principles. The new legislation reduces hours but does not completely protect the residents from a draconian workweek. A very important debate was held in June 2002 by the American Medical Association's Council on Medical Education in response to a groundswell of concern about sleep-deprived medical residents and related patient safety. The council's recommendations on resident work hours include the following:

- Total duty should not exceed 84 h per week, averaged over 2 weeks.

- Workdays that exceed 12 h should be defined as on-call, and scheduled on-call assignments should not exceed 24 h.

- Residents may remain on duty for up to 30 h to complete the transfer of patient care, but they may not be assigned new patients after 24 h.

The list continues for considerable length. Residents apparently still work very long hours.

EXTENDED WORK HOURS IN SURGERY. Work demands for residents, nurses, and physicians on surgery duty can be extremely great. As technology extends the possibilities for remedial surgery, the length of a surgical schedule can extend to 18 h or more (Figure 10.10). Transplant surgery has become relatively common in the past two decades. Successes include heart transplant, heart–lung transplant, kidney transplant, surgery on neonates, and even intrauterine surgery to correct heart defects. Clearly, the outcome is more promising if the lead surgeon and support team complete the procedure rather than rotating it to replacements. The continuity may become counterproductive, however, if a surgery session is not based on circadian principles.

Interstate trucking work schedules are very demanding[22,66]

The trucking industry has asserted for many years that long-haul trucking has an enormous impact on the

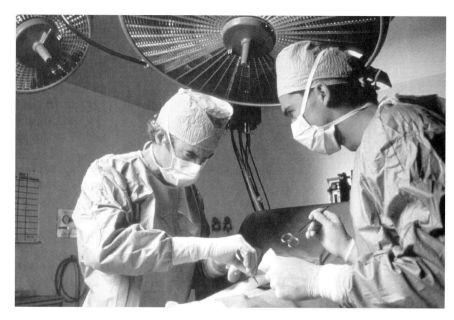

FIGURE 10.10 Surgery is the medical specialty with the most demanding hours.

American economy. Here are a few facts to support truckers' claims:

- Over 75% of U.S. communities depend solely on trucking for delivery of their goods and commodities.

- By 2008, the trucking industry will haul 9.3 billion tons of freight per year, or more than 64% of total U.S. tonnage.

- The industry involves 3.12 million truck drivers and over 500,000 trucking companies in the United States.

- A tractor-trailer truck may cost from $70,000 to $125,000, depending on the equipment.

- Over 2.3 million large trailer trucks are on U.S. highways at any one time.

- The trucking industry uses 43 billion gallons of diesel fuel and gasoline each year.

Interstate trucking routes, especially long transcontinental hauls, require several days of almost nonstop driving under a variety of conditions. The road task of a trucker traveling from a cheese factory in eastern New York State to Portland, Oregon, is continuously demanding. The geographic profile of the United States that the trucker must safely and speedily negotiate entails an eastern coastal plain; followed by the Appalachian Mountains, which present long, potentially dangerous grades for heavy vehicles; then a thousand miles of flat prairie; another very steep, hazardous mountain chain

FIGURE 10.11 Long-haul truckers face severe fatigue during extended work hours. The problems may be compounded by bad weather when crossing icy moutain passes.

in crossing the Rocky Mountains; a high plains area in Wyoming and Idaho; and then another very steep mountain chain in the Cascade Mountains and Coast Ranges of Oregon. The trucker travels at 70 to 80 miles per hour in 14 h days on the flattest and best-built interstate highways, making only hasty stops for gas, food, and a few hours sleep per night.

Minimum crossing time is 4 to 5 days to negotiate 3,100 miles in moderate weather, provided no highway construction delays or accidents occur. Much of the travel is night work. The travel demands are compounded by an order of magnitude in the event of ice and snow (Figure 10.11). In Colorado, for example, the snow alert signals that warn truckers to carry chains start flashing in early September and persist through May for one of the main crossing of the Rockies, via the Eisenhower Pass.

Several considerations with circadian overtones stand out. The cost of the vehicle is great enough that it falls into the category of requiring nearly continuous use to yield a profit. Economic considerations also dictate that more than one driver is usually not possible. The truck is very heavy, and as a missile in the hands of a fatigued driver, it has few equals for decimating passenger vehicles in a collision. Drivers formerly tended to drive extended hours, sometimes even driving continuously, nonstop, on a

coast-to-coast crossing. The accident rate rose to a level that caused a public outcry. Eventually pressure was placed on the U.S. Congress to pass legislation under the sanction of regulating interstate commerce.

Increasing regulation based on 24 h scheduling indicates growing awareness of circadian alertness and sleep–wake cycles relative to circadian body temperature cycles in humans. The interstate trucker has basically been categorized in new rules and regulations as a night-shift worker. For safe operation of a huge vehicle on the public highways, circadian rhythms must not be overlooked in attempts to keep U.S. highways from becoming dieways.

Power plant operator schedules require extreme diligence over long hours under potentially hazardous conditions[5]

In 1951 the first electric power from atomic sources was produced in an experimental power plant. The amount was sufficient to light four incandescent light bulbs. By 1953, nuclear power had been perfected enough to use in nuclear submarines. The achievement made possible extended tours of duty. Submarines could now prowl the seas for many months at a time without surfacing. Since 1955, commercial power plants have provided large-scale home and industrial electricity. Currently 103 commercial nuclear power plants operate at 64 sites in 31 states. The largest, Palo Verde Nuclear Generating Station in Arizona, generates more electricity per year than any other U.S. power plant of any kind, including natural gas, coal, oil, hydro, wind, or thermal plants. In 1999, this power plant generated over 32 million megawatt-hours of electricity.

Nuclear power plants are currently the second largest source of electricity in the United States, and they provide about 20% of total electricity each year. Vermont scores the highest percentage, with 67% of its total energy needs generated by nuclear power. South Carolina and Vermont both supply over half of their energy from nuclear plants. On a worldwide basis, nuclear power plants provided about 16% of total energy in the year 2000. The percentage of power coming from nuclear power plants in industrialized countries ranges from a high of 76% for France to a low of 36% for Switzerland.

Nuclear power plants were envisioned after 1951 as the ultimate energy answer for highly industrialized, fuel-hungry nations. They seemed to offer energy efficiency, freedom from smog, automated operation, and cost effectiveness. For example, nuclear power costs 1.76 cents per kilowatt hour compared to 1.79 cents for coal-burning plants, 5.28 cents for oil, and 5.69 cents for natural gas. However, these numbers are merely for routine operation and do not take into account construction costs or potential costs to the environment in the event of a nuclear disaster. Nuclear facilities are extremely expensive to build, which greatly increases the real cost per kilowatt hour. The production of nuclear energy does not release pollutants such as sulfur dioxide or carbon monoxide, but it does result in radioactive wastes such as spent fuel rods, the disposal of which is extremely problematic because of the long half-life of radioactive materials.

Within a decade after the first plant was built, some profound problems began to appear. Not the least of these was the immense power that these plants held within their concrete containment buildings and cooling towers, which needed constant monitoring by highly trained personnel 24 h a day. The control room of a nuclear power plant consists of a complicated panel board for each reactor containment building and its associated cooling system (Figure 10.12). Banks of warning lights, indicators, switches, and monitor screens fill the panel board and help operators detect problems and keep the plant running smoothly. Both 8 h and 12 h shifts have been used in different plants. Operators are often required to take two consecutive work shifts, or to work extended hours during outages. In March 1987, the Nuclear Regulatory Commission (NRC) received notice that several control room operators at the Peach Bottom atomic power plant had been found sleeping on the job. One week later, for the first time in U.S. nuclear power plant history, the NRC took the unprecedented step of shutting down the facility for 2 years to overhaul its management policies.

Personnel are required to be alert and responsive throughout their shifts, to watch the information on the control panels at all times, and to carry out prescribed checks and safety routines continuously. The routineness of the tasks and lack of physical exercise tend to induce sleepiness, particularly during the night shift. Studies of tasks that require high levels of alertness and vigilance have repeatedly found that human error increases after about half an hour on such a job. Not surprisingly, maintenance of attention span is one of the major objectives for control room operators. Part of the solution involves varying the tasks of the operators as much as possible. Experts recommend changing tasks every 2 h. Breaks in routine include walking to adjacent buildings to monitor cooling facilities, coffee breaks, or short exercise bouts away from the monitor screen.

The facts described here indicate the widespread use of nuclear power in the United States. A total of 103 commercial energy plants, plus 17 research and weapons fuel plants, exist today. All are licensed and tightly supervised by the federal Nuclear Regulatory Commission. Nevertheless, the degree of public concern about their safety is immense. The average age of U.S. plants is currently 19 years; their expected life span is 40 years, with an option to renew for another 20 years. When poorly managed, targeted by terrorists, rocked by earthquakes, or merely senescent, their potential for mass destruction is monumental. Mistakes or neglect by operators could cause incalculable damage to humanity, as the Chernobyl and Three Mile Island incidents have shown. Although nuclear power plant operators may seem at first glance to be merely another example of factory shift workers, their great responsibility to society and the margin for catastrophic accidents set this group of workers apart in a circadian context.

FIGURE 10.12 Control panel at a nuclear power plant. Vigilance must be maintained in spite of monotonous sedentary tasks.

Military operations offer another example of extended work hours under high stress[51]

Many military assignments pose duty schedules that lead to severe circadian disruption, sleep deficit, and other circadian stresses that decrease performance. Routine training operations often require extended hours and night work (Figure 10.13), but the worst scenario is actual combat assignment. Operation Desert Storm, for example, illustrated the advanced technology that made possible dramatic night attacks for many weeks. Prolonged exposure to long, highly stressful work hours, however, leads inevitably to sleepiness and fatigue, particularly at times of circadian body temperature low and efficiency low.

FIGURE 10.13 Military maneuvers for heavy-duty equipment operators.

The military has unique operations and work settings. Many military assignments require continuous manning, including medical, security, communication, ship navigation, transportation system operations, and command. In general, the military works in three modes:

1. In routine operations or for peacetime training, regular shift schedules are employed with adequate opportunities for rest between work assignments.

2. In the continuous operation mode, the workload is unceasing, but demands on the individual are low.

3. In the sustained operation mode required during simulated or actual combat, individuals must perform at peak levels often for 12 hours or more at a stretch under high psychological and physical stress, without adequate opportunities for rest. In these so-called surge conditions, performance degradation is often the limiting factor in the mission.

In addition to workload, translocation can involve jet-lag factors. Rapid deployment by plane to overseas assignments may induce jet lag of 5 to 12 h, depending on the locale. Acute short-term dysfunction may result from the dislocation, and long-term chronic circadian disruption may be caused by the shifting work schedules.

Covering more than the briefest details of military aspects of circadian dysfunction is outside the scope of this textbook. Only a few of the interesting aspects of extended manning in the four U.S. military services—army, air force, navy, and marines—will be considered.

In the army, no special guidelines exist for work schedules except for pilots. Individual commanders must regulate assignments to ensure that their units are rested and fit for duty. Regular deployments of units are usually scheduled far in advance, but they may include transmeridian air transport with concomitant jet lag. Surge operations, such as a show of force to display policy in a place of tension, are particularly demanding. The combined elements of jet lag, rapid deployment, and immediate duty on prolonged schedules may introduce much circadian disruption. Another category, special operations, covers a variety of classified, inherently irregular assignments. The special forces personnel may have very unpredictable requirements to respond immediately to a large range of tasks from counterterrorism to search and rescue. Circadian disruption can be extreme.

The air force workload centers around tactical forces for manning fighter planes, bombers, and helicopters, and a second set of transport forces for deploying tanker and transport planes. Unlike most army operations, air force personnel are regulated by several assignment codes. These restrictions cover maximum flight duty periods, maximum monthly and quarterly flying hours for air crews, and minimum crew rest periods. One special workforce is termed Strategic Forces. The job of this elite group is to deter nuclear attack and maintain the missile shield. Strategic Forces crews operate on a carefully designed schedule intended to maintain peak performance.

Naval nonshipboard operations differ little from army or nonflight air force assignments in their scheduling. Workloads are intended to provide a 40 h work-week with minimal round-the-clock watches. On shipboard, however, many tasks must be performed continuously. The guiding principle on shipboard is that the ship must be self-sufficient at all times when under way, from the points of view of housekeeping, maintenance, and the military mission of the ship.

General work such as maintenance and repair is usually carried out in 12 h shifts, but assignments on a specific piece of equipment or at a station are carried out on a watch schedule. The watch schedules rotate members of a watch team with frequent changing of the work hours, thus generating great circadian disruption. The result is often sleep dysfunction, fatigue, and degraded performance.

As a ship goes into higher alert status toward combat status, it progresses through several designated stages. Condition I battle readiness is equivalent to sustained operations. Maintenance tasks are discontinued, all combat systems are manned, the crew has no rest period, and the crew is expected to be able to endure for

24 continuous hours. Condition II battle readiness allows 4 to 6 h of rest per person per day over an extended period of days.

Naval submariners have some of the most challenging schedules of any service group. Essential to a submarine's mission is to cruise underwater undetected for long periods of time in a high state of vigilance. Submarines in one category carry ballistic missiles with an assignment comparable to the air force Strategic Forces unit. They cruise in critical need areas ready to deploy missiles on command. Attack submarines carry out such tasks as defense of surface ships, surveillance, and attacks on enemy ships. Almost all submarines use 18 h watches: 6 h on duty, 12 h off for sleep, meals, and relaxation. Average sleep time is 4 h per 18 h watch. The mariner's watch is the circadian equivalent of a daily swing shift. As a result, submarine crew are almost continuously sleep deprived.

In summary, military operations by their very nature demand that most personnel engage in work at all hours, often on an extended duty schedule, particularly in wartime or on high alert status. The disruption of circadian rhythms can be severe, and the stress of high alert or combat modes of operation can result in marked performance dysfunction.

Space flight poses unique scheduling and performance demands[19,39]

Unparalleled conditions accompany the astronauts who carry out space missions. Their task is in many ways at the outer edge of human performance capacity. By comparison with U.S. military alert stages, U.S. astronauts are on continuous-duty, maximum alert/combat status for extended periods of time. The mission is to package the highest-caliber engineer/scientist staff after years of intensive training into a minuscule capsule on top of a highly explosive depot of fuel, then blast the crew into orbit around Earth for periods of days to weeks (Figure 10.14).

In outer space the crew is expected to carry out complex tasks in an extraordinarily hostile atmosphere. After defying weightlessness for the entire period, solving all manner of crises that may have arisen, and completing their assignments, they must navigate their craft back into Earth's stratosphere and safely land either in the ocean or on the salt flats of Nevada.

Although shift work is involved, the conditions are almost beyond the imagination of the average Earth-bound citizen. A brief verbal snapshot will give only an inkling of life on board. Space shuttle flights are relatively short in duration. The crew usually ranges from three to five. Once the high tension of liftoff is over, the spacecraft settles into its orbital circuits, and tasks fall

FIGURE 10.14 Launch of the space shuttle *Discovery*. Highly skilled shift work is required of astronauts.

into a daily work routine (Figure 10.15). However, the celestial rhythm is anything but routine. The view of Earth and other heavenly bodies is magnificent. The sun rises every 90 minutes during the mission.

Work proceeds according to plan during daytime, and at day's end an 8 h sleep period is scheduled. A few differences from sleeping on Earth must be considered, however. Because of the lack of gravity, weightlessness prevails and the crew members literally float around freely. Crew members must attach themselves to a bunk bed, seat, or wall to avoid drifting around or bumping into equipment or each other while they sleep. The shuttle has four narrow bunk shelves. If more than four crew members are aboard, they can sleep in the commander's seat, the pilot's seat, or crew seats, or taped to a wall. Usually the crew sleep in standard sleeping bags.

The International Space Station consists of modules from the various participating nations; therefore it may vary considerably in length and configuration from one expedition to the next. During Expedition Two, the space station consisted of the huge Destiny laboratory at one end, linked successively to the Unity node, the Zarya control module, and the Zvezda service module, which was docked to the Soyuz module. The length of the Station during that mission was 52 m.

FIGURE 10.15 Shift work in space entails huge challenges for astronauts.

Unpleasant digestive feelings may include loss of appetite and indigestion with the sensation of bloating after food intake. Cognitive distress may involve decrease of ability to concentrate and loss of motivation, increases in irritability, and occurrence of headaches. The time zone–induced symptoms are often compounded by long waits in airports between connections without an adequate place to rest. As with night work, sleep disturbances are a prime source of complaints for transmeridian travelers and flight crews alike.

CIRCADIAN BASIS OF JET LAG. Jet lag is the result of a temporary loss of synchrony between an abruptly shifted sleep–wake cycle and the local time. The endogenous and exogenous components are mismatched. Until the internal clock is gradually reentrained and the phases of the body temperature and sleep–wake cycle are locked on to local time, the discomfort of jet lag persists. The unadjusted rhythm is phase-delayed relative to local time after an eastward flight, and phase-advanced after a flight to the west (Figure 10.16). Advance adjustment of the eastward flight is slower but essentially complete by day 8. The phenomenon does not occur after long-haul flights in a north–south direction in which no time-zone changes are experienced.

After a westward flight across eight time zones from Los Angeles to Hong Kong, for example, a passenger feels tired at 4:00 P.M. (16:00) local time, which is equivalent to midnight (24:00) in the time zone just left, the time zone to which the internal clock is still adjusted. The traveler begins to wake up at midnight local time because this corresponds to 8:00 A.M. (08:00) subjective internal time. In the same way, after an eastward flight across eight time zones from Los Angeles to London, the individual does not feel tired at midnight by local time. Instead, he is ready to sleep as the new day dawns at 8:00 A.M. For most people, the symptoms last longer after a flight to the east than one to the west, partly because the internal clock, with an intrinsic period greater than 24 h, naturally tends to phase-delay rather than phase-advance.

Some individuals suffer from jet lag more than others, for reasons that are unclear. Supposedly the young, the fit, and the flexible sleepers suffer less, but the supporting evidence is weak. Morning types, individuals who awaken energized and who typically function well at the start of the day, will have less trouble with eastward flights than evening types because an advance of

The space station has two small crew cabins, each large enough for one person. Remaining crew stake out sleeping positions in other parts of the craft, such as the Destiny lab or the Zvezda service module. Motion sickness, weightlessness, and general excitement over being in space may disrupt circadian sleep–wake patterns. Sleeping in the shuttle cabin has especially severe effects because the rising sun every 90 minutes shines through the cockpit window. The added warmth and light may be disruptive unless an astronaut wears a sleep mask. Equally damaging to consolidated sleep are the crowded conditions of many crew members sleeping in close quarters. As a result, astronauts may become sleep deprived during the mission.

The National Aeronautics and Space Administration (NASA) has employed timed exposure to bright light to help both space shuttle astronauts and ground crew adjust to night shifts. This bright light phase-shifting procedure has been so successful that it has become a permanent part of the space shuttle program.

Jet lag varies directly with the number of time zones crossed[2,4,17,58]

SYMPTOMS OF JET LAG. Following transmeridian travel, many air passengers and crew suffer from jet lag for several days. The set of symptoms is similar to night worker's malaise. The exact symptoms of jet lag, as well as their severity and duration, depend on the individual, the direction of flight, and the number of time zones crossed. The malaise generally increases with the number of time zones crossed in excess of two or three.

Symptoms may include fatigue during the new daytime but inability to sleep satisfactorily at nighttime.

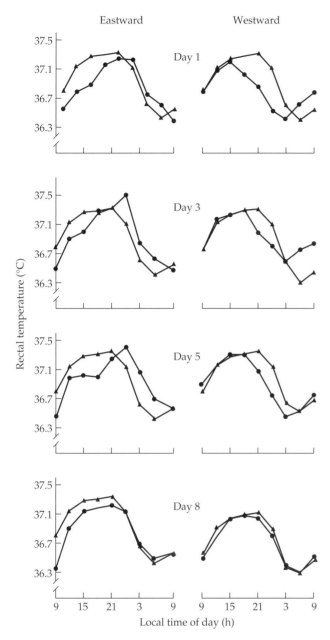

FIGURE 10.16 Adjustment of the body temperature rhythm to an eastward or westward flight across six time zones, shown for days 1, 3, 5, and 8 after flight and plotted on the new local time. Lines with triangles represent the control rhythm after final adjustment to the time zone; lines with circles represent observed temperature rhythms on the days indicated. (From Klein et al., 1972.)

problem is different. Often their duty period requires them to return home soon after the outward flight. In this case they should try to retain a sleep–wake cycle that is as appropriate for their home time zone as their duty periods will allow. In addition, long-haul flights often take place during the crew's subjective night, and opportunities for naps can be shared by the air crew members during the course of the flight. In very long-haul flights exceeding 8 h, cabin crews generally alternate duty assignments and take opportunities to sleep in crew bunks. The result of these complexities is a difficulty in distinguishing jet-lag problems from night-work problems.

TREATMENT OF JET LAG. Effects of time-zone transitions may persist for many days. Even with remedial measures, adjustment is not immediate but takes up to about 10 days after an eastward transition of 10 time zones, and slightly less after a similar transition to the west. The symptoms of jet lag fade as the internal clock adjusts to the new time zone through entrainment. For short stays in the new time zone before returning home, full adjustment is not possible, and the traveler is advised to time appointments and lifestyle to coincide as close as possible with daytime on home time.

For longer stays in the new site, specific remedial measures are recommended. In line with circadian principles of photic phase shifting, scheduled exposure to bright light should alleviate jet-lag discomfort by accelerating reentrainment of the internal clock to the new time zones. Portable light sources, in the form of battery-operated visors, are available, which allow a traveler to provide the required light, even during the flight. A pair of strong sunglasses is also important to avoid light at appropriate times. However, few detailed reports of the effectiveness of the light/sunglasses regime have been published.

The old saying, "When in Rome, do as the Romans do," has merit after a westward flight because light exposure in the new time zone will promote a phase delay. In contrast, after a flight to the east the advice is unsound. Eastbound travelers should avoid light by staying indoors in the morning and seek light only in the afternoon in the new time zone. For further details of jet-lag treatment, see the section on chronopharmacology later in this chapter.

Circadian Physiology Is Important in Human Health and Disease[36,38]

Many human health parameters are circadian in nature[61]

Virtually every bodily function has been shown to exhibit a circadian rhythm with remarkable precision and stability in healthy individuals. At a physiological

the internal clock is required. The opposite should occur with westward flights, where a phase delay is required, but the evidence again is weak.

Some evidence indicates that jet lag is less if the time interval between the last full sleep in the time zone just left and the first full sleep in the new time zone is made as short as possible. Coordinating such timing requires careful planning of travel arrangements. For the air crew, the level, body temperature, heart rate, and blood pressure are generally highest in the afternoon, while hearing and pain sensitivity are more acute in the evening. Some hormones, such as cortisol and testosterone, are highest in the morning when a person arises; others, such as gastrin, insulin, and renin, are highest in the afternoon or early evening, while melatonin, prolactin, and growth hormone reach their peak during sleep. Thus circadian rhythms are part of normal physiology.

The onset of several acute medical conditions is correlated with circadian phase[6,15,43,71]

Some illnesses show considerable daily variation, either in onset or in the severity of symptoms. Cerebral strokes and heart attacks occur more frequently between 6:00 A.M. and noon than in any other 6 h period, and the incidence of asthma attacks is greatest at night. People with rheumatoid arthritis tend to experience their worst symptoms on waking in the morning. The increased incidence of cardiovascular attacks and death in the morning can be explained in terms of normal circadian physiology (Figure 10.17). In early morning, cardiac activity rises after awakening. Sympathetic tone to the vasculature increases, and systemic blood pressure rises. All of these changes place greater demands on the heart. Blood platelets also stick together more, and the mechanism for breaking down clots is less active. In suscep-

tible individuals, these combined factors may tilt the balance toward a cardiovascular crisis.

Similarly, the evening peak in asthmatic symptoms coincides with falling levels of both adrenaline and cortisol. Adrenaline is a bronchodilator and cortisol is a natural immunosuppressant hormone. Airways constrict naturally in the evening, but asthmatics have hypersensitive airways. The evening constriction is exaggerated and can lead to the discomfort of breathlessness, or even to fatal asthmatic attacks. With rheumatoid arthritis, the low evening plasma levels of cortisol lead to increased autoimmune responses in the joints after a delay of some hours. In the morning, therefore, the joints are swollen, stiff, and painful.

A knowledge of chronobiology helps in making some medical diagnoses and in timing the treatment of some illnesses[31]

DIAGNOSIS. A simple example illustrating the value of choosing the correct time for diagnosis is provided by the timing of blood samples. Cortisol and growth hormone show large-amplitude circadian rhythms (see Figure 10.2). As a result, judging the normality of either hormone depends on the time of measurement. Taking the time of sampling into consideration can prevent misinterpretation of the data.

Another example of circadian diagnosis comes from HIV infection. The alteration of normal daily cycling of

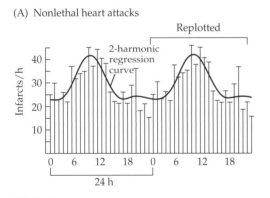

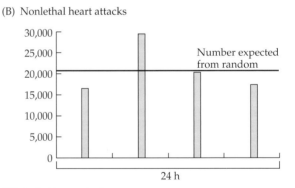

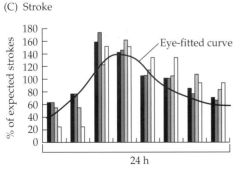

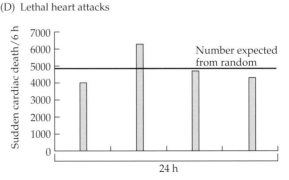

FIGURE 10.17 Circadian occurrence of cardiovascular dysfunctions. (A) Nonlethal heart attacks (myocardial infarcts). (B) Statistical analysis of 83,929 patients. (C) Stroke. (D) Sudden cardiac death for 19,390 patients. Horizontal dark lines in B–D indicate the number expected to occur randomly. (A from Muller, 1994; B and D from Cohen et al., 1997; C from Elliott, 2001.)

blood constituents is one of the earliest detectable signs of disease in HIV-infected patients. In healthy individuals, circulating blood cells such as CD4+ lymphocytes display a robust circadian rhythm with the highest level at night and the lowest level in the morning. In HIV-positive asymptomatic patients, the CD4+ lymphocyte circadian cycle flattens and the circadian phase angle between circulating lymphocyte and plasma cortisol level is altered significantly. Serum neopterin, a marker of T-cell activity, normally displays a stable circadian rhythm with the highest level during the late night and early morning, but some HIV-infected patients have higher levels during the day than at night, indicating a disease-related alteration of circadian rhythm in this variable. All of these circadian disorders occur in the very early stages of HIV infection before other methods of detection are available.

TIMING TREATMENT.   Certain drugs are more effective at a particular circadian phase. The standard treatment for patients with hypertension or rheumatoid arthritis was formerly a single morning dose of a beta-blocker such as atenolol for hypertension and nonsteroidal immunosuppressive drugs for the latter. Atenolol is effective in the daytime because it significantly reduces heart rate, systolic blood pressure, and diastolic blood pressure. However, the effect of the drug wears off during the night. As a result, the early-morning rises in systolic and diastolic blood pressure become even more marked than without treatment (Figure 10.18). Recently clinicians have started to prescribe these therapeutic treatments in divided doses, taken morning and evening. In addition, slow-release formulations, calcium channel blockers, and ACE (angiotensin-converting enzyme) inhibitors have been developed to provide effective 24 h treatment.

A different problem is encountered in hormone replacement therapy for patients with pituitary malfunctions that lead to delayed puberty or stunted growth. If hormone levels are maintained at a constant high level by frequent regular injections, the target organs become refractory to the effects of treatment. Duplicating the normal rhythm of hormone concentration in the plasma with one dose per day at bedtime overcomes the problem. The evening dose maintains high overnight values and low daytime values. Treatment is also cheaper because less hormone needs to be given.

USE OF LD SCHEDULES FOR INTENSIVE CARE PATIENTS AND THE STAFF WHO TEND THEM.   A limited amount of data suggests that the circadian system functions poorly in intensive care patients. Circadian rhythms of these patients are absent, low in amplitude, or irregularly phased. One possible explanation is that sedation affects the circadian clock or its output signals. Another possibility also has chronobiological overtones. The patients are usually kept in an almost constant environmental setting. Usually the facilities have no windows, and

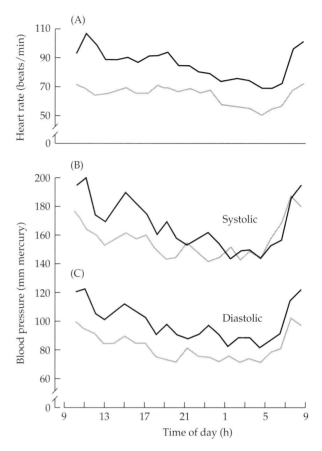

FIGURE 10.18 Responses to a single morning dose of atenolol. (A) Heart rate. (B) Systolic blood pressure. (C) Diastolic blood pressure. Black lines represent rates before treatment; gray lines, rates after treatment. (From Raftery et al., 1981.)

because of the constant care, the rooms are always brightly lighted and busy.

The benefit of emphasizing normal day–night changes in the intensive care unit has not been systematically explored. Because improvement of patients is often correlated with increasing circadian rhythmicity, a study seems warranted. Benefits could extend not only to the patients, but also to nursing staff, who often complain of a loss of the sense of time in the constant environment.

Sleep and Mood Disorders Belong in the Realm of Chronopathology[3,7,12,14,29,46,57,65,70,72,74,75]

Phase disorders of the sleep–wake cycle include several different syndromes[7,24,35,64,65]

ARRHYTHMICITY OF THE SLEEP–WAKE CYCLE.   Some patients have irregular or fully arrhythmic sleep–wake patterns (Figure 10.19). Several different causes are possible. Generally tumors in the SCN region are responsible. Alternatively, disturbances in sleep regulation may be at fault, such as the total lack of sleep that character-

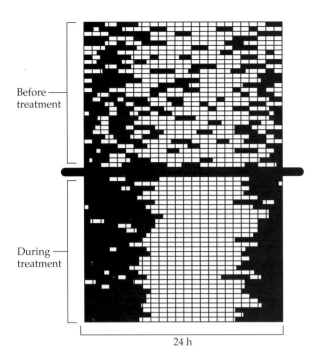

Before treatment

During treatment

24 h

FIGURE 10.19 Examples of the effect of melatonin therapy on the sleep times in young patients with irregular sleep–wake cycles. (From Jan et al., 1994.)

izes fatal familial insomnia. The arrhythmicity is very disruptive to family life and to successful employment. Melatonin administration has been used to entrain patients to a normal day.

FREE-RUNNING SLEEP–WAKE RHYTHM. Non-24 h sleep–wake syndrome occurs when a sighted subject cannot entrain to the external day–night cycle, in spite of attempts to do so. The data appear at first to be free-running, but closer inspection indicates relative coordination with variation of the period. When sleep falls at the subjectively expected time, the observed circadian period length is closer to 24 h than when sleep is in the middle of the day. The variable period length indicates interaction between the environment and the sleep–wake cycle. The resultant partial entrainment may be due to social entraining agents.

ADVANCED SLEEP PHASE SYNDROME. Advanced sleep phase syndrome (ASPS) is characterized by early sleep onset and offset. The syndrome is rare, having a frequency in the population of less than one in 10,000. The main interest in the syndrome comes from its genetic basis, and the insight into sleep–wake regulation that it provides. The trait is conferred by a single autosomal dominant gene that resides on the end of the short arm of chromosome 2.

Information arising from wholesale genomic sequencing of the human genome had suggested that a homolog of the *Drosophila per* gene, *hPer2* (see Chapter 7), resided in that region of the genome. An educated guess was made on the basis of the association of circadian clock dysfunction and sleep disorders. Further investigation led to the association of the ASPS allele and the *hPer2* gene. The final work was technically demanding but straightforward. Genomic DNA was isolated from the affected individuals, and the ASPS allele was shown to be the result of a serine-to-glycine mutation in the hPER2 protein. The mutation lies within the part of the protein bound by casein kinase 1ε, a kinase that normally phosphorylates hPER2 and regulates its degradation. The misregulation of hPER2 results in disruption of the circadian clock and in turn an advance in the sleep–wake cycle.

This relationship of a single circadian gene to a specific behavioral output is the clearest example of such a connection in the human chronobiological literature. Treatment of the syndrome is simple. Late-evening light exposure from 9:00 to 11:00 P.M. is effective in delaying the circadian system sufficiently to entrain sleep.

DELAYED SLEEP PHASE SYNDROME. Patients with delayed sleep phase syndrome (DSPS) cannot get to sleep before 2:00 or 3:00 A.M., and they have difficulty waking before late morning. The temperature, melatonin, and sleep rhythms of DSPS subjects are very similar to normal subjects, but the phase is delayed (Figure 10.20A,B). The body temperature trough for 64 subjects ranged from only slightly phase delayed to almost 12 h phase delayed (Figure 10.20C). Obviously the disturbances of a chronic case of delayed sleep phase syndrome make life very difficult for a person trying to fit productively into society.

In one interesting case, a student suffered from severe delayed sleep phase syndrome. Getting up at 6:00 A.M. on school days severely curtailed the individual's sleep, but sleep loss was made up partially on weekends, when sleep lasted from about 4:00 A.M. to 3:00 P.M. During vacation, the late hours of sleep were not a problem. Therapy consisted of delaying sleep times until sleep occurred at the desired time, from 9:00 P.M. to 6:00 A.M. Thereafter it was possible to maintain the new phasing of the sleep–wake cycle.

The sleep-shifting protocol is not effective for all patients. Furthermore, the use of light to shift the rhythm may actually be deleterious for patients suffering from DSPS. Once they initiate phase delaying as a treatment, they may not be able to reentrain at all. Although light treatment is simple in concept, case management is often complicated. Many of these individuals are evening types, or so-called owls, and are

(A)
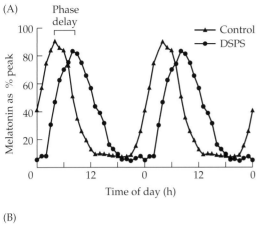

(B)
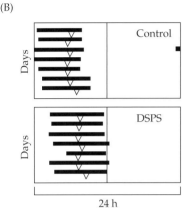

(C)
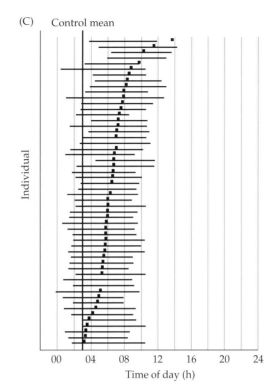

FIGURE 10.20 Delayed sleep phase syndrome. (A) Doubleplot of the melatonin rhythm of normal subjects and DSPS patients, shown as a percentage of melatonin peak. (B) Actograph of the sleep time and body temperature trough ("V") of an intact control subject (above) and a DSPS subject (below) for 1 week. (C) Summary of phase of the sleep period (bars) and urinary sulfatoxymelatonin peaks (squares), averaged over 7 days for 54 DSPS subjects. The mean phase of urinary sulfatoxymelatonin in normal subjects is indicated by the vertical line. (A from Shibui et al., 1999; B from Uchiyama et al., 2000; C from Cole et al., 2002.)

reluctant to shift earlier. Most patients with DSPS subsequently fail to comply with the recommended light schedule and allow themselves to relapse. Some are able to maintain a normalized sleep phase without maintenance for periods of up to several months; others drift back toward the delayed sleep phase within days. The use of melatonin for treatment of this syndrome is still controversial, in spite of some reports of successful treatment with melatonin timed to advance the phase.

IRREGULAR SLEEP–WAKE PATTERNS CHARACTERISTIC OF STUDENTS. The responsibilities of a full-time job and a family force most adults to conform to a fairly regular sleep–wake pattern. For 80% of the full-time employed workers in the United States, the schedule is a routine 8 h of work for 5 days a week, about 48 weeks of the year. After several years, the time scheduling of daily life for most of these individuals becomes very predictable and routine. The 9-to-5 grind is another name for such time restriction. Even for the other 20% of adults who work

full time on night work, a certain predictability in scheduling prevails.

Quite the opposite is the scheduling for most students in colleges, universities, and technical schools, and to some extent in advanced professional programs, such as medical school, law school, and graduate school. The few who work to help support themselves may fit in the category of homeowner and provider for a family. Most, however, are supported by parents and are unmarried. Personal responsibilities are accordingly directed toward obtaining a degree with creditable grades. The timetable is directed by scheduling of classes in an irregular distribution of lectures, labs, and recitations dictated by the registration schedule.

Some studies of the sleep–wake cycles of students have shown considerable daily variation. Times of going to bed may vary over a range of 4 h. Anecdotal evidence is available that some students go to bed and rise progressively later until they miss an entire night's sleep. The scheduling, however, indicates social and work considerations rather than malfunction of the circadian system. In the absence of meetings and other commitments, and with the opportunity to plan their own habits, students have an opportunity to ignore conventional synchronizers.

Duration disorders in the sleep–wake cycle include several loosely related syndromes[18,52,58,62]

A PRIMER ON AFFECTIVE DISORDER. Mood disorders, or so-called affective disorders, have long been considered a consequence of circadian pathology. The best that can be said, however, is that the timing and severity of behavioral changes that occur in these disorders are strongly modulated by circadian and seasonal rhythms. These apparent links have led researchers to investigate the behavioral mechanisms that regulate circadian and seasonal rhythms in afflicted patients.

One of the most important first steps was the identification of a seasonal form of mood disorder, called seasonal affective disorder (SAD). A novel treatment, based on circadian principles, employs light to improve the condition. Unfortunately, few studies of circadian rhythms in patients with mood disorders have been conducted in a constant routine. Consequently, many uncertainties in the interpretation of mood disorder results remain.

Affective disorders are characterized by periods of depression and/or periods of excitation called mania. Patients typically experience several episodes of depression during their lives. These episodes vary in length. They usually last several months and then terminate spontaneously. In endogenous depression, patients lose appetite and weight, and they sleep less. In so-called atypical depression, patients oversleep, overeat, crave carbohydrates, and gain weight. The pattern of recurrences can be highly irregular. However, in approximately 30% of cases, symptoms recur with some regularity. In half of those cases, symptoms return on a seasonal basis.

Many of the changes that occur in depression and mania very superficially resemble changes in behavior and physiology that occur in connection with seasonal rhythms in animals. Depression has been likened by many clinicians to a state of hibernation. However, depression is an inherited, long-lasting medical illness in sentient, self-aware human beings. In these respects, depression has few true parallels with animal hibernation. Human depression is characterized by numerous changes in behavior and physiology that go far beyond sadness as a normal human emotional response. In fact, some depressed patients are not even sad, and others experience sadness only secondarily, as an emotional reaction to the impairment they feel during a depressive episode.

When depressed, a patient feels a loss of energy, a loss of initiative and drive, and a pervasive loss of interest in surroundings. The capacity to enjoy normally pleasurable activities is absent, and interest in sex disappears. The ability for mental work is reduced. Thoughts come slowly, and concentration is difficult. Any sparks of creativity are lacking. Because of these deficiencies, patients suffering from depression find difficulty in meeting the demands of work. They are preoccupied with imagined illnesses. This pessimism, loss of interest, and inward turning causes patients to withdraw from their social and natural environment. They avoid company, take leave from work, and occasionally become bedridden. Patients may experience a feeling of failure and a lessening of self-worth, and they may feel guilty about failing to meet obligations to family, colleagues, and friends. In these cases, guilt and low self-esteem appear to be secondary, psychological complications of illness.

In many respects, mania is the opposite of depression. Patients exhibiting mania are in an excited state with increased energy and drive. Their thoughts race, and they are full of ideas. The manic patient is impulsive, spends money recklessly, exercises poor judgment, and is talkative and intrusive. Patients exhibiting mania are unrealistically optimistic with an inflated self-opinion. This behavior eventually alienates most family, friends, and colleagues. The patient goes forth into the social and physical environment with enthusiasm and seeks out experiences and activities. The need for sleep is decreased, often markedly so. Mild mania can be an extraordinarily productive and creative state. Unfortunately, mania usually escalates out of control and can destroy relationships and lead to financial ruin.

THE LINK BETWEEN CIRCADIAN OR SEASONAL RHYTHMS AND AFFECTIVE DISORDERS. Diurnal variation in symptom severity is one of the characteristic symptoms of depression. Patients feel worst early in the day, improve during the course of the day, and by evening feel almost well. The following day, the cycle repeats itself. Diurnal variation in severity of depressive symptoms appears to be a true circadian rhythm that persists during a 40 h constant routine protocol. An endogenous rhythm of mood, driven by the circadian pacemaker, is even documented in control subjects. The circadian process that modulates depressive symptoms does not appear to be specific to depression, however. Diurnal mood variation has been detected in other psychiatric illnesses and in healthy human subjects.

Depression and mania are also subject to seasonal influences. Although episodes of these disorders usually recur at irregular intervals, they are more likely to begin in the spring and in the fall than at other times of year. Symptoms of winter depression characteristically include oversleeping, overeating, and carbohydrate craving. Symptoms of summer depression tend to be opposite: insomnia and loss of appetite and weight. As with circadian variation in mood, the processes responsible for increased sleep and appetite in winter do not seem to be specific to depression. Similar seasonal patterns have been described for mood, sleep, appetite, carbohydrate intake, and weight in healthy human subjects.

PHASE CHANGES DURING DEPRESSION. The most systematic investigations of circadian rhythm phase disturbances in depression have been carried out in patients suffering from seasonal affective disorder (SAD). The syndrome may be a consequence of a phase-delayed circadian system, although other circadian hypotheses relate SAD abnormalities to a disturbed photoperiodic response or to a diminished circadian amplitude. Various studies of the phase of the melatonin rhythm in SAD have used either dim-light melatonin onset (see Chapter 9) or the entire rhythm measured during a constant routine as a phase marker. Most patients show a small phase delay.

A second important circadian marker, the core temperature rhythm, has been measured in a constant routine for SAD patients and was also found to be slightly delayed. An abnormal circadian phase in winter may trigger the depressive phase in SAD. Remedial treatment is discussed in the section on chronopharmacology a little later in this chapter.

CHANGED AMPLITUDE IN DEPRESSION. In major depression, amplitudes of daily rhythms have frequently been found to be reduced. For example, the nightly decline in core temperature is diminished and the nocturnal rise in melatonin is often blunted. An amplitude reduction was seen only when patients were studied in more or less naturalistic conditions and allowed to sleep and engage in other routine activities, but not in patients in constant routine conditions. Apparently the abnormalities of rhythm amplitude in depression arise from exogenous masking processes connected with the sleep–wake cycle.

Sleep disorders may be severe in those who are blind[55,58]

Legal blindness is defined as acuity for visual image formation of not greater than 20/200 in the better eye with best correction, or a visual field of less than 20 degrees. A person with normal visual acuity can see an object clearly at 200 feet; a person who is legally blind must be 20 feet or closer to see the same object. In a circadian sense, however, the issue is more complex and requires a clear understanding of the structure of the eye and its photic and nonphotic functions (see Chapter 6).

In the better-known imaging function of the human eye, resolution is of highest importance. Definitions of legal blindness refer strictly to visual acuity. The photic input to the SCN for circadian entrainment uses highly specialized retinal ganglion cells for luminance detection (see Chapter 6). Therefore, even a legally blind individual may detect sufficient light to entrain. Sleep disturbance in the blind is usually associated with complete lack of light perception and, specifically, the transport of light information to the SCN via the retinohypothalamic tract. In a recent epidemiological study, only 60% of registered blind subjects in the United Kingdom appeared to suffer from sleep disorder, and a greater prevalence was exhibited among individuals with no conscious light perception at all. Similar statistics have been found in smaller studies in other countries. Presumably those able to entrain have some circadian photoreception.

Sleep disorders in people who are blind usually involve free-running rhythms. Melatonin therapy is often effective in stabilizing the free-running rhythm. In one instance, the circadian period length before treatment was about 24.3 h (Figure 10.21A), was stabilized during treatment with melatonin (Figure 10.21B, days 39–57), and reverted to free-running when treatment stopped (Figure 10.21B, days 57 and following).

Sleep in the elderly is often disrupted[38]

Insomnia is a major complaint among the elderly. The syndrome is characterized not by difficulty in initiating sleep, but rather by problems of staying asleep, particularly in the second half of the night. One of the most consistent observations in elderly individuals is a decline in the amplitude of the melatonin rhythm. The eyes begin to lose the capacity to transmit light with the aging process, and the entrainment signal weakens. In circadian terms, this means an inadequate coupling to the external light–dark cycle. In some cases the primary defect may also be at the SCN level, as suggested by postmortem analysis in older humans that has revealed decreased cell numbers and lowered vasopressin concentrations in the SCN. The result is an almost constant feeling of fatigue throughout the day, and a decrement in performance.

As a result of the ubiquity of sleep disturbances in the elderly, an inordinate proportion of sleeping pill prescriptions goes to people over 65 years of age. Because of the substantial problems associated with the use of hypnotics in the older population, however, efforts have been made to develop treatments not involving drugs for age-related sleep disturbance. Instead, timed exposure to bright light and replacement melatonin admin-

(A) Before treatment

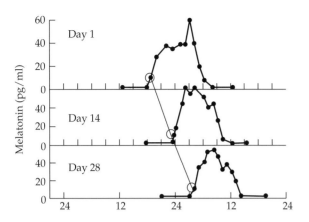

(B) During and after treatment

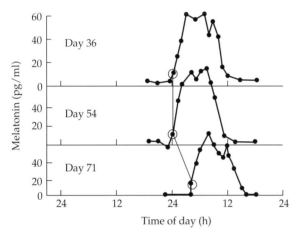

Time of day (h)

FIGURE 10.21 Melatonin therapy for subjects who are blind. Melatonin profiles of a single subject taken at about 2-week intervals. Dim-light melatonin profiles are shown by circles. (A) Before treatment. (B) During and after treatment. For further explanation, see the text. (From Sack et al., 2000.)

istration are being recommended. For example, in one study exposure to bright light in the evening for 12 days reduced nocturnal wakefulness by 1 hour per night in sleep-disturbed older subjects.

The extreme photic environments of the arctic far north and Antarctica may cause circadian problems, particularly for nonnatives[25,53]

NATIVE GROUPS OF THE ARCTIC. For over 10,000 years Eskimos, Aleuts, and other native peoples have adapted to life in one of the coldest and harshest regions of the world, the Arctic. Summers are short and cool; winters are extremely severe. The average temperature in most parts of the region rises above freezing for only 2

or 3 months of the year. During the coldest months, from December to March, temperatures fall to –20 or –30°F. Snow covers the ground from September until June, and winds howl across the treeless, open tundra. From a circadian point of view, the most unique feature is the continuous darkness of winter and constant light of summer (see Figure 3.19 in Chapter 3). With the coming of European whaling ships in the 1800s, the way of life for Eskimos changed greatly, and most native arctic peoples today live in houses with electric appliances much like their neighbors to the south.

The fascinating story of the original arctic lifestyle, different from that of any other people on Earth, has many important circadian implications. A thumbnail sketch will provide the backdrop for the circadian details. No other group developed the unique food, clothing, homes, transportation, hunting methods, language, or social structure that made possible survival in the harsh arctic climate (Figure 10.22).

Every aspect of life revolved around survival. Individual cultural groups were geographically determined, marked by great isolation. Eskimos had no laws, but rather rules of conduct that required everyone to help in the struggle to survive, and to live peacefully with others in the group. Food consisted of the local available meat, chiefly seal and caribou, but also fish, birds, whales, and plant products in summer. Clothing was made of animal skins, often the light, warm skin of

FIGURE 10.22 The life of a traditional Eskimo during an arctic winter is harsh.

caribou. Transportation was by dogsled in winter, and by foot or by boat in summer. Shelter in summer was normally a tent; winter homes were more permanent. In most regions, winter sod houses were countersunk into the ground and topped with sod. In more northerly regions, igloos or snow houses were constructed of ice blocks cleverly stacked into comfortable dome-shaped shelters.

Although Eskimos stayed in very small shelters for many months at a time during the winter darkness, no indication of seasonal affective disorder is apparent in the records of Caucasian visitors. Instead, the diaries of arctic explorers emphasize the cheerfulness and good humor of Eskimos in midwinter. Apparently Eskimos loved to have fun, especially in winter, when storms and darkness kept them inside. They sang, told stories, played games, enjoyed their children, and danced to the throbbing beat of native drums. They exercised with games of wrestling among the men. The happy, balanced, stable life of Eskimos before the arrival of white men gives impressive testimony to purely natural, non-pharmacological maintenance of sleep patterns and mood in an extreme environment with prolonged periods of continuous darkness.

NONNATIVE GROUPS OF THE ARCTIC. In northern Alaska, residents face subzero temperatures and continuous darkness from October to May. During the ensuing summer season, continuous light prevails. Over 25% of all nonnative Alaskans experience mild seasonal affective disorder symptoms during winter. As pointed out already, the symptoms in winter are sleepiness even in daytime, increased irritability, weight gain, and less interest in social activities.

The long winters are thought to disrupt normal cycles of melatonin. Normally the onset of daylight inhibits melatonin production. In the continuous darkness of winter, untimely amounts of melatonin are produced, causing sleepiness throughout the day. One approach is to use as much outdoor daylight as possible during the shortened days. Another approach is to provide bright artificial light in the home or workplace.

ANTARCTIC RESEARCH STATIONS. No native peoples have lived closer to Antarctica than at Tierra del Fuego, the southernmost tip of South America. The hostile continent of Antarctica experiences some of the fiercest weather known in the world, with temperatures dropping at times to −100°F. Native cultures cannot survive the winters.

One of the greatest adventure dramas of all time was also played out on the frozen ice sheet of Antarctica, when Sir Ernest Shackleton's scientific expedition of 1914–1916 failed in its attempt to make the first crossing of the antarctic continent by dogsled. His ship and crew were unexpectedly trapped for 16 months in the hard-fast ice of the Weddell Sea in an early winter before ever landing on the continent. Their heroic escape involved overwintering near their ice-imprisoned ship until she was finally cracked by the ice, then dragging their small dinghies to open water and navigating across surging open ocean water to the nearest land, Elephant Island.

In a final race, a single open boat and crew of six fought their way through titanic storm-tossed seas for 16 days to reach the whaling station at South Georgia Island. After a pinpoint landing on the back side of the island, Shackleton and his mate scaled the towering ice cliffs and crossed the glaciers of the island in order to reach rescue help at the station. The life of the crew in the constant dark of winter and continuous light of summer gives much insight into human physiology and psychology under these natural constant lighting regimes.

In the latter half of the twentieth century, major permanent research stations were established by Russians, Americans, and British for studying many physical and biological aspects of the extreme environment and adaptive features of the biota of the antarctic ecosystem. Access to the continent for most personnel and cargo is by ship and for bases with an airstrip also by airlift. The picture for extended periods is vast ice-covered plains and mountains, frozen sea, and unimaginable cold. The British base of Halley, for example, is isolated for 10 months of the year.

Under these conditions, crews of research staff work in shifts that, in a temporal sense, are not much different from factory work in Pennsylvania or any other temperate part of the world. During winter, circadian rhythms are clearly either delayed or free-run, but timed light treatment can restore a summer phase position. Personnel adapt readily to night shifts, especially in winter, and have more problems readapting to day-shift schedules. Problems of shift work and sleep deprivation, compounded by possible seasonal affective disorder, demand solutions.

Treatment of Circadian Dysfunction by Chronopharmacology Is Still Largely in the Experimental Stages[12]

Several approaches can be used to achieve circadian well-being[3,13]

For both healthy individuals and those with particular circadian malfunctions, a variety of regimes are available for use singly or in combination. Behavioral therapy can sometimes be very effective. Treatment by this means can be therapeutic particularly in sleep and mood disorders. The routine includes promoting a patient's daytime social or physical activities, and discouraging daytime sleep or naps. At night, rest or sleep is encouraged. The technique is very effective in elderly subjects, both for healthy individuals and for those with sleep and mood disorders. It also helps young children with sleep problems.

The major chronopharmacological method is light therapy. Many different syndromes benefit from carefully timed light supplements. Light therapy is widely used to ameliorate shift-work problems across a wide range of occupations, including factory work and nursing, as well as astronauts' adaptation to space travel, jet lag, and sleep or mood disorders.

The final category of chronopharmacology is drug therapy. Chronomedicine considers how a knowledge of circadian rhythms is of value in a clinical context, and chronopharmacology directly applies the knowledge of circadian variations in susceptibility to drugs. The usual doctor's prescription must now be modified from "three times a day after meals" to a new circadian version: "Medication can be given with the maximum efficacy and minimal side effects at specific circadian times of day."

Highest on the list of treatments is melatonin and its analogs. Cases range from control of free-running rhythms in the blind and control of jet-lag symptoms to treatment of sleep and mood disorders.

Human phase response curves to both light and melatonin provide the theoretical basis for timing administration of a synchronizer[33]

Timed exposure to bright light, behavioral modification, or melatonin can be used to reset the circadian clock and achieve an appropriate phase relationship between the circadian timing system and habitual sleep time. Time of day, however, is not equivalent to circadian phase. Individual differences in endogenous period, together with socially determined sleep–wake timing lead to a wide range of circadian phases in any group of individuals. Furthermore, normal phase relationships and intact circadian rhythms may no longer be present in many illnesses.

Care should be taken to make sure that a timed treatment is given at the appropriate circadian phase. Phase assessment based on timing of the rest–activity cycle is the usual method for circadian sleep disorders. Measurement of dim-light melatonin onset in plasma or saliva, of the level of 6-sulfatoxymelatonin in urine, or of the temperature minimum during a constant routine protocol are more accurate methods but are expensive. The urinary method is widely used to assess rhythms in the blind or in shift workers.

In a clinical setting the exact cause of an abnormality in humans is occasionally clearly detected. Some children suffer from bed-wetting. This distressing syndrome is often due to an inadequate concentration of antidiuretic hormone in the blood. Levels normally rise at night, reducing urine production at this time. An injection of the hormone or analog each evening is effective. Similarly, the free-running circadian rhythm in totally blind people is almost certainly due to their inability to respond to the light–dark cycle as a synchronizer.

In contrast, the cause of a circadian dysfunction is frequently not known. A variety of factors could be at fault, including perception of the entraining agent, transmission of the entraining agent to the clock, faults in the pacemaker itself, or output to the body's systems. The deterioration of circadian rhythms in old age is an illustration. Many aspects of the circadian system may be inefficient, involving poor vision, decreased amplitude of the pacemaker, or reduced hormonal output.

The observation that the circadian system is acting abnormally doesn't mean the system itself is at fault. The primary defect may be elsewhere, and the effects on the circadian system may be only one of several abnormalities produced by the disorder. Treating the circadian system might improve an individual's sleep at night and activity during the daytime, but it does not address the underlying cause of the problem. These warnings are not a problem for those who wish to help patients, but they are a stumbling block for those who wish to develop a rationale for treatment. Fortunately, the chronobiological basis for many disorders is becoming better known. As a result, the rationale for treatment can often be based on chronobiological knowledge. Much of the rest of this chapter will deal with specific examples.

Behavioral treatment has many applications[45]

Behavioral therapy has been used for various age groups. Usually the protocol is simple encouragement to maintain a normal circadian sleep–wake schedule. Cases include the elderly with sleep consolidation problems, patients with Alzheimer's, and people with other dementia problems. The technique is also very useful in combating seasonal affective disorder depressions. In addition, it has been used with children having phase problems of the sleep–wake cycle. Data from a patient with Alzheimer's disease illustrates the striking improvement that is possible with the simple behavioral therapy of eliminating daytime naps.

Light treatment is often very effective[8,63,67]

Although years of research involving both animals and humans have demonstrated unequivocally that timed exposure to bright light is an effective means of manipulating the circadian timing system, many more years will apparently be needed to develop effective, reliable, and practical treatment strategies using bright light (Figure 10.23). Some issues may differ depending on the targeted application and population. Questions involve

FIGURE 10.23 Light therapy for treatment of SAD and other circadian dysfunctions. (Courtesy of EnviroMed and the Lightbox Company of New England.)

optimal intensity, spectral composition, duration and timing of light exposure, individual differences in response to light, and age effects. Issues of compliance must also be addressed.

Light treatment can be used to treat a variety of circadian dysfunctions. Light therapy increases the consolidation of nocturnal sleep in patients with senile dementia (Figure 10.24). Light has often been given in the morning, but increasing light intensity indoors throughout the daytime has also been successful. Improvement in the sleep–wake cycle is often associated with a decrease in the severity of any preexisting symptoms of disorientation and general confusion.

All the light considerations that apply to helping adaptation to night work apply in principle to adaptation to time-zone changes, and many simulation studies are appropriate to both situations. However, the differences between jet lag and night work must be taken into account. Whereas shift workers are always fighting against the natural synchronizers, travelers do not have this problem. All entraining agents in the new time zone will act in concert to promote adjustment to new local time after jet lag.

Melatonin used in physiological doses is a very effective chronotherapeutic agent[56,58,73]

Melatonin generally is taken in capsule form about 3 h before sleep is desired. The current dosage is based on physiological

levels of less than 5 mg per day. In some cases a dose of 0.5 mg per day is effective. With very few side effects, melatonin has been used successfully to alleviate jet-lag problems. An overall 50% reduction of subjective self-rated jet lag and improved sleep quality has been registered after melatonin treatment. In addition, melatonin treatment helps the internal clock adapt to local time more rapidly.

Federal Legislation Has Been an Important Part of Regulating Work Schedules in the Public Interest Sector[68]

Governments worldwide are beginning to recognize the significance of circadian timing for human performance in critical industries[49,54]

THE U.S. "DECADE OF THE BRAIN": 1990–2000. On July 25, 1989, the U.S. Congress took the unprecedented step of legally setting aside a decade to advance understanding of vital neural processes. The importance of the proclamation lay in the recognition of circadian function in humans, and the need to legally regulate work hours in situations where circadian abuses were endangering the lives of large numbers of citizens. Public Law 101-58 recognized that "50 million Americans are affected each year by disorders and disabilities involving the brain including major mental illnesses, inherited and degenerative diseases, stroke, epilepsy, addictive disorders, injury from prenatal events, environmental neurotoxins and trauma, as well as speech, language, hearing, and other cognitive disorders." It also noted that the cost of brain disorders represented a total economic burden of $305 billion annually in 1989.

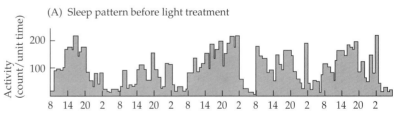

(A) Sleep pattern before light treatment

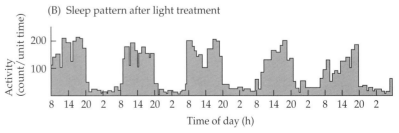

(B) Sleep pattern after light treatment

Time of day (h)

FIGURE 10.24 Effects of light therapy on an Alzheimer's patient with sleep consolidation problems. (From Van Someren et al., 1997.)

Part of the work that started immediately was the publication of three books in a series called *New Developments in Neuroscience*. The third book in the series was *Biological Rhythms: Implications for the Worker*, published in 1991. This book was a sweeping review of circadian rhythms, with case studies of nuclear power plant operators, of nurses and residents in hospitals, and of factory shift workers. Soon regulatory legislation based on circadian principles was under way in Congress.

Circadian rhythms are related to public workplace safety and efficiency[34]

POLICY ISSUES AND LEGISLATION OPTIONS FOR CONGRESSIONAL ACTION. Many occupations must schedule nonstandard work hours in order to carry out production for the full 24 h day. Included are manufacturing facilities, utility power plants, and protective and health services, such as police, fire, and rescue departments, and hospitals. Others that must schedule nonstandard work hours include transportation by airlines, trucking, railroads, and shipping. Major construction projects, such as dams, bridges, tunnels, and interstate highway construction, also use nonstandard work hours and have public safety issues. Military operations and many services, such as financial institutions, entertainment, and recreation, also are needed on a 24 h basis, as described at the beginning of the chapter.

The federal government works in two main ways to protect workers and the greater public in areas of nonstandard work hours. The first is by enacting national laws in Congress. These laws can be modified only by legislative action of Congress. For example, Congress enacted the landmark Fair Labor Standards Act in 1938, which established a standard 40 h workweek and dictated a minimum wage. The Federal Hours of Service Act was passed in 1907 to impose a limitation on work hours in the railroad industry.

The second function of the federal government is to delegate regulation to an administrative agency. For example, the federal government has authority to monitor and regulate working safety and health conditions for workers through the Occupational Safety and Health Administration (OSHA) of the Department of Labor. The Nuclear Regulatory Commission (NRC) is the watchdog for the nuclear power industry, and the Federal Railroad Administration oversees railroads. Other federal regulations are imposed through agencies such as the Department of Transportation and the Federal Highway Administration for drivers, and the Federal Aviation Administration for airline crews. At present, all federal regulations concern the total hours worked in a specified period of time, not the timing of the shifts or the proportion of day and night work. Any new regulations in the United States would be derived from additional Congressional laws or from regulation by federal administrative agencies.

EFFECTIVENESS OF GOVERNMENT REGULATION. Only the briefest of comments can be made on the impact of regulation on human circadian health and welfare. In the arena of international regulation of shift work, the United States is considered a leader in safety and health regulation, but its political system hinders it from enacting a national standard. Many of the regulations are under the jurisdiction of individual states.

Regulatory means differ from nation to nation, largely in a history-dependent fashion based on the type of government that has prevailed. As a result, some countries have extensive regulation for shift workers and others practically none. Most regulations concern work hours and pay. Only a few countries regulate leisure facilities, housing, and retirement programs for workers. Few nations, however, have a standard workweek exceeding 50 h.

SUMMARY

Even though the circadian clock of healthy humans is entrained to a period of about 24 h, the exact timing varies day by day in any individual and between individuals. These differences are rarely a source of inconvenience, but for some individuals the deviations are marked enough to cause difficulties in dealing with a conventional lifestyle. At a certain point they must be considered clinical abnormalities. Symptoms often appear as difficulties with sleep, such as delayed sleep phase syndrome, advanced sleep phase syndrome, and free-running rhythms. Free runs are especially common in people who are blind. Another group of affective disorders, particularly seasonal affective disorder, has been linked to circadian rhythm disturbance. Some older people also experience sleep difficulties, generally arising from poor sleep habits.

One group of problems exists in normal, healthy subjects who make substantial changes to their sleep–wake cycles during night work or after transmeridian travel. Sleep and other problems arise that may have important implications for the general public workplace.

Chronobiological treatment generally consists of strengthening exposure to appropriate circadian synchronizers, and basing therapy on the phase response curves for light and melatonin. Timed melatonin ingestion enhances adaptation to simulated and real night shifts, and it improves subjective sleep quality, circadian adaptation, and subjectively perceived jet lag in field studies. It can also stabilize sleep onset in blind subjects without necessarily entraining all circadian rhythms. Melatonin has been used successfully to synchronize sleep in some cases of delayed sleep phase syndrome or irregular sleep–wake disorders, and to increase duration and quality of sleep in the elderly.

Light therapy has been similarly effective with all these groups, except the profoundly blind, and it has the advantage of often being much cheaper. In addition, light therapy carries with it no legal requirement to perform long-term toxicology tests, unlike the case of melatonin or its analogs, for which no results are yet available. Melatonin and light may act in concert to maintain entrainment of the circadian timing system, and their combined use should provide optimal phase-shifting strategies.

Knowledge of chronobiology plays a small role in a clinical setting, enabling better diagnosis and treatment regimens to be developed. U.S. federal enforcement of a reasonable circadian policy in the workplace is applied by direct laws of the U.S. Congress, and by federal administrative regulatory agencies. Additional regulation is provided by state statutes and agencies.

CONTRIBUTORS

James M. Waterhouse and Patricia J. DeCoursey wish to thank Josephine Arendt, Scott Campbell, William J. M. Hrushesky, Björn Lemmer, Alfred J. Lewy, Michael Terman, Thomas A. Wehr, and Anna Wirz-Justice for their material contributions to Chapter 10.

STUDY QUESTIONS AND EXERCISES

1. The cause of seasonal affective disorder (SAD) and the use of light therapy for treating it have been considered in this chapter. The following exercise will illustrate some of the scientific problems that can arise in considering the cause and treatment of any complex disorder in humans, who have subjective feelings and preconceptions about treatment. What follow here are presumably true statements (S) related to SAD, followed by questions (Q) to consider:

    a. **S:** The incidence of SAD in the population increases as groups living farther away from the equator are studied. **Q:** Is SAD an effect of light or temperature?

    b. **S:** The severity of SAD is proportional to a percentage reduction of total light in a particular year. **Q:** What would you predict about the incidence of SAD relative to cloud cover during the fall months of a particular year? About the incidence of SAD in those who work outdoors compared to those who work indoors? About the incidence of SAD in those who wear sunglasses indoors and

outside as a fashion accessory? About the incidence of SAD in those who work indoors near a window compared with those who work indoors but away from a window?

    c. **S:** The incidence of SAD is directly related to decreased day length in winter. **Q:** What predictions can be made regarding the effectiveness of light therapy in the morning? In the middle of the day? In the evening?

2. Chapter 10 may be the most relevant chapter in the book for the reader personally because it portrays, at least in part, the myriad ways that circadian timing systems affect humans in their daily lives. Review Table 10.1, which is a summary of this chapter. Then carry out the following two exercises:

    a. Add as many more areas to the table as possible.

    b. Design a poster for a minisymposium that will be called "The Importance of Biological Pacemaking Systems on Human Well-Being" covering one topic or part of a topic. On the appointed day, be prepared to ask questions of other students on their topics and to answer questions about yours.

3. How unique are humans in comparison with all other species? Try a definition of a human being that includes the large category plus the unique feature(s) that distinguish humans from all other members of that category. (*Hint:* Humans are members of the order Primates, which differ uniquely and exclusively from all other primates in the following three feature[s]: _____, _____, and _____.) World-class scientists have had problems trying to define humanity. Humans certainly differ in degree by orders of magnitude in their sentient or self-aware nature, and in their reasoning power, as well as their ability to communicate in spoken and written ways about complex ideas. Using the unique or outstanding features of humanness, develop a 5-minute presentation on a specific topic for a minisymposium titled "The Effects of Human Reasoning and Other Uniquely Human Traits on Personal and Societal Chronobiological Performance." Use illustrations such as slides for part of the presentation.

4. Placebo trials are a very controversial topic in present-day medicine. Conventional methods in pharmacology are to give a placebo pill or injections to part of a group of patients in a trial, who act as controls for evaluating the effectiveness of a

treatment. The placebo procedures are indistinguishable from the real treatment because the pill that contains no drug or the injection that consists of only the vehicle looks exactly like its potent counterpart. Placebo issues can be very difficult to deal with when patients' preconceptions about treatment are concerned. For example, in light therapy the analogy would be to place the patient in front of a light box, but not to turn it on. Who would be fooled by this approach, since light therapy is widely thought by patients to be effective?

    a. What problems are connected with giving light as a placebo—that is, at a time predicted to be ineffective or even detrimental? What are the ethical considerations?

    b. What problems might occur in a second alternative to "involve or excite" patients by an amount equaling their involvement with or excitement from light therapy? How could this be done? (Get them to dance? Play music to them?)

5. Chapter 10 has drawn attention to the clash that exists for the shift worker, particularly the night worker, between living a normal social life and the chronobiological requirements of working abnormal hours. The list that follows gives several examples of the types of changes to shift schedules that shift workers often wish to make, and their reasons. For each change, give the chronobiological evidence related to the requested changes and your chronobiological evaluation of advisability.

    a. Desired change: Work 12 h shifts rather than 8 h shifts. **Reason:** 12 h shifts will cover the entire 24 h day, but the number of shifts for an individual and the related travel time will decrease.

    b. Desired change: Use a rapidly rotating shift system rather than a slowly rotating one. **Reason:** The rapid rotation every 1 or 2 days means that at least some days are normal in each week. In contrast, shifts that rotate slowly, with a week or longer on every shift before changing, result in some weeks that are a total loss as far as normal family life is concerned.

    c. Desired change: Reverse the usual swing-shift order and instead rotate from afternoon to morning to night to rest days. **Reason:** The reverse order has the advantage that the new shift cycle starts with an afternoon shift rather than a morning shift, thus allowing an extra final day of sleeping late.

    d. Desired change: Make the night shift later, from 11:00 P.M. to 7:00 A.M. rather than 10:00 P.M. to 6:00 A.M. **Reason:** The change would give more opportunity for social and domestic activities before the night shift. (*Hint for evaluation*: Consider problems of transport availability and safety, especially for women, of traveling at this time.)

REFERENCES

1. Åkerstedt, T., and J. Fröberg. 1975. Work hours and 24 h temporal patterns in sympathetic-adrenal medullary activity and self-rated activation. In: Experimental Studies of Shiftwork, P. Colquhoun, S. Folkard, and P. Knauth (eds.), pp. 78–93. Westdeutscher, Opladen, Germany.

2. Arendt, J. 1999. Jet-lag and shift work: Therapeutic use of melatonin. J. R. Soc. Med. 92: 402–405.

3. Arendt, J. 2000. Melatonin, sleep and circadian rhythms [Editorial]. N. Engl. J. Med. 343: 1114–1116.

4. Arendt, J., B. Stone, and D. J. Skene. 2000. Jet lag and sleep disruption. In: Principles and Practice of Sleep Medicine, 3rd Edition, M. Kryger, T. Roth, and W. C. Dement (eds.), pp. 591–599. Saunders, Philadelphia.

5. Bobko, N., A. Karpenko, A. Gerasimov, and V. Chernyuk. 1998. The mental performance of shift workers in nuclear and heat power plants of Ukraine. Int. J. Indust. Ergon. 21: 333–340.

6. Cohen, M. C., K. M. Rohtla, C. E. Lavery, J. E. Muller, and M. A. Mittleman. 1997. Meta-analysis of the morning excess of acute myocardial infarction and sudden cardiac death. Am. J. Cardiology 79: 1512–1515.

7. Cole, R. J., J. S. Smith, Y. C. Alcala, J. A. Elliott, and D. F. Kripke. 2002. Bright-light mask treatment of delayed sleep phase syndrome. J. Biol. Rhythms 17: 89–101.

8. Czeisler, C. A., and D-J. Dijk. 1995. Use of bright light to treat maladaptation to night shift work and circadian rhythm sleep disorders. J. Sleep Res. 4(Suppl. 2): 70–73.

9. Czeisler, C. A., R. E. Kronauer, J. S. Allan, J. F. Duffy, M. E. Jewett, et al. 1989. Bright light induction of strong (type 0) resetting of the human circadian pacemaker. Science 244: 1328–1333.

10. Czeisler, C. A., G. Richardson, R. Coleman, J. Zimmerman, M. Moore-Ede, et al. 1981. Chronotherapy: Resetting the circadian clocks of patients with delayed sleep phase insomnia. Sleep 4: 1–21.

11. Daan, S., and J. Aschoff. 2001. The entrainment of circadian rhythm. In: Handbook of Behavioral Neurobiology Volume 12: Circadian Clocks, J. S. Takahashi, F. W. Turek, and R. Y. Moore (eds.), pp. 7–43. Kluwer Academic/Plenum, New York.

12. Duncan, W. C. J. 1996. Circadian rhythms and the pharmacology of affective illness. Pharmacol. Ther. 71: 253–312.

13. Eastman, C., and S. Martin. 1999. How to use light and dark to produce circadian adaptation to night shift work. Ann. Med. 31: 87–98.

14. Elliott, A. L., J. N. Mills, and J. M. Waterhouse. 1970. A man with too long a day. J. Physiol. 212: 30P–31P.

15. Elliott, W. J. 2001. Cyclic and circadian variations in cardiovascular events. Am. J. Hypertension 14: 291S–295S.

16. Folkard, S., and T. Åkerstedt. 1991. A three-process model of the regulation of alertness and sleepiness. In: Sleep, Arousal, and Performance: Problems and Promises, R. Ogilvie and R. Broughton (eds.), pp. 11–26. Birkhauser, Boston.

17. Gander, P. H., D. Nguyen, M. R. Rosekind, and L. J. Connell. 1993. Age, circadian rhythms, and sleep loss in flight crews. Aviat. Space. Environ. Med. 64: 189–195.

18. Goodwin, F. K., and K. R. Jameson. 1990. Manic-Depressive Illness. Oxford University Press, Oxford, UK.

19. Gundel, A., V. Polyakov, and J. Zully. 1997. The alteration of human sleep and circadian rhythms during spaceflight. J. Sleep Res. 6: 1–8.

20. Hamelin, P. 1987. Lorry drivers' time habits in work and their involvement in traffic accidents. Ergonomics 30: 1323–1333.

21. Hastings, J. W., B. Rusak, and Z. Boulos. 1991. Circadian rhythms: The physiology of biological timing. In: Neural and Integrative Animal Physiology, C. L. Prosser (ed.), pp. 435–545. Wiley, New York.

22. Horne, J., and L. Reyner. 1995. Driver sleepiness. J. Sleep Res. 4(Suppl. 2): 23–29.

23. Jan, J., H. Espezel, and R. Appleton. 1994. The treatment of sleep disorders with melatonin. Dev. Med. Child Neurol. 36: 97–107.

24. Jones, C. R., S. S. Campbell, S. E. Zone, F. Cooper, A. DeSano, et al. 1999. Familial advanced sleep-phase syndrome: A short-period circadian rhythm variant in humans. Nature Med. 5: 1062–1065.

25. Kennaway, D. F., and C. F. Van Dorp. 1991. Free running rhythms of melatonin, cortisol, electrolytes and sleep in humans in Antarctica. Am. J. Physiol. 260: R1137–R1144.

26. Klein, D. C., R. Y. Moore, and S. M. Reppert. 1991. Suprachiasmatic Nucleus: The Mind's Clock. Oxford University Press, New York.

27. Klein, K., H. Wegmann, and B. Hunt. 1972. Desynchronization of body temperature and performance circadian rhythm as a result of outgoing and homegoing transmeridian flights. Aerospace Med. 43: 119–132.

28. Knutsson, A., and H. Boggild. 2000. Shiftwork and cardiovascular disease: Review of disease mechanisms. Rev. Environ. Health 15: 359–372.

29. Kokkoris, C., E. D. Weitzman, C. Pollack, A. Spielman, C. A. Czeisler, et al. 1978. Long-term ambulatory temperature monitoring in a subject with a hypernychthemeral sleep–wake cycle disturbance. Sleep 1: 177–190.

30. Krieger, D. T., W. Allen, F. Rizzo, and H. P. Krieger. 1971. Characterization of the normal pattern of plasma corticosteroid levels. J. Clin. Endocrinol. Metab. 32: 266–284.

31. Lemmer, B. 1995. Clinical chronopharmacology: The importance of time in treatment. In: Circadian Clocks and Their Adjustment (CIBA Foundation Symposium 183), D. Chadwick and K. Ackrill (eds.), pp. 235–247. Wiley, Chichester, UK.

32. Lemmer, B., and K. Witte. 2000. Biologische Rhythmen und kardiovaskulare Erkrankungen. Uni-med, Bremen, Germany.

33. Lewy, A. J., V. Bauer, S. Ahmed, K. Thomas, N. Cutler, et al. 1998. The human phase response curve (PRC) to melatonin is about 12 hours out of phase with the PRC to light. Chronobiol. Int. 15: 71–83.

34. Liskowsy, D. R. (Project Director). 1991. Biological Rhythms: Implications for the Worker. Congress of the United States, Office of Technology Assessment, Washington, DC.

35. Machado, E., B. Viviane, and M. Andrade. 1998. The influence of study schedules and work on the sleep-wake cycle of college students. Biol. Rhythm Res. 29: 578–584.

36. Minors, D. S., and J. M. Waterhouse. 1981. Circadian Rhythms and the Human. Wright, Bristol, UK.

37. Mitler, M. M., M. A. Carskadon, C. A. Czeisler, W. C. Dement, D. F. Dinges, et al. 1988. Catastrophes, sleep, and public policy: Consensus report. Sleep 11: 100–109.

38. Monk, T. H., and D. Kupfer. 2000. Circadian rhythms in healthy aging: Effects downstream from the pacemaker. Chronobiol. Int. 17: 355–368.

39. Monk, T. H., D. Buysse, B. Billy, K. Kennedy, and L. Willrich. 1998. Sleep and circadian rhythms in four orbiting astronauts. J. Biol. Rhythms 13: 188–201.

40. Moore, R. Y., and R. K. Leak. 2001. Suprachiasmatic nucleus. In: Handbook of Behavioral Neurobiology Volume 12: Circadian Clocks, J. S. Takahashi, F. W. Turek, and R. Y. Moore (eds.), pp. 141–179. Kluwer Academic/Plenum, New York.

41. Moore-Ede, M. C. 1993. The Twenty-Four Hour Society: Understanding Human Limits in a World That Never Stops. Addison-Wesley, Reading, MA.

42. Moore-Ede, M. C., F. M. Sulzman, and C. A. Fuller. 1982. The Clocks That Time Us. Harvard University Press, Cambridge, MA.

43. Muller, J. E. 1999. Circadian variations in cardiovascular events. Am. J. Hypertension 12: 35S–42S.

44. Novak, R., and S. Auvil-Novak. 1996. Focus group evaluation of night nurse shiftworking difficulties and coping strategies. Chronobiol. Int. 13: 457–463.

45. Okawa, M., K. Mishima, Y. Hishikawa, S. Hozumi, H. Hori, et al. 1991. Circadian rhythm disorders in sleep-waking and body temperature in elderly patients with dementia and their treatment. Sleep 14: 478–485.

46. Oren, D., E. Turner, and T. A. Wehr. 1995. Abnormal circadian rhythms of plasma melatonin and body temperature in the delayed sleep phase syndrome. J. Neurol. Neurosurg. Psychiatry 58: 379–394.

47. Parker, D. C., L. G. Rossman, and D. F. Kripke, W. Gibson, and K. Wilson. 1979. Rhythmicities in human growth hormone concentrations in plasma. In: Endocrine Rhythms, D. Krieger (ed.), pp. 143–173. Raven, New York.

48. Raftery, E., M. Millar-Craig, S. Mann, and V. Balasubramanian. 1981. Effects of treatment on circadian rhythms of blood pressure. Biotelemetry Patient Monitoring 8: 113–120.

49. Rajaratnam, S. M. W., and J. Arendt. 2001. Health in a 24-h society. Lancet 358: 999–1005.

50. Ribeiro, D. C. O., S. M. Hampton, L. Morgan, S. Deacon, and J. Arendt. 1998. Altered postprandial responses in a simulated shift work environment. J. Endocrinol. 158: 305–310.

51. Rosekind, M., R. Smith, D. Miller, E. Co, K. Gregory, et al. 1995. Alertness management: Strategic naps in operational settings. J. Sleep Res. 4(Suppl. 2): 62–66.

52. Rosenwasser, A. M., and A. Wirz-Justice. 1996. Circadian rhythms and depression: Clinical and experimental models. In: Handbook of Experimental Pharmacology: Physiology and Pharmacology of Biological Rhythms, P. H. Redfern and B. Lemmer (eds.), pp. 457–486. Springer, Berlin.

53. Ross, J. K., J. Arendt, J. Horne, and W. Haston. 1995. Night-shift work in Antarctica: Sleep characteristics and bright light treatment. Physiol. Behav. 57: 1169–1174.

54. Roth, T. 1995. An overview of the report of the commission on sleep disorders. Eur. Psychiatry 10(Suppl. 3): S109–S113.

55. Sack, R. L., R. W. Brandes, A. R. Kendall, and A. J. Lewy. 2000. Entrainment of free-running circadian rhythms by melatonin in blind people. N. Engl. J. Med. 343: 1070–1077.

56. Sharkey, K., and C. Eastman. 2002. Melatonin phase shifts human circadian rhythms in a placebo-controlled simulated night study. Am. J. Physiol.: Regul. Integr. Comp. Physiol. 282: R454–R463.

57. Shibui, K., M. Uchiyama, and M. Okawa. 1999. Melatonin rhythms in delayed sleep phase syndrome. J. Biol. Rhythms 14: 72–76.

58. Skene, D. J., S. W. Lockley, and J. Arendt. 1999. Use of melatonin in the treatment of phase shift and sleep disorders. Adv. Exp. Med. Biol. 467: 79–84.

59. Smith, R., and C. Kushida. 2000. Risk of fatal occupational injury by time of day. Sleep 23: A110–A111.

60. Smith-Coggins, R., M. Rosekind, K. Buccino, D. Dinges, and R. Moser. 1997. Rotating shiftwork schedules: Can we enhance physician adaptation to night shifts? Acad. Emerg. Med. 4: 951–961.

61. Smolensky, M. 1995. Present achievements and future prospects for clinical chronobiology. In: Circadian Clocks and Their Adjustment (CIBA Foundation Symposium 183), D. Chadwick and K. Ackrill (eds.), pp. 253–271. Wiley, Chichester, UK.

62. Souetre, E., E. Salvati, and J-L. Belugou. 1989. Circadian rhythms in depression and recovery: Evidence for blunted amplitude as the main chronobiological abnormality. Psychiatry Res. 28: 263–278.

63. Terman, M. 2000. Light therapy. In: Principles and Practice of Sleep Medicine, 3rd Edition, M. H. Kryger, T. Roth, and W. C. Dement (eds.), pp. 1258–1274. Saunders, Philadelphia.

64. Toh, K. L., C. R. Jones, Y. He., E. J. Eide, W. A. Hinz, et al. 2001. An hPer2 phosphorylation site mutation in familial advanced sleep phase syndrome. Science 291: 1040–1043.

65. Uchiyama, M., M. Okawa, K. Shibui, K. Kim, and H. Tagaya, et al. 2000. Altered phase relation between sleep timing and core body temperature rhythm in delayed sleep phase syndrome and non-24-hour sleep-wake syndrome in humans. Neurosci. Lett. 294: 101–104.

66. VanHemel, S., and S. Rogers. 1998. Survey of truck drivers' knowledge and beliefs regarding driver fatigue. Transport. Res. Record 1640: 65–73.

67. Van Someren, E., A. Kessler, M. Mirmiran, and D. Swaab. 1997. Indirect bright light improves circadian rest-activity rhythm disturbances in demented patients. Biol. Psychiatry 41: 955–963.

68. Waterhouse, J. M., S. Folkard, and D. S. Minors. 1992. Shiftwork, Health and Safety: An Overview of the Scientific Literature 1978–1990. HMSO, London.

69. Waterhouse, J., T. Reilly, and G. Atkinson. 1997. Jet lag. Lancet 350: 1611–1616.

70. Wehr, T. A. 1990. Effects of wakefulness and sleep on depression and mania. In: Sleep and Biological Rhythms, J. Montplaisir and R. Godbout (eds.), pp. 42–86. Oxford University Press, London.

71. White, W. 2001. Cardiovascular risk and therapeutic intervention for the early morning surge in blood pressure and heart rate. Blood Press. Monit. 6: 63–72.

72. Wirz-Justice, A. 1995. Biological rhythms in mood disorders. In: Psychopharmacology: The Fourth Generation of Progress, F. E. Bloom and D. J. Kupfer (eds.), pp. 999–1017. Raven, New York.

73. Wirz-Justice, A., and S. M. Armstrong. 1996. Melatonin: Nature's soporific? J. Sleep Res. 5: 137–141.

74. Wirz-Justice, A., and R. H. Van den Hoofdakker. 1999. Sleep deprivation in depression: What do we know, where do we go? Biol. Psychiatry 46: 445–453.

75. Zisapel, N. 2001. Circadian rhythm sleep disorders: Pathophysiology and potential approaches to management. CNS Drugs 15: 311–328.

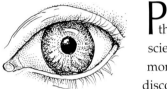

Proust's lucid epigraph at the start of this chapter recognizes that progress in science often lies in new views about commonplace objects. His idea applies brilliantly to the discovery of biological pacemakers. Although the suprachiasmatic nucleus of the mammalian hypothalamus had been noticed since the early 1800s, nothing was known about its function beyond idle speculation. Only half a century ago, a few isolated scientists apologetically voiced their views that undiscovered internal biological timers played a key role in regulating rhythmic activity of a wide variety of animals and plants. The prompt reply of most of the scientific community was "Show me an internal clock!"

One of the first living clocks to be found was in the optic lobes of insects, and for molluscs like *Bulla,* literally in the eyes. Figuratively, new eyes have been opened in a meteoric trajectory of discovery based on new ideas and techniques. Chapter 11 opens with a consideration of human cultural evolution and the consequent awareness of both spatial and temporal organization. The rapid progress of the field of chronobiology in the past decade is discussed briefly, and the difficulty of accurately projecting the course of chronobiology research findings in the future is mentioned. The significance of biological timing in our present-day world is portrayed in six themes emphasizing recent advances in chronobiology research.

11

Looking Forward

The real voyage of discovery consists not in seeking new landscapes but in having new eyes.

—Marcel Proust

Introduction: Accurate Predictions about the Future of Chronobiology Are Hard to Make[41,61]

A glance back at the history of human interest in timekeeping will sharpen the focus of any forward projections[8,10,12,13,21,25,33]

HUMAN TEMPORAL SELF-AWARENESS. This book began its journey with a look back 5,100 years ago to the important archeological site at Newgrange, Ireland. The ancient burial site features a 25 m tunnel leading to an underground crypt. Only at sunrise on the day of the winter solstice does the sun's alignment with the tunnel allow a mote of sunlight to illuminate the crypt. Present-day humans can only guess at the significance of the light mystically touching the stone, but the work of constructing the monument suggests that highly significant time-related rituals were involved. Even more significant was the epoch of megalithic stone monument building at Stonehenge. Starting about 5,000 years ago and continuing over a period of almost 2,500 years, a good deal of community energy must have been involved. Time allocated to moving the gigantic stones, raising them into position, and then carrying out the solar and lunar ceremonies must have been considerable.

The driving force behind these events is best understood in the framework of a key development in human evolution that had begun about a million years earlier. Advanced learning capacity and self-awareness were emerging slowly out of the former restraints of stereotyped, rigidly programmed behavior typical of nonhuman animals. At some point early humans evolved the capacity for viewing themselves in a time realm: past, present, and future. That unique endowment was a major factor in the cultural evolution that was to follow. Clearly evident is a sudden new human interest in time measurement devices and their application to human everyday practical life, as well as spiritual life.

THE RATE OF HUMAN CULTURAL EVOLUTION. The millennia-long evolution in human understanding of biological timekeeping is cogently relevant to predictions of developments that the next century might bring to the field of chronobiology. The gradually emerging understanding of the complexity and adaptiveness of biological timekeeping has not been a static progression because the rate of increase of human intellectual awareness has been neither linear nor uniform. Two different factors have been at work. First, human cultural evolution proceeds in general at an ever-increasing rate. Second, brilliant steplike advances are interjected into long-lasting plateaus.

The logarithmic increase in the rate of cultural change is analogous to the opening and page turning of a huge book such as an unabridged dictionary. When the book is opened, the extent of the vast compendium of word-filled pages becomes evident. Each page stands figuratively for a major advance in human intellectual understanding. The pages were turned slowly at first, when knowledge of scientific processes was minimal and most decisions were based on superstition. Soon, however, the rate of turning quickened, as new concepts opened unexpected horizons for exploration, and eventually the fin-

(A)

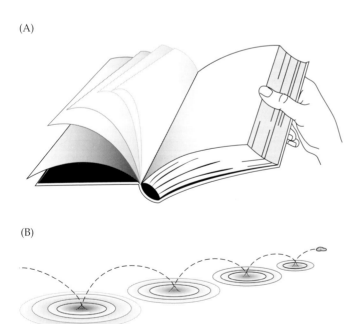

(B)

FIGURE 11.1 Figurative depiction of the rate of increase of human culture as a factor in predicting future progress in chronobiology. (A) The turning pages of a large book represent the predictable logarithmic rate of cultural change. (B) Stones skipping on water signify the unpredictable, serendipitous scientific discoveries that result in cultural change.

gers could no longer turn the pages fast enough to keep up with the pace of scientific progress. Now the ever increasing tempo of turning causes the pages to spin past and dissolve into a blur. (Figure 11.1A). The process is illustrated concretely by the escalation in the annual number of scientific papers published and in the increase in the number of levels of operation of circadian clocks studied, from the societal and community levels down to the molecular level.

The second factor, the alternation of advances and plateaus, resembles the impact of a flat, smooth stone skillfully pitched across the glassy smooth water of a lake. The stone skips gracefully across the surface. Each time it touches the surface ephemerally, the stone generates a ring of ripples that fan out and interact with preceding ring sets (Figure 11.1B). As a metaphor for human cultural evolution, each touchdown and new ring of expanding ripples signifies a major advance.

Often in the history of science, an advance is the de novo discovery of a new scientific concept or a new tool. The building of a simple microscope, hardly more than a magnifying glass, by the Dutchman Antoni van Leeuwenhoek in the late 1600s changed the course of biology. Similarly, Charles Darwin's brief stay in the Galápagos Islands, just 1 month long, gave rise to his monumental book *On the Origin of Species by Natural Selection*. The timely quote at the beginning of this chapter by Marcel Proust captures these ideas well.

THE ROLE OF NEW IDEAS AND NEW TOOLS IN ADVANCING CHRONOBIOLOGY KNOWLEDGE. Certainly one of the touchdowns of the skipping stone for chronobiology was the recognition of internal timing capacity. Clocks and calendars were a pervasive force in regulating practically every aspect of the rhythmic life of early human cultures. Whether early observations were truly astronomical or merely astrological is anyone's guess. Although the true origins of temporal self-awareness will never be known, some clues can be found.

Many instances of the adaptiveness of temporal programming undoubtedly were apparent to early humans in antiquity. The pithy quotes that start the chapters and other parts of this book reflect powerfully the interest of humans from all walks of life in the inevitable and never-ending march of time. Plutarch wrote 2,000 years ago, "Time is the wisest of counselors." From Chapter 2 comes an ancient proverb that almost certainly predates written language. "Time and tide waiteth for no one" was first recorded in one of the earliest European written records, a British book of proverbs, but it surely had been handed down by word of mouth for millennia previously. No one could have imagined at Newgrange what 5 millennia would bring on the scientific scene. For that matter, no one could have foreseen what a century of progress would produce in understanding internal time.

Awareness of the adaptive importance of biological timing has grown rapidly in the past decade[61]

The pace of discovery in the field of chronobiology has been very rapid in the past decade. Symbolic of the progress are accolades by the prestigious scientific journal *Science*. Read each week by about half a million scientists, it is one of the foremost forums in the world for reporting new scientific advances. For the past few years, *Science* has featured an article at year's end that aims to predict the six fields that will make scientific headlines in the coming year and critiques the accuracy of the preceding year's choices. An accompanying disclaimer states: "But forecasting is no exact science; here we rate whether our crystal ball was cloudy or clear."

A second part of the annual *Science* survey recognizes about nine noteworthy scientific discoveries "that transform our ideas about the natural world and also offer potential benefits to society." *Science* applauded breakthroughs in chronobiology in three of the past six years. In December 1997, the breakthrough discovery entailed the cloning of several new circadian genes, including *white collar-1* and *-2* in *Neurospora*, and *Clock* in mice. A gene similar to *per* was also found in mice and humans. Not only was the *per* gene active in the brain pacemaker of the fruit fly but also in many parts of the fly's body. In December 1998, the first runner-up status was awarded for further genetic advances in chronobiology. At the close of 2002, the field of chronobiology was recognized by *Science* for the localization of previously unknown mammalian circadian retinal ganglion photoreceptors.

The premier association for those who study biological clocks in the United States is the Society for Research on Biological Rhythms, and its literary voice to the scientific community of the world is the *Journal of Biological Rhythms* (JBR; see Figure 1.13 in Chapter 1). On the occasion of the 1998 announcement in *Science*, JBR's Editor, Fred W. Turek of Northwestern University, wrote an editorial entitled "A remarkable year for clocks based on many years of interdisciplinary research." He stated that

> Much of what was referred to as the '. . .double-quick barrage of discoveries. . .' that '. . .exceeded all expectations. . .' involved the discovery that flies and mice separated by nearly 700 million years [of evolution] share the same timekeeping genes and proteins. Similarly, organisms separated by billions of years, while having different clock genes, use similar feedback loops to generate 24 h rhythm. While the degree of similarity may have come as a surprise to many circadian biologists, the assumption that similarities would indeed be found has been a driving force in the field for years. Investigators working in organisms as diverse as bacteria, plants, invertebrates and vertebrates shared their findings on the regulation of biological rhythms at scientific meetings and in interdisciplinary journals such as this one in anticipation that there would indeed be similarities at the formal and mechanistic levels across most if not all living

organisms. . . . What other unexpected relationships between rhythmic processes across the living world will be discovered and where will the next cross-species breakthroughs be made?

Turek's words serve both as an inspiration for the future and as a warning about the dangers of making predictions about progress too far into the future. Chapter 11 will make no attempt to predict what the twenty-second or twenty-third century, much less the thirtieth, will bring. Instead it will extend some of the breakthrough ideas developed in the proceeding ten chapters into likely directions of the next decade.

Six forward-looking themes have been chosen to close Chronobiology: Biological Timekeeping[17,27,54,63,66,69]

The preceding section emphasized the difficulty of predicting scientific progress over long time spans. More realistic estimates can be made for shorter periods. Rather than predict the course of the whole chronobiology field, a simpler goal has been chosen. From a multitude of possible topics, six themes will be highlighted as cameo portraits of important recent breakthrough events in chronobiology that may help channel the directions of circadian research in the next decade.

1. Progress in sleep–wake research and its application to human welfare

2. The use of nonhuman vertebrate models in circadian sleep research

3. Growing use of invertebrate model species for circadian molecular research

4. Peripheral pacemakers in animals

5. Use of SCN *Clock* mutant chimeras in understanding circadian pacemaker mechanisms

6. The consequences of ecosystem degradation relative to entrainment of circadian systems

Theme 1: Sleep Research in Humans Is an Important Area for Future Study[2,3,7,14,19,31,35,63,64,65]

Numerous theories postulate the function and mechanism of sleep in humans[63,65]

Because sleep for most humans is an obvious and personally interesting aspects of the circadian clock, questions about it will clearly be a focus of future research. The topic of the function and the mechanism of sleep in humans is still debated, although a number of persuasive theories have been proposed. As common sense suggests, most theories propose a recovery function for sleep. The behavioral features of sleep include prolonged immobility, increased arousal threshold, and homeostatic drive to make up lost sleep. If mere energy conservation were the only requirement, simple immobility would suffice. Even less apparent is the value of reducing awareness of the environment if the purpose of sleep were simply to reduce metabolic rates. In fact, reduced responsiveness to the environment and the overwhelming need to sleep can be highly maladaptive in some situations. A sleeping rabbit in an open field or a drowsy human driving a car is obviously an individual at risk.

Sleep may affect somatic as well as central nervous system (CNS) function, but the elaborate nature of multiple states of mammalian sleep has directed attention to CNS issues. Several theories propose that neural function is enhanced by sleep-related neural plasticity. In the adult, promotion of long-term memory consolidation during sleep could optimize adaptation to a changing environment. In the developing organism, sleep may foster neural maturation. In an aging animal or one subject to neural malfunctions, sleep may promote healing or at least reduce the damage of disease, trauma, and degenerative processes. As part of the mechanism for neural restoration, sleep may specifically regulate energy and metabolism in neural tissues. In this way, neurons may be protected from energy depletion due to the toxic effects of high rates of metabolism or fluctuating oxygen availability during an organism's active period.

Sleep research and circadian research are increasingly collaborative[38,64]

Until about 1995 interaction between the sleep and circadian rhythm research communities was quite minimal. Sleep disorders such as insomnia, or the more severe affective sleep disorders (see Chapter 10), were treated by physicians as medical problems. Few circadian physiologists dealt with humans as research subjects, primarily because of the large costs and ethical considerations. Another reason for the lack of overlap of the sleep and circadian fields was their two different views of sleep. Scientists in the sleep field believed that the clock just told the organism when to sleep but said nothing related to function, or the drive to sleep, while circadian scientists considered the sleep–wake cycle to be just another output of the clock, merely analogous to feeding, body temperature, or hormone rhythms. As a result, connections between the two disciplines of human sleep and circadian rhythms were very limited.

The outlook for rewarding collaborations of the two fields in the future is promising. A growing number of joint circadian–sleep society programs have been held at national and international meetings. Synergistic interactions of the two fields are fostered with increasing frequency by the view that circadian rhythms, the sleep–wake cycle, and the metabolic budget of the human body have all coevolved since the beginning of life on this planet.

Chronic partial sleep loss is a major health issue in industrialized countries[35,38,39,62,63,65]

No longer do most physicians scoff at the idea of timely administration of pharmaceutical drugs, sleep therapy, or light treatment for affective disorders (see Chapter 10). The reason is clear: Sleep deficit affects millions of people. At least 65% of people over 65 years of age, for instance, have sleep problems severe enough to result in sleep deficits, and a majority of nonprescription pharmaceuticals bought by the elderly are for coping with sleep problems. Sleep deficit has major implications for optimal daily performance at home and on the job. Chronic sleep loss is directly related to industrial and highway accident rates. In addition, sleep deficiency has major overtones for long-term chronic health problems.

Sleep duration is declining in America. Average sleep duration has decreased since 1900 from 9 h for adults to 7.5 h in 1975. Several factors are involved. Since 1969, the amount of time spent at work has increased by 158 h per year. A 24 h society, shift work, and moonlighting jobs are all in part to blame. The insatiable demand of Americans for "the good life" has also reduced sleep time. Greater income, greater recreational opportunities, and increased television watching have changed the face of society from the typical farm life in rural communities in 1900 to the bustling, nerve-jangling, peripatetic urban lifestyle of the twenty-first century.

One convincing statistic of the detrimental effect of sleep deficit on performance comes from recording short-term memory lapses relative to normal function, sleep debt, and sleep recovery periods. Without a doubt, cognitive function is seriously impaired by sleep deficit. Although some effect is seen in young adults, the severe disruption in the elderly gives a disquieting view for the future, given our aging population.

A state of sleep debt in young adults is associated with endocrine and metabolic alterations that are similar to those observed in normal aging. The changes include decreased glucose tolerance, increased evening cortisol levels, and increased sympathetic tone. Chronic sleep loss is an increasingly common condition in industrialized countries. This sleep loss is likely to increase the severity of widespread age-related chronic conditions such as obesity, diabetes, and hypertension.

Sleep deficit aggravates many health conditions of the elderly[39,49,63,64,65]

Normal sleep in most human cultures entails a uni-modal period of consolidated sleep lasting 8–9 hours with a regular predictable succession of cycling stages of REM and non-REM sleep. Sleep can be measured by electroencephalography (Figure 11.2A), which portrays the characteristic brainwave patterns of the specific sleep stages (Figure 11.2B). Generally the temperature cycle is tightly linked with the sleep–wake cycle, thus ensuring that the temperature minimum falls shortly after the middle of the sleep period. With aging, some of the characteristics of the sleep cycle change and various pathologies result. The tendency is often aggravated by napping during the day as a result of fatigue induced by sleep dysfunction. A vicious circle may result that can spill over from simple fatigue to a major impact on many physiological functions and medical conditions.

(A)

(B)

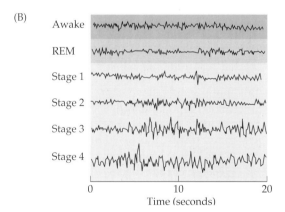

FIGURE 11.2 **Summary of the myographic, encephalographic, and ocular features of the normal sleep cycle of humans.** (A) Recording the sleep cycle in a human subject. (B) Sample short-term EEG recording for the brain waves of the stages of sleep. In practice, diagnosing the stages and their normality requires considerable skill and experience. (A courtesy of the Sleeplab, Department of Health Science and Technology, Aalborg University, Denmark; B from Purves et al., 2001.)

(A)

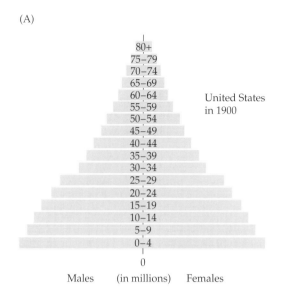

(B)

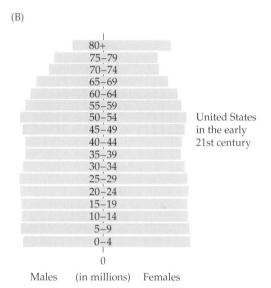

FIGURE 11.3 **Age structure changes in the United States between 1900 (A) and 2000 (B).** (From Olshansky, 1997.)

As life span increases in industrialized countries, age-related sleep problems become a matter of national concern. Age structuring in the United States has changed dramatically in the past century because of better health care and the resulting increase in longevity. A comparison of longevity in 1900 and 2000 illustrates the gains in life span (Figure 11.3). Today many elderly individuals continue to enjoy an active physical and social life well into their 80s and even 90s, but sleep deficit problems are widespread in the elderly. These sleep–wake disorders jeopardize good physical and mental health. Chronic sleep deficit adversely affects memory, performance capabilities, and the general quality of life. Although research on sleep continues actively on human subjects, many constraints limit progress.

Limitations on sleep research in humans will lead increasingly to the use of animal models in the future[37,47,61]

Sleep research with humans is expensive and entails many ethical considerations. Consequently, alternative rodent models are very desirable. In the future, rodents will probably supplant human subjects for many research projects, particularly for experimental approaches in the laboratory. Some rodent models are already in the pipeline.

The circadian *Clock* mutant mouse, for example, is proving a highly valuable model for human research. Since the *Clock* mutant is defective in one critical circadian gene, it has been used to tease apart the circadian components from the sleep–wake cycle components. In the homozygous state, individuals initially have very long free-running periods in DD. The free runs deteriorate gradually into arrhythmicity. The gene is affiliated strongly with SCN-regulated functions in mammals. As a result, the mutant mouse offers many opportunities for examining gene control of sleep. One example concerns the role of the SCN, which has been very controversial in mammals, including humans. Little is known about the consolidation of sleep and wake states by the generation of an arousal signal by the SCN. If such a signal exists, nothing is known yet about the mechanism of signaling. The extent to which the circadian system influences sleep recovery in humans is also unknown. One representative data set will be presented to exemplify the many new facts obtained from sleep studies in the *Clock* mutant mouse.

When homozygous *Clock* mice are subjected to stress by physical restraint, fatigue is induced. Various measures of sleep can be recorded in the subsequent recovery period. Measurement of the total amount of sleep, non-REM sleep, and REM sleep in the 12 h dark period reveals highly significant differences between wild-type and homozygous, arrhythmic mutant mice (Figure 11.4).

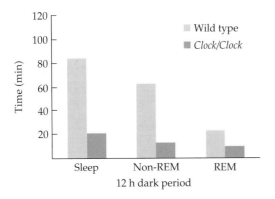

FIGURE 11.4 Relative effect of restraint stress on sleep recovery in wild-type control mice and *Clock* mutant mice. (Courtesy of A. E. Easton and F. Turek.)

A summary of features of the arrhythmic mice relative to wild-type controls indicated increased wakefulness, decreased amplitude of the sleep–wake cycle, and reduced sleep recovery after sleep deprivation or restraint stress.

The next theme will elaborate other ways to learn about sleep in humans. Comparative studies of sleep processes in nonhuman vertebrates promise to give many insights into the phylogenetic origin of sleep and its functions.

Theme 2: Wild, Free-Living Vertebrate Species Are Being Used to Understand the Circadian Basis of Sleep[1,11,34,48,54,55,58,59]

Studies of nonhuman vertebrate species, especially wild mammals and birds, are a recent dynamic development in sleep research[1,34,54,55,58,59]

CHALLENGES. Research has emphasized human sleep until very recently for good reasons. In humans the evolutionary development of sleep has reached its pinnacle. Generally, a unimodal sleep pattern coinciding with the hours of darkness is present, and a distinctive cycling between stages of sleep over a discrete 8 to 9 h period per day occurs. The behavioral stages correlate well with brain and muscle recording. In addition, sleep intensity and recovery have been relatively easy to document in humans. In contrast, sleep patterns are more complex and more difficult to record and interpret in nonhuman vertebrates, and the very existence of sleep has been questioned for most invertebrates.

Only recently have the invaluable opportunities for ecological and physiological research in nonhuman vertebrate species been recognized and aggressively explored. Free-living wild species, of course, present some disadvantages in research. Wild creatures are generally very secretive during the vulnerable quiescent rest or sleep period. They are often difficult to capture, and they may show many sleep artifacts during observation and recordings in captivity. Nevertheless, a surge of enthusiasm for studies using wild species promises rapid progress in the near future. Two trends aimed at clarifying functions of sleep are evident: a phylogenetic approach and an ecological approach.

PHYLOGENETIC APPROACHES The thrust of the comparative phylogenetic approach is to record fundamental sleep properties of key species for as many classes and orders of mammals, birds, and lower vertebrates as possible. The hope is to reveal the first appearance of sleep properties from simple to complex in order to gain a better handle on the mechanism of sleep (Table 11.1). A potential pitfall is that a function involving multiple brain centers will certainly not be a simple straight-chain

TABLE 11.1   Occurrence of sleep and sleeplike properties in vertebrate groups

Property	Placental mammals	Marsupials	Monotremes	Birds	Reptiles	Amphibians	Fish
EEG difference between vigilance states	+	+	+	+	+ spikes	+/−	+/−
High-voltage slow-wave sleep	+	+	+	+	−	− tendency	− tendency
Low-voltage fast-wave sleep	+	+	?	+	?	−	−
Behavioral sleep	+	+	+	+	+	+/−	+/−

Source: Tobler, 1983.
Note: +, present; −, absent; ?, questionable.

sequence, but rather an interwoven web of features and mechanisms that may be difficult to tease apart.

Comparative questions about sleep in vertebrates have attracted a surprising number of scientists. As a result, sleep has recently been examined in at least one species of most mammalian and avian orders (see Plates 2 and 4). In the comparative study of sleep mechanisms and functions, several fundamental properties have been considered. Behavioral sleep is relatively easy to document. One aspect of behavioral sleep concerns the distribution of sleep relative to the environmental light–dark cycle. Other features are physiological, including total REM and non-REM sleep, the intensity of sleep and sleep recovery, and the occurrence of the phenomenon of unihemispheric sleep.

Several generalizations about sleep stages for mammals and lower vertebrates are now possible. In contrast to humans, most other vertebrates do not have a strongly unimodal, consolidated period of sleep. Instead, many have complex multiphasic sleep patterns. Even if they are highly nocturnal or diurnal in their activity patterns, they may alternate brief sleep and wake periods throughout the inactive period of the day. Behavioral sleep, along with a clear difference in the EEG between sleep and wake states, is evident in all vertebrate classes from reptiles to mammals, and is seen even in a few species of amphibians and fish. Slow-wave sleep is well developed in birds and mammals. REM sleep is developed in most mammals and birds. Studies in mammals have not shown correlations between sleep parameters and body temperature, presence of hibernation, or life span.

A large proportion of the research has involved the phylogenetic distribution of REM sleep in mammals, because this stage of the sleep–wake cycle has been considered a feature of higher vertebrates. Data from numerous species have allowed phylogenetic comparisons between mammal groups. An overlap in the total amount of REM sleep is apparent across orders, ranging from 0 h to a maximum of 8.0 h (Table 11.2). However, more than 100% variation occurs within particular orders. REM sleep has been documented in all mammals except dolphins, though the data about dolphins may be an artifact of the recording method, which requires recording a restrained dolphin. Other cetaceans have very low amounts of REM sleep, and also very small amounts of total sleep. Similarly, the reported lack of REM sleep in the echidna, a monotreme, may be an artifact. The echidna exhibited sleep properties different from typical recording in other mammals, which made interpretation of the data ambiguous.

The issue of REM sleep in echidnas is important in light of recent careful studies of the only other monotreme, the duck-billed platypus. In this unusual species, the total quiet time resembling sleep ranged from 5.8 to 8 h, with sleep resembling REM constituting most of the sleep period (Figure 11.5). The motivation for intense study of the monotremes lies in the question of evolutionary origin of the fundamental properties of sleep. Early anatomical and physiological studies of the brain suggested that the pons was phylogenetically an old part of the brain, and that it was also a vital part of

TABLE 11.2   Amount of REM sleep in several groups of mammals

Group	Example(s)	REM sleep (number of hours per 24 h)
Monotremes	Platypus	5.8–8.0 (of total 8 h quiet state/day)
Marsupials	spp.	1.5–4.0
Rodents	Ground squirrel	0.8–3.4
Carnivores	Cat (domestic)	1.3–3.2
Primates	Owl monkey	0.7–1.9
Edentates	Sloth, armadillo	3.1–6.1
Artiodactyls	Cattle, giraffe, pig	0.6–2.4
Cetaceans	Dolphin	0.0

Source: Data from J. M. Siegel.

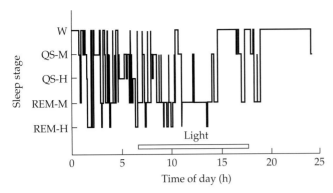

FIGURE 11.5 Sleep stage structure in the duck-billed platypus. W, wake; QS-M, quiet sleep, moderate voltage; QS-H, quiet sleep, high voltage; REM-M, REM sleep, moderate voltage; REM-H, REM sleep, high voltage. (From Siegel et al., 1999.)

REM sleep. These facts argued for the early evolution of REM sleep. Arguments against this theory are the occurrence of vivid dreams in humans, suggesting a late origin of REM sleep. Evidence for the late derivation of REM sleep also comes from the lack of its occurrence in reptiles and amphibians.

ECOLOGICAL APPROACHES IN MAMMALS. The second trend has been an ecological perspective that covers field observation and laboratory recording plus experimentation for as many species as possible. Close members of a taxonomic group with extreme ecology and behavior patterns have been emphasized. Presumably the correlation of sleep characteristics with ecological habitats will give a better understanding of the function of sleep. Much emphasis has been placed on mammalian REM sleep as the measurement parameter.

Several consistent ecological correlations are seen with respect to mammalian REM sleep. Generally, the amount of REM sleep parallels the amount of non-REM sleep, and the amount of REM sleep time is correlated with total sleep time. A high correlation is also seen between non-REM sleep time or total sleep time and body size. The correlation is negative: the larger the animal, the less it sleeps.

Another correlation concerns the relationship of amount of REM sleep in mammals relative to the security of the sleep site. This link reflects the ecology of a species, especially predator–prey status. The more secure the sleep site is, the greater the amount of REM sleep that is shown. Consequently, predators and cryptic sleepers with concealed sleep sites can apparently afford extended periods of sleep. Lions are the proverbial sleepers of the animal kingdom, and they exhibit large amounts of total, as well as REM, sleep. Similarly, most cats, including tigers and domestic house cats, have large amounts of total and REM sleep. In contrast,

animals with unprotected sleep sites, such as cattle, giraffes, and other large herbivores, have relatively low amounts of total sleep, including REM sleep.

A positive correlation is also found in the relationship of the amount of REM sleep and the degree of helplessness at birth. Altricial animals, which are born immature and helpless, such as marsupials, exhibit relatively large amounts of REM sleep. The amount of REM sleep decreases in the adult but still is high compared to advanced or precocial mammals. Humans, with the most altricial birth state of any primate, exhibit the highest amount of REM sleep, 1.9 h per day. A close runner-up is the owl monkey, with 1.8 h per day. No correlation has been found between REM sleep and brain size, ratio of brain to body weight, or circadian activity type relative to the day–night cycle.

DIFFERENCES IN REM SLEEP BETWEEN BIRDS AND MAMMALS. Birds contradict the mammalian ecological correlations. All birds exhibit REM sleep. The cycling, however, is simpler than in higher mammals. As with nonprimate mammals, substages of non-REM sleep cannot be distinguished. Compared to mammals, birds in general are small. Many are predators. Most predatory birds are day-active and seek secure sleeping places under cover of darkness. The great majority of species have extremely altricial young that hatch from a vulnerable egg. Exceptions are the precocial ground-nesting species, such as most waterfowl and almost all shorebirds, gulls, and terns.

On all these counts, birds might be expected to show large amounts of REM sleep, but the data come as a surprise. Most birds average less than 30 minutes of REM sleep per day. Even more surprising is the fact that the largest birds show the greatest amount of REM sleep. The generalizations from mammals are diametrically opposite to bird results.

UNIHEMISPHERIC SLEEP IN MAMMALS AND BIRDS. Another correlation concerns the relationship of unihemispheric sleep with the need for constant vigilance both during daytime and nighttime. A few species of mammals and birds can sleep with one brain hemisphere while the other hemisphere maintains a vigilant state. Without a doubt, the ecological function of unihemispheric sleep is protection, because it occurs primarily in animals with specialized sleep vigilance needs. Best known of these species are cetaceans such as dolphins. During periods of rest, dolphins may keep one eye closed, on the side opposite the awake hemisphere. Similar results have been reported in pinnipeds such as seals. Most of the dolphin research has involved behavioral observations. Some doubt exists about the data on eyelid position because of problems in recording it while dolphins are actively swimming.

Excellent new data on unihemispheric slow-wave sleep and the state of the eyes have very recently become available for the beluga, or white, whale. This species is a high-arctic toothed whale with a number of extreme behavioral and physiological adaptations for surviving in icy waters, including the ability to sleep underwater for extended periods of time without sounding to the surface for breathing. Eye position, swimming motion, and brain wave activity were simultaneously recorded over a period of 4 days. For the most part, the results supported the hypothesis of eye closure correlated with unilateral sleep of the opposite brain hemisphere (Figure 11.6). An overall relationship between the sleeping hemisphere and eyelid position was observed, but rapid changes in eyelid position were not always followed by immediate changes in EEG. Eye position and EEG were highly correlated over long periods of time, but they could be independent over short periods of time lasting less than 1 minute. This capability allows both beluga whales and dolphins to breathe, sleep, swim, and monitor their environment simultaneously. The ability may be very important for a species that must constantly use swimming movements to maintain a stable position in a hydrodynamically changing environment.

Similar observations of unihemispheric sleep have been made in aquatic birds. Particularly interesting are data from social birds such as ducks in flocks. Like cetaceans, ducks can apparently keep one eye scanning while sleeping with the opposite brain hemisphere. A scanning duck in behavioral sleep position can react to a threatening visual stimulus in a fraction of a second. Data support the hypothesis that this type of sleeping is

an antipredator adaptation. Birds more exposed to predation use such a sleep pattern more often than birds in the inside of a group, and the eye that stays open most is the one facing the outside of the group.

Although birds and wild mammals are good alternatives to humans for circadian studies such as sleep, other even simpler models may be necessary in future circadian research[26,27]

Extensive studies of the CNS of mammals have yielded a relatively detailed understanding of the anatomical structures regulating sleep and other circadian factors. Much is also known about systemic humoral and metabolic factors that influence states of arousal, and about the neurotransmitter systems involved in sleep and vigilance. A well-organized array of changes occurs in neural firing and in neural networks. The electrophysiological changes that accompany normal daily changes in consciousness have also been extensively investigated. However, the identification of a single gene related to sleep, such as the gene regulating advanced sleep phase syndrome (ASPS), is a quite recent event (see Chapter 10). Even though it has been identified, the role of this gene in normal sleep–wake regulation is not understood. Thus, the molecular basis of sleep regulation remains a frontier.

The application of modern tools of molecular biology to an array of animal models strongly favors the use of simple models such as the nematode or fruit fly. Since the genetics of these species are relatively well known, genetic engineering approaches are possible. In addition, the central nervous systems of these organisms, particularly fruit flies, are relatively simple. In contrast, functions in the complex brains of vertebrates are much more difficult to study. For example, the reticular system of the human brain extends throughout the central core of the brain and regulates levels of vigilance and the sleep–wake cycle. A part of the reticular system called the reticular activating system maintains the state of wakefulness by high neuronal firing rates. The midbrain, as well as the pons and medulla, is involved. Consequently, finding single genes with major effects on the sleep–wake cycle and tracing their mode of action is much more challenging than in simpler organisms.

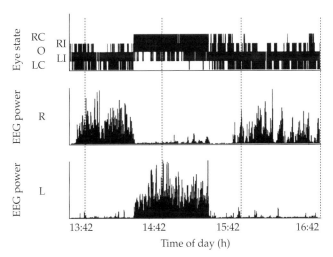

FIGURE 11.6 Relationship between EEG and state of the eyes in a beluga whale during hemispheric sleep for 3 h. The top trace shows data for the state of the eyelids (RC, right eye closed; RI, right eye intermediate; LC, left eye closed; LI, left eye intermediate; O, eyes open). The middle trace is the EEG for the right hemisphere; the left-hemisphere EEG trace is at the bottom. (From Lyamin et al., 2002.)

Theme 3: Simple Invertebrates Hold Out Promise for Studying Molecular Aspects of Sleep–Wake Rhythms[26,27]

Many circadian questions about sleep remain unanswered[7,11,30,52,65,70]

PROBLEMS IN SLEEP REGULATION STUDIES. The difficulty of moving beyond interesting theories to pinpoint the genes involved in regulating any circadian process such as sleep is that sleep is a very complex behavioral state. Traditional studies have investigated sleep in mammals and birds. Episodes of sleep in these species are readily recognizable because they all share the major characteristics of human sleep. In reality, the behavioral hallmarks of prolonged immobility, reduced responsiveness, and homeostatic rebound after deprivation are not restricted to birds and mammals, but are found in a number of species throughout the animal kingdom. Innovative research on circadian mutant models is already helping speed up the rate of success in sleep research.

Two phylogenetically simple animals are possible models for the molecular genetic study of sleep because even relatively simple animals share a large number of their genes with humans, including major classes of genes involved in neural function in human diseases. Consequently, widening the scope of animal sleep models could be a useful strategy. Research in the future will make more and more use of animals in which genetic research is relatively simple, productive, and inexpensive (Figure 11.7). Although a simpler model system sacrifices the obvious relevance of studying mammals, it allows for very important gains in efficiency and comprehensiveness. The point of studying a simpler animal that exhibits a sleeplike state is to approach the question of whether conserved genetic mechanisms exist for regulating the states of arousal. Broadly conserved functions for the state may exist.

ADVANTAGES OF *DROSOPHILA*. Much of the sleep research with relatively simple invertebrate species has focused on the nematode worm, *Caenorhabditis elegans*, and the fruit fly, *Drosophila melanogaster*. Historically, fruit flies have provided an excellent system for understanding genetics. Fruit flies are effective models for studying genetic control of not only development and metabolism, but also such complex behavior patterns as courtship, learning, and circadian rhythms.

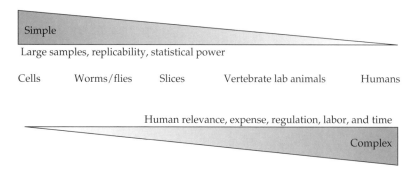

FIGURE 11.7 Factors affecting the choice of a model system for sleep research. (Courtesy of J. Hendricks.)

Drosophila offers excellent genetic databases, tools for manipulating genes, and the benefits inherent in any simpler system. Automated methods for studying locomotion are also well established and can easily be modified to analyze the data for rest, and also for activity. Adult flies have a life span of several weeks, during which they exhibit an array of complex behavior patterns. In contrast, worms live for only a few days and relatively little circadian information is known about the worm of choice, *C. elegans*.

Fruit flies have a circadian rest phase with behavioral sleep features that include a preferred posture and location, and an increased response threshold. Prior to entering their circadian rest phase, they adopt a particular posture, face down on the floor of the container, and then remain immobile, except for respiratory movements, for periods up to 2.5 h during an average total consolidated rest period of 7 to 8 h. Fruit flies also show evidence of needing sleep. If flies are not allowed to rest during the usual consolidated nighttime rest period, they exhibit a subsequent rest rebound with increased rest during the ensuing recovery period.

Studies have been carried out with Drosophila *on an important cellular signaling pathway*[28]

Most eukaryotic cells contain signaling pathways that allow extracellular changes to affect gene expression in the nucleus. Such pathways allow cells to change their complement of proteins and thereby allow animals to adapt to environmental changes. In a common signaling pathway, binding of small molecules to receptors on the cell surface results in increased production of a modified nucleotide called cyclic adenosine monophosphate (cyclic AMP or cAMP for short). Cyclic AMP in turn binds to a protein complex, resulting in the release of an active kinase, PKA, that moves to the nucleus and phosphorylates a protein called CREB. This phosphorylated CREB then has the ability to activate gene expression. The cAMP–PKA–CREB pathway is important for learning and for consolidation of long-term memories. An initial question was whether the process was also important for regulating sleep and rest. Because long-term memory consolidation is apparently fostered by sleep in mammals, changing the activity of genes in the pathway might change sleep and waking.

Varieties of flies are available with mutations and transgenes that change the activity of components of the pathway. Experiments showed convincingly that activity of the cAMP–PKA–CREB pathway is negatively correlated with the amount of daily rest. In the six genotypes examined, the more cAMP activity in the

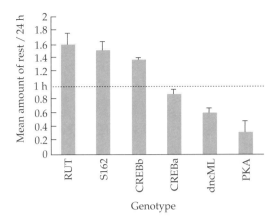

FIGURE 11.8 Relationship of cAMP activity to rest time in six mutant or transgenic genotypes of fruit flies. When the genotypes are arranged from left to right in terms of increasing mean amount of cAMP activity, a negative correlation is seen between the amount of rest per 24 h and the cAMP activity in controls, which is indicated by the horizontal line. (From Hendricks et al., 2001.)

cAMP pathway, the less rest time the animals experienced (Figure 11.8). Blocking CREB activity produced animals that had great difficulty recovering from rest deprivation. These animals did not return to normal amounts of waking even 3 days after rest was disrupted for 6 h. Because of the nature of this pathway, this discovery should improve an understanding of which CREB-related genes foster wakefulness. Modern tools can help identify genes that are regulated by CREB. Microarrays can be used to examine the genes whose expression is changed when CREB activity is altered. Further comparative studies of such candidates in fruit flies and mammals could show how this molecular pathway changes the ability of animals to stay awake and to recover from rest deprivation.

A second productive line of investigation focuses on the role of Clock genes in regulating rest[27,53]

Selective disruptions of clock function in flies have been carried out as described in Chapter 7. Because the clock regulates rest–activity timing, these disruptions should change the timing of rest and activity. Surprisingly, they also changed the total *amount* of rest needed. Moreover, when mutants of all of the central clock genes were studied, flies with a null mutation of the *cyc* gene showed a pronounced and unique phenotype of changes in rest regulation that are different in males and females. Females with a null mutation responded to rest deprivation with an enormously exaggerated rest rebound. Some of these flies died when rest deprivation was prolonged. To test whether the rest deprivation was a general stressor, the effect of other stressful manipulations was examined. The flies were not generally frag-

ile, but instead seemed specifically susceptible to rest deprivation. Clearly these mutants are abnormally vulnerable to rest deprivation.

The mutants also failed to increase expression of a group of genes normally induced by thermal stress, those that encode heat shock proteins. These genes generally respond to stress, including rest deprivation. Through the use of a transgene that increased the expression of specific heat shock proteins, the abnormal rest rebound and the lethal effect of rest deprivation could be counteracted in the mutant flies. Thus a possible link has been established between rest need and stress tolerance.

The identification of specific candidate genes linking rest deprivation to a lethal outcome is an important finding because it can be used to drive genetic selection for genes that encode the proteins involved in this regulation. Further studies will focus on understanding how rest deprivation produces these changes, and the generality of this mechanism. Investigations of gender differences in *cycle* mutants will add an additional dimension. Studies of gender differences in sleep need are just starting in mammals. If a gender dimorphism is found in fruit flies, it may allow for rapid progress in understanding the molecular basis of this difference.

Like CREB, CYCLE is a protein that regulates gene transcription. CYCLE acts in the *Drosophila* core circadian feedback loop with its partner CLOCK to make a dimeric complex that activates expression of several genes in the circadian cycle, including *per, tim,* and *vri* (see Chapter 7). CYCLE–CLOCK is the *Drosophila* counterpart of the PAS–PAS heterodimeric transcription activator found in other eukaryotic clock feedback loops. Its mammalian counterpart is BMAL1. The use of molecular biological tools to identify which *cycle*-regulated genes are involved in the rest phenotype should be straightforward.

Because both the cAMP–PKA–CREB pathway and the core circadian feedback loop involving CYCLE are conserved in mammals, new findings about rest in *Drosophila* should suggest fruitful approaches in mammals. The number of mutants and transgenic tools available to investigate genetic pathways is generally greater in flies than in mammals. In addition, studies in mammals are much more time-consuming and expensive. For these reasons it is easier to try experiments that use the homologous genes and pathways in flies before attempting them in mammals. General regulatory mechanisms are often conserved along with the precise roles of genes. This was the case with the circadian feedback loop.

Can single-gene mutants affecting sleep be identified?[53]

Predicting the molecular mechanisms that mediate waking and sleep is difficult. If states of consciousness were regulated like circadian rhythms, then a particular molecular pathway, or a small group of interacting genes, might produce wakefulness and sleep. Currently available data makes that hypothesis seem unlikely. In the SCN of mammals, a small collection of neurons in a circumscribed location is necessary and sufficient for circadian rhythms in locomotor activity, but a comparable sleep center has never been found. Most of the mammalian brain normally participates in changes in consciousness. Global interaction of multiple genes is probably also the case in fruit flies.

Whereas mammals can survive reasonably well without a circadian clock, sleep may be crucial for survival. If sleep is truly a vital function, complete abolition of sleep may not be achievable in a viable mutant. A vital function may also be mediated by multiple, redundant genes. As a result, knocking out any single gene or genetic pathway may not abolish sleep. Finally, the genes that encode the CYCLE and CREB proteins and are known to participate in sleep are also involved in other important aspects of cell function. Manipulating the function of such genes may not have a specific effect on sleep. Consequently, drawing conclusions about causation may be difficult.

Although the answer is likely to be complex, the search should not stop. Even in elaborate biological systems, meaningful insights can be gained from studies of the roles of specific pathways. Computational and genomics tools are available to examine both the specific genes involved in behavior patterns and the quantitative levels of expression of the genes in the genome during different behavioral states. Such a nuanced, quantitative approach will be needed to dissect the multiple interacting systems involved in sleep as well as in other circadian behavioral processes. The eventual schematic of the molecular biology of sleep may not be a linear pathway. Instead, it may be a network with a rank order of the various cell signaling and metabolic pathways having different but partially overlapping roles. These roles may also show tissue and species specificity, and perhaps gender dimorphism. Even developmental and aging differences may be important.

In conclusion, the fruit fly model illustrates the power of comparative biology. Fruit flies are small, efficient, and inexpensive to use. They also provide a broader phylogenetic base for understanding circadian processes.

Theme 4: Another Breakthrough Discovery Is the Widespread Distribution of Peripheral Pacemakers[4,51,56,69]

One of the first mammalian pacemakers to be discovered was the retinal clock of golden hamsters[60,68]

For many years after the discovery of the SCN in 1972, the mammalian circadian system appeared to be monolithic

in having only one chief central pacemaker. The central driving oscillator of the SCN was viewed as a major evolutionary change from the multiple-oscillator and multiple-photic-input arrangements seen in all other vertebrate groups. Then intriguing new data in mammals suggested that independent local oscillatory systems existed outside the SCN. In 1996 a surprising new circadian mammalian pacemaker was announced, residing locally in the retina. By culturing whole retinas in a flow-through chamber, investigators could follow the melatonin synthesis and secretion rhythm for as long as 5 days. The properties of this system resembled those of the SCN itself.

In retinas of wild-type hamsters, the free-running period in constant darkness was close to 24 h; in *tau* mutant hamsters, the period was shortened, having values similar to those of free-living intact mutants. The retinal rhythms could be phase-shifted by light and entrained to LD cycles. The fact that the rhythms persisted in isolated retina and could be phase-shifted as well as entrained indicated that both light input and the oscillator itself were located in the eye.

Because of the diversity of cells in mammalian retina, no information is available for the hamster retinal clock on localization of the circadian functions in particular types of retinal cells. Little is known of the cell physiology of such peripheral oscillators. However, very recent data suggest that the retinal clock is not merely a peripheral local pacemaker within the eye, but that it actually interacts with the SCN to regulate its rhythmic output. This first indication of extra-SCN autonomous oscillators in mammals foreshadowed the recognition of many other independent or at least local self-sustaining oscillators.

One recent promising development in circadian research has been the application of transgenic technology to cell culture of insect and mammalian pacemaker tissues[44,50,56,67,69]

OVERVIEW. Transgenic marker techniques had previously been used successfully for the unicellular cyanobacterium *Synechococcus* and for the mustard plant *Arabidopsis*. The adaptation of the general reporter technique proved more difficult for advanced invertebrates and vertebrate species. Then in 1997, the development of reporter systems in *Drosophila* gave new impetus to the study of peripheral oscillators in arthropod invertebrates and mammals (see also Chapters 5, 6, and 7). The bioluminescent luciferase gene *luc* was fused to the *per* gene of *Drosophila* to form a *per–luc* construct. By this means a rhythmic visible reporter system was made available that could easily be tracked over considerable periods of time.

MULTIPLE PERIPHERAL CLOCKS IN *DROSOPHILA* REVEALED BY *PER–LUC* TRANSGENE TECHNOLOGY.   The great advantage of the *per–luc* construct is that it is an excellent longitudinal marker for following the phase and period of circadian rhythms with minimal disturbance to the organism. Bioluminescent light emission is recorded photographically.

An alternative method is the use of *per*-driven green fluorescent protein as the reporter. Using *Drosophila*, scientists examined *per* oscillations in whole insects and in cultured body segments, including head, thorax, and abdomen. Rhythmicity was expressed in DD and LD by the intact fruit fly. The surprising fact, however, was that all three segments in culture also showed robust and widespread rhythmicity. In DD the rhythms free-ran but with relatively low amplitude. In LD the rhythms entrained to a shifted light cycle. Other body parts, such as proboscises and antennae, were also oscillatory in LD and DD (Figure 11.9). The expression was found at nearly single-cell level in legs and wings. Furthermore, scientists made the major discovery that every oscillatory tissue in *Drosophia* is photoreceptive.

Why hadn't the phenomenon of self-sustained oscillatory activity in multiple parts of the body been noticed previously for more animals? Probably the reason involved the method of recording rhythmicity. The initial evidence for central pacemakers had come primarily from behavioral data of intact organisms. Locomotor activity of individual flies and eclosion rhythms of populations were the preferred overt rhythms measured. Only when transgenic reporter methods were used did the tissue- and cell-level rhythms become apparent in *Drosophila*.

PERIPHERAL CLOCKS IN CULTURED RODENT TISSUES REVEALED BY TRANSGENIC MARKERS. The prevailing idea for several decades in chronobiology research had been that multiple oscillators were prevalent only in nonmammalian groups. The overlapping roles of pineal gland, retinal pacemaker, and SCN pacemaker of reptiles and birds were well-known examples. An initial break in that unified facade was the demonstration of an autonomous circadian clock in hamster retina, as described in the previous section.

At about this time, one group of scientists made an important discovery. Fibroblasts that had been in culture for a quarter of a century could be induced to become rhythmic again. The identification of this phenomenon was possible because clock genes and proteins, in this case *Per1* and *Per 2*, were known, and their expression could be followed without a detailed understanding of the rhythmic biology of these cells. One important extrapolation was obvious. If fibroblasts have a clock, why not liver cells, heart cells, hair follicles, or other cells that cannot perceive light?

Soon another piece of evidence for independent peripheral oscillators came from *per–luc* transgenic rats.

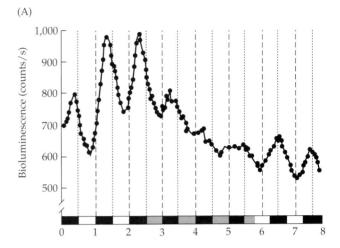

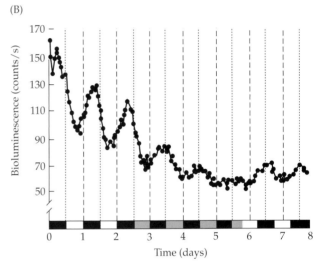

FIGURE 11.9 Free-running circadian bioluminescence rhythms in *Drosophila* tissues showing expression of *per*-driven green fluorescent protein. (A) A single isolated proboscis. (B) Antennae in culture. LD on days 1–2 and 6–7; DD on days 3–5. (From Plautz et al., 1997.)

Much as in the case for *per–luc Drosophila*, the transgenic rats express the *Per* gene activity visibly by means of a bioluminescent luciferin marker tag linked to the *Per* gene. When SCN tissue was explanted from a *Per1–luc* transgenic rat and maintained in constant conditions, a circadian rhythm with a near 24 h period persisted for 32 days in vitro. Even more surprising is the persistence of rhythmicity in cultured pieces of liver, lung, and skeletal muscle taken from the transgenic rats (Figure 11.10).

Further discoveries soon materialized. Peripheral clocks appeared regulated loosely by the SCN but reentrained independently after abrupt phase shifts of the light schedule. Using the transgenic rat model, researchers studied effects of competitive light cycles and food availability as entraining agents in intact rats.

(A)

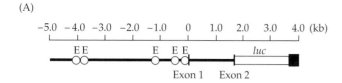

(B)

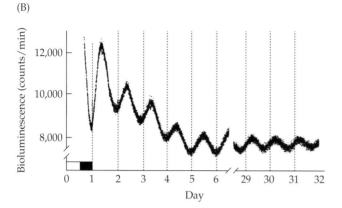

(C)

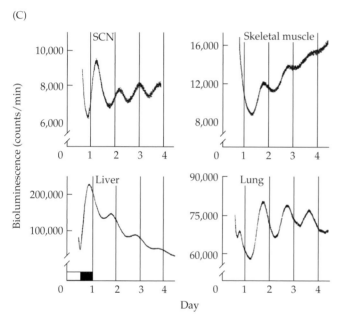

FIGURE 11.10 Rhythmicity of cultured transgenic liver cells. Bioluminescent reporter genes were used to record circadian activity. (A) Gene map. kb, kilobases. (B) SCN in DD for 32 days. (C) Comparison of persistent rhythmicity in four types of tissue of *Per–luc* transgenic rats. (From Yamazaki et al., 2000.)

The liver clocks could be entrained by restricted feeding schedules, much as intact rats entrain to restricted feeding schedules. Food entrainment occurred even when a light cycle was provided that was out of phase with the feeding schedule. Although SCN rhythmicity remained linked to the LD regime, the liver was rapidly entrained by restricted feeding. Shifts as large as 10 h were accomplished in 2 days. These results challenge the well-estab-

lished concept of a mammalian circadian hierarchy dominated by the SCN as a central pacemaker.

In fact, many types of cells do have clocks. The old view of a dominant clock in one tissue like the SCN has given way to a new view of a clock shop of highly interconnected oscillators. The biology of most of these oscillators and the spectrum of rhythmically regulated activities that they control are still largely unknown. In recent years this shift in thinking has resulted in a name change from *circadian clock* to *circadian program*. The new term denotes the distinction between a single pacemaker dictating time for the organism and a group of intercoupled clocks collectively providing an internal measure of time correctly coordinated with dawn and dusk through seasonal changes in day length. The program is a clock shop of coupled oscillators, some self-sustaining and some damped.

Theme 5: Single Independently Oscillating SCN Clock Cells Are Coordinated into a Multicellular Regulator of Behavioral Output Rhythms[32,57]

Chimera technology is providing answers about the control of circadian behavior by the ensemble of oscillators in the SCN[32,57]

BACKGROUND. Recently single pacemaker cells have been shown to oscillate independently in culture, with a cell-specific free-running period (see Chapter 6). Obviously the differing periods of many clock cells must somehow be modulated to yield an integrated output command to the higher organizational levels and eventually to behavioral output. One approach to understanding the interactions of a multiplicity of SCN oscillators has been the chimera paradigm. In this demanding technology, chimeric mice are produced, which have varying proportions of mutant SCN cells and wild-type SCN cells. In the experiments described in the next section, the differing proportions of cells across a series of chimeric individuals gave clues to the process of circadian regulation of behavior. In 2002, Sharon Low-Zeddies was a co-recipient of the Donald B. Lindsley Prize in Behavioral Neuroscience from the Society for Neuroscience for outstanding Ph.D. thesis of the year in recognition of this elegant study.

METHODS. The idea is simple; the procedures are very complex. In 1994 a semidominant mutation called *Clock* was induced in mice that resulted in prominent modification of the circadian period and amplitude of the running-wheel activity rhythms, as well as phase shifting by light. The wild-type period was about 23.7 h, with a robust circadian rhythm and a normal phase response curve. Heterozygous mice differ by showing lengthened

FIGURE 11.11 A study of mammalian SCN pacemaker function using *Clock* chimeras. (A) Chimeric blastocysts implanted in foster mothers gave rise to *Clock* chimeric offspring (top photo) whose circadian phenotype was then measured quantitatively in terms of coat color, wheel-running activity, and SCN histology (bottom photo). (B) Two component strains were used to construct the chimeras. The wild-type strain (left) had a pigmented coat and a LacZ marker for SCN cells, which stained dark blue. The albino homozygous *Clock* mutant strain (right) had a colorless coat and did not stain for LacZ. (C) Double-plotted activity records indicate the circadian activity pattern consecutively in LD 12:12, in DD, in response to a 6 h light pulse (horizontal arrows), and in a final LD schedule for wild-type mice (left), genetic control F1 *Clock/+* mice (middle), and *Clock/Clock* chimeras (right). (D) Fourier analyses of circadian amplitude for the three groups as relative power spectral density (rPSD) for the 20 day DD interval between the light pulses. (E) Periodogram for the 20 day DD period between light pulses, demonstrating the dominant circadian period in wild-type and genetic control F1 mice, and arrythmicity in *Clock/Clock* chimeras. (Modified from Low-Zeddies and Takahashi, 2001.)

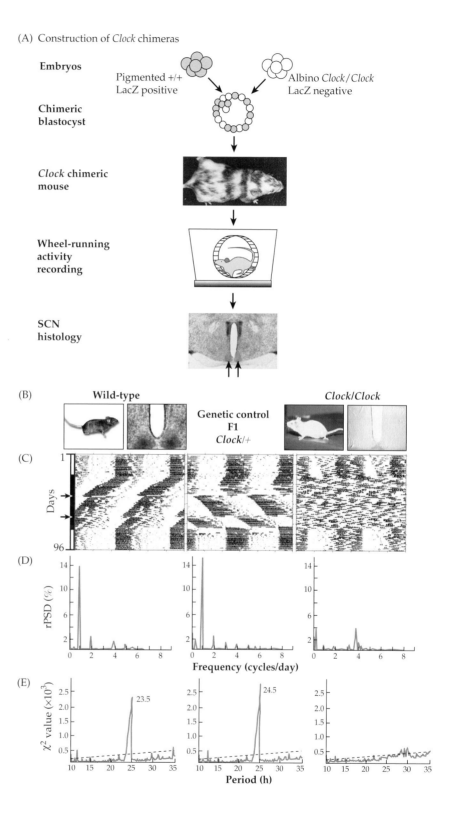

free-running periods of 24.5 h, decreased amplitude of rhythm, and a heightened response to phase shifting by light signals. Homozygous mice were usually initially weakly rhythmic, with an exaggerated lengthening of period between 27 and 29 h under constant conditions, but they quickly became arrhythmic (see Figure 11.11B). The period has now been shown to exist in single oscillating pacemaker cells from wild-type and *Clock* mutant mice under in vitro culture conditions. These features were utilized in the chimera experiment.

Two strains of mice were first generated to provide suitable embryos for creating the chimeras (Figure 11.11A and B). One was a wild-type strain with coat color as an overt marker and a LacZ cell marker for the SCN cells. This strain exhibited a normal circadian period for running-wheel activity. The other strain was an albino homozygous mutant mouse with arrhythmic circadian period that lacked the LacZ marker. Pairs of wild-type and mutant eight-cell stage

Clock/Clock chimera circadian behavior

Wild-type
behavior

Mutant
behavior

◀ **FIGURE 11.12 Wheel-running records for 137** *Clock* **chimeras.** The data are arranged from left to right, top to bottom, grading from typical wild-type wheel-running patterns at top left to typical homozygous *Clock* mutant wheel-running patterns at lower right. (From Low-Zeddies and Takahashi, 2001.)

embryos were aggregated in vitro to form chimeric blastocysts. The composite embryos were then transferred into surrogate mothers to continue development.

Random intermingling of wild-type and mutant cells takes place within a chimeric embryo during development such that each chimeric individual carries a unique combination of the two different cell types. A total of 137 *Clock* chimeras were born and used for data collection. After weaning they were placed in running wheels, and locomotor activity was recorded for 96 days. Rhythms were recorded first in LD conditions, then in DD conditions with two incidences of exposure to a phase-shifting light signal, and finally the mice were returned to LD conditions (see Figure 11.11B). Coat color was recorded. Histological analysis of the SCN took advantage of the LacZ marker, which is bright blue for wild-type cells and colorless for *Clock* mutant cells. In addition to the experiments described already, studies of several control groups were necessary to avoid any ambiguity in interpretation of the results.

A graded range of phenotypes is found in the Clock *chimeras, from typical wild-type circadian behavior to homozygous mutant behavior*[32,57]

ACTIVITY RHYTHMS. The clean design of the chimera experiments allows quantitative evaluation of the effects of the chimeric mixing with a representative range of wild-type and mutant clock oscillator cells. In terms of activity rhythms, it was possible to arrange the patterns in an orderly sequence. At the top left point of the series shown in Figure 11.12, LD entrainment, free-running period, and phase shifting were purely wild type in pattern. The effect is modulated in the progression from wild type to initial minor change in pattern, to moderate change in pattern, to mutant actographs. Finally, in the lower part of the composite figure, the circadian activity patterns appear purely homozygous in the *Clock* mutant.

COAT COLOR. An ordering of coat color paralleled the activity composite. Individuals with the darkest coat colors, at the top of the composite color chart, also had the most confirmed wild-type actographs. Individuals further down the series were paler and paler until finally they were almost entirely the white albino mutant color. The degree of correlation between coat color and activity pattern was very high.

HISTOLOGY. Similarly, the histology series paralleled coat color and activity pattern in a graded progression. The more wild type the circadian phenotype, the more wild-type cells were present in the SCN, as indicated by degree of LacZ staining. A clear majority of either wild-type or mutant SCN cells was required to dominate the behavioral phenotype. It was not the case that a small number of either wild-type or mutant cells could "rescue" or drive behavior.

About a third of the chimeras, appearing in the middle of the sequence, expressed an intermediate degree of mutant properties. Intermediate behavior results when the number of mutant and wild-type cells are approximately balanced. For example, some chimeras behaved like *Clock* heterozygotes, which is surprising because they were composed solely of wild-type and homozygous *Clock* mutant cells. Other intermediate phenotypes further suggested that circadian period and circadian amplitude are determined by different complements of cells.

In no cases did the analyses of wheel running show any evidence of multiple components of circadian rhythmicities of differing frequencies. The multiple periods of the individual cellular oscillators were smoothly integrated into one behavioral output frequency. The period range is a figurative dose response curve of cell-type representation. The authors concluded that intermediate behavioral periods indicated that SCN cells interact to determine the realized circadian period.

Theme 6: The Pressing Global Issue of Ecosystem Integrity and Biodiversity Has Circadian Aspects[4,9,18,23,42,43]

Ecosystems are relatively stable configurations of biomass and species diversity coexisting in a particular set of climatic and physiographic conditions[9,46,66]

SPECIFICITY OF ENVIRONMENTAL ENTRAINING AGENTS. One major circadian breakthrough of 2002, the localization of retinal photoreceptors in mammals (see Chapter 6), may have surprising ecological connotations. Unexpected was the finding that mammalian retinal cells other than the imaging rod and cone cells could contain a photopigment and be light sensitive. The discovery highlights the important fact that extraretinal photoreceptors occur widely in nonmammalian animals. Many of these photoreceptors appear finely tuned by natural selection over eons of time to carry out functions other than visual imaging. Most are radiance detectors with properties other than short-duration, high-resolution response. They are often located near the target organ and can have the properties precisely needed for a specific task.

One of the greatest ecological threats of all time is the massive elimination of biodiversity on the planet by habitat destruction, overharvesting, and various types of pollution. If evolutionary processes have honed circadian photoreceptors for such specialized purposes, what will be the effect on species survival of tinkering with environmental lighting and temperature entraining signals? The question is particularly important for species already on the brink of extinction.

The following section will look briefly at rhythmic aspects of ecosystems. It will then present data on the correlation of photic entraining thresholds with particular environmental lighting conditions. Finally, field data about daily activity patterns of the desert iguana and its ant prey will demonstrate the potential effects of slight temperature changes on the window for predation.

ECOSYSTEM FEATURES. Although ecosystems are very diverse, they can usually be characterized by three or four diagnostic features. Daily and annual temperature patterns are important. These may be determined by latitudinal site, from equator to the poles, or by vertical elevation from sea level to the tops of highest mountains. Water availability in terms of annual rainfall or snowfall throughout the year is another key feature. Substrate is often critically important. Photic environment, in terms of absolute intensity as well as the duration and amplitude of the LD cycle, also plays a role. Different combinations of these factors define specific ecosystems (see Chapter 2; see also Plate 1).

An ecosystem is balanced and stable in terms of biomass and species diversity. Each species is delicately poised in a web of relationships with its physical and biotic environment in such a way that competition within the ecosystem is minimized. The optimization of resource utilization by lack of competition is an important part of the great stability of each ecosystem. An often overlooked factor is that rhythmic environmental features can also play a vital role in species survival within a habitat. The glaring sunlight pattern of a desert is as vital for photoentrainment of the circadian rhythms of a desert ground squirrel as the shadowy light of a rain forest for a nocturnal rodent.

CHANGES IN ECOSYSTEMS. Most small disturbances do not seriously damage an ecosystem because of its great inherent stability. A severe windstorm may blow down a few trees in a redwood forest and create a few openings, but seedlings grow rapidly toward the newly available light and soon close the gap. Major calamities, however, may overwhelm an ecosystem, and it may not be able to withstand truly cataclysmic changes. Subsidence of coasts may result in saltwater flooding and eventual inundation of coastal forests that may lead to long-lasting effects or permanent extinction of the forest.

On a global scale, Earth has sustained catastrophic changes, including collisions with giant meteorites and periodic coolings that have changed the face of the planet and altered the very course of evolution of species. When considered on a geological timescale, these almost unimaginable events have a certain regularity. Five such events have been so titanic as to earn the name *major extinctions*. Another is now in progress.

The geological record of life on Earth bears witness to five worldwide major extinctions during the past half-billion years[9,20,46]

THE FIVE PREVIOUS EXTINCTIONS. A glance at a geological timescale with the fate of its fauna superimposed on it can be sobering (Figure 11.13). At the beginning of the

FIGURE 11.13 Major biotic extinctions on Earth.

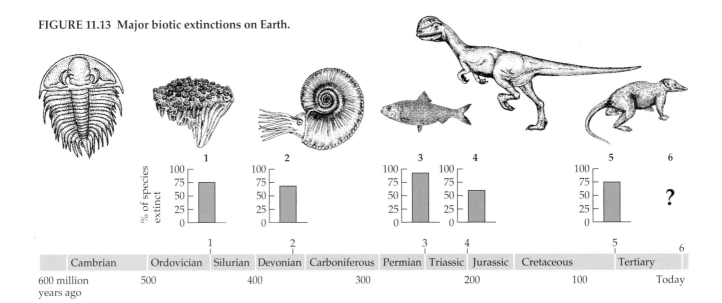

Silurian era, prolonged glaciation decimated almost three-quarters of all families of organisms, especially marine invertebrates. Near the end of the Devonian, 70% of invertebrates and most fish families perished. In the Permian the greatest extinction of all time took place, and almost 95% of all families were eradicated. At the end of the Triassic, 75% of marine invertebrates vanished, perhaps in a meteorite-caused catastrophe. The Cretaceous–Tertiary (K-T) extinction again eliminated 75% of all families, including the dinosaurs.

THE HANDWRITING ON THE WALL: A CURRENT, SIXTH, HUMAN-CAUSED EXTINCTION? The extinctions caused by geological factors have a certain inevitability about them. More disturbing is the mounting realization that humans are causing the sixth and possibly final mass extinction of life on Earth. In approximately one million years, humans have fanned out from sites of origin and populated most of the habitable parts of the planet. Nonhuman species have gone extinct from fairly obvious causes. Hunting eliminated almost all the species of very large grassland herbivores. Farming and logging destroyed forests.

Currently the litany of insults to the planet is growing at an ever increasing rate. Urban sprawl, industrialization, and interstate highway systems all create air and water pollution. Land clearing, marginal farming in Third World countries, and overlogging and overgrazing devastate pristine habitats. Overuse of water results in desertification. Excessive recreational hunting and fishing decimate rare and endangered species. These factors have been emphasized many times and in many ways, but one important factor in species survival has rarely been addressed: circadian aspects of temperature and light pollution.

The changing of ecosystem parameters may impact circadian entrainment[2,5,6,14,15,16,22,24,29,36,45]

LIGHT POLLUTION IN ECOSYSTEMS. A link can easily be seen between the total destruction of the vegetation of a ecosystem and the immediate elimination of all living species in a restricted area. Examples of such events are the explosive eruption of Krakatau in 1883 in the Sumatra Straits, the calamitous eruption of Mount Saint Helens in 1981, and the 500,000-acre Biscuit fire in Oregon in 2002. Far more subtle and insidious, however, are the small changes that erode away the integrity of an ecosystem by a few percentage points per year over long periods of time. Scientists have growing concerns about the increasing impact of light pollution and thermal pollution (discussed in a later section). Nighttime pictures of Earth from satellites provide galvanizing evidence of the creeping progress of human illumination into ecosystems at night.

The phrase *light pollution* has been in use for a long time. In 1917 a California ornithologist, Carlos Lastreto, investigated reports that migrating birds were dying by crashing into the beacon lights of lighthouses. Lighthouses have now been largely decommissioned, but modern analogs have taken their place. Research on industrial lighting has led to the development of efficient, inexpensive night lighting for homes and businesses. Halogen and metal-halide lights deliver intense light in smaller packages than their incandescent predecessors, and they are cheaper to run. During the past decade home builders have increased their use of large coach lights to illuminate front door entrances or driveways, and they now frequently employ large, unshielded floodlights to illuminate garage entrances and decks.

RADIANCE DETECTORS AND PHOTIC ENTRAINMENT. Light pollution is of special concern given the newfound role of radiance detectors in photic entrainment. The discovery of mammalian retinal ganglion cells (see Chapter 6) emphasized the multiple roles that photoreceptors play in visual imaging and in luminance measurement in organisms. Luminance detection gives long-term information on the average amount of light available over the entire time of illumination. The ability of mammalian retinal ganglion cells to detect luminance makes them ideal for circadian entrainment. Perhaps these same photoreceptors play a role in melatonin suppression by light and in this way help regulate the vital photoperiodic processes of many organisms.

GROWING PUBLIC AWARENESS. The harmful effects of artificial night light have been documented with growing frequency. A recent article in *Science* entitled "Lighting's Dark Side" warned of the growing danger of round-the-clock light pollution to natural habitats. Another article in 2002 in the Pittsburgh *Post-Gazette* was entitled "Lighting a path to extinction?"; it alerted readers to the dangers of human-made bright nights in altering habitats and the behavior of nocturnal creatures. Many biologists agree that although night light may not be killing animals outright, it is affecting some aspects of their circadian behavior that could increase mortality. Acute effects on endangered sea turtles, for example, have been noted with alarm. Both the adult females coming ashore to lay eggs in the dunes and the hatchlings scrambling across the beach to gain the safety of the ocean water are negatively affected. Millions of migrating birds die each year, piled up in heaps at the base of brightly lighted high-rise buildings. Apparently the lights lure the birds, perhaps since the moon is used by many small avian migrants as their navigational compass. Equally damaged are night-flying insects that are attracted phototactically to street lamps.

CIRCADIAN PHOTIC EFFECTS OF HABITAT CHANGE. The photic entrainment mechanism of an animal such as a flying squirrel ensures that it wakes up and becomes active at an optimal time each day (see Figure 2.6 in Chapter 2). For many nocturnal species the start of activity time is signaled at an intensity of a few lux of light, shortly after sunset. It is the time of transition from daytime cone vision to nighttime rod vision and also the time when most diurnal predators abruptly cease activity.

Data are now available for the phase-shifting thresholds of circadian photoreceptors for entrainment of several species living in different photic habitats. The threshold is important because of the role it plays in light sampling during photoentrainment (see Figures 2.3B and 2.24D in Chapter 2). If light intensity is above

threshold during light sampling, the animal's clock is phase delayed, and it returns to its den to nap for periods up to several hours. Although the time of day for the start of nocturnal activity may be almost exactly the same time for animals in habitats with different light characteristics, clearly the threshold for triggering the start of circadian activity must differ greatly. A comparative study documented photic phase-shifting thresholds for two desert species and two dense-forest species of small nocturnal rodents. A dose response curve for a single flying squirrel can be plotted as the mean of many test points at a given intensity ± the standard deviation of the response (Figure 11.14A). The response is remarkably precise, allowing the threshold to be delineated as the point at which a phase-shifting response greater than 10 minutes occurs. The results are very revealing.

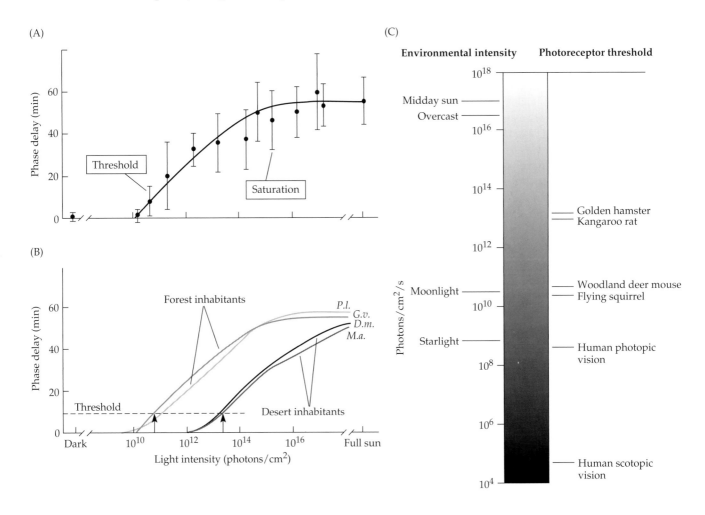

FIGURE 11.14 Threshold for circadian phase shifting in nocturnal rodents based on intensity response curves. (A) Data from isolated single pulses for one free-running flying squirrel in DD. Pulses of dim light lasted 15 minutes each. The mean and standard deviation are shown for each test time. (B) Summary dose response (fluence) curves for two desert species and two forest species, with horizontal dashed line indicating threshold level. (C) Environmental light intensity scale with photic phase shifting thresholds indicated for the two desert species (top labels) and the two forest species. *D.m., Dipodomys merriami,* Merriam's kangaroo rat; *G.v., Glaucomys volans,* flying squirrel; *M.a., Mesocricetus auratus,* golden hamster; *P.l., Peromyscus lelucopus,* woodland deer mouse. (From DeCoursey, 1990.)

FIGURE 11.15 Lizard and ant interaction and potential effects of changed entraining conditions. (A) Lizard activity range. (B) Ant activity range. (C) Superimposed activity patterns of lizard and ant throughout the year. Black indicates the overlap of activity of lizards and ants, which would allow predation. (From Porter et al., 1973.)

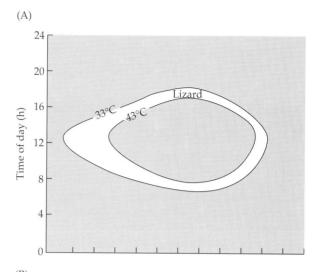

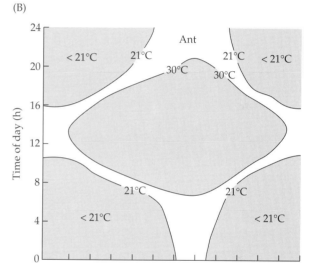

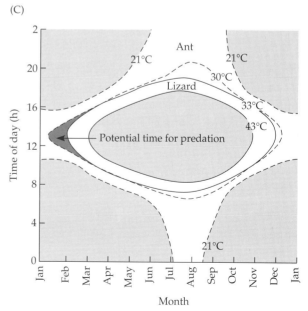

Both desert species have quite similar, overlapping curves. They resemble closely the curves for the two forest species, but the threshold of the desert species was almost a thousand times higher than that of the forest species. The results are even more cogent when the data are plotted on a logarithmic scale of environmental light intensity (Figure 11.14C). If a forest were thinned by 50%, many severe changes in the photic parameters of the habitat would result. The changes in habitat light intensity could affect the timing of activity start and thereby influence many aspects of a species' behavior patterns.

PREDATOR–PREY IMPLICATIONS OF THERMAL HABITAT CHANGES. Forest thinning could also modify thermal profiles. The changes in turn could influence circadian activity rhythms of poikilotherms that use temperature signals for entrainment. One persevering scientist measured seasonal changes in environmental profiles relative to a real predator, the desert iguana, and its predicted ant prey. The data consist of careful measurements from simulated models and live lizards in the Mojave Desert, as well as in a wind tunnel in the laboratory.

The lizards were active aboveground only between 33°C and 43°C (Figure 11.15A). Seasonal behavior patterns for the lizard show that it hibernates from November until early February, then is active in spring in full sun at midday. Gradually as the surface temperatures rise, the lizard seeks partial cover and becomes bimodal in summer with activity peaks in the cool morning and late afternoon. The activity of a hypothetical ant is restricted to the temperature range of 21 to 30°C (Figure 11.15B). The predicted behavior pattern shows a single activity period at midday in winter, a bimodal activity pattern in spring and fall, and a nocturnal pattern in winter. When the two activity patterns are superimposed, the surprising fact emerges that ants will fall prey to desert lizards for only a few weeks in February.

Now imagine what would happen if the temperature patterns were modified by all the desert scrub and brush being burned off, allowing sunlight to warm the surface a few degrees and shift the timing of particular temperature patterns. Shifts in activity pattern might greatly enhance or just as likely reduce chances of the lizard's success and the ant's mortality (Figure 11.15C).

If either of the two species were rare or endangered, the shifts in ambient temperature could tip the balance in survival of one species and potentially push it over the brink into extinction.

Closure: A Remarkable Island Symbolizes Adaptive Rhythmicity

Biological timing is widespread in plants and animals

The fact that nearly all organisms possess biological timing systems suggests that internal temporal organization is important for survival. The majority of terrestrial animals have evolved striking daily behavioral rhythms by virtue of living on a revolving planet. The biosphere is dramatically rhythmic in its environmental features for practically every available niche. Most environmental rhythms have great potential for acting on living organisms as ultimate evolutionary selective factors to prune and shape internal timing systems. As a result, behavioral and physiological rhythms of organisms are not mere passive reactions to environmental cycles. In almost all cases, a physiological pacemaker is corrected to local time by a cue such as the 24 h solar cycle.

An allegory will portray the power of biological timing

Somewhat like a jigsaw puzzle of many pieces, the structure of any textbook must rather arbitrarily divide a whole into fragments of individual chapters. In the course of a semester of study, some of the earlier fragments have misted over or have been forgotten, and by the very nature of learning they have been assimilated piecemeal. The trick is now to integrate quickly in one final example a scintillating, lucid picture of the whole and re-create, in a lightning sweep, the beauty and excitement of that whole.

The story of Mont-St.-Michel will stand as symbolic closure of this book. The exquisite monastery of Mont-St.-Michel on the Normandy coast remains today, after more than a thousand years, one of the architectural jewels of France. The locale has long been revered for its natural beauty and its panoramic view. Reputedly it was the site of prehistoric Druid temples. Later conquering Romans may have built a temple on the site to honor their god Jupiter.

A legend relates that a vision of St. Michael came to Bishop Aubert in 708 and instructed him to build a church on the promontory in the forest. Before he could carry out his task that year, a massive earthquake shook the ground that surrounded the pinnacle. The earth heaved and groaned and with a shattering noise began to sink. After the tremors subsided, an even more ominous event began. A giant tidal wave as tall as the trees rushed straight for the towering rock. Everything in its path was devastated: trees and farmhouses, as well as farmers and their animals. Astonishingly, the wall of water stopped at the base of the granite spire. The water then receded, leaving a freestanding island and bare earth stretching several kilometers to the ocean.

Work on building the monastery began in 709. Over the years the first small chapel became ever wealthier and grander. Word of the shrine soon spread throughout Europe, and pilgrims came in droves. Soon, however, the shrine attracted visitors of a less desirable kind: robbers, plunderers, and warring armies who wished to steal its treasures. Many attempts were made over 13 centuries to breach the rampart walls, but no attack or siege has ever been successful.

Twice each day since the earthquake, a 15 m high tidal bore has raced toward the island and rushed out to sea again. Until the building of an access causeway from the island to the mainland starting in 1874, the tides at

FIGURE 11.16 The temporal symbolism of Mont-St.-Michel, France, at ebb tide. Mont-St.-Michel is invulnerable, impregnable, protected by its rhythmic environment.

Mont-St.-Michel were among the highest and most ferocious in the world. The water didn't drain sedately away during the 6 h ebb or return gently during the 6 h flood tide. It roared out to sea and back to the base of Mont-St.-Michel with the speed of a galloping horse (Figure 11.16).

No longer can invading armies lay siege to the monastery. Forever after that fateful day in 709, attackers would have to run for their lives every 12 h to avoid being decimated by the smashing flood tides. No army or plunderer has scaled the walls of the monastery. The rhythmic temporal organization of Mont-St.-Michel stands as a symbol for the adaptiveness of temporal programming of the life processes of all living organisms.

SUMMARY

Six themes have portrayed some recent important discoveries in chronobiology research that may influence the course of the field in the near future. The first two themes considered sleep research in humans and in a phylogenetic sampling of vertebrate species. Information about the mechanism and function of sleep in humans is increasing, partly due to the use of surrogate model species of vertebrates. Some features of sleep as defined by EEG criteria and behavioral states are found in all mammal and bird groups, as well as in lower vertebrates. The more restricted features, such as REM sleep, occur only in mammals and birds. Possibly REM sleep originated relatively early in the advanced reptilian ancestors of mammals.

The third theme considered invertebrate species as alternate models for circadian research. Several very simple, inexpensive models, such as the fruit fly circadian system, may accelerate the rate of progress of research on sleep and other circadian behavior patterns at the molecular level.

Two themes dealt with properties of pacemakers at the cellular level: peripheral clocks in culture and chimeric analysis. Recognition of the importance of peripheral oscillators has changed the view of a central dominant pacemaker to a clock shop of multiple oscillators. In the *Clock* chimera experiment, a graded series of chimeric mice was used to dissect SCN function at a cellular level. The identification of a range of resultant circadian phenotypes supports the idea that an integrated consensus periodicity results from an ensemble of interacting SCN cellular oscillators. The realized circadian behavioral phenotype is dependent nearly quantitatively on the representation of the two cell types.

The final theme explored the degradation of the photic and thermal entraining signals of the natural environment, which is proceeding at an alarming rate. These changes in ecosystems may have deleterious effects on circadian processes in organisms, which could seriously threaten survival, especially for endangered species.

CONTRIBUTORS

Patricia J. DeCoursey wishes to thank Joan Hendricks and Fred W. Turek for their material contributions to Chapter 11.

REFERENCES

1. Alcock, J. 2001. Animal Behavior, 6th Edition. Sinauer, Sunderland, MA.

2. Aschoff, J. 1989. Temporal orientation: Circadian clocks in animals and humans. The Niko Tinbergen Lecture, 1986. Anim. Behav. 37: 881–896.

3. Aschoff, J. 1998. Circadian parameters as individual characteristics. J. Biol. Rhythms 13: 123–131.

4. Balsalobre, A., F. Damiola, and U. Schibler. 1998. A serum shock induces circadian gene expression in mammalian culture cells. Cell 93: 929–937.

5. Barinaga, M. 2002. How the brain's clock gets daily enlightenment. Science 295: 955–957.

6. Berson, D. A., F. A. Dunn, and M. Takao. 2002. Phototransduction by retinal ganglion cells that set the circadian clock. Science 295: 1070–1073.

7. Borbely, A. A., D-J. Dijk, P. Achermann, and I. Tobler. 2001. Processes underlying the regulation of the sleep-wake cycle. In: Handbook of Behavioral Neurobiology Volume 12: Circadian Clocks, J. Takahashi, R. W. Turek, and R Y. Moore (eds.), pp. 458–479. Kluwer Academic/Plenum, New York.

8. Brearley, H. C. 1919. Telling Time through the Ages. Ingersoll, New York.

9. Bush, M. 2003. Ecology of a Changing Planet. Prentice Hall, Upper Saddle River, NJ.

10. Campbell, B. 1974. Human Evolution: An Introduction to Man's Adaptations, 2nd Edition. Aldine, Chicago.

11. Campbell, S., and I. Tobler. 1984. Animal sleep: A review of sleep duration across phylogeny. Neurosci. Biobehav. Rev. 8: 269–300.

12. Castleden, R. 1987. The Stonehenge People: An Exploration of Life in Neolithic Britain 4700–2000 B.C. Routledge and Kegan Paul, New York.

13. Cunliffe, B. W. 1997. The Ancient Celts. Oxford University Press, New York.

14. Daan, S., and J. Aschoff. 2001. The entrainment of circadian rhythm. In: Handbook of Behavioral Neurobiology Volume 12: Circadian Clocks, J. Takahashi, R. W. Turek, and R. Y. Moore (eds.), pp. 7–43. Kluwer Academic/Plenum, New York.

15. DeCoursey, P. J. 1989. Photoentrainment of circadian rhythms: An ecologist's viewpoint. In: Circadian Clocks and Ecology, T. Hiroshige and K-I. Honma (eds.), pp. 187–206. University of Hokkaido Press, Sapporo, Japan.

16. DeCoursey, P. J. 1990. Circadian photoentrainment in nocturnal mammals: Ecological overtones. Biol. Behav. 15: 213–238.

17. Dunlap, J. C. 1999. Molecular bases of circadian clocks. Cell 96: 271–290.

18. Ehrenfeld, D. W. 1970. Biological Conservation. Holt, Rinehart and Winston, New York.

19. Gillin, J. C. 2002. How long can humans stay awake? Sci. Am. July: 95.

20. Gore, R., and J. Blair. 1989. Extinctions. Natl. Geogr. 175: 662–699.

21. Griffin, D. R. 1976. The Question of Animal Awareness. Rockefeller University Press, New York.

22. Grigione, M. M., P. Burman, V. C. Bleich, and B. M. Pierce. 1999. Identifying individual mountain lions Felis concolor by their tracks: Refinement of an innovative technique. Biol. Conserv. 88: 25–32.

23. Hammond, A. L. Ed. 1995. World Resources: A Guide to the Global Environment. Oxford University Press, New York.

24. Hastings, J. W., B. Rusak, and Z. Boulos. 1991. Circadian rhythms: The physiology of biological timing. In: Integrative Animal Physiology, C. L. Prosser (ed.), pp. 435–546. Wiley, New York.

25. Hawkins, G. S. 1965. Stonehenge Decoded. Doubleday, Garden City, NY.

26. Hendricks, J. C., S. M. Finn, K. A. Panckeri, J. Chavkin, J. A. Williams, et al. 2000a. Rest in Drosophila is a sleep-like state. Neuron 25: 129–138.

27. Hendricks, J. C., A. Sehgal, and A. I. Pack. 2000b. The need for a simple animal model to understand sleep. Prog. Neurobiol. 61: 339–351.

28. Hendricks, J. C., J. A. Williams, K. Panckeri, D. Kirk, J. C-P. Yin, et al. 2001. A non-circadian role for cAMP signaling and CREB activity in waking and rest homeostasis in Drosophila melanogaster. Nature Neurosci. 4: 1108–1115.

29. Holden, C. 2002. Lighting's dark side. Science 295: 1227.

30. Kilduff, T. S., and C. Peyron. 2000. The hypocretin/orexin ligand receptor system: Implications for sleep and sleep disorders. Trends Neurosci. 23: 359–365.

31. Long, M. E., and L. Psihoyos. 1987. What is this thing called sleep? Natl. Geogr. 172: 787–820.

32. Low-Zeddies, S., and J. S. Takahashi. 2001. Chimera analysis of the Clock mutation in mice shows that complex cellular integration determines circadian behavior. Cell 105: 25–42.

33. Lumsden, C. J., and E. O. Wilson. 1983. Promethean Fire: Reflections on the Origin of Mind. Harvard University Press, Cambridge, MA.

34. Lyamin, O. I., L. M. Mukhametov, J. M. Siegel, E. A. Nazarenko, I. G. Polyakove, et al. 2002. Unihemispheric slow wave sleep and the state of the eyes in a white whale. Behav. Brain Res. 129: 125–129.

35. Meerlo, P., M. Koehl, K. van der Borght, and F. W. Turek. 2002. Sleep restriction alters the hypothalamic-pituitary-adrenal response to stress. J. Neuroendocrinol. 14: 397–402.

36. Menaker, M., and G. Tosini. 1995. The evolution of vertebrate circadian systems. In: Circadian Organization and Oscillatory Coupling, K-I. Honma and S. Honma (eds.), pp. 39–52. University of Hokkaido Press, Sapporo, Japan.

37. Menaker, M., and M. A. Vogelbaum. 1993. Mutant circadian period as a marker of suprachiasmatic nucleus function. J. Biol. Rhythms 8: S93–S98.

38. Minors, F., and J. Waterhouse. 1981. Circadian Rhythms and the Human. Wright, Bristol, UK.

39. Moore-Ede, M. 1993. The Twenty-Four Hour Society: Understanding Human Limits in a World That Never Stops. Addison-Wesley, Reading, MA.

40. Olshansky, S. J. 1997. The demography of aging. In: Geriatric Medicine, 3rd Edition, C. K. Cassel, H. J. Cohen, E. B. Larson, et al. (eds.), pp. 229–236. Springer, New York.

41. Pittendrigh, C. S. 1960. Circadian rhythms and the circadian organization of living things. In: Cold Spring Harb. Symp. Quant. Biol. 25: 159–184.

42. Pittendrigh, C. S. 1961. On temporal organization in living systems. Harvey Lect. 56: 93–125.

43. Pittendrigh, C. S. 1993. Temporal organization: Reflections of a Darwinian clockwatcher. Annu. Rev. Physiol. 55: 17–54.

44. Plautz, J. D., M. Kaneko, J. C. Hall, and S. A. Kay. 1997. Independent photoreceptive circadian clocks throughout Drosophila. Science 278: 1632–1635.

45. Porter, W. P., J. W. Mitchell, W. A. Beckman, and C. B. DeWitt. 1973. Behavioral implications of mechanistic ecology: Thermal and behavioral modeling of desert ectotherms and their microenvironment. Oecologia 13: 1–54.

46. Purves, W. K., D. Sadava, G. H. Orians, and H. C. Heller. 2001. Life: The Science of Biology, 6th Edition. Sinauer, Sunderland, MA.

47. Ralph, W. R., and M. H. Vitaterna. 2001. Mammalian clock genetics. In: Handbook of Behavioral Neurobiology Volume 12: Circadian Clocks, J. Takahashi, R. W. Turek, and R. Y. Moore (eds.), pp. 433–453. Kluwer Academic/Plenum, New York.

48. Rattenborg, N. C., S. L. Lima, and C. J. Amlaner. 1999. Facultative control of avian unihemispheral sleep under risk of predation. Behav. Brain Res. 105: 163–172.

49. Rowe, J. W., and R. L. Kahn. 1998. Successful Aging. Pantheon Books, New York.

50. Rutter, J., M. Reick, and S. L. McKnight. 2002. Metabolism and the control of circadian rhythms. Annu. Rev. Biochem. 71: 307–331.

51. Schibler, U., J. A. Ripperger, and S. A. Brown. 2001. Chronobiology—Reducing time. Science 293: 437–438.

52. Shaw, P. J., C. Cirelli, R. J. Greenspan, and G. Tononi. 2000. Correlates of sleep and waking in Drosophila melanogaster. Science 287: 1834–1837.

53. Shaw, P. J., G. Tononi, R. J. Greenspan, and D. F. Robinson. 2002. Stress response genes protect against lethal effects of sleep deprivation in Drosophila. Nature 417: 287–291.

54. Siegel, J. M. 1995. Phylogeny and the function of REM sleep. Behav. Brain Res. 69: 29–34.

55. Siegel, J. M., P. R. Manger, R. Nienhuis, H. M. Fahringer, T. Shalita, et al. 1999. Sleep in the platypus. Neuroscience 91: 391–400.

56. Stokkan, K-A., S. Yamasaki, H. Tei, Y. Sakaki, and M. Menaker. 2001. Entrainment of the circadian clock in the liver by feeding. Science 291: 490–493.

57. Takahashi, J. S., L. H. Pinto, and M. H. Vitaterna. 1994. Forward and reverse genetic approaches to behavior. Science 264: 1724–1732.

58. Tobler, I. 1983. Introduction: Phylogenetic approaches to the function of sleep. In: Proceedings of the 6th European Congress on Sleep Research, pp. 126–146. Karger, Basel, Switzerland.

59. Tobler, I. 1995. Is sleep fundamentally different between mammalian species? Behav. Brain Res. 69: 35–41.

60. Tosini, G., and M. Menaker. 1996. Circadian rhythms in cultured mammalian retina. Science 272: 419–421.

61. Turek, F. W. 1999. Editorial: A remarkable year for clocks based on many years of interdisciplinary research. J. Biol. Rhythms 14: 83.

62. Turek, F. W. 2002. Sleep restriction alters the hypothalamic-pituitary-adrenal response to stress. J. Endocrinol. 14: 397–402.

63. Turek, F. W., and P. C. Zee (Eds.) 1999. Regulation of sleep and circadian rhythms. Dekker, New York.

64. Turek, F. W., P. Penev, Y. Zhang, O. van Reeth, and P. Zee. 1995. Effects of age on the circadian system. Neurosci. Biobehav. Rev. 19: 53–58.

65. Waterhouse, J., D. Minors, M. Waterhouse, T. Riley, and G. Atkinson. 2002. Keeping in Time with Your Body Clock. Oxford University Press, Oxford, UK.

66. Wilson, E. O. 1999. The Biodiversity of Life. Norton, New York.

67. Wilson, T., and J. W. Hastings. 1998. Bioluminescence. Annu. Rev. Cell Dev. Biol. 14: 197–230.

68. Yamazaki, S., V. Alones, and M. Menaker. 2002. Interaction of the retina with suprachiasmatic pacemakers in the control of circadian behavior. J. Biol. Rhythms 17: 315–329.

69. Yamazaki, S., R. Numano, M. Abe, A. Hida, R-I. Takahashi, et al. 2000. Resetting central and peripheral circadian oscillators in transgenic rats. Science 288: 682–685.

70. Zhdanova, I. V., S. Y. Wang, O. J. Leclai, and N. P. Danilova. 2001. Melatonin promotes sleep-like state in zebrafish. Brain Res. 903: 263–268.

Appendix

Species Lists

Multiple species of a genus are indicated by "sp."

Latin to English

Acronicta rumicis   knotgrass moth
Aedes sp.   mosquito
Albizia sp.   leguminous tree or shrub
Amphioxus sp.   lancelet
Angraecum sesquipedale   Darwin's orchid
Anolis sp.   anole lizard
Anolis carolinensis   Carolina chameleon; Carolina
    anole
Anopheles sp.   mosquito
Antheraea polyphemus   silk moth
Antheraea pernyi   silk moth
Anthrenus sp.   dermestid beetle
Apanteles glomeratus   parasitic wasp
Aplysia californica   sea hare
Arabidopsis thaliana   mustard plant; mouse ear cress,
    thale cress
Bulla gouldiana   clouded bubble snail
Caenorhabditis elegans   nematode
Canavalia ensiformis   jackbean
Cattleya sp.   cattleya orchid
Cereus hildmannianus   night-blooming cactus, the
    "Queen of the Night"
Clunio marinus   marine midge
Cnemidophorus uniparens   teiid lizard
Coleus sp.   variegated herbaceous plant
Culex quinquefasciatus   mosquito
Danio rerio   zebra fish
Diatraea grandiosella   southwestern corn borer moth
Dipodomys sp.   kangaroo rat
Dipodomys merriami   Merriam's kangaroo rat
Dipsosaurus dorsalis   desert iguana
Donax sp.   butterfly clam
Drosophila auraria   fruit fly

Drosophila melanogaster   fruit fly
Drosophila pseudoobscura   fruit fly
Emerita talpoida   beach hopper crustacean
Euglena gracilis   flagellate protist
Excirolana   beach hopper crustacean
Fringilla coelebs   chaffinch
Glaucomys volans   eastern flying squirrel
Gonyaulax polyhedra   unicellular dinoflagellate; "red
    tide" (note that it has recently been renamed
    Lingulodinium polyedrum)
Gymnorhamphichthys sp.   electric knife fish; gymnotid
    fish
Gymnotus carapo   banded carapo; gymnotid electric
    fish
Iguana iguana   green iguana lizard
Kalanchoe fedtschenkoi   lavender scallops, "air plant"
Lacerta sp.   lizard
Lampetra japonica   lamprey
Leucophaea maderae   cockroach
Manduca sexta   tobacco hornworm moth
Mesocricetus auratus   golden hamster; Syrian hamster
Mimosa pudica   sensitive plant
Mus musculus   house mouse; lab mouse
Myotis sp.   insectivorous bat
Nasonia vitripennis   jewel wasp
Neurospora crassa   bread mold
Odontosyllis enopla   Atlantic fireworm
Paramecium   protozoan
Peromyscus leucopus   white-footed mouse
Peromyscus maniculatus   deer mouse
Petromyzon marinus   lamprey
Pieris brassica   cabbage butterfly
Podarcis sicula   ruin lizard

Pterygophora californica   stalked kelp
Rana esculenta   European green frog
Samia cynthia   silk moth
Sarcophaga argyrostoma   flesh fly
Sceloporus occidentalis   fence lizard
Sceloporus olivaceus   Texas spiny lizard
Sceloporus virgatus   striped plateau lizard
Sinapsis alba   white mustard plant
Sylvia borin   garden warbler
Synchlidium sp.   beach hopper crustacean

Synechococcus elongatus   cyanobacterium; blue-green alga
Taphozous sp.   bat
Teleogryllus commodus   black field cricket
Tetrahymena sp.   unicellular protozoan or ciliate protozoan
Uca sp.   fiddler crab
Vibrio fischeri   bacterial symbiont
Xenopus laevis   African clawed frog

English to Latin

African clawed frog   *Xenopus laevis*
African stonechat   *Saxicola torquata axillaris*
air plant   *Kalanchoe fedtschenkoi*
alga, blue-green   *Synechococcus elongatus*
anole lizard   *Anolis carolinensis*
aphid, green vetch   *Megoura viciae*
Atlantic fireworm   *Odontosyllis enopla*
bacterial symbiont   *Vibrio fischeri*
banded carapo   *Gymnotus carapo*
bat   *Taphozous* sp. See also *horseshoe-nosed bat* and *insectivorous bat.*
beach hopper crustacean   *Emerita talpoida, Excirolana* sp., *Synchlidium* sp.
bear (brown or black)   *Ursus* sp.
beaver   *Castor canadensis*
beetle, ground   *Pterostichus nigrita*
blowfly   *Calliphora vicina*
blue-green alga   *Synechococcus elongatus*
bread mold   *Neurospora crassa*
butterfly, cabbage   *Pieris brassica*
butterfly clam   *Donax* sp.
cabbage butterfly   *Pieris brassica*
cactus, night-blooming   *Cereus hildmannianus*
carapo, banded   *Gymnotus carapo*
Carolina chameleon or anole   *Anolis carolinensis*
cattleya orchid   *Cattleya* sp.
cave crayfish   *Orconectes pellucidus*
chaffinch   *Fringilla coelebs*
chameleon, Carolina   *Anolis carolinensis*
chicken   *Gallus* sp.
chipmunk, eastern   *Tamias striatus*
cichlid fish, convict   *Cichlasoma nigrofasciatum*
clam, butterfly   *Donax* sp.
clouded bubble snail   *Bulla gouldiana*
cocklebur plant   *Xanthium strumarium*
cockroach   *Leucophaea maderae, Periplaneta americana*
collared flycatcher   *Ficedula albicollis*
convict cichlid fish   *Cichlasoma nigrofasciatum*

crab   See *fiddler crab* and *horseshoe crab.*
crayfish, cave   *Orconectes pellucidus*
cress   *Arabidopsis thaliana*
cricket   *Gryllus bimaculatus, Teleogryllus commodus*
cyanobacterium   *Synechococcus elongatus*
dark-eyed junco   *Junco hyemalis*
Darwin's orchid   *Angraecum sesquipedale*
deer, sika   *Cervus nippon*
deer mouse   *Peromyscus maniculatus*
dermestid beetle   *Anthrenus* sp.
desert iguana   *Dipsosaurus dorsalis*
dinoflagellate   *Gonyaulax polyhedra*
eastern chipmunk   *Tamias striatus*
eastern flying squirrel   *Glaucomys volans*
European green frog   *Rana esculenta*
European stonechat   *Saxicola torquata rubicola*
fence lizard   *Sceloporus occidentalis*
fiddler crab   *Uca* sp.
finch   See *house finch* and *weaver finch.*
fireworm, Atlantic   *Odontosyllis enopla*
flagellate protist   *Euglena gracilis*
flesh fly   *Sarcophaga* sp.
fly   See *blowfly, flesh fly,* and *fruit fly.*
flycatcher   See *collared flycatcher* and *pied flycatcher.*
flying squirrel, eastern   *Glaucomys volans*
four-o'clock flower   *Mirabilis jalapa*
frog   *Rana* sp. See also *African clawed frog, European green frog*
fruit fly   *Drosophila* sp.
garden warbler   *Sylvia borin*
golden hamster   *Mesocricetus auratus*
golden-mantled ground squirrel   *Spermophilus lateralis*
green iguana   *Iguana iguana*
green vetch aphid   *Megoura viciae*
ground beetle   *Pterostichus nigrita*
ground squirrel   See *white-tailed antelope ground squirrel* and *golden-mantled ground squirrel.*

gymnotid fish   *Gymnorhamphichthys* sp., *Gymnotus carapo*

hamster   See *golden hamster, Siberian hamster*, and *Syrian hamster.*

honeybee   *Apis mellifera*

hops   *Humulus lupulus*

horseshoe crab   *Limulus polyphemus*

horseshoe-nosed bat   *Rhinolophus ferrum-equinum*

house finch   *Passer domesticus*

house mouse   *Mus musculus*

human   *Homo sapiens*

iguana   See *desert iguana* and *green iguana.*

insectivorous bat   *Myotis* sp.

jackbean   *Canavalia ensiformis*

Japanese quail   *Coturnix japonica*

jewel wasp   *Nasonia vitripennis*

jumping mouse   *Zapus hudsonicus*

junco, dark-eyed   *Junco hyemalis*

kangaroo rat, Merriam's   *Dipodomys merriami*

knife fish   *Gymnorhamphichthys* sp.

knotgrass moth   *Acronicta rumicis*

lab mouse   *Mus musculus*

lamprey   *Lampetra japonica, Petromyzon marinus*

lancelet   *Amphioxus* sp.

lavendar scallops   *Kalanchoe fedtschenkoi*

leguminous tree or shrub   *Albizia* sp.

linden bug   *Pyrrhocoris apterus*

lizard   *Lacerta* sp. See also *anole lizard, fence lizard, ruin lizard, striped plateau lizard, teiid lizard,* and *Texas spiny lizard.*

marijuana   *Cannabis sativa*

marmot   *Marmota* sp.

Merriam's kangaroo rat   *Dipodomys merriami*

midge   *Clunio marinus*

mole-rat, naked   *Heterocephalus glaber*

morning glory   *Ipomoea purpurea*

mosquito   *Aedes* sp., *Anopheles* sp., *Culex* sp.

moth   See *knotgrass moth, silk moth,* and *tobacco hornworm moth.*

mouse   See *deer mouse, house mouse, jumping mouse, lab mouse,* and *white-footed mouse.*

mustard plant   *Arabidopsis thaliana*

naked mole-rat   *Heterocephalus glaber*

nematode   *Caenorhabditis elegans*

night-blooming cactus   *Cereus hildmannianus*

orchid   See *cattleya orchid* and *Darwin's orchid.*

parasitic wasp   *Apanteles glomeratus*

pied flycatcher   *Ficedula hypoleuca*

prairie vole   *Microtus ochrogaster*

protozoan   *Paramecium* sp., *Tetrahymena* sp.

quail, Japanese   *Coturnix japonica*

rainbow trout   *Salmo gairdneri*

rat   *Rattus norvegicus.* See also *Merriam's kangaroo rat* and *naked mole-rat.*

red-backed vole   *Clethrionomys gapperi*

"red tide"   *Gonyaulax polyhedra*

ruin lizard   *Podarcis sicula*

scorpion   *Androctonus australis*

sea hare   *Aplysia californica*

sensitive plant   *Mimosa pudica*

sheep   *Ovis* sp.

Siberian hamster   *Phodopus sungorus*

sika deer   *Cervus nippon*

silk moth   *Antheraea pernyi, Antheraea polyphemus, Bombyx mori, Samia cynthia*

skunk, spotted   *Spilogale putorius*

snail, clouded bubble   *Bulla gouldiana*

sparrow   See *tree sparrow* and *white-crowned sparrow.*

spider mite   *Tetranychus urticae*

spotted skunk   *Spilogale putorius*

squirrel   See *white-tailed antelope ground squirrel, eastern flying squirrel,* and *golden-mantled ground squirrel.*

stalked kelp   *Pterygophora californica*

starling   *Sturnus vulgaris*

stonechat   *Saxicola torquata*

striped plateau lizard   *Sceloporus virgatus*

Syrian hamster   *Mesocricetus auratus*

teiid lizard   *Cnemidophorus uniparens*

Texas spiny lizard   *Sceloporus olivaceus*

tobacco hornworm moth   *Manduca sexta*

tree sparrow   *Spizella arborea*

trout, rainbow   *Salmo gairdneri*

vole   See *prairie vole* and *red-backed vole.*

warbler, garden   *Sylvia borin*

wasp, parasitic   *Apanteles glomeratus*

weasel   *Mustela* sp.

weaver finch   *Euplectes hordeaceus*

white-crowned sparrow   *Zonotrichia leucophrys*

white-footed mouse   *Peromyscus leucopus*

white mustard plant   *Sinapsis alba*

white-tailed antelope ground squirrel   *Ammospermophilus leucopus*

zebra fish   *Danio rerio*

Glossary

action spectrum (pl. spectra) A graph of the magnitude of a light-dependent biological process versus the wavelength of incident light.

actogram, actograph A graph of daily activity.

advanced sleep phase syndrome (ASPS) A sleep disorder characterized by early sleep onset and offset. Compare *delayed sleep phase syndrome*.

affective disorder A chronic emotional or mood disorder often associated with abnormal sleep patterns.

aftereffect An effect on circadian period length arising from a previous entrainment schedule.

aliquot (n) An exact portion or fraction of the whole. (v) To measure out into equal parts.

allele A particular variant of a gene, distinguishable from other variants of the same gene.

allostery Regulation of the activity of a protein by the binding of a nonsubstrate small molecule to the protein at a site other than the catalytic site. Generally allostery is assumed to be the result of a change in the shape or structure of the protein as a result of binding the small molecule.

alpha The portion of a daily rest–activity cycle corresponding to activity. Compare *rho*.

amplitude The extent of an oscillatory movement, measured from mean to extreme value.

anterograde transport Transport of substances, generally in a neuron, away from the cell body. Compare *retrograde transport*.

anuran Pertaining to frogs or toads.

ASPS See *advanced sleep phase syndrome*.

axilla (pl. axillae) Armpit or an analogous part of anatomy.

bistability In circadian entrainment by skeleton photoperiods, the phenomenon in which the pacemaker can adopt either of two alternative stable phase relationships with respect to the entraining cycle, the choice being dependent on starting conditions.

bouton The distal terminal of a neuron that secretes the neurochemical for crossing a synapse.

bregma (pl. bregmata) Position on the roof of the mammalian skull at the midline intersection of the sutures of frontal and parietal bones. The bregma is an important landmark in SCN lesioning surgery.

chromophore The light-absorbing molecule in a photoreceptor protein.

chronobiology The study, at all levels of organization, of adaptations evolved by living organisms to cope with regularly occurring environmental cycles.

circadian rhythm A biological rhythm that persists under conditions of constant light, temperature, and other environmental factors with a period length of about a day, whose phase can be reset by a brief interruption in the constant regimen, and whose period length is relatively independent of temperature within the physiological range of normal growth.

circadian time (CT) Subjective internal organism time in which one circadian period length is divided into 24 equal parts, each a circadian hour. By convention, CT 0 corresponds to subjective dawn and CT 12 to subjective dusk.

circalunar Having a period length of about a month.

circannual Having a period length of about a year.

circatidal Having a period length of about one tidal cycle, usually 12.4 h.

civil twilight The time when the sun's disk is 6 degrees below the horizon. Civil twilight is the time at which many nocturnal animals start their evening activity.

clepsydra A water thief or water clock for measuring daily time.

clock An assembly of components used for measuring time.

clock-controlled gene (*ccg*) A gene whose expression is rhythmically regulated by a clock.

cone A photoreceptor cell of the retina. Compare *rod*.

constant routine A protocol used for measuring free-running period lengths in humans, whereby the subject is kept constantly employed in routine tasks.

crepuscular Active predominantly in the early evening or twilight.

cryptochrome A blue-light photoreceptor using a pterin and flavin as chromophores.

CT See *circadian time*.

DD The light–dark cycle abbreviation that represents constant darkness. Compare *LD* and *LL*.

delayed sleep phase syndrome (DSPS) A sleep disorder in which the onset of sleep is chronically delayed. Compare *advanced sleep phase syndrome*.

diapause A state of suspended animation in insects akin to hibernation in mammals.

diurnal (1) Pertaining to a rhythm that recurs on a daily basis under entrainment conditions in the real world but that may not, or has not been shown to, recur under constant conditions. (2) Referring to a rhythmic behavior or process that peaks in the daytime rather than at night.

DNA Deoxyribonucleic acid, the coding material of genes.

domain A part of a protein whose structure is associated with a particular function.

double plot The graphing of an actogram in which activity records of 2 days are plotted next to one another, such that the record of day *n* + 1 is plotted to the right of day *n*.

driver oscillator An oscillator that drives or entrains another oscillator. Compare *slave oscillator*.

DSPS See *delayed sleep phase syndrome*.

eclosion Hatching of an insect pupa into an adult.

ecology The study of the relationship of an organism to its environment.

ectothermic Cold-blooded. Synonymous with *poikilothermic*.

entrainment The process by which an environmental rhythm such as the day–night cycle regulates the period and phase relationship of a self-sustained biological pacemaker.

entraining agent An environmental cycle that controls the period and phase relationship of a self-sustained oscillator.

florigen A substance that elicits flowering.

fluence Usually referring to light, the rate at which photons strike a surface.

forced desynchronization The process in which two mutually entrained oscillators assume different period lengths and move out of entrainment as a result of exposure to an entrainment regimen. Compare *spontaneous desynchronization*.

free run The state of an oscillator when not influenced by any external time cues.

free-running period (FRP) The period length of a biological oscillator. Also called *tau*.

FRP See *free-running period*.

genotype The complement of genes that an organism possesses. Compare *phenotype*.

hibernaculum (pl. hibernacula) The winter den in the laboratory or field in which an animal hibernates.

infradian Having a period length of greater than one day. Compare *ultradian*.

irradiance Usually referring to light, the total amount of light incident upon a surface.

LD The light–dark cycle abbreviation that represents an entrainment regimen composed of alternating periods of light and dark. Compare *DD* and *LL*.

LL The light–dark cycle abbreviation that represents constant light. Compare *DD* and *LD*.

luminance The amount of emitted or reflected light from a surface in a given direction per unit of projected area.

masking The phenomenon in which an external factor interferes with the expression of a rhythm or with observation of the behavior of the pacemaker by directly affecting expression of the overt rhythm.

melanopsin A putative photoreceptor pigment in the vertebrate eye.

melatonin N-Acetyl-5-methoxytryptamine, which is the same as 5-hydroxytryptamine. A hormone produced by the pineal gland that contributes to entrainment of the circadian clock in mammals.

military time A time-reporting convention in which midnight is designated as 00:00, and noon as 12:00. The afternoon hours continue past 12:00 by increments of 1 each hour up to 24:00, or midnight. The addition of "A.M." and "P.M." is therefore not necessary. Military time is used for most scientific work and data records because the morning and afternoon hours are clearly differentiated.

monochromatic Pertaining to light of a single- or narrow-band wavelength.

mustelid A mammal of the family Mustelidae, which includes minks, otters, weasels, and badgers.

neuroendocrine enabler A substance within an organism that is not directly involved in timekeeping but that provides a neuroendocrine link between the clock and the overt rhythmic output.

nonphotic entrainment Entrainment of a free-running rhythm by environmental cycles other than light. Compare *photoentrainment.*

ocellus (pl. ocelli) The simple eye of an arthropod.

open reading frame (ORF) A series of triplets within the sequence of bases in DNA that can be decoded by use of the genetic code without any STOP codons, and that is potentially translatable into a protein.

ORF See *open reading frame.*

oscillator A system of components whose interaction produces a metric that varies between alternative extremes with a definable period length. In chronobiology, the term *oscillator* refers to components within a cell whose action and regulatory interaction are sufficient to produce a rhythm. A *circadian oscillator* is a set of components whose action and regulatory interaction are sufficient to produce a circadian rhythm.

overt rhythm A rhythm in an observable characteristic that is directly or indirectly linked to and controlled by the actual pacemaker.

pacemaker A localizable, functional anatomical region capable both of sustaining its own oscillations and of entraining other oscillators.

parameter In oscillator theory, a variable that must be defined in order for the oscillatory state of the system to be known, but that need not itself oscillate.

pelage The coat of an animal, made up of its hair, wool, or fur.

period The time after which a defined phase of an oscillation recurs.

phase The instantaneous state of an oscillation within a period.

phase angle The difference between an identifiable phase in one oscillation and the corresponding phase point in another oscillation, such as the difference expressed in hours or in degrees of arc between the peak in a driving oscillator and the peak in a driven or entrained oscillator.

phase response curve (PRC) A map of phase-dependent resetting—that is, the phase-dependent response of a circadian clock to an entraining agent delivered at different times through a circadian day.

phase shift The steady-state change in phase brought about by the action of an entraining agent.

phenotype The observable characteristics of an individual as they have developed under the combined influence of the organism's genotype and environment. Compare *genotype.*

photoentrainment Entrainment brought about by the action of a light cycle. Compare *nonphotic entrainment.*

photoperiod The time of light in a light–dark cycle.

photoperiodic time measurement The detection of changes in day length by living organisms.

photoperiodism The use of changes in the day length on an annual basis, and to regulate seasonal behavioral or physiological processes.

photophil The portion of an organism's daily cycle in which light is expected; subjective daytime. Compare *scotophil.*

photoreceptor Either the protein that bears a chromophore and detects light, or the anatomical structure that houses the light-detecting proteins.

photorefractory Nonresponsive to light.

phototactic Referring to attraction (+) or repulsion (–) of an organism by light.

phototaxis Behavior in which an organism moves toward or away from a source of light.

phytochrome A photoreceptor protein that mediates many light-regulated processes, using a linear tetrapyrrole as a chromophore. Phytochromes can exist in active and inactive forms; the switch between the two states is typically initiated by absorption of light in the red or far-red range of the light spectrum.

pineal The neuroendocrine gland of vertebrates in which melatonin is synthesized and from which melatonin is subsequently released into the circulatory system.

poikilothermic The phenomenon of maintaining body temperature close to environmental temperature; cold-blooded. Synonymous with *ectothermic.*

polyvoltine Usually referring to insects or mammals, producing multiple broods or litters of offspring each year. Compare *univoltine.*

PRC See *phase response curve.*

promoter The part of a gene responsible for initiating and regulating expression of the transcript.

recrudescence Shrinkage, absorption, or regression of gonads.

retrograde transport Transport of substances, generally in a neuron, toward the cell body. Compare *anterograde transport.*

rhabdom A photoreceptor unit of an arthropod compound eye.

rho The portion of a rest–activity cycle corresponding to rest or lack of activity. Compare *alpha.*

rhythm A nonrandom series of events without any statement of causation.

rod A photoreceptor cell of the vertebrate retina. Compare *cone*.

SCN The suprachiasmatic nucleus of the ventral hypothalamus; the chief mammalian circadian pacemaker.

scotophil The portion of an organism's day in which darkness is expected; subjective nighttime. Compare *photophil*.

skeleton photoperiod A photoperiod in which the transitions from dark to light and light to dark are represented by brief exposures to light in an otherwise dark environment.

slave oscillator An oscillator that is driven or entrained by another oscillator. Compare *driver oscillator*.

spontaneous desynchronization The phenomenon in which two oscillators that had been mutually entrained spontaneously move out of phase with one another. Compare *forced desynchronization*.

state variable In oscillator theory, a variable that must be defined in order for the oscillatory state of the system to be known, and that must itself oscillate. More generally, a variable whose value defines the state of a system.

subjective day The portion of a circadian day in constant darkness corresponding to the day phase in a light–dark cycle.

subjective night The portion of a circadian day in constant darkness corresponding to the night phase in a light–dark cycle.

suprachiasmatic nucleus (pl. nuclei) See *SCN*.

synchronizer An agent that promotes synchrony between or among oscillators.

synchrony The state in which two or more oscillators are oscillating in phase with the same period length.

syncytium (pl. syncytia) A growth mass in which more than one nucleus is shared within a common cytoplasm.

T The period or cycle time of a time giver.

tau See *free-running period*.

T-cycle The cycle of an entraining agent, or time giver.

T-cycle experiment An experimental protocol in which the organism is exposed to T-cycles of varying length to demonstrate entrainment characteristics.

temperature compensation A defining characteristic of circadian rhythms, in which the period length of the rhythm under constant conditions is shown to vary relatively little (generally less than 20%) over a 10°C range in temperature. A Q_{10} close to or equal to 1.

time giver See *zeitgeber*.

torpid Dormant, hibernating.

transient A cycle in an overt rhythm observed after a phase shift but before the rhythm has reached a new steady-state phase.

Type 0 PRC A phase response curve showing strong resetting of a circadian system, in which, on average, the new phase is the same regardless of when in the cycle a standard entraining agent acted.

Type 1 PRC A phase response curve showing weak resetting of a circadian system, in which, on average, the new phase is a function primarily of the phase at which a standard time giver acted.

ultradian Having a period length of significantly less than a day. Compare *infradian*.

ungulate (a) Having hooves. (n) An organism characterized by hooves, including horses, cattle, swine, deer, giraffes, hippos, and elephants.

univoltine Usually referring to insects or mammals, producing a single litter or egg batch each year. Compare *polyvoltine*.

untranslated region (UTR) The portion of a messenger RNA molecule that does not provide information to encode the primary sequence of a protein.

UTR See *untranslated region*.

zeitgeber Entraining agent.

Photo Credits

Chapter 1 *Harrison's watch:* © National Maritime Museum, London. **1.1:** © Painet. **1.5:** Jeff Saward/Labyrinthos. **1.6:** © Painet.

Chapter 2 *squirrel:* © Painet. **2.12:** Ursula Schleicher-Benz. From *Lindauer Bilderbogen* no. 5, edited by Friedrich Boer, Jan Thorbecke Verlag, Sigmaringen, Germany. **2.18:** Courtesy of the National Library of Medicine. **2.24B:** © Painet.

Chapter 4 *deer:* © Painet.

Chapter 5 **5.1A:** © Painet.

Chapter 9 *sleeping child:* © Painet. **9.2A:** © Painet. **9.2B:** Courtesy of Commander J. Bortniak, NOAA Corps.

Chapter 10 *factory:* © Painet. *airplane:* © Painet. **10.1:** © Reuters NewMedia Inc./Corbis. **10.3, 10.5B, 10.10:** © Painet. **10.11:** © Royalty-Free/Corbis. **10.13:** © Painet. **10.14, 10.15:** Courtesy of NASA. **10.22:** Courtesy of the U.S. Fish and Wildlife Service.

Chapter 11 **11.11A:** Courtesy of B. Mintz. **11.2A:** © Royalty-Free/CORBIS.

Color Plates

Plate 1 *Earth:* Courtesy of NASA. **A, B, C, D:** Courtesy of P. DeCoursey. **E, F:** © Painet.

Plate 2 **A, B:** Courtesy of P. DeCoursey. **C, D, E:** © Painet. **F:** Courtesy of P. DeCoursey. **G:** Courtesy of M. L. Springer. **H:** Courtesy of M. Grace. **I:** Courtesy of A. Morgan. **J:** Courtesy of P. DeCoursey.

Plate 3 **A:** Courtesy of M. Grace. **B:** Courtesy of P. DeCoursey. **C:** Courtesy of C. Cook. **D, E:** Courtesy of R. Vogt; transgenic fly in **E** gift of Dean Smith. **F:** Courtesy of I. Provencio and József Czégé. **G:** Courtesy of the U.S. Fish and Wildlife Service.

Plate 4 **A:** © Painet. **B:** Courtesy of W. Hall, University of Delaware. **C:** Courtesy of P. DeCoursey. **D:** © L. J. Friesen. **E:** © Painet. **F:** Courtesy of M. Van Woert, NOAA NESDIS, ORA. **G:** © Painet. **H:** Courtesy of G. Zahm/U.S. Fish and Wildlife Service. **I:** © Painet.

Index